Probability, Statistics and Random Processes

Third Edition

About the Author

T Veerarajan is currently heading the Department of Science and Humanities, Sree Sowdambika College of Engineering Aruppukotai, Tamil Nadu. A Gold Medalist from Madras University, he has had a brilliant academic career all through. He has 47-years of teaching experience at undergraduate and postgraduate levels in various established engineering colleges in Tamil Nadu including Anna University, Chennai.

He is the author of several textbooks. Some of them are titled as:

Engineering Mathematics for first year

Engineering Mathematics for III Semester

Discrete Mathematics

Numerical Methods (Sigma Series)

Numerical Methods: With programs in C

All these books which are primarily meant for BE/B. Tech courses, are published by Tata McGraw-Hill Publishing Company, New Delhi. His prime area of interest includes Random Processes which has inspired him to write this book on Probability, Statistics and Random Processes.

Probability, Statistics and Random Processes

Third Edition

T Veerarajan
Professor of Mathematics and Head
Science and Humanities Department
Sree Sowdambika College of Engineering
Aruppukottai, Tamil Nadu

Tata McGraw-Hill Publishing Company Limited
NEW DELHI

McGraw-Hill Offices
New Delhi New York St Louis San Francisco Auckland Bogotá Caracas
Kuala Lumpur Lisbon London Madrid Mexico City Milan Montreal
San Juan Santiago Singapore Sydney Tokyo Toronto

Tata McGraw-Hill

Published by the Tata McGraw-Hill Publishing Company Limited,
7 West Patel Nagar, New Delhi 110 008.

ISBN 13: 978-0-07-069956-4
ISBN 10: 0-07-069956-9

Managing Director: *Ajay Shukla*

General Manager: Publishing—SEM & Tech Ed: *Vibha Mahajan*
Sponsoring Editor: *Shukti Mukherjee*
Editorial Executive: *Suman Sen*
Executive—Editorial Services: *Sohini Mukherjee*
Senior Production Assistant: *Stephen Manuvel*

General Manager: Marketing—Higher Education & School: *Michael J Cruz*
Product Manager: SEM & Tech Ed: *Biju Ganesan*

Controller—Production: *Rajender P Ghansela*
Asst. General Manager—Production: *B L Dogra*

Published by the Tata McGraw-Hill Publishing Company Limited, 7 West Patel Nagar, New Delhi 110 008, typeset at Bukprint India, B-180A, Guru Nanak Pura, Laxmi Nagar-110 092 and printed at Adarsh Printers, C-50-51, Mohan Park, Naveen Shahdara, Delhi 110 032

Cover: SDR Printers

RCLACRQXRBLRA

The McGraw-Hill Companies

Contents

Preface to the Third Edition

I am extremely gratified by the overwhelming response shown for the first two editions of this book.

Though the previous editions catered to the requirements of engineering students as per the Anna University syllabus, a recent survey of the syllabi followed by many universities across the country made me revise the book. The two topics 'Reliability Engineering' and 'Statistical Quality Control' have been removed, as the mechanical engineering students study them along with other subjects. The topic 'Statistical Averages' has been thoroughly revised—more concepts about basics of statistical measures have been included and more number of worked examples and exercise problems have been added. Various chapters have been rearranged.

The Online Learning Centre of the book can be accessed at http://www.mhhe.com./veerarajan/psrp3e and contains the following material for students:

- useful web links for further reading
- Chapters on
  —Reliability Engineering
  —Statistical Quality Control

I am extremely grateful to the Management Chockalingapuram Devangar Varthaga Sangam, Aruppukottai, which has sponsored Sree Sowdambika College of Engineering, Aruppukottai, in which I am presently working, for the support extended to me in this project.

I wish to express my thanks to Prof. M Jegan Mohan, Principal, Sree Sowdambika College of Engineering for the keen interest shown and constant encouragement given to me while preparing this revised edition of the book.

While preparing this edition of the book, care has been taken to completely eliminate a few errors which were present in the previous edition.

Critical evaluation and suggestions for further improvement of the book are most welcome.

T VEERARAJAN

Preface to the First Edition

This book is intended for a one-semester course in Probability, Statistics and Random Processes for undergraduate level Electrical/Electronics/Computer Engineering and Information Technology students. The various chapters also cover the syllabus content of 'Probability and Statistics' introduced recently for the BE (CSE and IT) course by the Madurai Kamaraj University of Tamil Nadu.

Most engineering students, who are used to a deterministic outlook of Physics and Engineering problems, find the theory of probability unreliable, vague and difficult. This is due to inadequate understanding of the basic concepts of probability theory and the wrong impression that the subject is an advanced branch of Mathematics.

The book is written in such a manner that beginners may develop an interest in the subject and may find it useful to pursue their studies. Basic concepts and proofs of theorems are explained in as lucid a manner as possible. Although the theory of probability is developed rigorously based on measure theory, it is developed in this book by a simple set-theory approach.

As engineering students find it easier to generalize specific results and examples than to specialize general results, considerable attention is devoted to working of problems. Nearly 300 problems including those with applications to communication theory are worked out in various chapters. Unless the students become personally involved in solving exercises, they cannot really develop an understanding and appreciation of the ideas and a familiarity with the pertinent techniques. Hence, in addition to a large number of short-answer questions under Part-A, over 350 problems have been given under Part-B of the Exercises in various chapters. Answers are provided at the end of every chapter.

Though chapters 7 and 8 are meant for Electrical/Electronics Engineering students, the other chapters that deal with probability theory, random variables, probability distributions and statistics may be useful to students of any discipline, such as those doing MCA and M.Sc courses who have Mathematical Statistics as a subject of their study.

I am sure that the students and the faculty will find this book very useful.

Critical evaluation and suggestions for improvement of the book will be highly appreciated and gratefully acknowledged.

I am extremely grateful to Dr K V Kuppusamy, Chairman, and Mr K Senthil Ganesh, Managing Trustee, RVS Educational Trust, Dindigul, for the support extended to me in this project.

I wish to express my thanks to Dr K M Karuppannan, Principal, RVS College of Engineering and Technology, Dindigul, for the appreciative interest shown and constant encouragement given to me while writing this book.

I am thankful to my publishers, Tata McGraw-Hill Publishing Company Limited, New Delhi, for their painstaking efforts and cooperation in bringing out this book in a short span of time.

I would also like to thank Dr A Rangan, Professor, Department of Mathematics, IIT Madras; Dr S Leela Devi, Professor and Head, Department of Mathematics J J College of Engineering and Technology, Tiruchirapalli; and Mr Sitharselvan and Mr Muthuraman of Bannari Amman Institute of Technology for reviewing and providing valuable suggestions during the developmental stages of the book.

I have great pleasure in dedicating this book to my beloved students, past and present.

T VEERARAJAN

1 Probability Theory

Probability theory had its origin in the analysis of certain games of chance that were popular in the seventeenth century. It has since found applications in many branches of Science and Engineering and this extensive application makes it an important branch of study. Probability theory, as a matter of fact, is a study of random or unpredictable experiments and is helpful in investigating the important features of these random experiments.

Random Experiment

An experiment whose outcome or result can be predicted with certainty is called a deterministic experiment. For example, if the potential difference E between the two ends of a conductor and the resistance R are known, the current I flowing in the conductor is uniquely determined by Ohm's law, $I = \frac{E}{R}$.

Although all possible outcomes of an experiment may be known in advance, the outcome of a particular performance of the experiment cannot be predicted owing to a number of unknown causes. Such an experiment is called a random experiment.

Whenever a fair 6-faced cubic dice is rolled, it is known that any of the 6 possible outcomes will occur, but it cannot be predicted what exactly the outcome will be, when the dice is rolled at a point of time.

Although the number of telephone calls received in a board in a 5-min. interval is a non-negative integer, we cannot predict exactly the number of calls received in the next 5-min. In such situations we talk of the chance or the probability of occurrence of a particular outcome, which is taken as a quantitative measure of the likelihood of the occurrence of the outcome.

Mathematical or Apriori Definition of Probability

Let S be the sample space (the set of all possible outcomes which are assumed equally likely) and A be an event (a sub-set of S consisting of possible outcomes) associated with a random experiment. Let $n(S)$ and $n(A)$ be the number of elements of S and A. Then the probability of event A occurring, denoted as $P(A)$, is defined by

$$P(A) = \frac{n(A)}{n(S)} = \frac{\text{Number of cases favourable to } A}{\text{Exhaustive number of cases in } S}$$

For example, the probability of getting an even number in the die tossing experiment is 0.5, as $S = \{1, 2, 3, 4, 5, 6\}$, $E = \{2, 4, 6\}$, $n(S) = 6$ and $n(E) = 3$.

Statistical or Aposteriori Definition of Probability

Let a random experiment be repeated n times and let an event A occur n_A times out of the n trials. The ratio $\frac{n_A}{n}$ is called the relative frequency of the event A. As n increases, $\frac{n_A}{n}$ shows a tendency to stabilise and to approach a constant value. This value, denoted by $P(A)$, is called the probability of the event A, i.e., $P(A) = \lim_{n \to \infty} \frac{n_A}{n}$.

For example, if we want to find the probability that a spare part produced by a machine is defective, we study the record of defective items produced by the machine over a considerable period of time. If, out of 10,000 items produced, 500 are defective, it is assumed that the probability of a defective item is 0.05.

Note *From both the definitions, it is obvious that $0 \le P(A) \le 1$. If A is an impossible event, $P(A) = 0$. Conversely, if $P(A) = 0$, then A can occur in a very small percentage of times in the long run. On the other hand, if A is a certain event, $P(A) = 1$. Conversely, if $P(A) = 1$, then A may fail to occur in a very small percentage of times in the long run.*

Axiomatic Definition of Probability

Let S be the sample space and A be an event associated with a random experiment. Then the probability of the event A, denoted by $P(A)$, is defined as a real number satisfying the following axioms.

(i) $0 \le P(A) \le 1$

(ii) $P(S) = 1$

(iii) If A and B are mutually exclusive events, $P(A \cup B) = P(A) + P(B)$ and

(iv) If $A_1, A_2, \ldots, A_n, \ldots$ are a set of mutually exclusive events, $P(A_1 \cup A_2 \cup \ldots \cup A_n \ldots) = P(A_1) + P(A_2) + \ldots + P(A_n) + \ldots$

The term *mutually exclusive* used in the above definition can be explained as follows. A set of events is said to be mutually exclusive if the occurrence of any one of them excludes the occurrence of the others. Two events A and B are mutually exclusive if A occurs and B does not occur and vice versa. In other words, A and B cannot occur simultaneously, i.e., $P(A \cap B) = 0$.

In the development of the probability theory, all results are derived directly or indirectly using only the axioms of probability, as can be seen from the following theorems.

Theorem 1 The probability of the impossible event is zero, i.e., if ϕ is the subset (event) containing no sample point, $P(\phi) = 0$.

Proof The certain event S and the impossible event ϕ are mutually exclusive.

Hence $P(S \cup \phi) = P(S) + P(\phi)$ [Axiom (iii)].

But $S \cup \phi = S$.

$\therefore$ $P(S) = P(S) + P(\phi)$

$\therefore$ $P(\phi) = 0$

Theorem 2 If $\overline{A}$ is the complementary event of A, $P(\overline{A}) = 1 - P(A) \le 1$.

Proof A and $\overline{A}$ are mutually exclusive events, such that $A \cup \overline{A} = S$.

$\therefore$ $P(A \cup \overline{A}) = P(S)$

$= 1$ [Axiom (ii)]

i.e., $P(A) + P(\overline{A}) = 1$ [Axiom (iii)]

$\therefore$ $P(\overline{A}) = 1 - P(A)$

Since $P(A) \ge 0$, it follows that $P(\overline{A}) \le 1$.

Theorem 3 If A and B are any 2 events, $P(A \cup B) = P(A) + P(B) - P(A \cap B) \le P(A) + P(B)$

Proof A is the union of the mutually exclusive events $A\overline{B}$ and AB and B is the union of the mutually exclusive events $\overline{A}B$ and AB.

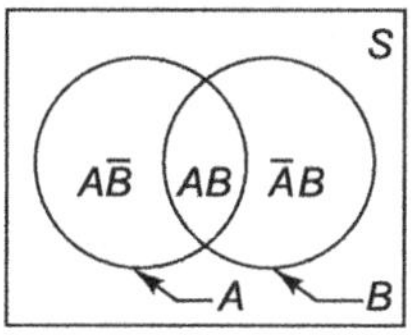

$\therefore$ $P(A) = P(A\overline{B}) + P(AB)$ [Axiom (iii)]

and $P(B) = P(\overline{A}B) + P(AB)$ [Axiom (iii)]

$\therefore$ $P(A) + P(B) = [P(A\overline{B}) + P(AB) + P(\overline{A}B) + P(AB)]$

$= P(A \cup B) + P(A \cap B)$

The result follows. Clearly, $P(A) + P(B) - P(A \cap B) \le P(A) + P(B)$]

Theorem 4 If $B \subset A$, $P(B) \le P(A)$.

Proof B and $A\overline{B}$ are mutually exclusive events such that $B \cup A\overline{B} = A$.

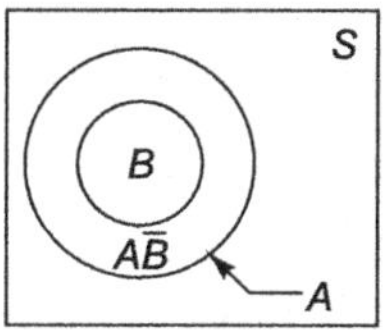

$\therefore$ $P(B \cup A\overline{B}) = P(A)$

i.e. $P(B) + P(A\overline{B}) = P(A)$ [Axiom (iii)]

$\therefore$ $P(B) \le P(A)$

Note *In probability theory developed using the classical definition of probability, theorem 3 above is termed as Addition theorem of probability as applied to any 2 events. The theorem can be extended to any 3 events A, B and C as follows:*

$$P(A \cup B \cup C) = P(\text{at least one of } A, B \text{ and } C \text{ occurs})$$
$$= P(A) + P(B) + P(C) - P(A \cap B) - P(B \cap C) - P(C \cap A) + P(A \cap B \cap C)$$

In the classical approach, probability axiom (iii) is termed as addition theorem of probability as applied to 2 mutually exclusive events, which is proved in the following way.

Let the total number of cases (outcomes) be n, of which n_A are favourable to the event A and n_B are favourable to the event B.

Therefore the number of cases favourable to A or B, i.e., $A \cup B$ is $(n_A + n_B)$, since the events A and B are disjoint.

$$\therefore \quad P(A \cup B) = \frac{n_A + n_B}{n} = \frac{n_A}{n} + \frac{n_B}{n} = P(A) + P(B)$$

Conditional Probability

The conditional probability of an event B, assuming that the event A has happened, is denoted by $P(B/A)$ and defined as

$$P(B/A) = \frac{P(A \cap B)}{P(A)}, \text{ provided } P(A) \neq 0$$

For example, when a fair dice is tossed, the conditional probability of getting '1', given that an odd number has been obtained, is equal to 1/3 as explained below:

$$S = \{1, 2, 3, 4, 5, 6\};\ A = \{1, 3, 5\};\ B = \{1\}$$

$$\therefore \quad P(B/A) = \frac{n(A \cap B)}{n(A)} = \frac{1}{3}$$

As per the definition given above, $P(B/A) = \dfrac{P(A \cap B)}{P(A)} = \dfrac{1/6}{1/2} = \dfrac{1}{3}$.

Rewriting the definition of conditional probability, we get $P(A \cap B) = P(A) \times P(B/A)$. This is sometimes referred to as *Product theorem of probability*, which is proved as follows.

Let n_A, n_{AB} be the number of cases favourable to the events A and $A \cap B$, out of the total number n of cases.

$$\therefore \quad P(A \cap B) = \frac{n_{AB}}{n} = \frac{n_A}{n} \times \frac{n_{AB}}{n_A} = P(A) \times P(B/A)$$

The product theorem can be extended to 3 events A, B and C as follows:

$$P(A \cap B \cap C) = P(A) \times P(B/A) \times P(C/A \text{ and } B)$$

The following properties are easily deduced from the definition of conditional probability:

1. If $A \subset B$, $P(B/A) = 1$, since $A \cap B = A$
2. If $B \subset A$, $P(B/A) \geq P(B)$, since $A \cap B = B$, and $\dfrac{P(B)}{P(A)} \geq P(B)$, as $P(A) \leq P(S) = 1$

3. If A and B are mutually exclusive events, $P(B/A) = 0$, since $P(A \cap B) = 0$
4. If $P(A) > P(B)$, $P(A/B) > P(B/A)$ (MKU — Apr. 96)
5. If $A_1 \subset A_2$, $P(A_1/B) \leq P(A_2/B)$ (BU — Apr. 96))

Independent Events

A set of events is said to be independent if the occurrence of any one of them does not depend on the occurrence or non-occurrence of the others.

When 2 events A and B are independent, it is obvious from the definition that $P(B/A) = P(B)$. If the events A and B are independent, the product theorem takes the form $P(A \cap B) = P(A) \times P(B)$. Conversely, if $P(A \cap B) = P(A) \times P(B)$, the events A and B are said to be independent (pairwise independent). The product theorem can be extended to any number of independent events: If $A_1, A_2, \ldots, A_n$ are n independent events.

$$P(A_1 \cap A_2 \cap \ldots \cap A_n) = P(A_1) \times P(A_2) \times \ldots \times P(A_n)$$

When this condition is satisfied, the events $A_1, A_2, \ldots, A_n$ are also said to be *totally independent*. A set of events $A_1, A_2, \ldots, A_n$ is said to be *mutually independent* if the events are totally independent when considered in sets of 2, 3, …, n events.

In the case of more than 2 events, the term 'independence' is taken as 'total independence' unless specified otherwise.

Theorem 1 If the events A and B are independent, the events $\overline{A}$ and B (and similarly A and $\overline{B}$) are also independent.

Proof The events $A \cap B$ and $\overline{A} \cap B$ are mutually exclusive such that $(A \cap B) \cup (\overline{A} \cap B) = B$.

$\therefore \quad P(A \cap B) + P(\overline{A} \cap B) = P(B)$ (by addition theorem)

$$\begin{aligned} \therefore \quad P(\overline{A} \cap B) &= P(B) - P(A \cap B) \\ &= P(B) - P(A)\,P(B) \text{ (by product theorem)} \\ &= P(B)\,[1 - P(A)] \\ &= P(\overline{A})\,P(B) \end{aligned}$$

Theorem 2 If the events A and B are independent, then so are $\overline{A}$ and $\overline{B}$.

(MU — Apr. 96)

Proof
$$\begin{aligned} P(\overline{A} \cap \overline{B}) &= P(\overline{A \cup B}) = 1 - P(A \cup B) \qquad (1) \\ &= 1 - [P(A) + P(B) - P(A \cap B)] \text{ (by addition theorem)} \\ &= 1 - P(A) - P(B) + P(A) \times P(B) \\ &\qquad \text{(since } A \text{ and } B \text{ are independent)} \\ &= [1 - P(A)] - P(B)\,[1 - P(A)] \\ &= P(\overline{A}) \times P(\overline{B}) \qquad (2) \end{aligned}$$

Note *From (1) and (2), if follows that when the events A and B are independent,* $P(A \cup B) = 1 - P(\overline{A}) \times P(\overline{B})$.

Worked Examples 1(A)

Example 1 *A fair coin is tossed 4 times. Define the sample space corresponding to this random experiment. Also give the subsets corresponding to the following events and find the respective probabilities:*

(a) More heads than tails are obtained.

(b) Tails occur on the even numbered tosses.

Solution S = {HHHH, HHHT, HHTH, HHTT, HTHH, HTHT, HTTH, HTTT, THHH, THHT, THTH, THTT, TTHH, TTHT, TTTH, TTTT}

(a) Let A be the event in which more heads occur than tails.

Then A = {HHHH, HHHT, HHTH, HTHH, THHH}

(b) Let B be the event in which tails occur in the second and fourth tosses.

Then B = {HTHT, HTTT, TTHT, TTTT}

$$P(A) = \frac{n(A)}{n(S)} = \frac{5}{16}; \ P(B) = \frac{n(B)}{n(S)} = \frac{1}{4}$$

Example 2 *There are 4 letters and 4 addressed envelopes. If the letters are placed in the envelopes at random, find the probability that (i) none of the letters is in the correct envelope and (ii) at least 1 letter is in the correct envelope, by explicitly writing the sample space and the event spaces.*

Solution Let the envelopes be denoted by A, B, C and D and the corresponding letters by a, b, c and d.

S = {(Aa, Bb, Cc, Dd), (Aa, Bb, Cd, Dc), (Aa, Bc, Cb, Dd), (Aa, Bc, Cd, Db), (Aa, Bd, Cb, Dc), (Aa, Bd, Cc, Db), (Ab, Ba, Cc, Dd), (Ab, Ba, Cd, Dc), (Ab, Bc, Ca, Dd), (Ab, Bc, Cd, Da), (Ab, Bd, Ca, Dc), (Ab, Bd, Cc, Da), (Ac, Ba, Cb Dd), (Ac, Ba, Cd, Db), (Ac, Bb, Ca, Dd), (Ac, Bb, Cd, Da), (Ac, Bd, Ca, Db), (Ac, Bd, Cb, Da), (Ad, Ba, Cb, Dc), (Ad, Ba, Cc, Db), (Ad, Bb, Ca, Dc), (Ad, Bb, Cc, Da), (Ad, Bc, Ca, Db), (Ad, Bc, Cb, Da)}

where 'Aa' means that the letter 'a' is placed in the envelope A.

Let E_1 denote the event in which none of the letters is in the correct envelope.

Then E_1 = {(Ab, Ba, Cd, Dc), (Ab, Bc, Cd, Da), (Ab, Bd, Ca, Dc), (Ac, Ba, Cd, Db), (Ac, Bd, Ca, Db), (Ac, Bd, Cb, Da), (Ad, Ba, Cb, Dc), (Ad, Bc, Ca, Db), (Ad, Bc, Cb, Da)}

Let E_2 denote the event in which at least one of the letters is in the correct envelope.

We note that E_2 is the complement of E_1. Therefore E_2 consists of all the elements of S except those in E_1.

$$P(E_1) = \frac{n(E_1)}{n(S)} = \frac{9}{24} = \frac{3}{8} \text{ and } P(E_2) = 1 - P(E_1) = \frac{5}{8}$$

Example 3 *A lot consists of 10 good articles, 4 with minor defects and 2 with major defects. Two articles are chosen from the lot at random (without replacement). Find the probability that (i) both are good, (ii) both have major defects, (iii) at least 1 is good, (iv) at most 1 is good, (v) exactly 1 is good, (vi) neither has major defects and (vii) neither is good.*

Solution Although the articles may be drawn one after the other, we can consider that both articles are drawn simultaneously, as they are drawn without replacement.

(i) $$P(\text{both are good}) = \frac{\text{No. of ways drawing 2 good articles}}{\text{Total no. of ways of drawing 2 articles}}$$

$$= \frac{10C_2}{16C_2} = \frac{3}{8}$$

(ii) P(both have major defects)

$$= \frac{\text{No. of ways of drawing 2 articles with major defects}}{\text{Total no. of ways}}$$

$$= \frac{2C_2}{16C_2} = \frac{1}{120}$$

(iii) P(at least 1 is good) = P(exactly 1 is good or both are good)

= P(exactly 1 is good and 1 is bad or both are good)

$$= \frac{10C_1 \times 6C_1 + 10C_2}{16C_2} = \frac{7}{8}$$

(iv) P(atmost 1 is good) = P(none is good or 1 is good and 1 is bad)

$$= \frac{10C_0 \times 6C_2 + 10C_1 \times 6C_1}{16C_2} = \frac{5}{8}$$

(v) P(exactly 1 is good) = P(1 is good and 1 is bad)

$$= \frac{10C_1 \times 6C_1}{16C_2} = \frac{1}{2}$$

(vi) P(neither has major defects)

= P(both are non-major defective articles)

$$= \frac{14C_2}{16C_2} = \frac{91}{120}$$

(vii) P(neither is good) = P(both are defective)

$$= \frac{6C_2}{16C_2} = \frac{1}{8}$$

Example 4 *From 6 positive and 8 negative numbers, 4 numbers are chosen at random (without replacement) and multiplied. What is the probability that the product is positive?*

Solution If the product is to be positive, all the 4 numbers must be positive or all the 4 must be negative or 2 of them must be positive and the other 2 must be negative.

No. of ways of choosing 4 positive numbers = $6C_4 = 15$.

No. of ways of choosing 4 negative numbers = $8C_4 = 70$.

No. of ways of choosing 2 positive and 2 negative numbers

$$= 6C_2 \times 8C_2 = 420.$$

Total no. of ways of choosing 4 numbers from all the 14 numbers

$$= 14C_4 = 1001.$$

P(the product is positive)

$$= \frac{\text{No. of ways by which the product is positve}}{\text{Total no. of ways}}$$

$$= \frac{15 + 70 + 420}{1001} = \frac{505}{1001}$$

Example 5 *A box contains tags marked 1, 2, ..., n. Two tags are chosen at random without replacement. Find the probability that the numbers on the tags will be consecutive integers.*

Solution If the numbers on the tags are to be consecutive integers, they must be chosen as a pair from the following pairs.

$$(1, 2); (2, 3); (3, 4); \ldots; (n - 1, n)$$

No. of ways of choosing any one pair from the above $(n - 1)$ pairs = $(n - 1)\, C_1 = n - 1$.

Total No. of ways of choosing 2 tags from the n tags = nC_2.

$\therefore$ Required probability $= \dfrac{n-1}{\dfrac{n(n-1)}{2}} = \dfrac{2}{n}$

Example 6 *If n biscuits are distributed at random among m children, what is the probability that a particular child receives r biscuits, where r < n?*

(MKU — Nov. 96)

Solution The first biscuit can be given to any one of the m children, i.e., in m ways.

Similarly the second biscuit can be given in m ways.

Therefore 2 biscuits can be given in m^2 ways.

Extending, n biscuits can be distributed in m^n ways. The r biscuits received by the particular child can be chosen from the n biscuits in nC_r ways. If this child has got r biscuits, the remaining $(n - r)$ biscuits can be distributed among the remaining $(m - 1)$ children in $(m - 1)^{n - r}$ ways.

$\therefore$ No. of ways of distributing in the required manner

$$= nC_r(m - 1)^{n - r}$$

$\therefore$ Required probability $= \dfrac{nC_r\,(m - 1)^{n - r}}{m^n}$

Example 7 *If $P(A) = P(B) = P(AB)$, show that $P(A\overline{B} + \overline{A}B) = 0$ [$AB \equiv A \cap B$].*

Solution By addition theorem,

$$P(A \cup B) = P(A) + P(B) - P(AB) \tag{1}$$

From the Venn diagram on page (3), it is clear that

$$A \cup B = A\overline{B} + \overline{A}B + AB$$

$$\therefore \quad P(A \cup B) = P(A\overline{B}) + P(\overline{A}B) + P(AB) \text{ (by probability axiom)} \tag{2}$$

Using the given condition in (1),

$$P(A \cup B) = P(AB) \tag{3}$$

From (2) and (3), $P(A\overline{B}) + P(\overline{A}B) = 0$

Example 8 *If A, B and C are any 3 events such that $P(A) = P(B) = P(C) = 1/4$, $P(A \cap B) = P(B \cap C) = 0$; $P(C \cap A) = 1/8$. Find the probability that at least 1 of the events A, B and C occurs.*

Solution P(at least one of A, B and C occurs) $= P(A \cup B \cup C)$

$$P(A \cup B \cup C) = P(A) + P(B) + P(C) - P(A \cap B) - P(B \cap C) - P(C \cap A) + P(A \cap B \cap C) \tag{1}$$

Since $P(A \cap B) = P(B \cap C) = 0$, $P(A \cap B \cap C) = 0$. Equation (1) becomes

$$P(A \cup B \cup C) = \frac{3}{4} - 0 - 0 - \frac{1}{8} = \frac{5}{8}$$

Example 9 *Solve Example 5, if the tags are chosen at random with replacement.*

Solution If the tag with '1' is chosen in the first draw, the tag with '2' must be chosen in the second draw. Probability for each $= 1/n$.

$\therefore$ P('1' in the first draw and '2' in the second draw) $= 1/n^2$ (Product theorem)

Similarly, P('n' in the first draw and '$n - 1$' in the second draw $= 1/n^2$.

If the number drawn first is '2', the number drawn second may be '1' or '3'.

Probability of drawing consecutive numbers in this case

$$= \frac{1}{n} \times \frac{2}{n} = \frac{2}{n^2}$$

Similarly, when the first number drawn is '3', '4', ..., '$(n - 1)$' probability of drawing consecutive numbers will be $2/n^2$.

All the above possibilities are mutually exclusive.

$$\therefore \quad \text{Required probability} = \frac{1}{n^2} + \frac{1}{n^2} + (n-2) \times \frac{2}{n^2} = \frac{2(n-1)}{n^2}$$

Example 10 *A box contains 4 bad and 6 good tubes. Two are drawn out from the box at a time. One of them is tested and found to be good. What is the probability that the other one is also good?*

Solution Let A = one of the tubes drawn is good and B = the other tube is good.

$$P(A \cap B) = P(\text{both tubes drawn are good})$$

$$= \frac{6C_2}{10C_2} = \frac{1}{3}$$

Knowing that one tube is good, the conditional probability that the other tube is also good is required, i.e., $P(B/A)$ is required.

By definition,

$$P(B/A) = \frac{P(A \cap B)}{P(A)} = \frac{1/3}{6/10} = \frac{5}{9}$$

Example 11 *Two defective tubes get mixed up with 2 good ones. The tubes are tested, one by one, until both defectives are found. What is the probability that the last defective tube is obtained on (i) the second test, (ii) the third test and (iii) the fourth test?*

Solution Let D represent defective and N represent non-defective tube.

(i) P(Second D in the II test) = P(D in the I test and D in the II test)

$$= P(D_1 \cap D_2), \text{ say}$$

$$= P(D_1) \times P(D_2) \text{ (by independence)}$$

$$= \frac{2}{4} \times \frac{1}{3} = \frac{1}{6}$$

(ii) P(second D in the III test) = $P(D_1 \cap N_2 \cap D_3$ or $N_1 \cap D_2 \cap D_3)$

$$= P(D_1 \cap N_2 \cap D_3) + P(N_1 \cap D_2 \cap D_3)$$

$$= \frac{2}{4} \times \frac{2}{3} \times \frac{1}{2} + \frac{2}{4} \times \frac{2}{3} \times \frac{1}{2}$$

$$= \frac{1}{3}$$

(iii) P(second D in the IV test) = $P(D_1 \cap N_2 \cap N_3 \cap D_4) + P(N_1 \cap D_2 \cap N_3 \cap D_4) + P(N_1 \cap N_2 \cap D_3 \cap D_4)$

$$= \frac{2}{4} \times \frac{2}{3} \times \frac{1}{2} \times 1 + \frac{2}{4} \times \frac{2}{3} \times \frac{1}{2} \times 1 + \frac{2}{4} \times \frac{2}{3} \times \frac{1}{2} \times 1$$

$$= \frac{1}{2}$$

Example 12 *In a shooting test, the probability of hitting the target is 1/2 for A, 2/3 for B and 3/4 for C. If all of them fire at the target, find the probability that (i) none of them hits the target and (ii) at least one of them hits the target.*

Solution Let $A \equiv$ Event of A hitting the target, and so on.

$$P(\bar{A}) = \frac{1}{2}, P(\bar{B}) = \frac{1}{3}, P(\bar{C}) = \frac{1}{4}$$

$$P(\bar{A} \cap \bar{B} \cap \bar{C}) = P(\bar{A}) \times P(\bar{B}) \times P(\bar{C}) \text{ (by independence)}$$

ie., $P(\text{none hits the target}) = \frac{1}{2} \times \frac{1}{3} \times \frac{1}{4} = \frac{1}{24}$

P(at least one hits the target)

$$= 1 - P(\text{none hits the target})$$

$$= 1 - \frac{1}{24} = \frac{23}{24}$$

Example 13 *A and B alternately throw a pair of dice. A wins if he throws 6 before B throws 7 and B wins if he throws 7 before A throws 6. If A begins, show that his chance of winning is 30/61.* (BU — Apr. 96)

Solution Throwing 6 with 2 dice ≡ Getting 6 as the sum of the numbers shown on the upper faces of the 2 dice.

$$P(\text{throwing 6 with 2 dice}) = \frac{5}{36}$$

$$P(\text{throwing 7 with 2 dice}) = \frac{1}{6}$$

Let $A \equiv$ Event of A throwing 6.

Let $B \equiv$ Event of B throwing 7.

A plays in the first, third, fifth, ..., trials.

Therefore A will win, if he throws 6 in the first trial or third trial or in subsequent (odd) trials.

$$\therefore \quad P(A \text{ wins}) = P(A \text{ or } \bar{A}\,\bar{B}\,A \text{ or } \bar{A}\,\bar{B}\,\bar{A}\,\bar{B}\,A \text{ or } \ldots)$$

$$= P(A) + P(\bar{A}\,\bar{B}\,A) + P(\bar{A}\,\bar{B}\,\bar{A}\,\bar{B}\,A) + \ldots \text{ (Addition theorem)}$$

$$= \frac{5}{36} + \left(\frac{31}{36} \times \frac{5}{6}\right)\frac{5}{36} + \left(\frac{31}{36} \times \frac{5}{6}\right)^2 \times \frac{5}{36} + \ldots \text{ upto } \infty$$

$$= \frac{5/36}{1 - (155/216)} \text{ (since the series is an infinite geometric series)}$$

$$= \frac{30}{61}$$

Example 14 *Show that $2^n - (n + 1)$ equations are needed to establish the mutual independence of n events.*

Solution n events are mutually independent, if they are totally independent when considered in sets of 2, 3, ..., n events.

Sets of r events can be chosen from the n events in nC_r ways.

To establish total independence of r events, say, $A_1, A_2, \ldots, A_r$ chosen in any one of the nC_r ways, we need one equation, namely, $P(A_1, A_2, \ldots, A_r) = P(A_1) \times P(A_2) \ldots \times P(A_r)$.

Therefore to establish total independence of all the nC_r sets, each of r events, we need nC_r equations.

Therefore the number of equations required to establish mutual independence

$$= \sum_{r=2}^{n} nC_r$$

$$= (nC_0 + nC_1 + nC_2 + \dots + nC_n) - (1 + n)$$
$$= (1 + 1)^n - (n + 1)$$
$$= 2^n - (n + 1)$$

Example 15 *Two fair dice are thrown independently. Three events A, B and C are defined as follows.*

(a) *Odd face with the first die*
(b) *Odd face with the second die*
(c) *Sum of the numbers in the 2 dice is odd. Are the events A, B and C mutually independent?* (MKU — Apr. 97)

Solution $P(A) = \frac{3}{6} = \frac{1}{2}$; $P(B) = \frac{3}{6} = \frac{1}{2}$

The outcomes favourable to the event C are (1, 2), (1, 4), (1, 6), (2, 1), (2, 3), (2, 5) and so on.

$\therefore$ $$P(C) = \frac{1}{2}$$

$$P(A \cap B) = P(B \cap C) = P(A \cap C) = \frac{1}{4}$$

$\therefore$ $P(A \cap B) = P(A)\ P(B)$, and so on

But $P(A \cap B \cap C) = 0$, since C cannot happen when A and B occur. Therefore $P(A \cap B \cap C) \neq P(A) \times P(B) \times P(C)$.

Therefore the events are pairwise independent, but not mutually independent.

Example 16 *If A, B and C are random subsets (events) in a sample space and if they are pairwise independent and A is independent of $(B \cup C)$, prove that A, B and C are mutually independent.* (MU — Nov. 96)

Solution Given:
$$P(AB) = P(A) \times P(B) \quad (1)$$
$$P(BC) = P(B) \times P(C) \quad (2)$$
$$P(CA) = P(C) \times P(A) \quad (3)$$
$$P[A\,(B \cup C)] = P(A) \times P(B \cup C) \quad (4)$$

Consider
$$P[A(B \cup C)] = P(AB \cup AC)$$
$$= P(AB) + P(AC) - P(AB \cap AC) \text{ (by addition theorem)}$$
$$= P(A) \times P(B) + P(A) \times P(C) - P(ABC) \text{ [by (1) and (3)]} \quad (5)$$

Therefore from (4) and (5), we get
$$P(ABC) = P(A) \times P(B) + P(A) \times P(C) - P(A) \times P(B \cup C)$$
$$= P(A) \times [P(B) + P(C) - P(B \cup C)]$$
$$= P(A) \times P(B \cap C) \text{ (by addition theorem)}$$
$$= P(A) \times P(B) \times P(C) \text{ [by (2)]} \quad (6)$$

From (1), (2), (3) and (6), the required result follows.

Exercise 1(A)

Part A (Short answer questions)

1. What is a random experiment? Give an example.
2. Give the apriori definition of probability with an example.
3. Give the aposteriori definition of probability with an example.
4. Give the relative frequency definition of probability with an example.
5. Define the sample space and an event associated with a random experiment with an example.
6. Give the axiomatic definition of probability.
7. State the axioms of probability.
8. What do you infer from the statements $P(A) = 0$ and $P(A) = 1$?
9. Define mutually exclusive events with an example.
   (**Example:** Getting an odd number and getting an even number when a 6-faced die is tossed are 2 mutually exclusive events.)
10. From a bag containing 3 red and 2 black balls, 2 balls are drawn at random. Find the probability that they are of the same colour.
11. When 2 cards are drawn from a well-shuffled pack of playing cards, what is the probability that they are of the same suit?
12. When A and B are 2 mutually exclusive events such that $P(A) = 1/2$ and $P(B) = 1/3$, find $P(A \cup B)$ and $P(A \cap B)$.
13. If $P(A) = 0.29$, $P(B) = 0.43$, find $P(A \cap \overline{B})$, if A and B are mutually exclusive.
14. When A and B are 2 mutually exclusive events, are the values $P(A) = 0.6$ and $P(A \cap \overline{B}) = 0.5$ consistent? Why?
15. Prove that the probability of an impossible event is zero (or prove that $P(\phi) = 0$.
16. Prove that $P(\overline{A}) = 1 - P(A)$, where $\overline{A}$ is the complement of A.
17. State addition theorem as applied to any 2 events. Extend it to any 3 events.
18. If $P(A) = 3/4$, $P(B) = 5/8$, prove that $P(A \cap B) \geq 3/8$.
19. A card is drawn from a well-shuffled pack of playing cards. What is the probability that it is either a spade or an ace?
20. The probability that a contractor will get a plumbing contract is 2/3 and the probability that he will get an electric contract is 4/9. If the probability of getting at least one contract is 4/5, what is the probability that he will get both?
21. If $P(A) = 0.4$, $P(B) = 0.7$ and $P(A \cap B) = 0.3$, find $P(\overline{A} \cap \overline{B})$.
22. If $P(A) = 0.35$, $P(B) = 0.75$ and $P(A \cup B) = 0.95$, find $P(\overline{A} \cup \overline{B})$.
23. Prove that $P(A \cup B) \leq P(A) + P(B)$. When does the equality hold good?
24. If $B \subset A$, prove that $P(B) \leq P(A)$.
25. Give the definitions of joint and conditional probabilities with examples.
26. Give the definition of conditional probability and deduce the product theorem of probability.

27. If $A \subset B$, prove that $P(B/A) = 1$.
28. If $B \subset A$, prove that $P(B/A) \geq P(B)$.
29. If A and B are mutually exclusive events, prove that $P(B/A) = 0$
30. If $P(A) > P(B)$, prove that $P(A/B) > P(B/A)$.
31. If $A \subset B$, prove that $P(A/C) \leq P(B/C)$.
32. If $P(A) = 1/3$, $P(B) = 3/4$ and $P(A \cup B) = 11/12$, find $P(A/B)$ and $P(B/A)$.
33. When are 2 events said to be independent? Give an example for 2 independent events.
34. What is the probability of getting atleast 1 head when 2 coins are tossed?
35. When 2 dice are tossed, what is the probability of getting 4 as the sum of the face numbers?
36. If the probability that A solves a problem is 1/2 and that for B is 3/4 and if they aim at solving a problem independently, what is the probability that the problem is solved?
37. If $P(A) = 0.65$, $P(B) = 0.4$ and $P(A \cap B) = 0.24$, can A and B be independent events?
38. 15% of a firm's employees are BE degree holders, 25% are MBA degree holders and 5% have both the degrees. Find the probability of selecting a BE degree holder, if the selection is confined to MBAs.
39. In a random experiment, $P(A) = 1/12$, $P(B) = 5/12$ and $P(B/A) = 1/15$, find $P(A \cup B)$.
40. What is the difference between total independence and mutual independence?
41. Can 2 events be simultaneously independent and mutually exclusive? Explain.
42. If A and B are independent events, prove that $\overline{A}$ and B are also independent.
43. If A and B are independent events, prove that A and $\overline{B}$ are also independent.
44. If $P(A) = 0.5$, $P(B) = 0.3$ and $P(A \cap B) = 0.15$, find $P(A/\overline{B})$.
45. If A and B are independent events, prove that $\overline{A}$ and $\overline{B}$ are also independent.
46. If A and B are independent events, prove that

$$P(A \cup B) = 1 - P(\overline{A}) \times P(\overline{B}).$$

47. A and B toss a fair coin alternately with the understanding that the one who obtains the head first wins. If A starts, what is his chance of winning?

Part B

48. Write the sample space associated with the experiment of tossing 3 coins at a time and the event of getting heads from the first 2 coins. Also find the corresponding probability.

49. Items coming off a production line are marked defective (D) or non-defective (N). Items are observed and their condition listed. This is continued until 2 consecutive defectives are produced or 4 items have been checked, whichever occurs first. Describe a sample space for this experiment.
50. An urn contains 2 white and 4 black balls. Two balls are drawn one by one without replacement. Write the sample space corresponding to this experiment and the subsets corresponding to the following events.
    (a) The first ball drawn is white.
    (b) Both the balls drawn are black.
    Also find the probabilities of the above events.
51. A box contains three 10-Ω resistors labelled R_1, R_2 and R_3 and two 50-Ω resistors labelled R_4 and R_5. Two resistors are drawn from this box without replacement. List all the outcomes of this random experiment as pairs of resistors. Also list the outcomes associated with the following events and hence find the corresponding probabilities.
    (a) Both the resistors drawn are 10-Ω resistors.
    (b) One 10-Ω resistor and one 50-Ω resistor are drawn.
    (c) One 10-Ω resistor is drawn in the first draw and one 50-Ω resistor is drawn in the second draw.
52. A box contains 3 white balls and 2 black balls. We remove at random 2 balls in succession. What is the probability that the first removed ball is white and the second is black? (BDU — Apr. 96)
53. An urn contains 3 white balls, 4 red balls and 5 black balls. Two balls are drawn from the urn at random. Find the probability that (i) both of them are of the same colour and (ii) they are of different colours.
54. One integer is chosen at random from the numbers 1, 2, 3, ..., 100. What is the probability that the chosen number is divisible by (i) 6 or 8 and (ii) 6 or 8 or both?
55. If there are 4 persons *A*, *B*, *C* and *D* and if *A* tossed with *B*, then *C* tossed with *D* and then the winners tossed. This process continues till the prize is won. What are the probabilities of each of the 4 to win? (MKU — Nov. 96)
56. Ten chips numbered 1 through 10 are mixed in a bowl. Two chips are drawn from the bowl successively and without replacement. What is the probability that their sum is 10?
57. A bag contains 10 tickets numbered 1, 2, ..., 10. Three tickets are drawn at random and arranged in ascending order of magnitude. What is the probability that the middle number is 5?
58. Two fair dice are thrown independently. Four events *A*, *B*, *C* and *D* are defined as follows:
    A: Even face with the first dice.
    B: Even face with the second dice.
    C: Sum of the points on the 2 dice is odd.

D: Product of the points on the 2 dice exceeds 20.
Find the probabilities of the 4 events.

59. A box contains 4 white, 5 red and 6 black balls. Four balls are drawn at random from the box. Find the probability that among the balls drawn, there is at least 1 ball of each colour.
60. Four persons are chosen at random from a group consisting of 4 men, 3 women and 2 children. Find the chance that the selected group contains at least 1 child.
61. A committee of 6 is to be formed from 5 lecturers and 3 professors. If the members of the committee are chosen at random, what is the probability that there will be a majority of lecturers in the committee?
62. Twelve balls are placed at random in 3 boxes. What is the probability that the first box will contain 3 balls?
63. If A and B are any 2 events, show that $P(A \cap B) \leq P(A) \leq P(A \cup B) \leq P(A) + P(B)$. (MU — Apr. 96)
64. A and B are 2 events associated with an experiment. If $P(A) = 0.4$ and $P(A \cup B) = 0.7$, find $P(B)$ if (i) A and B are mutually exclusive (ii) A and B are independent.
65. If $P(A + B) = 5/6$, $P(AB) = 1/3$ and $P(\overline{B}) = 1/2$, prove that the events A and B are independent.
66. If $A \subset B$, $P(A) = 1/4$ and $P(B) = 1/3$, find $P(A/B)$ and $P(B/A)$.
67. m objects are selected from n objects ($m < n$). What is the probability that the selection contains a particular object that was present in the n given objects?
68. What is the probability that there will be 53 sundays in (i) a leap year and (ii) a non-leap year?
69. If the probability that a communication system has high selectivity is 0.54 and the probability that it will have high fidelity is 0.81 and the probability that it will have both is 0.18, find the probability that (i) a system with high fidelity will also have high selectivity and (ii) a system with high selectivity will also have high fidelity.
70. An electronic assembly consists of two subsystems A and B. From previous testing procedures, the following probabilities are assumed to be known: $P(A \text{ fails}) = 0.20$, $P(A \text{ and } B \text{ fail}) = 0.15$ and $P(B \text{ fails alone}) = 0.15$. Evaluate (i) $P(A \text{ fails alone})$ and (ii) $P(A \text{ fails}/B \text{ has failed})$.
71. A consignment of 15 tubes contains 4 defectives. The tubes are selected at random, one by one, and examined. Assuming that the tubes tested are not put back, what is the probability that the ninth one examined is the last defective? (BDU — Apr. 97)
72. A card is drawn from a 52-card deck, and without replacing it, a second card is drawn. The first and second cards are not replaced and a third card is drawn.
    (a) If the first card is a heart, what is the probability of second card being a heart?

(b) If the first and second cards are hearts, what is the probability that the third card is the king of clubs? (BU — Nov. 96)

73. A pair of dice are rolled once. Let A be the event that the first die has a 1 on it, B the event that the second die has a 6 on it and C the event that the sum is 7. Are A, B and C independent? (BDU — Nov. 96)
74. A problem is given to 3 students whose chances of solving it are 1/2, 1/3 and 1/4. What is the probability that (i) only one of them solves the problem and (ii) the problem is solved. (MU — Nov. 96)
75. A and B alternately cut a pack of cards and the pack is shuffled after each cut. If A starts and the game is continued until one cuts a diamond, what is the chance that A wins at his second cut?
76. Players X and Y roll a pair of dice alternately. The player who rolls 11 first wins. If X starts, find his chance of winning.
77. Three persons A, B and C draw in succession from a bag containing 8 red and 4 white balls until a white ball is drawn. What is the probability that C draws the white ball?

Theorem of Total Probability

If $B_1, B_2, \ldots, B_n$ be a set of exhaustive and mutually exclusive events, and A is another event associated with (or caused by) B_i, then

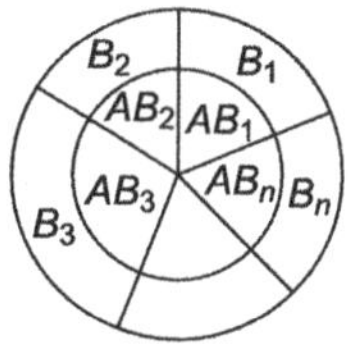

$$P(A) = \sum_{i=1}^{n} P(B_i)\, P(A/B_i)$$

Proof The inner circle represents the event A. A can occur along with (or due to) $B_1, B_2, \ldots, B_n$ that are exhaustive and mutually exclusive.

$\therefore$ $AB_1, AB_2, \ldots, AB_n$ are also mutually exclusive.

$$\therefore \quad A = AB_1 + AB_2 + \ldots + AB_n \text{ (by addition theorem)}$$

$$\therefore \quad P(A) = P(\Sigma\, AB_i)$$

$$= \Sigma\, P(AB_i) \text{ (since } AB_1, AB_2, \ldots, AB_n \text{ are mutually exclusive)}$$

$$= \sum_{i=1}^{n} P(B_i) \times P(A/B_i)$$

Bayes' Theorem or Theorem of Probability of Causes

If $B_1, B_2, \ldots, B_n$ be a set of exhaustive and mutually exclusive events associated with a random experiment and A is another event associated with (or caused by) B_i, then

$$P(B_i/A) = \frac{P(B_i) \times P(A/B_i)}{\sum_{i=1}^{n} P(B_i) \times P(A/B_i)}, \; i = 1, 2, \ldots, n$$

Proof $P(B_i \cap A) = P(B_i) \times P(A/B_i) = P(A) \times P(B_i/A)$

$$\therefore \quad P(B_i/A) = \frac{P(B_i) \times P(A/B_i)}{P(A)}$$

$$= \frac{P(B_i) \times P(A/B_i)}{\sum_{i=1}^{n} P(B_i) \times P(A/B_i)}, \; i = 1, 2, 3, \ldots, n$$

Worked Examples 1(B)

Example 1 *A bolt is manufactured by 3 machines A, B and C. A turns out twice as many items as B, and machines B and C produce equal number of items. 2% of bolts produced by A and B are defective and 4% of bolts produced by C are defective. All bolts are put into 1 stock pile and 1 is chosen from this pile. What is the probability that it is defective?*

Solution Let A = the event in which the item has been produced by machine A, and so on.

Let D = the event of the item being defective.

$$P(A) = \frac{1}{2}, \; P(B) = P(C) = \frac{1}{4}$$

$$P(D/A) = P(\text{an item is defective, given that } A \text{ has produced it})$$

$$= \frac{2}{100} = P(D/B)$$

$$P(D/C) = \frac{4}{100}$$

By theorem of total probability,

$$P(D) = P(A) \times P(D/A) + P(B) \times P(D/B) + P(C) \times P(D/C)$$

$$= \frac{1}{2} \times \frac{2}{100} + \frac{1}{4} \times \frac{2}{100} + \frac{1}{4} \times \frac{4}{100}$$

$$= \frac{1}{40}$$

Example 2 *An urn contains 10 white and 3 black balls. Another urn contains 3 white and 5 black balls. Two balls are drawn at random from the first urn and placed in the second urn and then 1 ball is taken at random from the latter. What is the probability that it is a white ball?*

Solution The two balls transferred may be both white or both black or 1 white and 1 black.

Let B_1 = event of drawing 2 white balls from the first urn, B_2 = event of drawing 2 black balls from it and B_3 = event of drawing 1 white and 1 black ball from it.

Clearly B_1, B_2 and B_3 are exhaustive and mutually exclusive events.

Let $\quad A$ = event of drawing a white ball from the second urn after transfer.

$$P(B_1) = \frac{10C_2}{13C_2} = \frac{15}{26};\ P(B_2) = \frac{3C_2}{13C_2} = \frac{1}{26};\ P(B_3) = \frac{10 \times 3}{13C_2} = \frac{10}{26},$$

$$\begin{aligned} P(A/B_1) &= P(\text{drawing a white ball/2 white balls have been transferred}) \\ &= P(\text{drawing a white ball/urn II contains 5 white and 5 black balls}) \\ &= \frac{5}{10} \end{aligned}$$

Similarly, $P(A/B_2) = \frac{3}{10}$ and $P(A/B_3) = \frac{4}{10}$.

By theorem of total probability,

$$\begin{aligned} P(A) &= P(B_1) \times P(A/B_1) + P(B_2) \times P(A/B_2) + P(B_3) \times P(A/B_3) \\ &= \frac{15}{26} \times \frac{5}{10} + \frac{1}{26} \times \frac{3}{10} + \frac{10}{26} \times \frac{4}{10} \\ &= \frac{59}{130} \end{aligned}$$

Example 3 *In a coin tossing experiment, if the coin shows head, 1 dice is thrown and the result is recorded. But if the coin shows tail, 2 dice are thrown and their sum is recorded. What is the probability that the recorded number will be 2?*

(BDU — Apr. 96)

Solution When a single dice is thrown, $P(2) = \frac{1}{6}$.

When 2 dice are thrown, the sum will be 2, only if each die shows 1.

$\therefore \quad P(\text{getting 2 as sum with 2 dice}) = \frac{1}{6} \times \frac{1}{6} = \frac{1}{36}$ (since independence)

By theorem of total probability,

$$\begin{aligned} P(2) &= P(H) \times P(2/H) + P(T) \times P(2/T) \\ &= \frac{1}{2} \times \frac{1}{6} + \frac{1}{2} \times \frac{1}{36} \\ &= \frac{7}{72} \end{aligned}$$

Example 4 *A bag contains 5 balls and it is not known how many of them are white. Two balls are drawn at random from the bag and they are noted to be white. What is the chance that all the balls in the bag are white?*

Solution Since 2 white balls have been drawn out, the bag must have contained 2, 3, 4 or 5 white balls.

Let B_1 = Event of the bag containing 2 white balls, B_2 = Event of the bag containing 3 white balls, B_3 = Event of the bag containing 4 white balls and B_4 = Event of the bag containing 5 white balls.

Let A = Event of drawing 2 white balls.

$$P(A/B_1) = \frac{2C_2}{5C_2} = \frac{1}{10},\ P(A/B_2) = \frac{3C_2}{5C_2} = \frac{3}{10}$$

$$P(A/B_3) = \frac{4C_2}{5C_2} = \frac{3}{5},\ P(A/B_4) = \frac{5C_2}{5C_2} = 1$$

Since the number of white balls in the bag is not known, B_i's are equally likely.

$$\therefore \quad P(B_1) = P(B_2) = P(B_3) = P(B_4) = \frac{1}{4}$$

By Bayes' theorem,

$$P(B_4/A) = \frac{P(B_4) \times P(A/B_4)}{\sum_{i=1}^{4} P(B_i) \times P(A/B_i)} = \frac{\frac{1}{4} \times 1}{\frac{1}{4} \times \left(\frac{1}{10} + \frac{3}{10} + \frac{3}{5} + 1\right)} = \frac{1}{2}$$

Example 5 *There are 3 true coins and 1 false coin with 'head' on both sides. A coin is chosen at random and tossed 4 times. If 'head' occurs all the 4 times, what is the probability that the false coin has been chosen and used?*

Solution $P(T) = P(\text{the coin is a true coin}) = \frac{3}{4}$

$P(F) = P(\text{the coin is a false coin}) = \frac{1}{4}$

Let A = Event of getting all heads in 4 tosses

Then $P(A/T) = \frac{1}{2} \times \frac{1}{2} \times \frac{1}{2} \times \frac{1}{2} = \frac{1}{16}$ and $P(A/F) = 1$.

By Bayes' theorem,

$$P(F/A) = \frac{P(F) \times P(A/F)}{P(F) \times P(A/F) + P(T) \times P(A/T)}$$

$$= \frac{\frac{1}{4} \times 1}{\frac{1}{4} \times 1 + \frac{3}{4} \times \frac{1}{16}} = \frac{16}{19}$$

Example 6 *For a certain binary, communication channel, the probability that a transmitted '0' is received as a '0' is 0.95 and the probability that a transmitted '1' is received as '1' is 0.90. If the probability that a '0' is transmitted is 0.4, find the probability that (i) a '1' is received and (ii) a '1' was transmitted given that a '1' was received.*

Solution Let A = the event of transmitting '1', $\overline{A}$ = the event of transmitting '0', B = the event of receiving '1' and, $\overline{B}$ = the event of receiving '0'.

Given: $P(\overline{A}) = 0.4$, $P(B/A) = 0.9$ and $P(\overline{B}/\overline{A}) = 0.95$

$\therefore$ $P(A) = 0.6$ and $P(B/\overline{A}) = 0.05$

By the theorem of total probability

$$P(B) = P(A) \times P(B/A) + P(\overline{A}) \times P(B/\overline{A})$$
$$= 0.6 \times 0.9 + 0.4 \times 0.05$$
$$= 0.56$$

By Bayes' theorem,

$$P(A/B) = \frac{P(A) \times P(B/A)}{P(B)} = \frac{0.6 \times 0.9}{0.56} = \frac{27}{28}$$

Exercise 1(B)

Part A (Short answer questions)

1. State the theorem of total probability.
2. Bag I contains 2 red and 1 black balls and bags II contains 3 red and 2 black balls. What is the probability that a ball drawn from one of the bags is red?
3. State Bayes' theorem on inverse probability.
4. Bag I contains 2 white and 3 black balls and bag II contains 4 white and 1 black balls. A ball chosen at random from one of the bags is white. What is the probability that it has come from bag I?
5. Five men out of 100 and 25 women out of 1000 are colour-blind. A colour-blind person is chosen at random. What is the probability that the person is a male? (Assume males and females are in equal numbers).

Part B

6. There are 2 bags one of which contains 5 red and 8 black balls and the other 7 red and 10 black balls. A ball is drawn from one or the other of the 2 bags. Find the chance of drawing a red ball.
7. In a bolt factory, machines A, B and C produce 25, 35 and 40% of the total output, respectively. Of their outputs, 5, 4 and 2%, respectively, are defective bolts. If a bolt is chosen at random from the combined output, what is the probability that it is defective? If a bolt chosen at random is found to be defective, what is the probability that it was produced by B or C? (MU — Apr. 96)
8. A box contains 2000 components of which 5% are defective. A second box contains 500 components of which 40% are defective. Two other boxes contain 1000 components, each with 10% defective components. We select at random one of the above boxes and remove from it at random a single component.
   (a) What is the probability that the component is defective?
   (b) Finding that the selected component is defective, what is the probability that it was drawn from box 2? (MSU — Apr. 96)

9. There are 4 candidates for the office of the highway commissioner; the respective probabilities that they will be selected are 0.3, 0.2, 0.4 and 0.1, and the probabilities for a project's approval are 0.35, 0.85, 0.45 and 0.15, depending on which of the 4 candidates is selected. What is the probability of the project getting approved? (MKU — Apr. 97)
10. In a binary communication system a '0' or '1' is transmitted. Because of noise in the system, a '0', can be received as a '1' with probability p and a '1' can be received as a '0' also with probability p. Assuming that the probability that a '0'is transmitted is p_0, and that a '1' is transmitted is q_0 $(= 1 - p_0)$ find the probability that a '1' was transmitted when a '1' is received.
11. A bag contains 7 red and 3 black marbles, and another bag contains 4 red and 5 black marbles. One marble is transferred from the first bag into the second bag and then a marble is taken out of the second bag at random. If this marble happens to be red, find the probability that a black marble was transferred.
12. The probability that a student passes a certain exam is 0.9, given that he studied. The probability that he passes the exam without studying is 0.2. Assume that the probability that the student studies for an exam is 0.75. Given that the student passed the exam, what is the probability that he studied?
13. Urn I has 2 white and 3 black balls, urn II has 4 white and 1 black balls and urn III has 3 white and 4 black balls. An urn is selected at random and a ball drawn at random is found to be white. Find the probability that urn I was selected. (MKU — Apr. 96)
14. Suppose that coloured balls are distributed in 3 boxes as follows:

	Box 1	*Box 2*	*Box 3*
Red	2	4	3
White	3	1	4
Blue	5	3	5

A box is selected at random from which a ball is selected at random and it is observed to be red. What is the probability that box 3 was selected? (MU — Nov. 96)
15. Three urns contain 3 white, 1 red and 1 black balls; 2 white, 3 red and 4 black balls; 1 white, 3 red and 2 black balls respectively. One urn is chosen at random and from it 2 balls are drawn at random. If they are found to be 1 red and 1 black ball, what is the probability that the first urn was chosen?
16. An urn contains 10 red and 3 black balls. Another urn contains 3 red and 5 black balls. Two balls are transferred from the first urn to the second urn, without noticing their colour. One ball is now drawn from the second urn and it is found to be red. What is the probability that 1 red and 1 black ball were transferred?

17. Box 1 contains 1000 bulbs of which 10% are defective. Box 2 contains 2000 bulbs of which 5% are defective. Two bulbs are drawn (without replacement) from a randomly selected box. (i) Find the probability that both bulbs are defective and (ii) assuming that both are defective, find the probability that they came from box 1.
18. The chance that a doctor A will diagnose a disease x correctly is 60%. The chance that a patient will die by his treatment after correct diagnosis is 40% and the chance of death by wrong diagnosis is 70%. A patient of doctor A, who had disease x, died. What is the chance that his disease was diagnosed correctly?
19. The chances of A, B and C becoming the general manager of a certain company are in the ratio 4:2:3. The probabilities that the bonus scheme will be introduced in the company if A, B and C become general manager are 0.3, 0.7 and 0.8 respectively. If the bonus scheme has been introduced, what is the probability that A has been appointed as general manager?

Bernoulli's Trials

Let us consider n independent repetitions (trials) of a random experiment E. If A is an event associated with E such that $P(A)$ remains the same for the repetitions, the trials are called *Bernoulli's trials*.

Theorem If the probability of occurrence of an event (probability of success) in a single trial of a Bernoulli's experiment is p, then the probability that the event occurs exactly r times out of n independent trials is equal to $nC_r\, q^{n-r} p^r$, where $q = 1 - p$, the probability of failure of the event.

Proof Getting exactly r successes means getting r sucesses and $(n - r)$ failures simultaneously.

$\therefore$ P(getting r successes and $n - r$ failures) $= p^r q^{n-r}$ (since the n trials are independent) (by product theorem).

The trials, from which the successes are obtained, are not specified. There are nC_r ways of choosing r trials for successes. Once the r trials are chosen for successes, the remaining $(n - r)$ trials should result in failures.

These nC_r ways are mutually exclusive. In each of these nC_r ways, P(getting exactly r successes) $= p^r q^{n-r}$.

Therefore, by the addition theorem, the required probability $= nC_r \times q^{n-r} \times p^r$.

De Moivre–Laplace Approximation

A result which is useful when a large number of terms of the form $nC_r\, q^{n-r} p^r$ is required to be summed up, is given below without proof.

If the probability of getting exactly r successes out of n Bernoulli's trials is denoted by $P_n(r)$, then

$$\sum_{r=r_1}^{r_2} P_n(r) = \sum_{r=r_1}^{r_2} nC_r\, q^{n-r} p^r \text{ is approximately equal to } \int_{r_1-\frac{1}{2}}^{r_2+\frac{1}{2}} y\, dx,$$

where $y = \frac{1}{\sqrt{2\pi npq}} e^{-(x-np)^2/2n/pq}$, which is the density of a normal distribution with mean np and variance npq.

As the reader is familiar with normal distribution, it can be easily seen that

$$\sum_{r=r_1}^{r_2} P_n(r) = \int_{t_1}^{t_2} \phi(t)dt$$

where $\phi(t)$ is the standard normal density and $t_1 = \frac{r_1 - np - 1/2}{\sqrt{npq}}$ and $t_2 = \frac{r_2 - np + 1/2}{\sqrt{npq}}$. Now $\int_{t_1}^{t_2} \phi(t)dt$ can be computed using the Table of areas under normal curve.

Generalisation of Bernoulli's Theorem Multinomial Distribution

If $A_1, A_2, \ldots, A_k$ are exhaustive and mutually exclusive events associated with a random experiment such that $P(A_i \text{ occurs}) = p_i$, where $p_1 + p_2 + \ldots + p_k = 1$, and if the experiment is repeated n times, then the probability that A_1 occurs r_1 times, A_2 occurs r_2 times, ..., A_k occurs r_k times is given by

$$P_n(r_1, r_2, \ldots, r_k) = \frac{n!}{r_1!\, r_2! \ldots r_k!} p_1^{r_1} \times p_2^{r_2} \ldots p_k^{r_k}$$

where $r_1 + r_2 + \ldots + r_k = n$.

Proof The r_1 trials in which the event A_1 occurs can be chosen from the n trials in nC_{r_1} ways. The remaining $(n - r_1)$ trials are left over for the other events.

The r_2 trials in which the event A_2 occurs can be chosen from the $(n - r_1)$ trials in $(n - r_1)C_{r_2}$ ways.

The r_3 trials in which the event A_3 occurs can be chosen from the $(n - r_1 - r_2)$ trials in $(n - r_1 - r_2)C_{r_3}$ ways, and so on.

Therefore the number of ways in which the events $A_1, A_2, \ldots, A_k$ can happen $= nC_{r_1} \times (n - r_1)C_{r_2} \times (n - r_1 - r_2)C_{r_3} \times \ldots \times (n - r_1 - r_2 \ldots - r_{k-1})C_{r_k}$

$= \frac{n!}{r_1!\, r_2! \ldots r_k!}$.

Consider any one of the above ways in which the events $A_1, A_2, \ldots, A_k$ occur.

Since the n trials are independent, $r_1, r_2, \ldots, r_k$ trials are also independent.

$\therefore$ $P(A_1 \text{ occurs } r_1 \text{ times}) = p_1^{r1}$

$P(A_2 \text{ occurs } r_2 \text{ times}) = p_2^{r_2}$, and so on.

$\therefore$ $P(A_1$ occurs r_1 times, A_2 occurs r_2 times, ..., A_k occurs r_k times$) = p_1^{r1} \times p_2^{r2} \times \ldots \times p_k^{rk}$

Since the ways in which the events happen are mutually exclusive, the required probability is given by

$$P_n(r_1, r_2, \ldots, r_k) = \frac{n!}{r_1!\, r_2! \ldots r_k!} p_1^{r1} \times p_2^{r2} \times \ldots \times p_k^{rk}$$

Worked Examples 1(C)

Example 1 *A coin is tossed an infinite number of times. If the probability of a head in a single toss is p, show that the probability that kth head is obtained at the nth tossing, but not earlier is* $(n-1)C_{k-1}\, p^k q^{n-k}$*, where* $q = 1 - p$.

Solution k heads should be obtained at the nth tossing, but not earlier.

Therefore $(k-1)$ heads must be obtained in the first $(n-1)$ tosses and 1 head must be obtained at the nth toss.

$$\therefore \quad \text{Required probability} = P[k-1 \text{ heads in } (n-1) \text{ tosses}] \times P(1 \text{ head in } 1 \text{ toss})$$
$$= (n-1)C_{k-1} p^{k-1} q^{n-k} \times p$$
$$= (n-1)C_{k-1} p^k q^{n-k}$$

Example 2 *Each of two persons A and B tosses 3 fair coins. What is the probability that they obtain the same number of heads?*

Solution $P(A$ and B get the same no. of heads)

= P(they get no head each or 1 head each or 2 heads each or 3 heads each)

= P(each gets 0 head) + P(each gets 1 head) + P(each gets 2 heads) + P(each gets 3 heads) (since the events are mutually exclusive)

= $P(A$ gets 0 head) $\times$ $P(B$ gets 0 head) + ... (since A and B toss independently)

$$= \left[3C_0\left(\frac{1}{2}\right)^3\right]^2 + \left[3C_1\left(\frac{1}{2}\right)^3\right]^2 + \left[3C_2\left(\frac{1}{2}\right)^3\right]^2 + \left[3C_3\left(\frac{1}{2}\right)^3\right]^2$$

$$= \frac{1}{64}(1 + 9 + 9 + 1) = \frac{5}{16}$$

Example 3 *A coin with P(head) = p = 1 − q is tossed n times. Show that the probability that the number of heads obtained is even is* $0.5[1 + (q-p)^n]$.

Solution P(even no. of heads are obtained)

$$= P(0 \text{ head or } 2 \text{ heads or } 4 \text{ heads or } \ldots)$$
$$= nC_0 q^n p^0 + nC_2 q^{n-2} p^2 + nC_4 q^{n-4} p^4 + \ldots \qquad (1)$$

Consider
$$1 = (q+p)^n = nC_0 q^n p^0 + nC_1 q^{n-1} p^1 + nC_2 q^{n-2} p^2 + \ldots \qquad (2)$$
$$(q-p)^n = nC_0 q^n p^0 - nC_1 q^{n-1} p^1 + nC_2 q^{n-2} p^2 - \ldots \qquad (3)$$

Adding (2) and (3), we get
$$1 + (q-p)^n = 2[nC_0 q^n p^0 + nC_2 q^{n-2} p^2 + nC_4 q^{n-4} p^4 + \ldots] \qquad (4)$$

Using (4) in (1), the required probability $= 0.5[1 + (q-p)^n]$.

Example 4 *If at least 1 child in a family with 2 children is a boy, what is the probability that both children are boys?* (MKU — Apr. 96)

Solution p = Probability that a child is a boy $= \frac{1}{2}$

$$\therefore \qquad q = \frac{1}{2} \text{ and } n = 2$$

P(at least one boy) = p(exactly 1 boy) + p(exactly 2 boys)

$$= 2C_1 \left(\frac{1}{2}\right)^2 + 2C_2 \left(\frac{1}{2}\right)^2 \qquad \text{(by Bernoulli's theorem)}$$

$$= \frac{3}{4}$$

P(both are boys/at least one is a boy)

$$= \frac{P(\text{both are boys} \cap \text{at least one is a boy})}{P(\text{at least one is a boy})}$$

$$= \frac{P(\text{both are boys})}{P(\text{at least one is a boy})} = \frac{\frac{1}{4}}{\frac{3}{4}} = \frac{1}{3}$$

Example 5 *Find the probability of getting at least 60 heads, when 100 fair coins are tossed.*

Solution Since the coins are fair, $p = q = \frac{1}{2}$.

$$n = 100$$

$$\therefore \qquad np = 50 \text{ and } \sqrt{npq} = 5$$

Required probability $P = \sum_{r=60}^{100} 100C_r \left(\frac{1}{2}\right)^{100}$

$$= \int_{1.9}^{10.1} \phi(t)\mathrm{d}t \simeq \int_{1.9}^{\infty} \phi(t)dt \text{ (by Demoivre–Laplace approximation)}$$

$$= 0.5 - \int_{0}^{1.9} \phi(t)dt = 0.5 - 0.4719 \text{ (from the Normal Table)}$$

$$= 0.0281$$

Example 6 *A fair dice is rolled 5 times. Find the probability that 1 shows twice, 3 shows twice and 6 shows once.*

Solution This problem is a direct application of Bernoulli's generalisation.
Here $n = 5$, $A_1 \equiv$ (getting 1), $A_3 \equiv$ (getting 3)

$A_6 =$ (getting 6), $k_1 = 2$, $k_3 = 2$ and $k_6 = 1$; $p_1 = p_3 = p_6 = 1/6$

$$\therefore \qquad \text{Required probability} = \frac{5!}{2!2!1!}\left(\frac{1}{6}\right)^2 \times \left(\frac{1}{6}\right)^2 \times \left(\frac{1}{6}\right)^1$$

$$= 0.0039$$

(It is to be noted that $k_2 = k_4 = k_5 = 0$, though $p_2 = p_4 = p_5 = \frac{1}{6}$.)

Exercise 1(C)

Part A (Short answer questions)

1. State Bernouilli's theorem on independent trials.
2. A fair coin is tossed 4 times. What is the probability of getting more heads than tails?
3. When 12 coins are tossed 256 times, how many times may one expect 8 heads and 4 tails?
4. If war breaks out on the average once in 25 years, find the probability that in 50 years at a stretch, there will be no war.
5. State the generalised form of Bernoulli's theorem on independent trials.
6. State De Moivre–Laplace theorem.

Part B

7. A binary number (composed only of the digits '0' and '1') is made up of n digits. If the probability of an incorrect digit appearing is p and that errors in different digits are independent of one another, find the probability of forming an incorrect number.
8. Suppose that twice as many items are produced (per day) by machine 1 as by machine 2. However 4% of the items from machine 1 are defective while machine 2 produces only about 2% defectives. Suppose that the daily output of the 2 machines is combined. A random sample of 10 items is taken from the combined output. What is the probability that this sample contains 2 defectives?
9. Binary digits are transmitted over a noisy communication channel in blocks of 16 binary digits. The probability that a received binary digit is in error because of channel noise is 0.1. If errors occur in various digit positions within a block independently, find the probability that the number of errors per block is greater than or equal to 5.
10. A company is trying to market a digital transmission system (modem) that has a bit error probability of 10^{-4}, and the bit errors are independent. The buyer will test the modem by sending a known message of 10^4 digits and checking the received message. If more than 2 errors occur, the modem will be rejected. Find the probability that the customer will buy the company's modem.
11. A fair coin is tossed 10,000 times. Find the probability that the number of heads obtained is between 4900 and 5100, using DeMoivre-Laplace approximation.
12. Over a period of 12h, 180 calls are made at random. What is the 'probability' that in a 4 h interval the number of calls is between 50 and 70? Use DeMoivre-Laplace approximation.

$$\left[\textbf{Hint: } P(\text{a particular call occurs in the 4-h interval}) = p = \frac{4}{12} = \frac{1}{3}\right].$$

13. A random experiment can terminate in one of 3 events A, B and C with probabilities 1/2, 1/4 and 1/4 respectively. The experiment is repeated 6 times. Find the probability that the events A, B and C occur once, twice and thrice respectively.
14. A throws 3 fair coins and B throws 4 fair coins. Find the chance that A will throw more number of heads than would B.
15. In a large consignment of electric bulbs 10% are defective. A random sample of 20 bulbs is taken for inspection. Find the probability that (i) exactly 3 of them are defective, (ii) at most 3 of them are defective and (iii) at least 3 of them are defective.
16. A lot contains 1% defective items. What should be the number of items in a random sample so that the probability of finding at least 1 defective in it is at least 0.95?
17. In a precision bombing attack there is a 50% chance that any one bomb will hit the target. Two direct hits are required to destroy the target completely. How many bombs must be dropped to give a 99% or more chance of completely destroying the target?

Answers

Exercise 1(A)

10. P(both balls are of the same colour)

$= P$(both balls are red or both are black)

$= P$(both are red) $+ P$(both are black)

$$= \frac{3C_2}{5C_2} + \frac{2C_2}{5C_2} = \frac{2}{5}$$

11. Required probability $= P$(both are spades) $+ P$(both are clubs) $+ P$(both are hearts) $+ P$(both are diamonds)

$$= 4 \times \frac{13C_2}{52C_2} = \frac{4}{17}$$

12. $P(A \cup B) = P(A) + P(B) = 5/6$; $P(A \cap B) = 0$, by definition.
13. When A and B are mutually exclusive, $P(A \cap \overline{B}) = P(A) = 0.29$
14. $P(A) = P(A \cap B) + P(A \cap \overline{B}) = 0 + P(A \cap \overline{B})$ i.e., $0.6 = 0.5$, which is absurd. Hence the given values are inconsistent.
18. P(any event) ≤ 1; $P(A \cup B) \le 1$; $P(A) + P(B) - P(A \cap B) \le 1$

$$\therefore P(A \cap B) \ge \frac{3}{4} + \frac{5}{8} - 1 \left(= \frac{3}{8} \right)$$

19. $P(S \cup A) = P(S) + P(A) - P(S \cap A)$

$$= \frac{13}{52} + \frac{4}{52} - \frac{1}{52} = \frac{4}{13}$$

20. $P(A \cup B) = P(A) + P(B) - P(A \cap B)$

$$\therefore P(A \cap B) = \frac{2}{3} + \frac{4}{9} - \frac{4}{5} = \frac{14}{45}$$

21. $P(\bar{A} \cap \bar{B}) = 1 - P(A \cup B)$

$$= 1 - [P(A) + P(B) - P(A \cap B)]$$
$$= 1 - [0.4 + 0.7 - 0.3] = 0.2$$

22. $P(\bar{A} \cup \bar{B}) = 1 - P(A \cap B)$

$= 1 - [P(A) + P(B) - P(A \cup B)] = 0.85$

23. Equality holds good, when A and B are mutually exclusive events.
25. The probability for the simultaneous occurrence of two events A and B is called the joint probability of A and B. Probability of getting 2 heads, when 2 coins are tossed, is an example of joint probability.

32. $P(A \cup B) = P(A) + P(B) - P(A \cap B)$; $P(A \cap B) = \frac{1}{3} + \frac{3}{4} - \frac{11}{12} = \frac{1}{6}$

$$P(A/B) = \frac{P(A \cap B)}{P(B)} = \frac{2}{9};\ P(B/A) = \frac{P(A \cap B)}{P(A)} = \frac{1}{2}$$

34. Required probability $= 1 - P$(both tails)

$$= 1 - \frac{1}{2} \times \frac{1}{2} = \frac{3}{4}$$

35. Required probability $= P$ (1 from I dice and 3 from II dice or 2 from 1 and 2 from II or 3 from I and 1 from II)

$$= \frac{1}{6} \times \frac{1}{6} + \frac{1}{6} \times \frac{1}{6} + \frac{1}{6} \times \frac{1}{6} = \frac{1}{12}$$

36. Required probability $= 1 - P$(the problem is not solved)

$$= 1 - \frac{1}{2} \times \frac{1}{4} = \frac{7}{8}$$

37. No, since $P(A \cap B) \neq P(A) \times P(B)$

38. $P(BE/MBA) = \dfrac{P(BE \cap MBA)}{P(MBA)} = \dfrac{0.05}{0.25} = 0.2$

39. $P(A \cap B) = P(A) \times P(B/A) = \dfrac{1}{12} \times \dfrac{1}{15} = \dfrac{1}{180}$

$P(A \cup \mathrm{B}) = P(A) + P(B) - P(A \cap B)$

$$= \frac{1}{12} + \frac{5}{12} - \frac{1}{180} = \frac{89}{180}$$

41. A and B are independent, if $P(A \cap B) = P(A) \times P(B)$. They are mutually exclusive, if $P(A \cap B) = 0$.

They are both independent and mutually exclusive if $P(A) \times PB) = 0$, i.e., if $P(A) = 0$ or $P(B) = 0$ or $P(A) = 0$ and $P(B) = 0$. The third case is trivial. Hence A and B can be both independent and mutually exclusive, provided either of the events is an impossible event.

44. $P(A \cap B) = P(A) \times P(B)$

$\therefore$ A and B are independent. Hence A and $\overline{B}$ are also independent.

$\therefore$ $P(A/\overline{B}) = P(A) = 0.5$.

47. $P = \frac{1}{2} + \frac{1}{2}\left(\frac{1}{2}\right)^2 + \frac{1}{2}\left(\frac{1}{2}\right)^4 + \ldots + \infty = \frac{\frac{1}{2}}{1 - \frac{1}{4}} = \frac{2}{3}$

48. $\frac{1}{4}$

49. $\{DD, NDD, DNDD, DNDN, DNND, DNNN, NDND, NDNN, NNDD, NNDN, NNND, NNNN\}$

50. (i) $\frac{1}{3}$ (ii) $\frac{2}{5}$

51. (i) $\frac{3}{10}$ (ii) $\frac{3}{5}$ (iii) $\frac{3}{10}$

52. $\frac{6}{25}$ (if the ball is replaced), $\frac{3}{10}$ (if the ball is not replaced).

53. (i) $\frac{19}{66}$ (ii) $\frac{47}{66}$

54. (i) $\frac{1}{5}$ (ii) $\frac{6}{25}$

55. $\frac{1}{4}$ 56. $\frac{4}{45}$ 57. $\frac{1}{6}$

58. $P(A) = \frac{1}{2} = P(B) = P(C);\ P(D) = \frac{1}{6}$

59. $\frac{48}{91}$ 60. $\frac{13}{18}$ 61. $\frac{9}{14}$ 62. $12\,C_3 \times 2^9/3^{12}$

64. (i) 0.3 (ii) 0.5

66. $\frac{3}{4}$ 67. $\frac{m}{n}$

68. (i) $\frac{2}{7}$ (ii) $\frac{1}{7}$

69. (i) $\frac{2}{9}$ (ii) $\frac{1}{3}$

70. (i) 0.05 (ii) 0.50

71. $\frac{8}{195}$

72. (a) $\frac{12}{51}$ (b) $\frac{1}{50}$

73. Pairwise independent, but not totally independent

74. (i) $\frac{11}{24}$ (ii) $\frac{3}{4}$

75. $\frac{9}{64}$ 76. $\frac{18}{35}$ 77. $\frac{7}{33}$

Exercise 1(B)

2. P (red ball) = P (red ball from bag I) + P (red ball from bag II)

$$= \frac{1}{2} \times \frac{2}{3} + \frac{1}{2} \times \frac{3}{5} = \frac{19}{30}$$

4. $$P(B_1/W) = \frac{P(B_1) \times P(W/B_1)}{P(B_1) \times P(W/B_1) + P(B_2) \times P(W/B_2)}$$

$$= \frac{\frac{1}{2} \times \frac{2}{5}}{\frac{1}{2} \times \frac{2}{5} + \frac{1}{2} \times \frac{4}{5}} = \frac{1}{3}$$

5. $P(M) = P(F) = \frac{1}{2}$; $P(B/M) = \frac{1}{20}$; $P(B/F) = \frac{1}{40}$

By Bayes' theorem, $P(M/B) = \dfrac{\frac{1}{2} \times \frac{1}{20}}{\frac{1}{2} \times \frac{1}{20} + \frac{1}{2} \times \frac{1}{40}} = \frac{2}{3}$

6. $\frac{88}{221}$

7. (i) $\frac{69}{2000}$ (ii) $\frac{44}{69}$

8. (i) $\frac{13}{80}$ (ii) $\frac{8}{13}$

9. 0.47

10. $\dfrac{(1-p)q_0}{(1-p)q_0 + pp_0}$

11. $\frac{12}{47}$

12. $\frac{27}{29}$

13. $\frac{14}{57}$

14. $\frac{5}{19}$

15. $\frac{3}{25}$

16. $\frac{20}{59}$

17. (i) 0.0062 (ii) 0.8005

18. $\frac{6}{13}$

19. $\frac{6}{25}$

Exercise 1(C)

2. Required probability = P (getting exactly 3 or 4 heads)

$$= 4C_3 \left(\frac{1}{2}\right)^4 + 4C_4 \left(\frac{1}{2}\right)^4 = \frac{5}{16}$$

3. Required number = $256 \times 12C_8 \left(\frac{1}{2}\right)^{12} \cong 31$

4. $p = \dfrac{1}{25}, q = \dfrac{24}{25}, n = 50; P = 50C_0 \left(\dfrac{1}{25}\right)^0 \left(\dfrac{24}{25}\right)^{50} = \left(\dfrac{24}{25}\right)^{50}$

7. $1-(1-p)^n$ 8. 0.0381 9. 0.017 10. 0.9197

11. 0.9545 12. 0.9876 13. 0.0293 14. $\dfrac{29}{128}$

15. (i) 0.1898 (ii) 0.8655 (iii) 0.3243

16. 299 17. 11

2 Random Variables

The outcomes of random experiments may be numerical or non-numerical in nature. For example, the number of telephone calls received in a board in 1 h is numerical in nature, while the result of a coin tossing experiment in which 2 coins are tossed at a time is non-numerical in nature. As it is often useful to describe the outcome of a random experiment by a number, we will assign a number to each non-numerical outcome of the experiment. For example, in the 2 coins tossing experiment we could assign the value 0 to the outcome of getting 2 tails, 1 to the outcome of getting 1 head and 1 tail and 2 to the outcome of getting 2 heads. Thus in any experimental situation we can assign a real number x to every element s of the sample space S. That is, the function $X(s) = x$ that maps the elements of the sample space into real numbers is called the random variable associated with the concerned experiment. A formal definition may be given as follows.

Definition: A random variable (abbreviatively RV) is a function that assigns a real number $X(s)$ to every element $s \in S$, where S is the sample space corresponding to a random experiment E.)

Note *Although we are expected to perform the random experiment E, we observe the outcome $s \in S$ and then evaluate $X(s)$ [i.e., assign a real number x to $X(s)$], the number $x = X(s)$ itself can be thought of as the outcome of the experiment and R_x as the sample space of the experiment. In this sense, we will hereafter talk about a random variable X taking the value x and $P(X = x)$. Actually, $P(X = x) = P\{s: X(s) = x\}$.*

Hereafter, R_x will be referred to as ***Range space***.
Similarly $\{X \le x\}$ represents the subset $\{s: X(s) \le x\}$ and hence an event associated with the experiment.

Discrete Random Variable

If X is a random variable (RV) which can take a finite number or countably infinite number of values, X is called a discrete RV. When the RV is discrete, the possible values of X may be assumed as $x_1, x_2, \ldots, x_n, \ldots$. In the finite case, the list of values terminates and in the countably infinite case, the list goes upto infinity.

For example, the number shown when a die is thrown and the number of alpha particles emitted by a radioactive source are discrete RVs.

Probability Function

If X is a discrete RV which can take the values $x_1, x_2, x_3, \ldots$ such that $P(X = x_i) = p_i$, then p_i is called the *probability function or probability mass function or point probability function*, provided p_i ($i = 1, 2, 3, \ldots$) satisfy the following conditions:

(*i*) $p_i \geq 0$, for all i, and

(*ii*) $\sum_i p_i = 1$

The collection of pairs $\{x_i, p_i\}$, $i = 1, 2, 3, \ldots$, is called *the probability distribution of the* RV X, which is sometimes displayed in the form of a table as given below:

$X = x_i$	$P(X = x_i)$
x_1	p_1
x_2	p_2
$\vdots$	$\vdots$
x_r	p_r
$\vdots$	$\vdots$

Continuous Random Variable

If X is an RV which can take all values (i.e., *infinite number* of values) in an interval, then X is called a *continuous* RV.

For example, the length of time during which a vacuum tube installed in a circuit functions is a continuous RV.

Probability Density Function

If X is a continuous RV such that

$$P\left\{x - \frac{1}{2}\mathrm{d}x \leq X \leq x + \frac{1}{2}\mathrm{d}x\right\} = f(x)\mathrm{d}x$$

then $f(x)$ is called the *probability density function* (shortly denoted as pdf) of X, provided $f(x)$ satisfies the following conditions:

(i) $f(x) \geq 0$, for all $x \in R_x$, and

(ii) $\int_{R_X} f(x)\mathrm{d}x = 1$

Moreover, $P(a \leq X \leq b)$ or $P(a < X < b)$ is defined as

$$P(a \leq X \leq b) = \int_a^b f(x)\mathrm{d}x.$$

The curve $y = f(x)$ is called the *probability curve of the* RV X.

Note *When X is a continuous RV*

$$P(X = a) = P(a \le X \le a) = \int_a^a f(x)\,dx = 0$$

This means that it is almost impossible that a continuous RV assumes a specific value. Hence $P(a \le X \le b) = P(a \le X < b) = P(a < X \le b) = P(a < X < b)$.

Cumulative Distribution Function (cdf)

If X is an RV, discrete or continuous, then $P(X \le x)$ is called the *cumulative distribution function* of X or *distribution function* of X and denoted as $F(x)$.
If X is discrete,

$$F(x) = \sum_{\substack{j \\ X_j \le x}} p_j$$

If X is continuous,

$$F(x) = P(-\infty < X \le x) = \int_{-\infty}^{x} f(x)\mathrm{d}x$$

Properties of the *cdf* F(x)

1. $F(x)$ is a non-decreasing function of x, i.e., if $x_1 < x_2$, then $F(x_1) \le F(x_2)$.
2. $F(-\infty) = 0$ and $F(\infty) = 1$.
3. If X is a discrete RV taking values $x_1, x_2, \ldots$, where $x_1 < x_2 < x_3 < \ldots < x_{i-1} < x_i < \ldots$, then $P(X = x_i) = F(x_i) - F(x_{i-1})$.
4. If X is a continuous RV, then $\frac{\mathrm{d}}{\mathrm{d}x} F(x) = f(x)$, at all points where $F(x)$ is differentiable.

Note *Although we may talk of probability distribution of a continuous RV, it cannot be represented by a table as in the case of a discrete RV. The probability distribution of a continuous RV is said to be known, if either its pdf or cdf is given.*

Special Distributions

The probability mass functions of some discrete RVs and the probability density functions of some continuous RV*s*, which are of frequent applications, are as follows:

Discrete Distributions

1. If the discrete RV X can take the values $0, 1, 2, \ldots, n$, such that $P(X = i) = nC_i\, p^i\, q^{n-i}$, $i = 0, 1, \ldots, n$, where $p + q = 1$, then X is said to follow a *binomial distribution* with parameters n and p,which is denoted a $B(n, p)$.

2. If the discrete RV X can take the values 0, 1, 2, ..., such that $P(x = i) = \frac{e^{-\lambda} \lambda^i}{i!}$, $i = 0, 1, 2, K$, then X is said to follow a Poisson distribution with parameter λ.
3. If the discrete RV X can take the values 0, 1, 2, ..., such that $P(X = i) = (n + i - 1)C_i p^n q^i$, $i = 0, 1, 2, \ldots$, where $p + q = 1$, then X is said to follow a *Pascal (or negative binomial) distribution* with parameter n.
4. A Pascal distribution with parameter 1 [i.e., $P(X = i) = pq^i$, $i = 0, 1, 2, \ldots$ and $p + q = 1$] is called a *geometric distribution*.

Continuous Distributions

5. If the pdf of a continuous RV X is $f(x) = \frac{1}{b-a}$ (a constant), $a \le x \le b$, then X follows a *uniform distribution* (*or rectangular distribution*).
6. If the pdf of a continuous RV X is $f(x) = \frac{1}{\sigma\sqrt{2\pi}} e^{-(x-\mu)^2/2\sigma^2}$, $-\infty < x < \infty$, then X is said to follow a *normal distribution* (or *Gaussian distribution*) with parameters μ and σ, which will be hereafter denoted as $N(\mu, \sigma)$.
7. If the pdf of a continuous RV X is $f(x) = \frac{1}{\Gamma(n)} e^{-x} x^{n-1}$, $0 < x < \infty$ and $n > 0$, then X follows a *gamma distribution* with parameter n. Gamma distribution is a particular case of *Erlang distribution*, the pdf of which is

   $$f(x) = \frac{c^n}{\Gamma(n)} x^{n-1} e^{-cx}, \; 0 < x < \infty, \; n > 0, \; c > 0.$$
8. An Erlang distribution with $n = 1$ [i.e., $f(x) = ce^{-cx}$, $0 < x < \infty$, $c > 0$] is called *an exponential* (or *negative exponential*) *distribution* with parameter c.
9. If the pdf of a continuous RV X is $f(x) = \frac{x}{\alpha^2} e^{-x^2/2\alpha^2}$, $0 < x < \infty$, then X follows a *Rayleigh distribution* with parameter α.
10. If the pdf of a continuous RV X is $f(x) = \frac{\sqrt{2}}{\alpha^3\sqrt{\pi}} x^2 e^{-x^2/2\alpha^2}$, $0 < x < \infty$, then X follows a *Maxwell distribution* with parameter α.
11. If the pdf of a continuous RV X is $f(x) = \frac{1}{2\lambda} e^{-|x-\mu|/\lambda}$, $-\infty < x < \infty$, $\lambda > 0$, then X follows a *Laplace* (or *double exponential*) *distribution* with parameters λ and μ.

12. If the pdf of a continuous RV X is $f(x) = \frac{\alpha}{\pi} \times \frac{1}{x^2 + \alpha^2}$, $\alpha > 0$, $-\infty < x < \infty$, then X follows a *Cauchy distribution* with parameter a.

Worked Examples 2(A)

Example 1 *From a lot containing 25 items, 5 of which are defective, 4 items are chosen at random. If X is the number of defectives found, obtain the probability distribution of X, when the items are chosen (i) without replacement and (ii) with replacement.*

Solution Since only 4 items are chosen, X can take the values 0, 1, 2, 3 and 4. The lot contains 20 non-defective and 5 defective items.

Case (i): When the items are chosen without replacement, we can assume that all the 4 items are chosen simultaneously.

$$\therefore \quad P(X = r) = P(\text{choosing exactly } r \text{ defective items})$$
$$= P(\text{choosing } r \text{ defective and } (4 - r) \text{ good items})$$
$$= \frac{5C_r \times 20C_{4-r}}{25C_4} \quad (r = 0, 1, \ldots, 4)$$

Case (ii): When the items are chosen with replacement, we note that the probability of an item being defective remains the same in each draw.

i.e., $$p = \frac{5}{25} = \frac{1}{5},\ q = \frac{4}{5} \text{ and } n = 4$$

The problem is one of performing 4 Bernoulli's trials and finding the probability of exactly r successes.

$$\therefore \quad P(X = r) = 4C_r \left(\frac{1}{5}\right)^r \left(\frac{4}{5}\right)^{4-r} \quad (r = 0, 1, \ldots, 4)$$

Example 2 *A shipment of 6 television sets contains 2 defective sets. A hotel makes a random purchase of 3 of the sets. If X is the number of defective sets purchased by the hotel, find the probability distribution of X.* *(MU — Apr. 96)*

Solution All the 3 sets are purchased simultaneously. Since there are only 2 defective sets in the lot, X can take the values 0, 1 and 2.

$$\therefore \quad P(X = r) = P(\text{choosing exactly } r \text{ defective sets})$$
$$= P[\text{choosing } r \text{ defective and } (3 - r) \text{ good sets}]$$
$$= \frac{2C_r \times 4C_{3-r}}{6C_3} \quad (r = 0, 1, 2)$$

The required probability distribution is represented in the form of the following table.

$X = r$	p_r
0	1/5
1	3/5
2	1/5
Total	1

Example 3 *A random variable X may assume 4 values with probabilities (1 + 3x)/4, (1 – x)/4, (1 + 2x)/4 and (1 – 4x)/4. Find the condition on x so that these values represent the probability function of X?*

Solution $P(X = x_1) = p_1 = (1 + 3x)/4;\ p_2 = (1 - x)/4;$

$$p_3 = (1 + 2x)/4;\ p_4 = (1 - 4x)/4$$

If the given probabilities represent a probability function, each $p_i \geq 0$ and

$$\sum_i p_i = 1.$$

In this problem, $p_1 + p_2 + p_3 + p_4 = 1$, for any x.

But $p_1 \geq 0$, if $x \geq -1/3$; $p_2 \geq 0$, if $x \leq 1$; $p_3 \geq 0$, if $x \geq -1/2$ and $p_4 \geq 0$, if $x \leq 1/4$.

Therefore, the values of x for which a probability function is defined lie in the range $-1/3 \leq x \leq 1/4$.

Example 4 *If the random variable X takes the values 1, 2, 3 and 4 such that 2P(X = 1) = 3P(X = 2) = P(X = 3) = 5P(X = 4), find the probability distribution and cumulative distribution function of X.*

Solution Let $P(X = 3) = 30K$. Since $2P(X = 1) = 30K$, $P(X = 1) = 15K$.

Similarly $P(X = 2) = 10K$ and $P(X = 4) = 6K$.

Since S $p_i = 1$, $15K + 10K + 30K + 6K = 1$.

$$\therefore \quad K = \frac{1}{61}$$

The probability distribution of X is given in the following table:

$X = i$	1	2	3	4
p_i	$\frac{15}{61}$	$\frac{10}{61}$	$\frac{30}{61}$	$\frac{6}{61}$

The cdf $F(x)$ is defined as $F(x) = P(X \leq x)$. Accordingly the cdf for the above distribution is found out as follows:

When $x < 1$, $F(x) = 0$

When $1 \leq x < 2$, $F(x) = P(X = 1) = \dfrac{15}{61}$

When $2 \leq x < 3$, $F(x) = P(X = 1) + P(X = 2) = \dfrac{25}{61}$

When $3 \leq x < 4$, $F(x) = P(X = 1) + P(X = 2) + P(X = 3) = \dfrac{55}{61}$

When $x \geq 4$, $F(x) = P(x = 1) + P(x = 2) + P(x = 3) + P(x = 4) = 1$.

Example 5 *A random variable X has the following probability distribution.*

x:	*– 2*	*– 1*	*0*	*1*	*2*	*3*
p(x):	*0.1*	*K*	*0.2*	*2K*	*0.3*	*3K*

(a) Find K, (b) Evaluate P(X < 2) and P(–2 < X < 2), (c) find the cdf of X and (d) evaluate the mean of X. (BU — Apr. 96)

Solution (a) Since S $P(x) = 1$, $6K + 0.6 = 1$

$$\therefore \quad K = \frac{1}{15}$$

∴ the probability distribution becomes

x	:	-2	-1	0	1	2	3
$p(x)$	:	1/10	1/15	1/5	2/15	3/10	1/5

(b) $P(X < 2) = P(X = -2, -1, 0 \text{ or } 1)$

$= P(X = -2) + P(X = -1) + P(X = 0) + P(X = 1)$

[since the events $(X = -2)$, $(X = -1)$ etc. are mutually exclusive]

$$= \frac{1}{10} + \frac{1}{15} + \frac{1}{5} + \frac{2}{15} = \frac{1}{2}$$

$P(-2 < X < 2) = P(X = -1, 0 \text{ or } 1)$

$= P(X = -1) + P(X = 0) + P(X = 1)$

$$= \frac{1}{15} + \frac{1}{5} + \frac{2}{15} = \frac{2}{5}$$

(c) $F(x) = 0$, when $x < -2$

$= \frac{1}{10}$, when $-2 \le x < -1$

$= \frac{1}{6}$, when $-1 \le x < 0$

$= \frac{11}{30}$, when $0 \le x < 1$

$= \frac{1}{2}$, when $1 \le x < 2$

$= \frac{4}{5}$, when $2 \le x < 3$

$= 1$, when $3 \le x$

(d) The mean of X is defined as $E(X) = \Sigma x p(x)$ (refer to Chapter 4)

$$\therefore \quad \text{Mean of } X = \left(-2 \times \frac{1}{10}\right) + \left(-1 \times \frac{1}{15}\right) + \left(0 \times \frac{1}{5}\right) + \left(1 \times \frac{2}{15}\right) + \left(2 \times \frac{3}{10}\right) + \left(3 \times \frac{1}{5}\right)$$

$$= -\frac{1}{5} - \frac{1}{15} + \frac{2}{15} + \frac{3}{5} + \frac{3}{5} = \frac{16}{15}$$

Example 6 *The probability function of an infinite discrete distribution is given by* $P(X = j) = 1/2^j$ $(j = 1, 2, \ldots, \infty)$. *Verify that the total probability is 1 and find the mean and variance of the distribution. Find also P(X is even),* $P(X \le 5)$ *and P(X is divisible by 3).*

Solution Let $P(X = j) = p_j$

$$\sum_{j=1}^{\infty} p_j = \frac{1}{2} + \frac{1}{2^2} + \frac{1}{2^3} + \cdots \infty, \text{ that is a geometric series.}$$

$$= \frac{\frac{1}{2}}{1 - \frac{1}{2}} = 1$$

The mean of X is defined as $E(X) = \sum_{j=1}^{\infty} jp_j$ (refer to Chapter 4).

$$\therefore \qquad E(X) = a + 2a^2 + 3a^3 + \ldots \infty, \text{ where } a = \frac{1}{2}$$

$$= a(1 + 2a + 3a^2 + \ldots \infty)$$

$$= a(1 - a)^{-2} = \frac{\frac{1}{2}}{\left(\frac{1}{2}\right)^2} = 2$$

The variance of X is defined as $V(X) = E(X^2) - [E(X)]^2$,

where $E(X^2) = \sum_{j=1}^{\infty} j^2 p^j$ (refer to Chapter 4).

$$E(X^2) = \sum_{j=1}^{\infty} j^2 a^j, \text{ where } a = \frac{1}{2}$$

$$= \sum_{j=1}^{\infty} [j(j+1) - j]a^j = \sum_{j=1}^{\infty} j(j+1)a^j - \sum_{j=1}^{\infty} ja^j$$

$$= a(1.2 + 2.3a + 3.4a^2 + \ldots \infty) - a(1 + 2a + 3a^2 + \ldots \infty)$$

$$= a \times 2(1 - a)^3 - a \times (1 - a)^{-2}$$

$$= \frac{2a}{(1-a)^3} - \frac{a}{(1-a)^2} = 8 - 2 = 6$$

$$\therefore \qquad V(X) = E(X^2) - \{E(X)\}^2 = 6 - 4 = 2$$

$P(X \text{ is even}) = P(X = 2 \text{ or } X = 4 \text{ or } X = 6 \text{ or etc.})$

$P(X = 2) + P(X = 4) + \ldots + \infty$

(since the events are mutually exclusive)

$$= \left(\frac{1}{2}\right)^2 + \left(\frac{1}{2}\right)^4 + \left(\frac{1}{2}\right)^6 + \cdots + \infty$$

$$= \frac{\frac{1}{4}}{1-\frac{1}{4}} = \frac{1}{3}$$

$$P(X \ge 5) = P(X = 5 \text{ or } X = 6 \text{ or } X = 7 \text{ or etc.})$$
$$= P(X = 5) + P(X = 6) + \ldots + \infty$$

$$= \frac{\frac{1}{2^5}}{1-\frac{1}{2}} = \frac{1}{16}$$

$$P(X \text{ is divisible by } 3) = P(X = 3 \text{ or } X = 6 \text{ or } X = 9 \text{ etc.})$$
$$= P(X = 3) + P(X = 6) + \ldots + \infty$$

$$= \left(\frac{1}{2}\right)^3 + \left(\frac{1}{2}\right)^6 + \left(\frac{1}{2}\right)^9 + \cdots + \infty$$

$$= \frac{\frac{1}{8}}{1-\frac{1}{8}} = \frac{1}{7}$$

Example 7 *A random variable X has the following probability distribution.*

x	:	0	1	2	3	4	5	6	7
$p(x)$	:	0	K	$2K$	$2K$	$3K$	K^2	$2K^2$	$7K^2 + K$

Find (i) the value of K, (ii) $P(1.5 < X < 4.5/X > 2)$ and (iii) the smallest value of λ for which $P(X \le \lambda) > 1/2$.

Solution $\Sigma\, p(x) = 1$

$\therefore$ $10K^2 + 9K = 1$

i.e., $(10K - 1)(K + 1) = 0$

$\therefore$ $K = \dfrac{1}{10}$ or -1.

The value $K = -1$ makes some values of $p(x)$ negative, which is meaningless.

$\therefore$ $K = \dfrac{1}{10}$

The actual distribution is given below:

x	:	0	1	2	3	4	5	6	7
$p(x)$	:	0	$\frac{1}{10}$	$\frac{2}{10}$	$\frac{2}{10}$	$\frac{3}{10}$	$\frac{1}{100}$	$\frac{2}{100}$	$\frac{17}{100}$

(i) $P(1.5 < X < 4.5 / X > 2) = P(A/B)$, say

$$= \frac{P(A \cap B)}{P(B)}$$

$$= \frac{P[(1\cdot5 < X < 4\cdot5) \cap (X > 2)]}{P(X > 2)}$$

$$= \frac{P(X=3) + P(X=4)}{\sum_{r=3}^{7} (X=r)} = \frac{\frac{5}{10}}{\frac{7}{10}} = \frac{5}{7}$$

(ii) By trials, $P(X \le 0) = 0$; $P(X \le 1) = \frac{1}{10}$; $P(X \le 2) = \frac{3}{10}$

$$P(X \le 3) = \frac{5}{10};\ P(X \le 4) = \frac{8}{10}$$

Therefore, the smallest value of λ satisfying the condition $P(X \le l) > 1/2$ is 4.

Example 8 *If* $p(x) = \begin{cases} x\,e^{-x^2/2} & x \ge 0 \\ 0 & x < 0 \end{cases}$

(a) show that p(x) is a pdf (of a continuous RV X.)
(b) find its distribution function P(x). (BU — Nov. 96)

Solution (a) If $p(x)$ is to be a pdf, $p(x) \ge 0$ and

$$\int_{R_X} p(x)\,dx = 1$$

Obviously, $p(x) = xe^{-x^2/2}$ 0, when $x \ge 0$

Now $\int_0^\infty p(x)dx = \int_0^\infty xe^{-x^2/2}\,dx = \int_0^\infty e^{-t}\,dt$ (putting $t = x^2/2$) = 1

$\therefore$ $p(x)$ is a legitimate pdf of a RV X.

$$F(x) = P(X \le x) = \int_0^x f(x)dx$$

$\therefore$ $F(x) = 0$, when $x < 0$

and $F(x) = \int_0^X xe^{-x^2/2}\,dx = 1 - e^{-x^2/2}$, when $x \ge 0$.

Example 9 *If the density function of a continuous RV X is given by*

$$\begin{aligned} f(x) &= ax, & 0 \le x \le 1 \\ &= a, & 1 \le x \le 2 \\ &= 3a - ax, & 2 \le x \le 3 \\ &= 0, & \text{elsewhere} \end{aligned}$$

(*i*) *find the value of a*
(*ii*) *find the cdf of X*
(*iii*) *If x_1, x_2 and x_3 are 3 independent observations of X, what is the probability that exactly one of these 3 is greater than 1.5?*

Solution (i) Since $f(x)$ is a pdf, $\int_{R_X} f(x)dx = 1$

i.e., $\int_0^3 f(x)dx = 1$

i.e., $\int_0^1 ax\,dx + \int_1^2 a\,dx + \int_2^3 (3a - ax)dx = 1$

i.e., $2a = 1$

$\therefore$ $a = \dfrac{1}{2}$

(ii) $F(x) = P(X \le x) = 0$, when $x < 0$

$$F(x) = \int_0^x \frac{x}{2}\,dx = \frac{x^2}{4}, \text{ when } 0 \le x \le 1$$

$$= \int_0^1 \frac{x}{2}\,dx + \int_1^x \frac{1}{2}\,dx = \frac{x}{2} - \frac{1}{4} \text{ when } 1 \le x \le 2$$

$$= \int_0^1 \frac{x}{2}\,dx + \int_0^1 \frac{1}{2}\,dx + \int_2^x \left(\frac{3}{2} - \frac{x}{2}\right)dx$$

$$= \frac{3}{2}x - \frac{x^2}{4} - \frac{5}{4}, \text{ when } 2 \le x \le 3$$

$$= 1, \text{ when } x > 3$$

(iii) $p(X > 1.5) = \int_{1.5}^3 f(x)\,dx$

$$= \int_{1.5}^2 \frac{1}{2}\,dx + \int_2^3 \left(\frac{3}{2} - \frac{x}{2}\right)dx = \frac{1}{2}$$

Choosing an X and observing its value can be considered as a trial and ($X > 1.5$) can be considered a success.

$\therefore$ $p = 1/2$, $q = 1/2$

As we choose 3 independent observations of X, $n = 3$.

By Bernoulli's theorem,

P(exactly one value > 1.5)

$$= P(1 \text{ success}) = 3C_1 \times (p)^1 \times (q)^2 = \frac{3}{8}$$

Example 10 *A continuous RV X that can assume any value between x = 2 and x = 5 has a density function given by f(x) = k(1 + x). Find P(X < 4).*

(MU — Apr. 96)

Solution By the property of pdf,

$\int_{R_x} f(x)\, dx = 1$. X takes values between 2 and 5.

$$\therefore \quad \int_2^5 k(1+x)dx = 1$$

$$\text{i.e.,} \quad \frac{27}{2}k = 1$$

$$\therefore \quad k = \frac{2}{27}$$

Now $p(X < 4) = p(2 < X < 4) = \int_2^4 k(1+x)dx = \frac{16}{27}$

Example 11 *A continuous RV X has a pdf $f(x) = kx^2e^{-x}$; $x \geq 0$. Find k, mean and variance.*

(MKU — Apr. 97)

Solution By the property of pdf,

$$\int_0^\infty kx^2e^{-x}dx = 1$$

$$\text{i.e.,} \quad 2k = 1$$

$$\therefore \quad k = \frac{1}{2}$$

Mean of X is defined as

$$E(X) = \int_{R_x} xf(x)dx$$

(refer to Chapter 4)

Variance of X is defined as

$$V(X) = E(X^2) - \{E(X)\}^2,$$

where $E(X^2) = \int_{R_x} x^2 f(x)dx$ (refer to Chapter 4)

$$\therefore \quad E(X) = \frac{1}{2}\int_0^\infty x^3e^{-x}dx$$

$$= \frac{1}{2}\left[x^3(-e^{-x}) - 3x^2(e^{-x}) + 6x(-e^{-x}) - 6(e^{-x})\right]_0^\infty$$

$$= 3$$

$$E(X^2) = \frac{1}{2}\int_0^\infty x^4 e^{-x} dx$$

$$= \frac{1}{2}\left[x^4(-e^{-x}) - 4x^3(e^{-x}) + 12x^2(-e^{-x}) - 24x(e^{-x}) + 24(-e^{-x})\right]_0^\infty$$

$$= 12$$

$$\therefore \quad V(X) = E(X^2) - \{E(X)\}^2 = 3$$

Example 12 *The probability that a person will die in the time interval (t_1, t_2) is given by*

$$p(t_1 \le t \le t_2) = \int_{t_1}^{t_2} a(t)dt.$$

Solution The function $a(t)$ is determined from long records and can be assumed to be

$$a(t) = \begin{cases} 3 \times 10^{-9} t^2 (100 - t)^2 & 0 \le t \le 100 \\ 0 & \text{elsewhere} \end{cases}$$

Determine (i) the probability that a person will die between the ages 60 and 70 and (ii) the probability that he will die between those ages, assuming he lived upto 60. (MSU — Apr. 96)

(i) $$P(60 < t < 70) = \int_{60}^{70} a(t)dt$$

$$= 3 \times 10^{-9} \int_{60}^{70} t^2(100 - t)^2 dt$$

$$= 0.1544$$

(ii) $$P(60 < t < 70/t \ge 60) = P(60 < t < 70/60 \le t \le 100)$$

$$= \frac{P(60 < t < 70)}{P(60 < t < 100)}$$

$$= \int_{60}^{70} a(t)dt \Big/ \int_{60}^{100} a(t)dt$$

$$= \frac{0 \cdot 15436}{0 \cdot 31744} = 0.4863$$

Example 13 *A continuous RV has a pdf $f(x) = 3x^2$, $0 \le x \le 1$. Find a and b such that*

(i) $P(X \le a) = P(X > a)$ and

(ii) $P(X > b) = 0.05$ (BDU — Nov. 96)

Solution

(i) $P(X \le a) = P(X > a)$

$$\therefore \quad \int_0^a 3x^2 dx = \int_a^1 3x^2 dx$$

i.e.,
$$a^3 = 1 - a^3$$

i.e.,
$$a^3 = \frac{1}{2}$$

$$\therefore \quad a = 0.7937$$

(ii) $P(X > b) = 0.05$

$$\int_b^1 3x^2 dx = 0.05$$

i.e.,
$$b^3 = 95$$

$$\therefore \quad b = 0.9830$$

Example 14 *The distribution function of a RV X is given by $F(x) = 1 - (1 + x)e^{-x}$, $x \ge 0$. Find the density function, mean and variance of X.* (MKU — Nov. 96)

Solution By the property of $F(x)$, the pdf $f(x)$ is given by $f(x) = F'(x)$ at points of continuity of $F(x)$.

The given cdf is continuous for $x \ge 0$.

$$\therefore \quad f(x) = (1 + x)e^{-x} - e^{-x} = xe^{-x}, x \ge 0$$

$$E(X) = \int_0^\infty x^2 e^{-x} dx = 2$$

$$E(X^2) = \int_0^\infty x^3 e^{-x} dx = 6$$

$$V(X) = E(X^2) - [E(X)]^2 = 2$$

Example 15 *The cdf of a continuous RV X is given by*

$$F(x) = 0, \; x < 0$$

$$= x^2, \; 0 \le x < \frac{1}{2}$$

$$= 1 - \frac{3}{25}(3 - x)^2, \; \frac{1}{2} \ge x < 3$$

$$= 1, \; x \ge 3$$

Find the pdf of X and evaluate $P(|X| \le 1)$ and $P(\frac{1}{3} \le X < 4)$ using both the pdf and cdf.

Solution The points $x = 0$, 1/2 and 3 are points of continuity

$\therefore$ $$f(x) = 0, x < 0$$

$$= 2x, 0 \le x < \frac{1}{2}$$

$$= \frac{6}{25}(3 - x), \frac{1}{2} \le x < 3$$

$$= 0, x ^3 3$$

Although the points $x = 1/2$, 3 are points of discontinuity for $f(x)$, we may assume that $f\left(\frac{1}{2}\right) = \frac{3}{5}$ and $f(3) = 0$.

$$P(|X| \le 1) = p(-1 \le x \le 1)$$

$$= \int_{-1}^{1} f(x)dx = \int_{0}^{1/2} 2x dx + \int_{1/2}^{1} \frac{6}{25}(3 - x)\,dx \text{ (using property of pdf)}$$

$$= \frac{13}{25}$$

If we use property of cdf

$$P(|X| \le 1) = P(-1 \le x \le 1) = F(1) - F(-1) = \frac{13}{25}$$

If we use the property of pdf

$$P(1/3 \le X < 4) = \int_{1/3}^{1/2} 2x dx + \int_{1/2}^{3} \frac{6}{25}(3 - x)dx = \frac{8}{9}$$

If we use the property of cdf

$$P(1/3 \le X < 4) = F(4) - F\left(\frac{1}{3}\right) = 1 - \frac{1}{9} = \frac{8}{9}.$$

Example 16 *If the RV k is uniformly distributed over (0, 5) what is the probability that the roots of the equation $4x^2 + 4kx + (k + 2) = 0$ are real?*

Solution The RV k is U (0, 5).

$\therefore$ $$\text{pdf of } k = \frac{1}{5}, 0 < k < 5$$

P(Roots of $4x^2 + 4kx + k + 2 = 0$ are real)

$= P$(Discriminant of the equation ≥ 0)

$= P(k^2 - k - 2 \ge 0) = P[(k - 2)(k + 1) \ge 0]$

$= P[(k \ge -1 \text{ and } k \ge 2) \text{ or } (k \le 2 \text{ and } k \le -1)]$

$= P(k \ge 2 \text{ or } k \le -1) = P(k \ge 2)$ [*since k takes values in* (0, 5)]

$$= \int_{2}^{5} f(k)dk = \frac{1}{5}(5 - 2) = \frac{3}{5}.$$

Example 17 *A point P is taken at random on a line AB of length 2a, all positions of the point being equally likely. Find the probability that the product* $(AP \times PB) > \dfrac{a^2}{2}$.

Solution Let $AP = X$.

$\therefore$ $PB = (2a - X)$

Since all positions of the point P are equally likely, $X(= AP)$ is uniformly distributed over $(0, 2a)$.

$$\therefore \quad \text{pdf of } X = \frac{1}{2a},\ 0 < x < 2a$$

$$P[AP \times PB > \frac{a^2}{2}] = P[X(2a - X) > \frac{a^2}{2}]$$

$$= P(2X^2 - 4aX + a^2 < 0)$$

$$= P\left[\left\{X - \left(1 - \frac{1}{\sqrt{2}}\right)a\right\}\left\{X - \left(1 + \frac{1}{\sqrt{2}}\right)a\right\} < 0\right]$$ [since the factors of $(2x^2 - 4ax + a^2)$ are $x - \left(1 - \frac{1}{\sqrt{2}}\right)a$ and $x - \left(1 + \frac{1}{\sqrt{2}}\right)a$]

$$= P\left[\left(1 - \frac{1}{\sqrt{2}}\right)a < X < \left(1 + \frac{1}{\sqrt{2}}\right)a\right]$$

$$= \int_{\left(1 - \frac{1}{\sqrt{2}}\right)a}^{\left(1 + \frac{1}{\sqrt{2}}\right)a} f(x)\,dx = \frac{1}{2a}\,[x]_{\left(1 - \frac{1}{\sqrt{2}}\right)a}^{\left(1 + \frac{1}{\sqrt{2}}\right)a} = \frac{\sqrt{2}\,a}{2a} = \frac{1}{\sqrt{2}}$$

Example 18 *If the continuous RV. X represents the time of failure of a system, that has been put into operation at t = 0, find the conditional density function of X, given that the system has survived upto time t. Deduce the same when X follows an exponential distribution with parameter λ.*

Solution The conditional distribution function of X, subject to the given condition, is given by

$$F(x/X > t) = \frac{P[X \le x \text{ and } X > t]}{P(X > t)}$$ [since unconditional $F(x) = P(X \le x)$]

$$= \frac{P[t < X \le x]}{P[t < X < \infty]}$$

$$= \frac{F(x) - F(t)}{1 - F(t)} \text{ for } x > t$$

$$= 0 \text{ for } x < t$$

Therefore, the conditional density function $f(x/X > t)$ is given by

$$f(x/X > t) = \frac{d}{dx} F(x/X > t)$$

$$= \frac{f(x)}{1 - F(t)}, \; x > t$$

For the exponential distribution with parameter λ,

$$f(x) = \lambda e^{-\lambda x}, \; x > 0, \text{ and } F(x) = \int_0^X \lambda e^{-\lambda x} dx = 1 - e^{-\lambda x}.$$

$$\therefore \quad f(x/X > t) = \frac{\lambda e^{-\lambda x}}{e^{-\lambda t}} = \lambda e^{-\lambda(x-t)} = f(x - t)$$

Example 19 *If $f(t)$ is the unconditional density of the time of failure of a system and $h(t)$ is the hazard rate (or conditional failure rate or conditional density of T, given $T > t$) find $f(t)$ in terms of $h(t)$. Deduce that T follows a Rayleigh distribution, when $h(t) = t$.*

Solution The conditional density of T, given $T > t$ or the hazard rate, is given by $h(t) = f(t/T > t)$

$$= \frac{f(t)}{1 - F(t)} = \frac{F'(t)}{1 - F(t)}$$

$$\therefore \quad \int_0^t h(t) dt = \int_0^t \frac{F'(t)}{1 - F(t)} dt$$

$$= [-\log\{1 - F(t)\}]_0^t$$

$= -\log\{1 - F(t)\}$ [since $F(0) = P(T \le 0) = 0$ as the system was put into operation at $t = 0$]

$$\therefore \quad F(t) = 1 - e^{-\int_0^t h(t) dt}$$

$$\therefore \quad f(t) = h(t) \times e^{-\int_0^t h(t) dt}$$

When $h(t) = t$,

$$f(t) = t \times e^{-\int_0^t t dt}$$

$= te^{-t^2/2}$, which is the pdf of a Raleigh distribution

Example 20 *If a continuous RV X follows N(0, 2), find* $P\{1 \le X \le 2\}$ *and* $P\{1 \le X \le 2/X \ge 1\}$.

Solution X follows $N(0, 2)$, the density function of which is $f(x) = \frac{1}{2\sqrt{2\pi}} e^{-x^2/8}$,

$-\infty < x < \infty$.

$$\therefore \quad P\{1 \le X \le 2\} = \int_1^2 f(x)dx$$

$$= \int_{0.5}^{1} \phi(t)dt, \text{ putting } t = \frac{x}{2}$$

where $\phi(t)$ is the standard normal density.

$$= \int_0^1 \phi(t)dt - \int_0^{0.5} \phi(t)dt$$

$= 0.3413 - 0.1915$ (from the normal tables)

$= 0.1498$

$$P\{1 \le X \le 2/X \ge 1\} = \frac{P\{(1 \le X \le 2) \text{ and } X \ge 1\}}{P\{X \ge 1\}}$$

$$= \frac{P\{1 \le X \le 2\}}{P\{1 \le X < \infty\}}$$

$$= \frac{0.1498}{\int_0^{\infty} \phi(t)\,dt - \int_0^{0.5} \phi(t)\,dt}$$

$$= \frac{0.1498}{0.5 - 0.1915} = 0.4856.$$

Exercise 2 (A)

Part A (Short answer questions)

1. Define a RV with an example.
2. Define a discrete RV with an example.
3. Define a continuous RV and give an example for the same.
4. Distinguish between a discrete RV and a continuous RV.
5. Define the probability mass function of a discrete RV.
6. Write down the probability distribution of the outcome when 2 fair dice are tossed.
7. Define the pdf of a continuous RV.
8. State the properties of the pdf of a continuous RV.
9. What is the probability curve of a continuous RV? Give an example.

10. Prove that it is almost impossible that a continuous RV assumes a specific value. (OR) If X is a continuous RV prove that $P(X = a) = 0$.
11. If X represents the total number of heads obtained, when a fair coin is tossed 5 times, find the probability distribution of X.
12. If the probability distribution of X is given as:

x:	1	2	3	4
p_x:	0.4	0.3	0.2	0.1

find $P\left(\frac{1}{2} < X < \frac{7}{2} / X > 1\right)$
13. Define the cdf of a RV. Explain how to find it for both kinds of RV.
14. Differentiate between the pdf and cdf of a RV.
15. State the properties of the cdf of a RV.
16. Verify whether $f(x) = \begin{cases} |x| & \text{in } -1 \le x \le 1 \\ 0, & \text{elsewhere} \end{cases}$

can be the pdf of a continuous RV.
17. If $f(x) = kx^2$, $0 < x < 3$, is to be a density function, find the value of k.
18. If the pdf of a RV X is given by

$f(x) = \begin{cases} 1/4 & \text{in } -2 < x < 2 \\ 0, & \text{elsewhere} \end{cases}$

find $P\{|X| > 1\}$.
19. Find the value of k, if $f(x) = \begin{cases} k\,x\,e^{-x} & x > 0 \\ 0 & \text{elsewhere} \end{cases}$

is the pdf of a RV X.
20. If the pdf of a RV X is $f(x) = \frac{x}{2}$ in $0 \le x \le 2$, find $P\{X > 1.5/X > 1)$.
21. The RV X has the following probability distribution:

x :	-2	-1	0	1
p_x:	0.4	k	0.2	0.3

find k and the mean value of X.
22. If X represents the outcome of the toss of a 6 faced dice, find $P(X \le x)$ as a function of x.
23. If the pdf of a RV X is $f(x) = 2x$, $0 < x < 1$, find the cdf of X.
24. If the cdf of a RV X is given by $f(x) = 1 - e^{-lx}$, when $x \ge 0$ and $= 0$, when $x < 0$, find the pdf of X.
25. If the cdf of a RV is given by $F(x) = 0$, for $x < 0$; $= x^2/16$ for $0 \le x < 4$ and $= 1$, for $4 \le x$, find $P(X > 1/X < 3)$.
26. Define binomial distribution. What are its mean and variance?
27. Give the probability law of Poisson distribution and also its mean and variance.
28. Define the exponential distribution.
29. If X follows an exponential distribution with parameter 1, find $P(|X| < 1)$.

30. Define Pascal distribution and define geometric distribution as a particular case of Pascal distribution.
31. Write down the pdf's of general normal distribution and standard normal distribution.
32. Define Erlang distribution. Deduce Gamma distribution as a particular case of Erlang distribution.
33. Deduce the pdf of an exponential distribution as a particular case of that of Erlang distribution.
34. Give the pdf of Raleigh distribution.
35. Define Maxwell distribution.
36. Write down the pdf of Laplace distribution.
37. Define Cauchy distribution.

Part B

38. Find the formula for the probability distribution of the number of heads, when a fair coin is tossed 4 times. (MU — Nov. 96)
39. A coin is known to come up heads 3 times as often as tails. This coin is tossed 3 times. Write down the probability distribution of the number of heads that appear and also the cdf. Make a sketch of both.
40. Consider the experiment of tossing a coin, the 2 events of the space being occurrence of head of tail. Assign probabilities p and q for head and tail respectively and define a random variable X by $X(h) = 1$ and $x(t) = 0$. Determine and plot the probability function $f(x)$ and the distribution function $F(x)$. (MSU — Apr. 96)
41. Two dice are tossed. If X is the sum of the numbers shown up, find the probability mass function of X.
42. Consider the experiment of tossing a fair coin 4 times. Define $X = 0$, if 0 or 1 head appears; $X = 1$, if 2 heads appear; $X = 2$, if 3 or 4 heads appear. Find the probability function, mean and variance of X.
43. A discrete RV X has the following probability distribution.

x:	0	1	2	3	4	5	6	7	8
$p(x)$:	a	$3a$	$5a$	$7a$	$9a$	$11a$	$13a$	$15a$	$17a$

Find the value of a, $P(X < 3)$, variance and distribution function of X. (MKU — Apr. 97)

44. The probability distribution of a RV X is given below:

x:	0	1	2	3
$p(x)$:	0.1	0.3	0.5	0.1

If $Y = X^2 + 2X$, find the probability distribution, mean and variance of Y.

45. The probability mass function of a RV X is defined as $P(X = 0) = 3C^2$, $P(X = 1) = 4C - 10C^2$ and $P(X = 2) = 5C - 1$, where $C > 0$ and $P(X = r) = 0$, if $r \neq 0, 1, 2$.
(i) Find the value of C, (ii) Find $P\{0 < X < 2/X > 0\}$, (iii) the distribution function of X, (iv) the largest value of X for which $F(x) < \frac{1}{2}$ and (v) the

smallest value of X for which $F(x) > \frac{1}{2}$.

46. If the probability mass function of a RV X is given by $P(X = r) = kr^3$; $r = 1, 2, 3, 4$, find (i) the value of k (ii) $P(1/2 < X < 5/2 / X > 1)$, (iii) the mean and variance of X and (iv) the distribution function of X.
47. Find the values of a for which $P(x = j) = (1 - a)a^j$, $j = 0, 1, 2$, K represents a probability mass function. Show also that for any 2 positive integers m and n

$$P(X > m + n/X > m) = P(X \geq n).$$

48. If a discrete probability distribution is given by $P(X = r) = k(1 - a)^{r-1}$, $0 < a < 1$, for $r = 1, 2, \ldots, \infty$, find the value of k and also the mean and variance of X.
49. If the probability distribution of a discrete RV X is given by $P(X = x) = ke^{-t}(1 - e^{-t})^{x-1}$, $x = 1, 2, \ldots, \infty$, find the value of k and also the mean and variance of X.
50. In a continuous distribution, the probability density is given by $f(x) = kx(2 - x)$, $0 < x < 2$. Find k, mean, variance and the distribution function. (MKU — Nov. 96)
51. The diameter of an electric cable X is a continuous RV with pdf $f(x) = kx(1 - x)$, $0 \leq x \leq 1$. Find (i) the value of k, (ii) cdf of X, (iii) the value of a such that $P(X < a) = 2P(X > a)$ and (iv) $P(X \leq 1/2 / 1/3 < X < 2/3)$.
52. X is a continuous RV with pdf given by $f(x) = kx$, in $0 \leq x \leq 2$; $= 2k$, in $2 \leq x \leq 4$, and $= 6k - kx$, in $4 \leq x \leq 6$. Find the value of k and $F(x)$.
53. The continuous RV X has pdf $f(x) = \frac{x}{2}$, $0 \leq x \leq 2$. Two independent determinations of X are made. What is the probability that both these determinations will be greater than 1? If 3 independent determinations had been made, what is the probability that exactly 2 of these are larger than 1?
54. A continuous RVX that can assume values between $x = 2$ and $x = 5$ has a density function given by $f(x) = 2(1 + x)/27$. Find $P(3 < X < 4)$. (MU — Nov. 96)
55. A continuous RV has the pdf $f(x) = kx^4$, $-1 < x < 0$. Find the value of k and also $P(X > -1/2 / X < -1/4)$.
56. Suppose that the life length of a certain radio tube (in hours) is a continuous RV X with pdf $f(x) = \frac{100}{x^2}$ $x > 100$ and $= 0$, elsewhere.
    (a) What is the probability that a tube will last less than 200 h, if it is known that the tube is still functioning after 150 h of service?
    (b) What is the probability that if 3 such tubes are installed in a set, exactly 1 will have to be replaced after 150 h of service?
    (c) What is the maximum number of tubes that may be inserted into a set so that there is a probability of 0.1 that after 150 h of service all of them are still functioning?

57. If the cdf of a continuous RV X is given by $F(x) = \frac{1}{2}e^{kx}$, $x \le 0$, and $F(x) = 1 - \frac{1}{2}e^{-kx}$, $x > 0$, find $P(|x| \le 1/k)$. Prove that the density function of X is $f(x) = \frac{k}{2}e^{-k|x|}$ $-\infty < x < \infty$, given that $k > 0$.
58. If the distribution function of a continuous RV X is given by $F(x) = 0$, when $x < 0$; $= x$, when $0 \le x < 1$ and $= 1$, when $1 \le x$, find the pdf of X. Also find $P(1/3 < X < 1/2)$ and $P(1/2 < X < 2)$ using the cdf of X.
59. A point is chosen on a line of length a at random. What is the probability that the ratio of the shorter to the longer segment is less than 1/4?
60. If the RV k is uniformly distributed over (1, 7) what is the probability that the roots of the equation $x^2 + 2kx + (2k + 3) = 0$ are real?
61. If $f(t)$ is the unconditional density of time to failure T of a system and $h(t)$ is the conditional density of T, given $T > t$, find $h(t)$ when (i) $f(t) = \lambda e^{-\lambda t}$ (ii) $f(t) = 1^2\, te^{-\lambda t}$, $t > 0$. Prove also that $h(t)$ is not a density function.
62. If the continuous RV X follows $N(1000, 20)$, find
 (i) $P(X < 1024)$, (ii) $P(X < 1024 / X > 961)$ and
 (iii) $P(31 < \sqrt{X} \le 32)$.

Two-Dimensional Random Variables

So far we have considered only the one-dimensional RV, i.e., we have considered such random experiments, the outcome of which had only one characteristic and hence was assigned a single real value. In many situations, we will be interested in recording 2 or more characteristics (numerically) of the outcome of a random experiment. For example, both voltage and current might be of interest in a certain experiment.

Definitions: Let S be the sample space associated with a random experiment E. Let $X = X(s)$ and $Y = Y(s)$ be two functions each assigning a real number to each outcomes s Î S. Then (X, Y) is called a *two-dimensional random variable.*

If the possible values of (X, Y) are finite or countably infinite, (X, Y) is called *a two-dimensional discrete* RV. When (X, Y) is a two-dimensional discrete RV the possible values of (X, Y) may be represented as (x_i, y_j), i = 1, 2, K, m, K; j = 1, 2, ..., n, ...

If (X, Y) can assume all values in a specified region R in the xy-plane, (X, Y) is called a *two-dimensional continuous* RV.

Probability Function of (X, Y)

If (X, Y) is a two-dimensional discrete RV such that $P(x = x_i, y = y_j) = p_{ij}$, then p_{ij} is called the *probability mass function* or simply the *probability function* of (X, Y) provided the following conditions are satisfied.

(i) $p_{ij} \ge 0$, for all i and j (ii) $\sum_j \sum_i p_{ij} = 1$

The set of triples $\{x_i, y_j, p_{ij}\}$, $i = 1, 2, ..., m, ..., j = 1, 2, K, n, ...$, is called *the joint probability distribution of (X, Y).*

Joint Probability Density Function

If (X, Y) is a two-dimensional continuous RV such that.

$P\left\{x - \frac{dx}{2} \le X \le x + \frac{dx}{2} \text{ and } y - \frac{dy}{2} \le Y \le y + \frac{dy}{2}\right\} = f(x, y)\,dx\,dy$, then $f(x, y)$ is called *the joint pdf* of (X, Y), provided $f(x, y)$ satisfies the following conditions.

(i) $f(x, y) \ge 0$, for all $(x, y) \in R$, where R is the range space.

(ii) $\iint_R f(x, y)\,dx\,dy = 1$.

Moreover if D is a subspace of the range space R, $P\{(X, Y) \in D\}$ is defined as

$$P\{(X, Y) \in D\} = \iint_D f(x, y)dx\,dy. \text{ In particular}$$

$$P\{a \le X \le b, c \le Y \le d\} = \int_c^d \int_a^b f(x, y)dx\,dy$$

Cumulative Distribution Function

If (X, Y) is a two-dimensional RV (discrete or continuous), then $F(x, y) = P\{X \le x \text{ and } Y \le y\}$ is called *the cdf of (X, Y).*

In the discrete case,

$$F(x, y) = \sum_{\substack{j \\ y_j \le y}} \sum_{\substack{i \\ x_i \le x}} p_{ij}$$

In the continuous case,

$$F(x, y) = \int_{-\infty}^{y} \int_{-\infty}^{x} f(x, y)dx\,dy$$

Properties of F(x, y)

(i) $F(-\infty, y) = 0 = F(x, -\infty)$ and $F(\infty,\infty) = 1$

(ii) $P\{a < X < b, Y \le y\} = F(b, y) - F(a, y)$

(iii) $P\{X \le x, c < Y < d\} = F(x, d) - F(x, c)$

(iv) $P\{a < X < b, c < Y < d\} = F(b, d) - F(a, d) - F(b, c) + F(a, c)$

(v) At points of continuity of $f(x, y)$

$$\frac{\partial^2 F}{\partial x\, \partial y} = f(x, y)$$

Marginal Probability Distribution

$$P(X = x_i) = P\{(X = x_i \text{ and } Y = y_1) \text{ or } (X = x_i \text{ and } Y = y_2) \text{ or etc.}\}$$
$$= p_{i1} + p_{i2} + \text{K} = \sum_j p_{ij}$$

$P(X = x_i) = \sum_j p_{ij}$ is called *the marginal probability function of* X. It is defined for $X = x_1, x_2, \ldots$ and denoted as P_{i*}. The collection of pairs $\{x_i, p_{i*}\}$, $i = 1, 2, 3, \ldots$ is called the *marginal probability distribution of* X.

Similarly the collection of pairs $\{y_j, p_{*j}\}$, $j = 1, 2, 3, \ldots$ is called *the marginal probability distribution of* Y, where $p_{*j} = \sum_i P_{ij} = P(Y = y_j)$.

In the continuous case,

$$P\{x - \frac{1}{2}\text{d}x \le X \le x + \frac{1}{2}\text{d}x, -\infty < Y < \infty\}$$
$$= \int_{-\infty}^{\infty} \int_{x - \frac{1}{2}\text{d}x}^{x + \frac{1}{2}\text{d}x} f(x, y)\, \text{d}x\, \text{d}y$$

$$= \left[\int_{-\infty}^{\infty} f(x, y)\, \text{d}y\right] \text{d}x \text{ [since } f(x, y) \text{ may be treated a constant in } (x - \frac{1}{2}\text{d}x, x + \frac{1}{2}\text{d}x)]$$

$$= f_X(x)\text{d}x, \text{ say}$$

$f_X(x) = \int_{-\infty}^{\infty} f(x, y)\text{d}y$ is called *the marginal density of* X.

Similarly, $f_Y(y) = \int_{-\infty}^{\infty} f(x, y)\text{d}x$ is called *the marginal density of* Y.

Note $P(a \le X \le b) = P(a \le X \le b, -\infty < Y < \infty)$

$$= \int_{-\infty}^{\infty} \int_a^b f(x, y)\text{d}x\, \text{d}y$$

$$= \int_a^b \left[\int_{-\infty}^{\infty} f(x, y)\, \text{d}y\right] \text{d}x = \int_a^b f_X(x)\text{d}x$$

Similarly, $P(c \le Y \le d) = \int_c^d f_Y(y)\text{d}y$

Conditional Probability Distribution

$P\{X = x_i / Y = y_j\} = \dfrac{P\{X = x_i\ Y = y_j\}}{P\{Y = y_j\}} = \dfrac{p_{ij}}{p_{*j}}$ is called *the conditional probability function of* X, given that $Y = y_j$.

The collection of pairs, $\left\{x_i, \dfrac{p_{ij}}{p_{*j}}\right\}$ $i = 1, 2, 3, \ldots,$

is called *the conditional probability distribution of* X*, given* $Y = y_j$.

Similarly, the collection of pairs, $\left\{Y_j, \dfrac{p_{ij}}{p_{i*}}\right\}, j = 1, 2, 3, \ldots,$ is called the *conditional probability distribution of* Y *given* $X = x_i$. In the continuous case,

$$P\left\{x - \frac{1}{2}\mathrm{d}x \le X < x + \frac{1}{2}\mathrm{d}x \,/\, Y = y\right\}$$

$$= P\left\{x - \frac{1}{2}\mathrm{d}x \le X \le x + \frac{1}{2}\mathrm{d}x \,/\, y - \frac{1}{2}\mathrm{d}y \le Y \le y + \frac{1}{2}\mathrm{d}y\right\}$$

$$= \frac{f(x, y)\,\mathrm{d}x\,\mathrm{d}y}{f_Y(y)\mathrm{d}y} = \left\{\frac{f(x, y)}{f_Y(y)}\right\}\mathrm{d}x.$$

$\dfrac{f(x, y)}{f_Y(y)}$ is called *the conditional density of* X*, given* Y, and is denoted by $f(x/y)$.

Similarly, $\dfrac{f(x, y)}{f_X(y)}$ is called *the conditional density of* Y, given X, and is denoted by $f(y/x)$.

Independent RVs

If (X, Y) is a two-dimensional discrete RV such that $P\{X = x_i / Y = y_j\} = P(X = x_i)$ i.e., $\dfrac{p_{ij}}{p_{*j}} = p_{i*}$, i.e., $p_{ij} = p_{i*} \times p_{*j}$ for all i, j then X and Y are said to be independent RVs.

Similarly if (X, Y) is a two-dimensional continuous RV such that $f(x, y) = f_X(x) \times f_Y(y)$, then X and Y are said to be independent RVs.

Random Vectors

Sometimes we may have to be concerned with Random experiments whose outcomes will have 3 or more simultaneous numerical characteristics. To study the outcomes of such random experiments we require knowledge of *n-dimensional random variables* or *random vectors*. For example, the location of a space vehicle in a cartesian co-ordinate system is a three-dimensional random vector.

Most of the concepts introduced above for the two-dimensional case can be extended to the n-dimensional one.

Definitions: A vector X: $[X_1, X_2, \ldots, X_n]$ whose components X_i are RVs is called a *random vector*. $(X_1, X_2, \ldots, X_n)$ can assume all values in some region R_n of the n-dimensional space. R_n is called the *range space*.

The joint distribution function of $(X_1, X_2, \ldots \ldots, X_n)$ is defined as $F(x_1, x_2, \ldots, x_n) = P[X_1 \le x_1, X_2 \le x_2, \ldots, X_n \le x_n]$

The joint pdf of $(X_1, X_2, \ldots, X_n)$ is defined as $f(x_1, x_2, \ldots, x_n)$

$$= \frac{\partial^n F(x_1, x_2 \cdots x_n)}{\partial x_1 \times \partial x_2 \cdots \partial x_n}$$ and satisfies the following conditions.

(i) $f(x_1, x_2, \ldots, x_n) \ge 0$, for all $(x_1, x_2, \ldots, x_n)$

(ii) $\displaystyle\iint\limits_{R_n}\int \cdots \int f(x_1, x_2, \ldots, x_n)\,dx_1 dx_2 \ldots dx_n = 1$

(iii) $P[(X_1, X_2, \ldots, X_n) \in D] = \displaystyle\iint\limits_D \ldots \int f(x_1, x_2, \ldots, x_n)\,dx_1 dx_2 \ldots dx_n$ where D is a subset of the range space R_n.

The marginal pdf of any subset of the n RVs $X_1, X_2, \ldots, X_n$ is obtained by "integrating out" the variables not in the subset. For example, if $n = 3$, then

$f_{X_1}(x_1) = \displaystyle\int_{-\infty}^{\infty}\int_{-\infty}^{\infty} f(x_1, x_2, x_3)\,dx_2 dx_3$ is the marginal pdf of the one-dimensional RV X_1 and $f_{X_1, X_2}(x_1, x_2) = \displaystyle\int_{-\infty}^{\infty} f(x_1\ x_2\ x_3)\,dx_3$ is the marginal joint pdf of the two-dimensional RV (X_1, X_2). The concept of independent RVs is also extended in a natural way. The RVs $(X_1, X_2, \ldots, X_n)$ are said to be independent, if

$f(x_1, x_2, \ldots, x_n) = f_{X_1}(x_1) \cdot f_{X_2}(x_2) \ldots f_{X_n}(x_n)$

The conditional density functions are defined as in the following examples.

If $n = 3$,

$$f(x_1, x_2 / x_3) = \frac{f(x_1, x_2, x_3)}{f_{X_3}(x_3)} \text{ and}$$

$$f(x_1 / x_2, x_3) = \frac{f(x_1, x_2, x_3)}{f_{X_2, X_3}(x_2, x_3)}.$$

Worked Examples 2(B)

Example 1 *Three balls are drawn at random without replacement from a box containing 2 white, 3 red and 4 black balls. If X denotes the number of white balls drawn and Y denotes the number of red balls drawn, find the joint probability distribution of (X, Y).*

Solution As there are only 2 white balls in the box, X can take the values 0, 1 and 2 and Y can take the values 0, 1, 2 and 3.

$$P(X=0, Y=0) = P(\text{drawing 3 balls none of which is white or red})$$
$$= P(\text{all the 3 balls drawn are black})$$
$$= 4C_3/9C_3 = \frac{1}{21}$$

$$P(X=0, Y=1) = P(\text{drawing 1 red and 2 black balls})$$
$$= \frac{3C_1 \times 4C_2}{9C_3} = \frac{3}{14}$$

Similarly, $P(X=0, Y=2) = \dfrac{3C_2 \times 4C_1}{9C_3} = \dfrac{1}{7}$; $P(X=0, Y=3) = \dfrac{1}{84}$

$P(X=1, Y=0) = \dfrac{1}{7}$; $P(X=1, Y=1) = \dfrac{2}{7}$; $P(X=1, Y=2) = \dfrac{1}{14}$;

$P(X=1, Y=3) = 0$ (since only 3 balls are drawn)

$P(X=2, Y=0) = \dfrac{1}{21}$; $P(X=2, Y=1) = \dfrac{1}{28}$; $P(X=2, Y=2) = 0$;

$P(X=2, Y=3) = 0$

The joint probability distribution of (X, Y) may be represented in the form of a table as given below:

X	Y			
	0	1	2	3
0	$\frac{1}{21}$	$\frac{3}{14}$	$\frac{1}{7}$	$\frac{1}{84}$
1	$\frac{1}{7}$	$\frac{2}{7}$	$\frac{1}{14}$	0
2	$\frac{1}{21}$	$\frac{1}{28}$	0	0

Note *Sum of all the cell probabilities = 1.*

Example 2 *For the bivariate probability distribution of (X, Y) given below, find* $P(X \le 1)$, $P(Y \le 3)$, $P(X \le 1, Y \le 3)$, $P(X \le 1/Y \le 3)$, $P(Y \le 3/X \le 1)$ *and* $P(X + Y \le 4)$.

X \ Y	1	2	3	4	5	6
0	0	0	1/32	2/32	2/32	3/32
1	1/16	1/16	1/8	1/8	1/8	1/8
2	1/32	1/32	1/64	1/64	0	2/64

Solution $P(X \le 1) = P(X = 0) + P(X = 1)$

$$= \sum_{j=1}^{6} P(X = 0, Y = j) + \sum_{j=1}^{6} P(X = 1, Y = j)$$

$$= \left(0 + 0 + \frac{1}{32} + \frac{2}{32} + \frac{2}{32} + \frac{3}{32}\right) + \left(\frac{1}{16} + \frac{1}{16} + \frac{1}{8} + \frac{1}{8} + \frac{1}{8} + \frac{1}{8}\right)$$

$$= \frac{1}{4} + \frac{5}{8} = \frac{7}{8}$$

$$P(Y \le 3) = P(Y = 1) + P(Y = 2) + P(Y = 3)$$

$$= \sum_{i=0}^{2} P(X = i, Y = 1) + \sum_{i=0}^{2} P(X = i, Y = 2)$$

$$+ \sum_{i=0}^{2} P(X = i, Y = 3)$$

$$= \left(0 + \frac{1}{16} + \frac{1}{32}\right) + \left(0 + \frac{1}{16} + \frac{1}{32}\right) + \left(\frac{1}{32} + \frac{1}{8} + \frac{1}{64}\right)$$

$$= \frac{3}{32} + \frac{3}{32} + \frac{11}{64} = \frac{23}{64}$$

$$P(X \le 1, Y \le 3) = \sum_{j=1}^{3} P(X = 0, Y = j) + \sum_{j=1}^{3} P(X = 1, Y = j)$$

$$= \left(0 + 0 + \frac{1}{32}\right) + \left(\frac{1}{16} + \frac{1}{16} + \frac{1}{8}\right) = \frac{9}{32}$$

$$P(X \le 1/Y \le 3) = \frac{P(X \le 1, Y \le 3)}{P(Y \le 3)} = \frac{9/32}{23/64} = \frac{18}{23}$$

$$P(Y \le 3/X \le 1) = \frac{P(X \le 1, Y \le 3)}{P(X \le 1)} = \frac{9/32}{7/8} = \frac{9}{28}$$

$$P(X + Y \leq 4) = \sum_{j=1}^{4} P(X = 0, Y = j) + \sum_{j=1}^{3} P(X = 1, Y = j) + \sum_{j=1}^{2} P(X = 2, Y = j)$$

$$= \frac{3}{32} + \frac{1}{4} + \frac{1}{16} = \frac{13}{32}$$

Example 3 *The joint probability mass function of (X, Y) is given by p(x, y) = k(2x + 3y), x = 0, 1, 2; y = 1, 2, 3. Find all the marginal and conditional probability distributions. Also find the probability distribution of (X + Y).*

Solution The joint probability distribution of (*X*, *Y*) is given below. The relevant probabilities have been computed by using the given law.

X	*Y*		
	1	2	3
0	$3k$	$6k$	$9k$
1	$5k$	$8k$	$11k$
2	$7k$	$10k$	$13k$

$$\sum_{j=1}^{3} \sum_{i=0}^{2} p(x_i, y_j) = 1$$

i.e., the sum of all the probabilities in the table is equal to 1.
i.e., $72k = 1$.

$$\therefore \qquad k = \frac{1}{72}$$

Marginal Probability Distribution of X: $\{i, p_{i*}\}$

$X = i$	$p_{i*} = \sum_{j=1}^{3} p_{ij}$
0	$p_{01} + p_{02} + p_{03} = \frac{18}{72}$
1	$p_{11} + p_{12} + p_{13} = \frac{24}{72}$
2	$p_{21} + p_{22} + p_{23} = \frac{30}{72}$
	Total = 1

Marginal Probability Distribution of Y: $\{j, p_{*j}\}$

$Y = j$	$p_{*j} = \sum_{i=1}^{2} p_{ij}$
1	15/72
2	24/72
3	33/72
	Total = 1

Conditional distribution of X, given $Y = 1$, is given by $\{i, P(X = i/Y = 1)\} = \{i, P(X = i, Y = 1)/P(Y = 1)\}$
$= \{i, p_{i1}/p_{*1}\}$, $i = 0, 1, 2.$
the tabular representation is given below:

$X = i$	p_{i1}/p_{*1}
0	$3k/15k = \frac{1}{5}$
1	$5k/15k = \frac{1}{3}$
2	$7k/15k = \frac{7}{15}$
	Total = 1

The other conditional distributions are given below:

C.P.D. of X, given $Y = 2$	
$X = i$	p_{i2}/p_{*2}
0	$\frac{6k}{24k} = \frac{1}{4}$
1	$\frac{8k}{24k} = \frac{1}{3}$
2	$\frac{10k}{24k} = \frac{5}{12}$
	Total = 1

C.P.D. of X, given $Y = 3$	
$X = i$	p_{i3}/p_{*3}
0	$\frac{9k}{33k} = \frac{3}{11}$
1	$\frac{11k}{33k} = \frac{1}{3}$
2	$\frac{13k}{33k} = \frac{13}{33}$
	Total = 1

C.P.D. of Y, given $X = 0$	
$Y = j$	p_{oj}/p_{o*}
1	$\frac{3k}{18k} = \frac{1}{6}$
2	$\frac{6k}{18k} = \frac{1}{3}$
3	$\frac{9k}{18k} = \frac{1}{2}$
	Total = 1

C.P.D. of Y, given $X = 1$	
$Y = j$	p_{1j}/p_{1*}
1	$\frac{5k}{24k} = \frac{5}{24}$
2	$\frac{8k}{24k} = \frac{1}{3}$
3	$\frac{11k}{24k} = \frac{11}{24}$
	Total = 1

C.P.D. of Y, given $X = 2$	
$Y = j$	p_{2j}/p_{2*}
1	$\frac{7k}{30k} = \frac{7}{30}$
2	$\frac{10k}{30k} = \frac{1}{3}$
3	$\frac{13k}{30k} = \frac{13}{30}$
	Total = 1

Probability distribution of $(X + Y)$	
$(X + Y)$	P
1	$p_{01} = \frac{3}{72}$
2	$p_{02} + p_{11} = \frac{11}{72}$
3	$p_{03} + p_{12} + p_{21} = \frac{24}{72}$
4	$p_{13} + p_{22} = \frac{21}{72}$
5	$p_{23} = \frac{13}{72}$
	Total = 1

Example 4 *A machine is used for a particular job in the forenoon and for a different job in the afternoon. The joint probability distribution of (X, Y), where X and Y represent the number of times the machine breaks down in the forenoon and in the afternoon respectively, is given in the following table. Examine if X and Y are independent RVs.*

X	Y		
	0	1	2
0	0.1	0.04	0.06
1	0.2	0.08	0.12
2	0.2	0.08	0.12

Solution X and Y are independent, if $P_{i*} \times P_{*j} = P_{ij}$ for all i and j. So, let us find $P_{i*}\ P_{*j}$ for all i and j.

$P_{0*} = 0.1 + 0.04 + 0.06 = 0.2;\ P_{1*} = 0.4;\ P_{2*} = 0.4$

$P_{*0} = 0.5;\ P_{*1} = 0.2;\ P_{*2} = 0.3$

Now $P_{0*} \times P_{*0} = 0.2 \times 0.5 = 0.1 = P_{00}$

$P_{0*} \times P_{*1} = 0.2 \times 0.2 = 0.04 = P_{01}$

$P_{0*} \times P_{*2} = 0.2 \times 0.3 = 0.06 = P_{02}$

Similarly we can verify that

$P_{1*} \times P_{*0} = P_{10};\ P_{1*} \times P_{*1} = P_{11};\ P_{1*} \times P_{*2} = P_{12};$

$P_{2*} \times P_{*0} = P_{20};\ P_{2*} \times P_{*1} = P_{21};\ P_{2*} \times P_{*2} = P_{22}$

Hence the RVs X and Y are independent.

Example 5 *The joint pdf of a two-dimensional RV (X, Y) is given by* $f(x, y) = xy^2 + \frac{x^2}{8}$, $0 \le x \le 2,\ 0 \le y \le 1$.

Compute $P(X > 1)$, $P\left(Y < \frac{1}{2}\right)$, $P(X > 1/Y < 1/2)$

$P\left(Y < \frac{1}{2}/X > 1\right)$, $P(X < Y)$ and $P(X + Y \le 1)$.

Solution Here the rectangle defined by $0 \le x \le 2$, $0 \le y \le 1$ is the range space R. $R_1, R_2, \ldots,$ are event spaces.

(i) $$P(X > 1) = \int\int_{\substack{R_1\\(x>1)}} f(x, y)\,dx\,dy$$

$$= \int_0^1\int_1^2 \left(xy^2 + \frac{x^2}{8}\right) dxdy = \frac{19}{24}$$

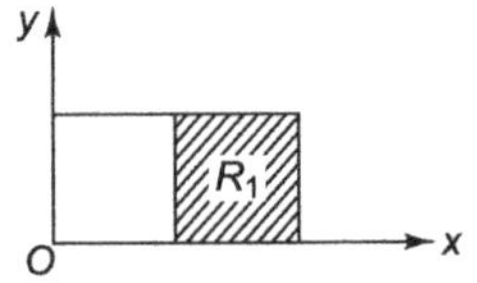

(ii) $$P(Y < 1/2) = \int_{\substack{R_2\\\left(y<\frac{1}{2}\right)}} \left(xy^2 + \frac{x^2}{8}\right) dx\,dy$$

$$= \int_0^{1/2}\int_0^2 \left(xy^2 + \frac{x^2}{8}\right) dx\,dy$$

$$= \frac{1}{4}$$

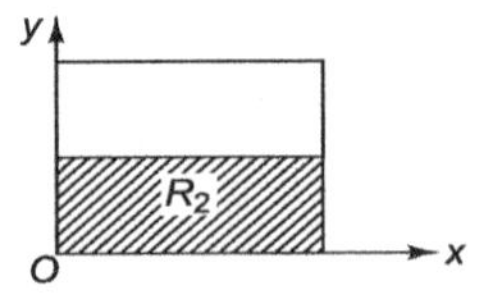

(iii) $P(X > 1, Y < 1/2) = \int\limits_{\substack{R_3 \\ \left(x>1\,\&\,y<\frac{1}{2}\right)}} \left(xy^2 + \frac{x^2}{8}\right) dx\,dy$

$$= \int_0^{1/2}\int_1^2 \left(xy^2 + \frac{x^2}{8}\right) dx\,dy = \frac{5}{24}$$

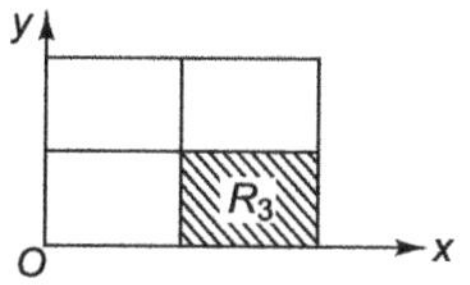

(iv) $P(X > 1/Y < \frac{1}{2}) = \dfrac{P\left(X>1, Y<\frac{1}{2}\right)}{P\left(Y<\frac{1}{2}\right)} = \dfrac{5/24}{1/4} = \dfrac{5}{6}$

(v) $P(Y < \frac{1}{2}/X > 1) = \dfrac{P\left(X>1, Y<\frac{1}{2}\right)}{P(X>1)} = \dfrac{5/24}{19/24} = \dfrac{5}{19}$

(vi) $P(X < Y) = \int\limits_{\substack{R_4 \\ (x<y)}}\int \left(xy^2 + \frac{x^2}{8}\right) dx\,dy$

$$= \int_0^1\int_0^y \left(xy^2 + \frac{x^2}{8}\right) dx\,dy = \frac{53}{480}$$

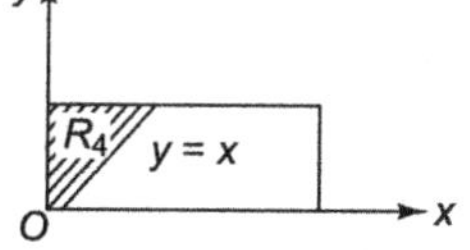

(vii) $P(X + Y \le 1) = \int\limits_{\substack{R_5 \\ (x+y\le 1)}}\int \left(xy^2 + \frac{x^2}{8}\right) dx\,dy$

$$= \int_0^1\int_0^{1-y} \left(xy^2 + \frac{x^2}{8}\right) dx\,dy = \frac{13}{480}$$

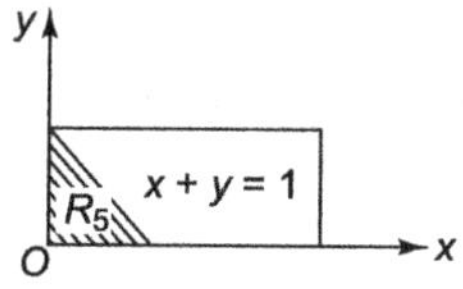

Example 6 *If the joint pdf of the RV (X, Y) is given by f(x, y)*

$= \dfrac{1}{2\pi\sigma^2}$ *exp* $\{-(x^2 + y^2)/2\sigma^2\}, -\infty < x, y < \infty$, *find* $P(X^2 + Y^2 \le a^2)$.

Solution Here the entire xy-plane is the range space R and the event-space D is the interior of the circle $x^2 + y^2 = a^2$.

$$P(X^2 + Y^2 \le a^2) = \iint_{x^2 + y^2 \le a^2} f(x, y)\, dx\, dy$$

Transform from cartesian system to polar system, i.e., put $x = r\cos\theta$ and $y = r\sin\theta$.

Then $dx\,dy = r d\,r d\,\theta$.

The domain of integration becomes $r \le a$.

$$\text{Then } P(X^2 + Y^2 \le a^2) = \int_0^{2\pi}\int_0^{a} \frac{1}{2\pi\sigma^2} e^{-r^2/2\sigma^2}\, r dr d\theta$$

$$= \frac{1}{2\pi}\int_0^{2\pi} \left(-e^{-r^2/2\sigma^2}\right)_0^a d\theta$$

$$= \frac{1}{2\pi}\int_0^{2\pi} \left(1 - e^{-a^2/2\sigma^2}\right) d\theta$$

$$= 1 - e^{-a^2/2\sigma^2}$$

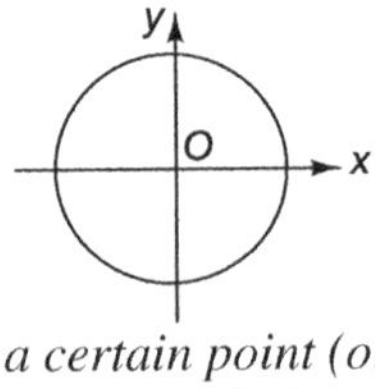

Example 7 *A gun is aimed at a certain point (origin of the co-ordinate system). Because of the random factors, the actual hit point can be any point (X, Y) in a circle of radius R about the origin. Assume that the joint density of X and Y is constant in this circle given by*

$$f_{xy}(x, y) = c, \quad \text{for } x^2 + y^2 \le R^2$$
$$= 0, \quad \text{otherwise}$$

(i) Compute c and (ii) show that

$$f_x(x) = \frac{2}{\pi R}\sqrt{1 - \left(\frac{x}{R}\right)^2}, \text{ for } -R \le x \le R$$

$= 0,$ *otherwise* (BDU — Nov. 96)

Solution Here the range space is the interior of the circle $x^2 + y^2 = R^2$. By the property of joint pdf,

$$\iint_{x^2 + y^2 \le R^2} f(x, y)\,dx\,dy = 1$$

i.e., $$\iint_{x^2 + y^2 \le R^2} c\,dx\,dy = 1$$

Changing over to polar co-ordinates, we have

$$\int_0^{2\pi}\int_0^{R} c\,r\,dr\,d\theta = 1$$

$\therefore$ $$c = \frac{1}{\pi R^2}$$

Note *We have defined earlier that* $f_X(x) = \int_{-\infty}^{\infty} f(x, y)dy$. *This definition holds good if the range space is the entire xy-plane. If the range space is different from the entire xy-plane* $f_X(x)$ *is given by* $\int f(x, y)dy$, *for which the limits are fixed as follows: Draw an arbitrary line parallel to y-axis (since x is to be treated as a constant). The y co-ordinates of the ends of the segment of such a line that lies within the range space are the required limits. These limits will be either constants or functions of x.*

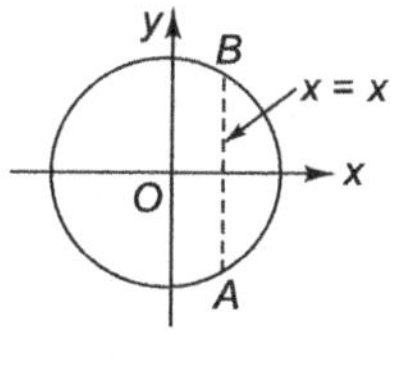

The point $A = \left\{x, -\sqrt{R^2 - x^2}\right\}$

and the point $B = \left\{x, \sqrt{R^2 - x^2}\right\}$

Now $$f_X(x) = \int_{-\sqrt{R^2 - x^2}}^{\sqrt{R^2 - x^2}} \frac{1}{\pi R^2}\,dy$$

$$= \frac{2}{\pi R^2}\sqrt{R^2 - x^2} = \frac{2}{\pi R}\sqrt{1 - \left(\frac{x}{R}\right)^2} \quad -R \le x \le R$$

Note *Whenever we are required to find the marginal and conditional density functions, the ranges of the concerned variables should also be specified.*

Example 8 *The joint pdf of the RV (X, Y) is given by* $f(x, y) = k\,x\,y\,e^{-(x^2+y^2)}$ $x > 0,\ y > 0$. *Find the value of k and prove also that X and Y are independent.*

Solution Here the range space is the entire first quadrant of the *xy*-plane.
By the property of the joint pdf

$$\iint_{x>0,\,y>0} kxy\,e^{-(x^2+y^2)}\,dx\,dy = 1$$

i.e.,
$$k\int_0^\infty ye^{-y^2}\,dy \int_0^\infty xe^{-x^2}\,dx = 1$$

i.e.,
$$\frac{k}{4} = 1$$

∴
$$k = 4$$

$$\text{Now } f_x(x) = \int_0^\infty 4x\,e^{-x^2} \times e^{-y^2}\,dy = 2x\,e^{-x^2},\ x > 0$$

Similarly, $f_y(y) = 2ye^{-y^2},\ y > 0$.

Now $f_x(x) \times f_y(y) = 4\,x\,y\,e^{-(x^2+y^2)} = f(x, y)$
∴ The RVs x and y are independent.

Note *If* $f(x, y)$ *can be factorised as* $f_1(x) \times f_2(y)$ *then X and Y will be independent.*

Example 9 *Given* $f_{xy}(x, y) = cx(x - y),\ 0 < x < 2,\ -x <, y < x$, *and 0 elsewhere, (a) evaluate c, (b) find* $f_x(x)$, *(c)* $f_{y/x}(y/x)$ *and (d)* $f_y(y)$. (BDU — Apr. 96)

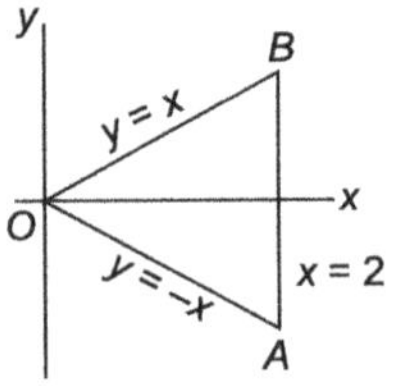

Solution Here the range space is the area within the triangle *OAB* (shown in the figure), defined by $0 < x < 2$ and $-x < y < x$.
(a) By the property of *jpdf*

$$\iint_{\Delta OAB} cx(x-y)\,dx\,dy = 1$$

$$\int_0^2\int_{-x}^{x} cx(x-y)\,dy\,dx = 1$$

i.e.,
$$8c = 1$$

∴
$$c = \frac{1}{8}$$

(b) $f_X(x) = \int_{-x}^{x} \frac{1}{8} x\,(x-y)\,dy$

$$= \frac{x^3}{4}, \text{ in } 0 < x < 2$$

(c) $f(y/x) = \frac{f(x,y)}{f_X(x)} = \frac{1}{2x^2}(x-y), \; -x < y < x$

(d) $f_Y(y) = \int_{-y}^{2} \frac{1}{8} x\,(x-y)\,dx, \text{ in } -2 \le y \le 0$

$$= \int_{y}^{2} \frac{1}{8} x\,(x-y)\,dx, \text{ in } 0 \le y \le 2$$

i.e., $$f_Y(y) = \begin{cases} \frac{1}{3} - \frac{y}{4} + \frac{5}{48} y^3, & \text{in } -2 \le y \le 0 \\ \frac{1}{3} - \frac{y}{4} + \frac{1}{48} y^3, & \text{in } 0 \le y \le 2 \end{cases}$$

Example 10 *Train X arrives at a station at random in the time interval (0, T) and stops for 'a'min. Train Y arrives independently in the same interval and stops for 'b' min.*

(i) *Find the probability P_1 that X will arrive before Y.*

(ii) *Find the probability P_2 that the two trains meet.*

(iii) *Assuming that they meet, find the probability P_3 that X arrived before Y.*

(MSU — Nov. 96)

Solution Let the trains X and Y arrive at the station at time instances X and Y respectively.

Then the lengths of the intervals $(0, X)$ and $(0, Y)$, namely X and Y are continuous RVs. Each of X and Y is uniformly distributed in $(0, T)$ (since the times of arrival are equally likely) with pdf $\frac{1}{T}$.

Since the 2 trains arrive independently, X and Y are independent RVs.

$\therefore$ The joint pdf of (X, Y) is given by

$$f(x, y) = f_X(x) \times f_Y(y) = \frac{1}{T^2};\; 0 \le x, y \le T$$

The range space is the square defined by $0 \le x \le T$ and $0 \le y \le T$.

(i) $P_1 = P(X < Y) = \iint\limits_{\substack{x<y \\ (\Delta OBC)}} f(x, y)\,dx\,dy$

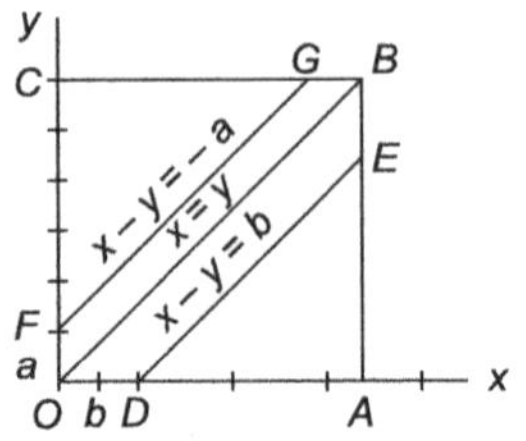

$$= \frac{1}{T^2} \iint_{\Delta OBC} \mathrm{d}x\, \mathrm{d}y = \frac{1}{T^2} \times \text{Area of } \Delta OBC$$

$$= \frac{1}{2}$$

(ii) If train X arrives first, the 2 trains will meet if $Y \le X + a$. If train Y arrives first, the 2 trains will meet if $X \le Y + b$.

$\therefore$ For the 2 trains to meet, $-a \le X - Y \le b$.

$$\therefore \quad P_2 = P(-a \le X - Y \le b) = \underset{\substack{-a \le x - y \le b \\ (ODEBGFO)}}{\iint} f(x, y)\mathrm{d}x\, \mathrm{d}y$$

$$= \frac{1}{T^2} \times \text{Area of the figure } ODEBGFO$$

$$= \frac{1}{T^2} \times (\text{Area of trapezium } ODEB + \text{that of } OBGF)$$

$$= \frac{1}{T^2} \times \left[\frac{1}{2} \times \frac{b}{\sqrt{2}}\left\{(T - b)\sqrt{2} + T\sqrt{2}\right\} + \frac{1}{2} \times \frac{a}{\sqrt{2}}\left\{(T - a)\sqrt{2} + T\sqrt{2}\right\}\right]$$

$$= \frac{1}{2T^2}\{a(2T - a) + b(2T - b)\}$$

$$= \frac{1}{2T^2}\{2(a + b)T - (a^2 + b^2)\}$$

(iii) $P_3 = P\{X < Y / -a \le X - Y \le b\}$

$$= \frac{P[X < Y \text{ and } -a \le X - Y \le b]}{P(-a \le X - Y \le b)}$$

$$= \frac{\frac{1}{T^2} \times \text{area of trapezium } OBGF}{P_2}$$

$$= \frac{a(2T - a)}{2(a + b)T - (a^2 + b^2)}$$

Example 11 *Two trains arrive at a station at random between 7 A.M. and 7:30 A.M. One train stops for 5 min and the other for x min. For what value of x, will the probability that the 2 trains meet be equal to* $\frac{1}{3}$*?*

Solution In the notation of the previous problem,

$$T = 30,\ a = 5,\ b = x \text{ and } P_2 = \frac{1}{3}$$

$$\therefore \qquad \frac{1}{2T^2}\{2(a+b)T - (a^2+b^2)\} = \frac{1}{3}$$

i.e., $$\frac{1}{1800}\{60(x+5) - (x^2+25)\} = \frac{1}{3}$$

i.e., $$x^2 - 60x + 325 = 0.$$

Solving, $x = 53.98$ (or) 6.02

As $x = 53.98$ is meaningless, $x = 6$ min (nearly).

Example 12 *The two-dimensional RV (X, Y) follows a bivariate normal distribution* $N(0, 0; \sigma_x, \sigma_y; r)$*. Find the marginal density function of X and the conditional density function of Y, given X.*

Solution The notation $N(0, 0; \sigma_x, \sigma_y; r)$ refers to a bivariate normal distribution with mean of X = mean of $Y = 0$, variance of $X = \sigma_x^2$, variance of $Y = \sigma_y^2$ and the co-efficient of correlation between X and $Y = r$.

The joint pdf of such a brivariate normal distribution is given by

$$f(x, y) = \frac{1}{2\pi\sigma_x\sigma_y\sqrt{1-r^2}} \exp\left\{-\frac{1}{2(1-r^2)}\left(\frac{x^2}{\sigma_x^2} - \frac{2rxy}{\sigma_x\sigma_y} + \frac{y^2}{\sigma_y^2}\right)\right\}$$

$$-\infty < x, y < \infty$$

The marginal density function of X is given

$$\text{by } f_x(x) = \int_{-\infty}^{\infty} f(x, y)\,dy$$

$$= A \int_{-\infty}^{\infty} \exp\left\{-\frac{1}{2(1-r^2)}\left(\frac{y}{\sigma_y} - \frac{rx}{\sigma_x}\right)^2\right\} \times$$

$$\exp\left(\frac{-x^2}{2\sigma_x^2}\right) dy, \text{ where } A = \frac{1}{2\pi\sigma_x\sigma_y\sqrt{1-r^2}}$$

$$= A \exp\left(\frac{-x^2}{2\sigma_x^2}\right) \times \sqrt{2}\,\sigma_y\sqrt{1-r^2} \int_{-\infty}^{\infty} e^{-t^2}\,dt,$$

by putting $\dfrac{1}{\sqrt{2(1-r^2)}}\left(\dfrac{y}{\sigma_y}-\dfrac{rx}{\sigma_x}\right)=t$

$$= A\exp\left(\frac{-x^2}{2\sigma_x^2}\right)\sqrt{2}\cdot\sigma y\sqrt{1-r^2}\,\Gamma\left(\frac{1}{2}\right)$$

$$= \frac{1}{2\pi\sigma_x\sigma_y\sqrt{1-r^2}}\exp\left(\frac{-x^2}{2\sigma_x^2}\right)\sqrt{2}\cdot\sigma y\sqrt{1-r^2}\sqrt{\pi}$$

$$= \frac{1}{\sigma_x\sqrt{2\pi}}\exp\left(\frac{-x^2}{2\sigma_x^2}\right),\ -\infty < x < \infty$$

which is the density function of a normal distribution $N(0, \sigma_x)$.

The conditional density function of Y given X is given by $f\left(\dfrac{y}{x}\right)=\dfrac{f(x, y)}{f_x(x)}$

$$\therefore\quad f\left(\frac{y}{x}\right)=\frac{\dfrac{1}{2\pi\sigma_x\sigma_y\sqrt{1-r^2}}\exp\left\{\dfrac{-1}{2(1-r^2)}\left(\dfrac{x^2}{\sigma_x^2}-\dfrac{2rxy}{\sigma_x\sigma_y}+\dfrac{y^2}{\sigma_y^2}\right)\right\}}{\dfrac{1}{\sigma_x\sqrt{2\pi}}\exp\dfrac{-x^2}{2\sigma_x^2}}$$

$$= \frac{1}{\sqrt{2\pi}\left(\sigma_y\sqrt{1-r^2}\right)}\exp\left\{-\frac{1}{2\sigma_y^2(1-r^2)}\left(y^2-2r\frac{\sigma_y}{\sigma_x}xy+\frac{r^2\sigma_y^2}{\sigma_x^2}\right)\right\}$$

$$= \frac{1}{\sqrt{2\pi}\left(\sigma_y\sqrt{1-r^2}\right)}\exp\left\{-\frac{1}{2\left(\sigma_y\sqrt{1-r^2}\right)^2}\left(y-\frac{r\sigma_y x}{\sigma_x}\right)^2\right\}$$

which is the density function of a normal distribution

$$N\left(\frac{r\sigma_y}{\sigma_x}x, \sigma_y\sqrt{1-r^2}\right)$$

Exercise 2(B)

Part A (Short answer questions)

1. Define a two-dimensional RV. Give an example for the outcome of a random experiment, that is a two-dimensional RV.
2. Define the joint pmf of a two-dimensional discrete RV.

3. Define the joint pdf of a two-dimensional continuous RV.
4. Write down the joint pdf of a bivariate normal distribution.
5. Define the cdf of a two-dimensional RV and write down the formulas for, finding the cdf of (X, Y), when (X, Y) is (i) a discrete RV and (ii) a continuous RV.
6. State the properties of the cdf of a two-dimensional RV (X, Y).
7. Define the marginal probability distributions of X and Y, when (X, Y) is a discrete RV.
8. Define the marginal probability density functions of X and Y, when (X, Y) is a continuous RV.
9. Define independence of 2 RVs X and Y, both in the discrete case and in the continuous case.
10. Define the conditional probability distributions of X and Y, given Y and X respectively, when (X, Y) is a discrete RV.
11. Define the conditional probability density functions of X and Y given Y and X respectively, when (X, Y) is a continuous RV.
12. Find the probability distribution of $(X + Y)$ from the bivariate distribution of (X, Y) given below.

X	*Y*	
	1	2
1	0.1	0.2
2	0.3	0.4

13. Find the marginal distributions of X and Y from the bivariate distribution of (X, Y) given in Q.12.
14. Find the conditional distribution of X, when $Y = 1$, from the bivariate distribution of (X, Y) given in Q.12.
15. Find the value of k, if $f(x, y) = k(1 - x)(1 - y)$, for $0 < x, y < 1$, is to be a joint density function.
16. If $f(x, y) = k(1 - x - y)$, $0 < x, y < \frac{1}{2}$, is a joint density function, find k.
17. If the joint pdf of (X, Y) is $f(x, y) = \frac{1}{4}$, $0 \le x, y \le 2$, find $P(X + Y \le 1)$.
18. If the joint pdf of (X, Y) is $f(x, y) = 6e^{-2x-3y}$, $x \ge 0$, $y \ge 0$, find the marginal density of X and conditional density of Y given X.
19. The j pdf of (X, Y) is given by $f(x, y) = e^{-(x+y)}$, $0 \le x, y < \infty$. Are X and Y independent? Why?
20. Define a random vector with an example.
21. Define the joint density and distribution functions of an n-dimensional RV. How are they related?

Part B

22. If X denotes the number of aces and Y the number of queens obtained when 2 cards are drawn at random (without replacement) from a deck of cards, obtain the joint probability distribution of (X, Y).
23. The joint probability function of two discrete RVs X and Y is given by $f(x, y) = c(2x + y)$, where x and y can assume all integers such that $0 \le x \le 2$ and $0 \le y \le 3$, and $f(x, y) = 0$ otherwise. (a) find the value of c and (b) find $P(X \ge 1, Y \le 2)$. (MKU — Apr. 96)

Note *$f(x, y)$ should not be mistaken as pdf, It is used instead of $p(x_i, y_i)$*

24. The joint probability distribution of a two-dimensional discrete RV (X, Y) is given below:

Y	X					
	0	1	2	3	4	5
0	0	0.01	0.03	0.05	0.07	0.09
1	0.01	0.02	0.04	0.05	0.06	0.08
2	0.01	0.03	0.05	0.05	0.05	0.06
3	0.01	0.02	0.04	0.06	0.06	0.05

(i) Find $P(X > Y)$ and $P\{\max (X, Y) = 3\}$ and
(ii) Find the probability distribution of the RV $Z = \min (X, Y)$.

25. The input to a binary communication system, denoted by a RV X, takes one of two values 0 or 1 with probabilities 3/4 and 1/4 respectively. Because of errors caused by noise in the system, the output Y differs from the input occasionally. The behaviour of the communication system is modeled by the conditional probabilities given below:
$$P(Y = 1/X = 1) = 3/4 \text{ and } P(Y = 0/X = 0) = 7/8$$
Find (i) $P(Y = 1)$, (ii) $P(Y = 0)$ and (iii) $P(X = 1/Y = 1)$.
26. The following table represents the joint probability distribution of the discrete RV (X, Y). Find all the marginal and conditional distributions.

Y	X		
	1	2	3
1	1/12	1/6	0
2	0	1/9	1/5
3	1/18	1/4	2/15

27. The joint distribution of X_1 and X_2 is given by $f(x_1, x_2) = \dfrac{x_1 + x_2}{21}$, $x_1 = 1, 2$ and 3; $x_2 = 1$ and 2. Find the marginal distributions of X_1 and X_2. (MU — Nov. 96).

28. If the joint pdf of a two-dimensional RV (X, Y) is given by

$$f(x, y) = x^2 + \frac{xy}{3}; 0 < x < 1, 0 < y < 2$$
$$= 0, \text{ elsewhere}$$

find (i) $P\left(X > \frac{1}{2}\right)$, (ii) $P(Y < X)$ and (iii) $P\left(Y < \frac{1}{2} / X < \frac{1}{2}\right)$.

29. If the joint pdf of a two-dimensional RV (X, Y) is given by

$$f(x, y) = k(6 - x - y); 0 < x < 2, 2 < y < 4$$
$$= 0, \text{ elsewhere}$$

find (i) the value of k, (ii) $P(X < 1, Y < 3)$,
(iii) $P(X + Y < 3)$ and (iv) $P(X < 1/Y < 3)$.

30. The joint density function of the RVs X and Y is given by

$$f(x, y) = 8xy; 0 < x < 1, 0 < y < x$$
$$= 0, \text{ elsewhere,}$$

find $P\left(Y < \frac{1}{8} / X < \frac{1}{2}\right)$. (MU — Nov. 96)

31. Given that the joint pdf of (X, Y) is

$$f(x, y) = e^{-y}; x > 0, y > x$$
$$= 0, \text{ elsewhere}$$

find (i) $P(X > 1/Y < 5)$ and (ii) the marginal distributions of X and Y.

32. If the joint pdf of a two-dimensional RV (X, Y) is given by

$$f(x, y) = 2; 0 < x < 1, 0 < y < x$$
$$= 0, \text{ otherwise}$$

find the marginal density function of X and Y. (BDU — Nov. 96)

33. If the joint pdf of (X, Y) is given by $f(x, y) = k$, $0 \le x < y \le 2$, find k and also the marginal and conditional density functions.

34. The joint density function of a RV (X, Y) is $f(x, y) = 8xy$, $0 < x < 1$, $0 < y < x$. find the conditional density function $f(y/x)$. (MU — Apr. 96)

35. The joint density function of a RV (X, Y) is given by $f(x, y) = axy$, $1 \le x \le 3$, $2 \le y \le 4$, and $= 0$, elsewhere.
Find (i) the value of a, (ii) the marginal densities of X and Y and (iii) the conditional densities of X and Y, given Y and X respectively.

36. Let X_1 and X_2 be two RVs with joint pdf given by $f(x_1, x_2) = e^{-(x_1 + x_2)}$; x_1, $x_2 \ge 0$, and $= 0$, otherwise. Find the marginal densities of X_1 and X_2. Are they independent? Also find $P[X_1 \le 1, X_2 \le 1]$ and $P(X_1 + X_2 \le 1)$. (BDU — Apr. 96)

37. The joint pdf of the RVs X and Y is given by $p(x, y) = xe^{-x(y+1)}$ where $0 \le x, y < \infty$. (i) Find $p(x)$ and $p(y)$ and (ii) Are the RVs independent? (BU — Nov. 96).

38. If the joint pdf of the RV (X, Y) is given by $f(x, y) = k(x^3y + xy^3)$, $0 \le x \le 2$, $0 \le y \le 2$, find (i) the value of k, (ii) the marginal densities of X and Y and (iii) the conditional densities of X and Y.

39. If the joint pdf of (X, Y) is given by

$$f(x, y) = \frac{9(1 + x + y)}{2(1 + x)^4 (1 + y)^4} \quad x > 0 \; y > 0$$

find the marginal densities of X and Y. Are they independent?

40. Trains X and Y arrive at a station at random between 8 A.M. and 8.20 A.M. Train A stops for 4 min and train B stops for 5 min. Assuming that the trains arrive independently of each other, find the probability that (i) X will arrive before Y, (ii) the trains will meet and (iii) X arrived before Y, assuming that they met.

41. If the two-dimensional RV (X, Y) follows a bivariate normal distribution $N(0, 0; \sigma_x, \sigma_y; r)$, find the marginal density function of Y and the conditional density function of X, given Y.

42. The two-dimensional RV (X, Y) has the joint density

$$f(x, y) = 8xy, \; 0 < x < y < 1$$
$$= 0, \text{ otherwise}$$

(i) Find $P(X < 1/2 \cap Y < 1/4)$,
(ii) Find the marginal and conditional distributions, and
(iii) Are X and Y independent? Give reasons for your answer.

(BDU — Apr. 97)

Answers

Exercise 2(A)

6. Assign the values 0, 1, 2 to X, when the outcome consists of 2 tails, 1 tail and 1 head and 2 heads respectively. Then the required probability distribution of X is

x:	0	1	2
p_x:	$\frac{1}{4}$	$\frac{1}{2}$	$\frac{1}{4}$

9. **Example** $y = \frac{1}{\sigma\sqrt{2\pi}} e^{-(x-\mu)^2/2\sigma^2}$, $-\infty < x < \infty$, is the equation of the normal curve, viz., the probability curve of the normal distribution.

11.

X:	0	1	2	3	4	5
p_x:	$\frac{1}{32}$	$\frac{5}{32}$	$\frac{10}{32}$	$\frac{10}{32}$	$\frac{5}{32}$	$\frac{1}{32}$

12. Required probability $= \dfrac{P\left\{\left(\frac{1}{2} < X < \frac{7}{2}\right) \cap (X > 1)\right\}}{P(X > 1)}$

$$= \frac{P(X = 2 \text{ or } 3)}{P(X = 2, 3 \text{ or } 4)} = \frac{0.5}{0.6} = \frac{5}{6}$$

16. $f(x) \geq 0; \int_{-1}^{1} |x| \, dx = 2 \int_{0}^{1} x \, dx = 1$

$\therefore$ $f(x)$ can be the pdf of continuous RV.

17. $\int_{0}^{3} kx^2 \, dx = 1; 9k = 1$

$\therefore \quad k = \frac{1}{9}$

18. $P\{|X| > 1\} = 1 - P\{|X| < 1\} = 1 - \int_{-1}^{1} \frac{1}{4} \, dx = \frac{1}{2}$

19. $\int_{0}^{\infty} kx \, e^{-x} \, dx = 1;$

$k[x(-e^{-x}) - 1 \times (e^{-x})]_0^\infty = 1$

$\therefore \quad k = 1$

20. Required probability $= \dfrac{P(X > 1.5)}{P(X > 1)} = \dfrac{\left(\frac{x^2}{4}\right)_{1.5}^{2}}{\left(\frac{x^2}{4}\right)_{1}^{2}} = \dfrac{7}{12}$

21. $\Sigma P_x = 1 \therefore k = 0.1; E(X) = -0.8 - 0.1 + 0 + 0.3 = -0.6$

22. The probability distribution of X is

X:	1	2	3	4	4	5	6
p_x:	$\frac{1}{6}$	$\frac{1}{6}$	$\frac{1}{6}$	$\frac{1}{6}$	$\frac{1}{6}$	$\frac{1}{6}$	$\frac{1}{6}$

$F(x) = P(X \leq x) = 0$, if $x < 1$; $= \frac{1}{6}$, if $1 \leq x < 2$;

$= \frac{2}{6}$ if $2 \leq x < 3$; $= \frac{3}{6}$, if $3 \leq x < 4$; $= \frac{4}{6}$, if $4 \leq x < 5$;

$= \frac{5}{6}$, if $5 \leq x < 6$ and $= 1$, if $6 \leq x$

23. When $x < 0$, $F(x) < 0$; when $0 \leq x < 1$, $F(x) = x^2$; when $1 \leq x$, $F(x) = 1$.

24. $f(x) = \dfrac{dF}{dx} = \lambda e^{-}\lambda^{x}$, when $x > 0$ and $= 0$ when $x < 0$.

25. $P(X > 1/X < 3)$

$$= \frac{P(1 < X < 3)}{P(0 < X < 3)} = \frac{F(3) - F(1)}{F(3) - F(0)} = \frac{\frac{9}{16} - \frac{1}{16}}{\frac{9}{16} - 0} = \frac{8}{9}$$

26. For the binomial distribution $B(n, p)$, mean $= np$ and variance $= npq$.

27. Mean = variance = λ for the Poisson distribution with parameter λ.

29. $P(|X| < 1) = P(-1 < X < 1)$

$$= \int_0^1 e^{-x}\, dx = \frac{e-1}{e}$$

Part B

38. $P(X = r) = 4\, C_r \left(\frac{1}{2}\right)^4$, $r = 0, 1, 2, 3, 4$

39. $P(X = 0) = \frac{1}{64}$, $P(X = 1) = \frac{9}{64}$, $P(X = 2) = \frac{27}{64}$, $P(X = 3) = \frac{27}{64}$

40. $P(X = 0) = q$, $P(X = 1) = p$; when $x < 0$, $F(x) = 0$; when $0 \le x < 1$, $F(x) = q$; when $1 \le x$, $F(x) = q + p = 1$

41. $P(X = r) = p(X = 14 - r) = \frac{(r-1)}{36}$, $r = 2, 3, 4, 5, 6, 7$

42. $P(X = 0) = \frac{5}{16}$, $P(X = 1) = \frac{6}{16}$, $P(X = 2)\ \frac{5}{16}$; $= E(X) = 1$; $V(X) = \frac{5}{8}$

43. $a = \frac{1}{18}$; $P(X < 3) = \frac{1}{9}$; $V(X) = 4.4719$; $F(x) = 0$ in $x < 0$, $F(x) = \frac{1}{81}$ in $0 \le x < 1$, $F(x) = \frac{4}{81}$ in $1 \le x < 2$, $F(x) = \frac{9}{81}$ in $2 \le x < 3$ etc. $F(x) = \frac{64}{81}$ in $7 \le x < 8$ and $F(x) = 1$ in $8 \le x$.

44. $P(Y = 0) = 0.1$; $P(Y = 3) = 0.3$; $P(Y = 8) = 0.5$; $P(Y = 15) = 0.1$; $E(Y) = 6.4$; $V(Y) = 16.24$.

45. $C = \frac{2}{7}$; $P = \frac{16}{37}$; $F(x) = 0$, when $x < 0$; $= \frac{12}{49}$, when $0 \le x < 1$; $= \frac{28}{49}$, when $1 \le x < 2$ and $= 1$, when $2 \le x$; $x = 0$; $x = 1$.

46. $k = \frac{1}{100}$; $P = \frac{8}{99}$; $E(X) = 3.54$; $V(X) = 0.4684$; $F(x) = 0$, when $x < 1$; $= \frac{1}{100}$, when $1 \le x < 2$; $= \frac{9}{100}$, when $2 \le x < 3$; $= \frac{36}{100}$, when $3 \le x < 4$ and $= 1$, when $4 \le x$.

47. $0 < a < 1$

48. $k = a$; $E(X) = \frac{1}{a}$; $V(X) = \frac{1}{a}\left(\frac{1}{a} - 1\right)$

49. $k = 1$; $E(X) = e^t$; $V(X) = e^t\,(e^t - 1)$.

50. $k = \frac{3}{4}$; $E(X) = 1$; $V(X) = \frac{1}{5}$; $F(x) = 0$, when $x < 0$; $= \frac{1}{4}\,(3x^2 - x^3)$, when $0 \le x < 2$; $= 1$, when $2 \le x$.

51. (i) $k = 6$;
(ii) $F(x) = 0$, when $x < 0$; $= 3x^2 - 2x^3$, when $0 \le x < 1$; $= 1$, when $1 \le x$;
(iii) the root of the equation $6a^3 - 9a^2 + 2 = 0$ that lies between 0 and 1;
(iv) $\frac{1}{2}$.

52. $k = \frac{1}{8}$; $F(x) = 0$, when $x < 0$; $= \frac{x^2}{16}$, when $0 \le x < 2$; $= \frac{1}{4}(x - 1)$, when $2 \le x < 4$; $= -\frac{1}{16}(20 - 12x + x^2)$, when $4 \le x < 6$; $= 1$, when $6 \le x$.

53. (i) $\frac{9}{16}$ (ii) $\frac{27}{64}$

54. $\frac{1}{3}$

55. $k = 5$; $P = \frac{1}{33}$

56. (a) $\frac{1}{4}$ (b) $\frac{4}{9}$ (c) 5

57. $P = 1 - \frac{1}{e}$; $f(x) = \frac{k}{2} e^{-k|x|}$

58. $f(x) = 1$ in $0 \le x \le 1$ and $= 0$, elsewhere; $\frac{1}{6}$; $\frac{1}{2}$.

59. $\frac{1}{5}$

60. $\frac{2}{3}$

61. (i) λ; (ii) $\frac{\lambda^2 t}{1 + \lambda t}$

62. (i) 0.8849 (ii) 0.8819 (iii) 0.8593

Exercise 2(B)

6. A

4. $f_{XY}(x, y) = \frac{1}{2\pi\sigma_x\sigma_y\sqrt{1-r^2}}$

$$\exp\left[-\frac{1}{2(1-r^2)}\left\{\frac{(x-\mu_x)^2}{\sigma_x^2} - \frac{2r(x-\mu_x)(y-\mu_y)}{\sigma_x\sigma_y} + \frac{(y-\mu_y)^2}{\sigma_y^2}\right\}\right],$$

$-\infty < x, y < \infty$

12.

$(X + Y)$:	2	3	4
p:	0.1	0.5	0.4

13.

X:	1	2	and	Y:	1	2
p_x:	0.3	0.7		p_y:	0.4	0.6

14.

X:	1	2
$p_{X/Y} = 1$:	0.25	0.75

15. $k \int_0^1 \int_0^1 (1 - x)(1 - y)\, dx\, dy = 1;$

$$k \left\{ \frac{(1-x)^2}{-2} \right\}_0^1 \left\{ \frac{(1-y)^2}{-2} \right\}_0^1 = 1$$

i.e., $\frac{k}{4} = 1$

$\therefore$ $k = 4$

16. $\int_0^{1/2} \int_0^{1/2} k(1 - x - y)\, dx\, dy = 1$

$$k \int_0^{1/2} \left(x - \frac{x^2}{2} - yx \right)_0^{\frac{1}{2}} dy = 1;$$

$$k \left(\frac{3}{8y} - \frac{y^2}{4} \right)_0^{\frac{1}{2}} = 1$$

$\therefore$ $k = 8$

17. $P = \int_0^1 \int_0^{1-y} \frac{1}{4}\, dx\, dy = \frac{1}{4} \int_0^1 (1 - y)\, dy = \frac{1}{8}$

18. $f_x(x) = 6 \int_0^{\infty} e^{-2x} e^{-3y}\, dy = 2e^{-2x};\ x \geq 0$

$$f_{y/x}(y) = \frac{f(x, y)}{f_x(x)} = 3e^{-3y};\ y \geq 0$$

19. $f_x(x) = e^{-x}$ and $f_y(y) = e^{-y}$

$f(x, y) = f_y(x), f_y(y)$

$\therefore$ X and Y are independent

22.

Y \ X	0	1	2
0	0.71	0.13	0.01
1	0.13	0.01	0
2	0.01	0	0

23. $c = \frac{1}{42}$; $P = \frac{4}{7}$

24. (i) 0.75, 0.21 (ii)

Z	0	1	2	3
P	0.28	0.30	0.25	0.17

25. (i) $\frac{9}{32}$ (ii) $\frac{23}{32}$ (iii) $\frac{2}{3}$

26. $\{i, p_{i*}\}$

$x = i$	p_{i*}
1	5/36
2	19/36
3	1/3

$\{j, p_{*j}\}$

$y = j$	p_{*j}
1	1/4
2	14/45
3	79/180

CPD of $X/Y = 1$

$x = i$	p_{i1}/p_{*1}
1	1/3
2	2/3
3	0

CPD of $X/Y = 2$

$x = i$	p_{i2}/p_{*2}
1	0
2	5/14
3	9/14

CPD of $X/Y = 3$

$x = i$	p_{i3}/p_{*3}
1	10/79
2	45/79
3	24/79

CPD of $Y/X = 1$

$y = j$	p_{1j}/p_{1*}
1	3/5
2	0
3	2/5

CPD of $Y/X = 2$

$x = j$	p_{2j}/p_{2*}
1	6/19
2	4/19
3	9/19

CPD of $Y/X = 3$

$y = j$	p_{3j}/p_{3*}
1	0
2	3/5
3	2/5

27.

M.D. of X_1

$X_1 = i$	pi^*
1	5/21
2	7/21
3	9/21

M.D. of X_2

$X_2 = j$	p^*j
1	9/21
2	12/21

28. (i) $\frac{5}{6}$ (ii) $\frac{7}{24}$ (iii) $\frac{5}{32}$

29. (i) 1/8 (ii) 3/8 (iii) 5/24 (iv) 3/5

30. $\frac{31}{256}$

31. (i) $\frac{e^4 - 5}{e^5 - 6}$ (ii) $f_x(x) = e^{-x}, x > 0; f_y(y) = y\,e^{-y}, y > 0.$

32. $f_x(x) = 2x$ in $0 < x < 1$; $f_y(y) = 2(1 - y)$ in $0 < y < 1$.

33. $k = 1/2$; $f_x(x) = \frac{1}{2}(2 - x), 0 \le x \le 2$; $f_y(y) = (1/2)\,y, 0 \le y \le 2$; $f(x/y) = 1/y; 0 < x < y$; $f(y/x) = \frac{1}{2 - x}, x < y < 2.$

34. $f(y/x) = 2y/x^2,\ 0 < y < x$
35. (i) $a = 1/24$
   (ii) $f_x(x) = x/4,\ 1 \le x \le 3;\ f_y(y) = y/6,\ 2 \le y \le 4$
   (iii) $f(x/y) = x/4,\ 1 < x < 3;\ f(y/x) = y/6,\ 2 < y < 4$
36. $f_{x1}(x_1) = e^{-x1};\ f_{x2}(x_2) = e^{-x2};\ X_1$ and X_2 are independent; $(1 - e^{-1})^2$; $(1 - 2e^{-1})$
37. $f_x(x) = e^{-x}, 0 < x < \infty;\ f_y(y) = (y + 1)^{-2}, 0 < y < \infty$; X and Y are not independent.
38. $K = 1/16;\ f_x(x) = \frac{1}{8}(x^3 + 2x),\ 0 \le x \le 2;\ f_y(y) = \frac{1}{8}(y^3 + 2y),\ 0 \le y \le 2;$

$$f(x/y) = x/2\,\frac{(x^2+y^2)}{(y^2+2)}\ 0 < x < 2;\ f(y/x) = y/2\,\frac{(x^2+y^2)}{(x^2+2)},\ 0 < y < 2$$

39. $f_x(x) = 3/4,\ \frac{2x+3}{(1+x)^4}\ x > 0$ and $f_y(y) = 3/4,\ \frac{2x+3}{(1+y)^4}\ y > 0;$

   Not independent.
40. (i) 1/2 (ii) 0.3988 (iii) 0.4514
41. $f_y(y) = \frac{1}{\sigma_y\sqrt{2\pi}}\exp\left(-\frac{y^2}{2\sigma_y^2}\right) - \infty < y < \infty$

$$f(x/y) = N\left(\frac{r\sigma_x}{\sigma_y}\cdot y,\ \sigma_x\sqrt{1-r^2}\right)$$

42. (i) 1/256
   (ii) $f_x(x) = 4x(1 - x^2)$ in (0, 1); $f_y(y) = 4y^3$ in (0, 1); $f(y/x)$

$$= \left|\frac{2y}{1-x^2}\right|\ f(x/y) = \frac{2x}{y^2}$$

   (iii) X and Y are not independent.

3 Functions of Random Variables

In the analysis of electrical systems, we will be often interested in finding the properties of a signal after it has been subjected to certain processing operations by the system, such as integration, weighted averaging, etc. These signal processing operations may be viewed as transformations of a set of input variables to a set of output variables. If the input is a set of random variables (RVs), then the output will also be a set of RVs. In this chapter, we deal with techniques for obtaining the probability law (distribution) for the set of output RVs when the probability law for the set of input RVs and the nature of transformation are known.

Function of One Random Variable

Let X be an RV with the associated sample space S_x and a known probability distribution. Let g be a scalar function that maps each $x \in S_x$ into $y = g(x)$. The expression $Y = g(X)$ defines a new RV Y. For a given outcome, $X(s)$ is a number x and $g[X(s)]$ is another number specified by $g(x)$. This number is the value of the RV Y, i.e., $Y(s) = y = g(x)$. The sample space S_y of Y is the set

$$S_y = \{y = g(x): x \in S_x)\}$$

How to find $f_y(y)$, when $f_x(x)$ is known

Let us now derive a procedure to find $f_y(y)$, the pdf of Y, when $Y = g(X)$, where X is a continuous RV with pdf $f_x(x)$ and $g(x)$ is a strictly monotonic function of x.

Case (i): $g(x)$ is a strictly increasing function of x.

$$\begin{aligned} F_y(y) &= P(Y \le y), \text{ where } F_y(y) \text{ is the cdf of } Y \\ &= P[g(X) \le y] \\ &= P[X \le g^{-1}(y)] \\ &= F_x(g^{-1}(y)] \end{aligned}$$

Differentiating both sides with respect to y,

$$f_y(y) = f_x(x)\,\frac{dx}{dy}, \text{ where } x = g^{-1}(y) \qquad (1)$$

Case (ii): $g(x)$ is a strictly decreasing function of x.

$$\begin{aligned} F_y(y) &= P(Y \le y) \\ &= P(g(X) \le y] \\ &= P[X \ge g^{-1}(y)] \\ &= 1 - P[X \le g^{-1}(y)] \\ &= 1 - F_x[g^{-1}(y)] \end{aligned}$$

$\therefore$
$$f_y(y) = -f_x(x)\,\frac{dx}{dy} \qquad (2)$$

Combining (1) and (2), we get

$$f_y(y) = f_x(x)\left|\frac{dx}{dy}\right|$$

i.e.,
$$f_y(y) = \frac{f_x(x)}{|g'(x)|}$$

Note *The above formula for $f_y(y)$ can be used only when $x = g^{-1}(y)$ is single valued.*

When $x = g^{-1}(y)$ takes finitely many values $x_1, x_2, \cdots x_n$, i.e., when the roots of the equation $y = g(x)$ are $x_1, x_2, \cdots, x_n$, the following extended formula should be used for finding $f_y(y)$:

$$f_y(y) = \frac{f_x(x_1)}{|g'(x_1)|} + \frac{f_x(x_2)}{|g'(x_2)|} + \cdots + \frac{f_x(x_n)}{|g'(x_n)|} \quad \text{or}$$

$$f_y(y) = f_x(x_1)\left|\frac{dx_1}{dy}\right| + f_x(x_2)\left|\frac{dx_2}{dy}\right| + \ldots + f_x(x_n)\left|\frac{dx_n}{dy}\right|$$

One Function of Two Random Variables

If a RV Z is defined as $Z = g(X, Y)$ where X and Y are RVs, we proceed to find $f_z(z)$ in the following way.

If z is a given number, we can find a region D_z in the xy-plane such that all points in D_z satisfy the condition $g(x, y) \le z$.

i.e.,
$$(Z \le z) = [g(X, Y) \le z] = [(X, Y) \in D_z)]$$

Now
$$F_z(z) = P(Z \le z) = P[(X, Y) \in D_z] = \iint_{D_z} f(x, y)\, dx\, dy$$

where $f(x, y)$ is the joint pdf of (X, Y). Thus, to find $F_z(z)$ it is sufficient to find the region D_z for every z and to evaluate the above integral. $f_z(z)$ is then found out as usual.

Theorem 1 If two RVs are independent, then the density function of their sum is given by the convolution of their density functions.

Proof Let the joint pdf of (X, Y) be $f(x, y)$

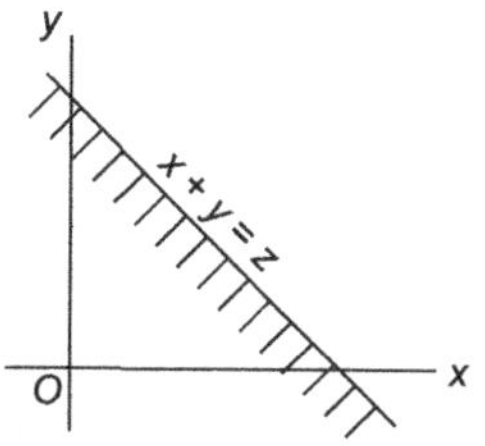

Let $$Z = X + Y$$

$$F_z(z) = P(X + Y \le z)$$

$$= \iint\limits_{(x+y\le z)} f(x, y)\, dx\, dy$$

$$= \int_{-\infty}^{\infty} \int_{-\infty}^{z-y} f(x, y)\, dx\, dy$$

Differentiating both sides with respect to z (Note that the upper limit for the inner integral is a function of z),

$$f_z(z) = \int_{-\infty}^{\infty} f(z - y, y)\, dy \tag{1}$$

Since X, Y are independent RVs

$$f(x, y) = f_x(x)\, f_Y(y)$$

$$\therefore \quad f(z - y, y) = f_x(z - y)\, f_y(y) \tag{2}$$

Using (2) in (1), we get

$$f_z(z) = \int_{-\infty}^{\infty} f_x(z - y)\, f_y(y)\, dy,$$ which is the convolution of $f_x(x)$ and $f_y(y)$.

Corollary

If $f_x(x) = 0$, for $x < 0$, and $f_y(y) = 0$, for $y < 0$, then $f_x(z - y)\, f_y(y) \neq 0$, only when $0 < y < z$.

$$\therefore \quad f_z(z) = \int_0^z f_x(z - y)\, f_y(y)\, dy,\ z > 0$$

Theorem 2 If two RVs X and Y are independent, find the pdf of $Z = XY$ in terms of the density functions of X and Y.

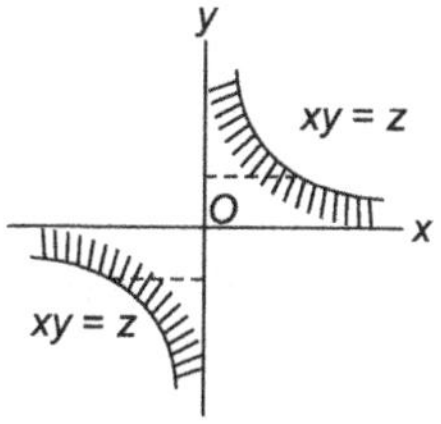

Let the joint pdf of (X, Y) be $f(x, y)$

$$F_z(z) = \iint\limits_{xy \le z} f(x, y)\, dx\, dy$$

Note *$xy = z$ is a rectangular hyperbola as shown in the figure.*

$$F_z(z) = \int_{-\infty}^{0} \int_{z/y}^{0} f(x, y)\, dx\, dy + \int_0^{\infty} \int_0^{z/y} f(x, y)\, dx\, dy$$

Differentiating both sides with respect to z,

$$f_z(z) = \int_{-\infty}^{0} -\frac{1}{y} f\left(\frac{z}{y}, y\right) dy + \int_0^{\infty} \frac{1}{y} f\left(\frac{z}{y}, y\right) dy$$

$$= \int_{-\infty}^{\infty} \left|\frac{1}{y}\right| f\left(\frac{z}{y}, y\right) \mathrm{d}y$$

$$= \int_{-\infty}^{\infty} \left|\frac{1}{y}\right| f_X\left(\frac{z}{y}\right) f_Y(y)\, \mathrm{d}y \quad \text{(since } X \text{ and } Y \text{ are independent)}$$

Theorem 3 If two RVs X and Y are independent, find the pdf of $Z = \dfrac{X}{Y}$ in terms of the density functions of X and Y.

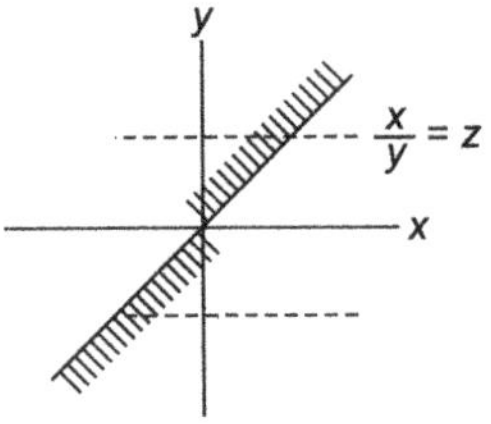

Let the joint pdf of (X, Y) be $f(x, y)$.

$$F_z(z) = \iint_{\frac{x}{y} \le z} f(x, y)\, \mathrm{d}x\, \mathrm{d}y$$

i.e.,

$$F_z(z) = \int_{-\infty}^{0} \int_{yz}^{\infty} f(x, y)\, \mathrm{d}x\, \mathrm{d}y + \int_{0}^{\infty} \int_{-\infty}^{yz} f(x, y)\, \mathrm{d}x\, \mathrm{d}y$$

Differentiating both sides with respect to z,

$$f_z(z) = \int_{-\infty}^{0} -yf(yz, y)\, \mathrm{d}y + \int_{0}^{\infty} yf(yz, y)\, \mathrm{d}y$$

$$= \int_{-\infty}^{\infty} |y| f(yz, y)\, \mathrm{d}y$$

$$= \int_{-\infty}^{\infty} |y| f_x(yz)\, f_y(y)\, \mathrm{d}y \quad \text{(since } X \text{ and } Y \text{ are independent)}$$

Two Functions of Two Random Variables

Theorem If (X, Y) is a two-dimensional RV with joint pdf $f_{xy}(x, y)$ and if $Z = g(X, Y)$ and $W = h(X, Y)$ are two other RVs, then the joint pdf of (Z, W) is given by

$$f_{zw}(z, w) = |J|\, f_{xy}(x, y), \text{ where } J = \frac{\partial(x, y)}{\partial(z, w)}$$

is called the Jacobian of the transformation and is given by

$$J = \begin{vmatrix} \dfrac{\partial x}{\partial z} & \dfrac{\partial x}{\partial w} \\ \dfrac{\partial y}{\partial z} & \dfrac{\partial y}{\partial w} \end{vmatrix}$$

Note *This theorem holds good, only if the equations $z = g(x, y)$ and $w = h(x, y)$ when solved, give unique values of x and y in terms of z and w.*

An alternative method to find the pdf of Z = g(X, Y)

Introduce a second RV $W = h(X, Y)$ and obtain the joint pdf of (Z, W), as suggested in the above theorem. Let it be $f_{zw}(z, w)$. The required pdf of Z is then obtained as the marginal pdf, i.e., $f_z(z)$ is obtained by simply integrating $f_{zw}(z, w)$ with respect to w.

i.e.,
$$f_z(z) = \int_{-\infty}^{\infty} f_{zw}(z, w)\, dw$$

Worked Examples 3

Example 1 *Find the distribution function of the RV $Y = g(X)$, in terms of the distribution function of X, if it is given that*

$$g(x) = \begin{cases} x - c & \text{for } x > c \\ 0 & \text{for } |x| \le c \\ x + c & x < -c \end{cases}$$

(MSU — Apr. 96)

Solution If $y < 0$,
$$F_y(y) = P(Y \le y)$$
$$= P(X + c \le y)$$

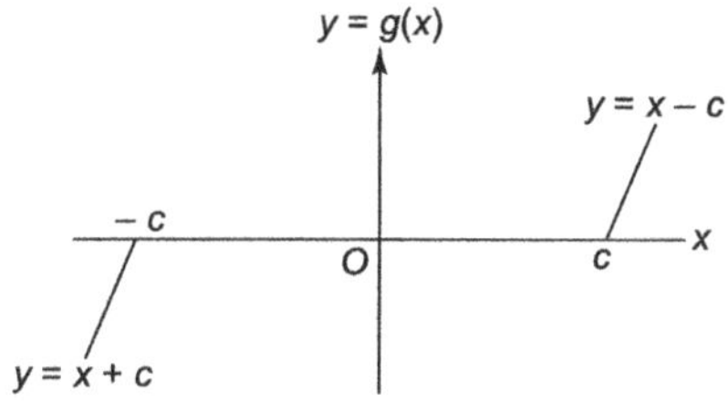

$$= P(X \le y - c)$$
$$= F_x(y - c)$$

If $y \ge 0$,
$$F_y(y) = P(X - c \le y)$$
$$= F_x(y + c)$$

Example 2 *The random variable Y is defined by $Y = \frac{1}{2}(X + |X|)$, where X is another RV. Determine the density and distribution function of Y in terms of those of X.*

(MSU — Nov. 96)

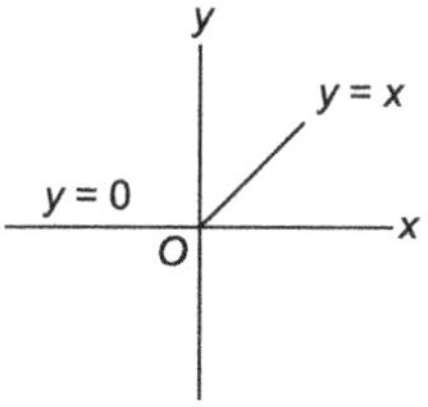

Solution When $X \ge 0$, $Y = X$

When $X < 0$, $Y = 0$

If $y < 0$, $F_y(y) = P(Y \le y) = 0$ (since there is no X, for which $Y \le y$)

If $y \ge 0$, $F_y(y) = P(Y \le y)$
$$= P(X \le y / X \ge 0)$$

$$= P(0 \le X \le y)/P(X \ge 0)$$

$$= \frac{F_x(y) - F_x(0)}{1 - F_x(0)}$$

$\therefore$ When $y < 0, f_y(y) = 0$ and

when $y \ge 0, f_y(y) = f_x(y)/[1 - F_x(0)]$

Example 3 *(a) Find the density function of $Y = aX + b$ in terms of the density function of X.*

(b) Let X be a continuous RV with pdf

$$f(x) = \frac{x}{12}, \text{ in } 1 < x < 5$$

$$= 0, \text{ elsewhere}$$

find the probability density function of $Y = 2X - 3$.

(MU — Apr. 96 and Nov. 96)

Solution (a) (i) Let $a > 0$

$$F_y(y) = P(Y \le y) = P(aX + b \le y)$$

$$= P\left(X \le \frac{y-b}{a}\right) \quad (\text{since } a > 0)$$

$$= F_x\left(\frac{y-b}{a}\right) \tag{1}$$

(ii) Let $a < 0$

$$F_y(y) = P(Y \le y) = P(aX + b \le y)$$

$$= P(aX \le y - b)$$

$$= P\left(X \ge \frac{y-b}{a}\right) \quad (\text{Since } a < 0)$$

$$= 1 - F_x\left(\frac{y-b}{a}\right) \tag{2}$$

From (1), $$f_y(y) = \frac{1}{a} f_x\left(\frac{y-b}{a}\right) \tag{3}$$

From (2), $$f_y(y) = -\frac{1}{a} f_x\left(\frac{y-b}{a}\right) \tag{4}$$

Combining (3) and (4),

$$f_y(y) = \frac{1}{|a|} f_x\left(\frac{y-b}{a}\right)$$

(b) $y = 2x - 3$, since $Y = 2X - 3$

$\therefore$ $x = \frac{1}{2}(y + 3)$, i.e., x is a single valued function of y

$$\therefore \quad f_y(y) = f_x(x)\left|\frac{dx}{dy}\right|$$

$$= \frac{x}{12} \times \frac{1}{2}$$

$$= \frac{1}{48}(y+3), \text{ in } -1 < y < 7$$

Note *The range of y is obtained from that of x (given in the problem) using the relation between x and y.*

Example 4 *If X is a continuous RV with some distribution defined over (0, 1) such that*

$$P(X \le 0.29) = 0.75, \text{ determine } k \text{ so that}$$
$$P(Y \le k) = 0.25, \text{ where } Y = 1 - X$$

Solution $\quad P(Y \le k) = 0.25$

i.e., $\quad P(1 - X \le k) = 0.25$

i.e., $\quad P(X \ge 1 - k) = 0.25$

$\therefore \quad P(X \le 1 - k) = 0.75 \quad (1)$

But it is given that $P(X \le 0.29) = 0.75 \quad (2)$

Comparing (1) and (2), $k = 0.71$

Example 5 *If $Y = X^2$, where X is a Gaussian random variable with zero mean and variance σ^2, find the pdf of the random variable Y.* (BU — Nov. 96)

Solution $\quad F_y(y) = P(Y \le y) = P(X^2 \le y)$

$$= P(-\sqrt{y} \le X \le \sqrt{y}), \text{ if } y \ge 0$$

$$= F_x(\sqrt{y}) - F_x(-\sqrt{y}). \quad (1)$$

and $F_y(y) = 0$, if $y < 0$ [since $X^2 = y$ has no roots, when $y < 0$]

Differentiating (1) with respect to y,

$$f_y(y) = \frac{1}{2\sqrt{y}}\left\{f_x(\sqrt{y}) + f_x(-\sqrt{y})\right\}, \text{ if } y \ge 0 \quad (2)$$

$$= 0, \text{ if } y < 0$$

It is given that X follows $N(0, \sigma)$.

$$\therefore \quad f_x(x) = \frac{1}{\sigma\sqrt{2\pi}} e^{-x^2/2\sigma^2} \quad -\infty < x < \infty$$

Using this value in (2), we get

$$f_y(y) = \frac{1}{\sigma\sqrt{2\pi y}} e^{-y/2\sigma^2} \quad y > 0$$

Example 6 *If the continuous RV X has pdf $f_x(x) = \frac{2}{9}(x + 1)$, in $-1 < x < 2$ and $= 0$, elsewhere, find the pdf of $Y = X^2$.*

Solution The transformation function $y = x^2$ is not monotonic in (–1, 2). So we divide the interval into two parts.
i.e., (– 1, 1) and (1, 2)
Since (–1, 1) is a symmetric interval, $f_y(y)$ is found out by using the formula (2) of the previous problem.
∴ When $-1 < x < 1$, i.e., $0 < y < 1$

$$f_y(y) = \frac{1}{2\sqrt{y}}\left\{\frac{2}{9}\left(1+\sqrt{y}\right)+\frac{2}{9}\left(1-\sqrt{y}\right)\right\}$$

$$= \frac{2}{9\sqrt{y}}$$

When $1 < x < 2$, i.e., $1 < y < 4$, $y = x^2$ is strictly increasing

$$\therefore \qquad f_y(y) = f_x(x)\left|\frac{dx}{dy}\right|$$

$$= \frac{2}{9}(x+1)\times\frac{1}{2x}$$

$$= \frac{1}{9}\left(1+\frac{1}{\sqrt{y}}\right)$$

Example 7 *According to the Maxwell-Boltzmann law of theoretical physics, the pdf of V, the velocity of a gas molecule is given by*

$$f_v(v) = \begin{cases} kv^2\, e^{-av^2}, & v > 0 \\ 0, & elsewhere \end{cases}$$

where a is a constant depending on its mass and the absolute temperature and k is an appropriate constant. Show that the kinetic energy $Y = \frac{1}{2}mV^2$ *is a random variable having Gamma distribution.*

Solution By the property of pdf,

$$k\int_0^{\infty} v^2\, e^{-av^2}\, dv = 1$$

i.e., $$\frac{k}{2a\sqrt{a}}\int_0^{\infty} t^{3/2-1}\, e^{-t}\, dt = 1, \text{ by putting } t = av^2$$

i.e., $$\frac{k}{2a\sqrt{a}}\,\Gamma\left(\frac{3}{2}\right) = 1$$

i.e., $$\frac{k}{2a\sqrt{a}}\times\frac{1}{2}\,\Gamma\left(\frac{1}{2}\right) = 1$$

∴ $$k = \frac{4a\sqrt{a}}{\sqrt{\pi}}, \text{ since } \Gamma\left(\frac{1}{2}\right) = \sqrt{\pi}$$

$$Y = \frac{m}{2} V^2$$

$$\therefore \qquad v = \pm \sqrt{\frac{2y}{m}}$$

since $v > 0$, $v = \sqrt{\frac{2y}{m}}$ is the only admissible value.

Now $$f_y(y) = f_v(v) \left| \frac{dv}{dy} \right|$$

$$= \frac{4a\sqrt{a}}{\sqrt{\pi}} \times \frac{2y}{m} e^{-2ay/m} \cdot \frac{1}{\sqrt{2my}}$$

$$= \frac{4\sqrt{2}\, a\sqrt{a}}{m\sqrt{m}\,\sqrt{\pi}} \times y^{1/2}\, e^{-2ay/m}$$

$$= \frac{(2a/m)^{3/2}}{\Gamma(3/2)} \times y^{3/2-1}\, e^{-(2a/m)y}, \quad y > 0$$

which is a 2-parameter Gamma distribution or Erlang distribution.

Example 8 *Given the RV X with density function*

$$f(x) = \begin{cases} 2x, & 0 < x < 1 \\ 0, & \text{elsewhere} \end{cases}$$

find the pdf of $Y = 8X^3$. (MU — Nov. 96)

Solution Since $y = 8x^3$ is a strictly increasing function in (0, 1),

$$f_y(y) = f_x(x) \left| \frac{dx}{dy} \right|, \text{ where } x = \frac{1}{2} y^{1/3}$$

$$= y^{1/3} \times \frac{1}{6} y^{-2/3}$$

$$= \frac{1}{6} y^{-1/3} \quad 0 < y < 8$$

Example 9 *If X is a Gaussian random variable with mean zero and variance* σ^2, *find the pdf of* $Y = |X|$.

$$F_y(y) = P(Y \le y) = P(|X| \le y)$$
$$= P\{-y \le X \le y\}$$
$$= F_x(y) - F_x(-y)$$

Solution Differentiating both sides with respect to y,

$$f_y(y) = f_x(y) + f_x(-y) \quad y > 0 \tag{1}$$

Now X follows $N(0, \sigma)$

$$\therefore \qquad f_x(x) = \frac{1}{\sigma\sqrt{2\pi}}\, e^{-x^2/2\sigma^2}, \quad -\infty < x < \infty \tag{2}$$

Using (2) in (1), we get

$$f_y(y) = \frac{2}{\sigma\sqrt{2\pi}}\, e^{-y^2/2\sigma^2}, \quad y > 0$$

Example 10 *(i) If X is a normal RV with mean zero and variance σ^2, find the pdf of $Y = e^X$.*
(ii) If X has an exponential distribution with parameter α, find the pdf of $Y = \log X$.

Solution (i) $f_y(y) = f_x(x)\left|\frac{dx}{dy}\right|$ (since $y = e^x$ is a monotonic function)

$$= \frac{1}{y} f_x(\log y)$$

$$= \frac{1}{\sigma y\sqrt{2\pi}}\, e^{-(\log y - \mu)^2/2\sigma^2}, \quad y > 0$$

which is the pdf of a lognormal distribution.

(ii) $f_y(y) = f_x(x)\left|\frac{dx}{dy}\right|$

$$= e^y \alpha\, e^{-\alpha e^y}, \quad -\infty < y < \infty$$

Example 11 *(i) If X has an exponential distribution with parameter 1, find the pdf of $Y = \sqrt{X}$.*
(ii) If X has a Cauchy's distribution with parameter α, prove that $Y = 1/X$ has also a Cauchy's distribution with parameter $1/\alpha$.

Solution (i) $f_y(y) = f_x(x)\left|\frac{dx}{dy}\right|$ (since $y = \sqrt{x}$ is m.i.)

$$= 2y\, e^{-y^2} \quad y > 0$$

(ii) $f_y(y) = f_x(x)\left|\frac{dx}{dy}\right|$ $\left(\text{since } y = \frac{1}{x} \text{ is m.d.}\right)$

$$= \frac{1}{y^2} \times \frac{\frac{\alpha}{\pi}}{\left(\frac{1}{y}\right)^2 + \alpha^2} \qquad \left(\text{since pdf of Cauchy's distribution } = \frac{\alpha/\pi}{x^2 + \alpha^2}\right)$$

$$= \frac{\frac{1}{\pi\alpha}}{y^2 + \frac{1}{\alpha^2}}, \quad -\infty < y < \infty$$

which is a Cauchy's distribution with parameter $1/\alpha$.

Example 12 *(i) If the RV X is uniformly distributed in* $(-\pi, \pi)$ *find the pdf of* $Y = a \sin(X + \alpha)$*, where* $a > 0$ *and* α *are constants.*

(ii) The horizontal range of a projectile is given by $R = \dfrac{v^2}{g} \sin 2\theta$*. If* θ *is uniformly distributed in* $(0, \pi/2)$ *and* $\dfrac{v^2}{g}$ *is a constant, find the pdf of R.*

Solution (i) If $|y| > a$, no solution exists for x

$\therefore$ $f_y(y) = 0$

If $|y| < a$, there exist only two values for x in $(-\pi, \pi)$.

Let them be $x_r = \sin^{-1}\left(\dfrac{y}{a}\right) - \alpha$; $r = 1, 2$

$$\therefore \quad \frac{dx_r}{dy} = \frac{1}{\sqrt{a^2 - y^2}}$$

Now,
$$f_y(y) = f_x(x_1)\left|\frac{dx_1}{dy}\right| + f_x(x_2)\left|\frac{dx_2}{dy}\right|$$

$$= \frac{1}{2\pi} \times \frac{1}{\sqrt{a^2 - y^2}} + \frac{1}{2\pi} \times \frac{1}{\sqrt{a^2 - y^2}}$$

$$\left(\text{since the pdf of } X = \frac{1}{2\pi}, \text{ as it is } U(-\pi, \pi)\right)$$

$$\therefore \quad f_y(y) = \frac{1}{\pi\sqrt{a^2 - y^2}} \quad |y| < a$$

(ii) θ is uniform in $(0, \pi/2)$

$\therefore$ $X = 2\theta$ is uniform in $(0, \pi)$ with pdf $1/\pi$

$$R = \frac{v^2}{g} \sin X = a \sin X, \text{ say } [R \text{ and } a > 0]$$

Therefore, when $r < a$, there exist only two values for x in $(0, \pi)$, given by $x_i = \sin^{-1} \dfrac{r}{a}$, $i = 1, 2$.

$$\therefore \quad \frac{dx_i}{dr} = \frac{1}{\sqrt{a^2 - r^2}}$$

Now
$$f_r(r) = f_x(x_1)\left|\frac{dx_1}{dr}\right| + f_x(x_2)\left|\frac{dx_2}{dr}\right|$$

$$= \frac{2}{\pi\sqrt{a^2 - r^2}}, \; 0 < r < a$$

$$= 0, \; r > a$$

Example 13 *(i) If X is uniformly distributed in $(-\pi/2, \pi/2)$ find the pdf of $Y = \tan X$.*

(ii) If X has the Cauchy's distribution with parameter 1, find the pdf of $Y = \tan^{-1} X$.

Solution (i) X is $U(-\pi/2, \pi/2)$

$$\therefore \quad f_X(x) = \frac{1}{\pi}$$

$$y = \tan x$$

Therefore, $x = \tan^{-1} y$, which is single valued in $(-\pi/2, \pi/2)$ i.e., for a given value of y, there exists only one value of $\tan^{-1} y$ in $(-\pi/2, \pi/2)$.

$$f_Y(y) = f_X(x) \left| \frac{dx}{dy} \right|$$

$$= \frac{\frac{1}{\pi}}{1 + y^2} \qquad -\infty < y < \infty$$

which is the pdf of a Cauchy's distribution.

(ii) $y = \tan^{-1} x$ is a monotonic increasing function

$$\therefore \quad f_Y(y) = f_X(x) \left| \frac{dx}{dy} \right|$$

$$= \frac{1}{\pi(1 + x^2)} \sec^2 y \quad (\text{since } x = \tan y)$$

$$= \frac{1}{\pi}, \quad \frac{-\pi}{2} < y < \frac{\pi}{2}$$

i.e., Y is uniformly distributed in $(-\pi/2, \pi/2)$.

Example 14 *(i) If X has an arbitrary distribution function $F_X(x)$, find $g(x)$ so that the random variable $Y = g(X)$ may be uniformly distributed in (0, 1).*

(ii) If X is uniformly distributed in (0, 1), find $g(x)$, so that the random variable $Y = g(X)$ may have an arbitrary distribution with cdf $F_Y(y)$.

(iii) If X is uniformly distributed in (–1, 1), find $g(x)$, so that the random variable $Y = g(X)$ may have the density function $f_Y(y) = 2e^{-2y}$, $y > 0$.

Solution (i) Y is to be uniform in (0, 1)

$$\therefore \quad f_Y(y) = 1 \quad \text{and} \quad F_Y(y) = \int_0^y f_Y(y)\, dy = y$$

$$\therefore \quad F_Y\{g(x)\} = g(x) \tag{1}$$

Now $$F_y(y) = P(Y \le y) = P[g(X) \le y]$$
$$= P[X \le g^{-1}(y)]$$
$$= P(X \le x) \qquad \text{[since } y = g(x) \text{ and hence } x = g^{-1}(y)]$$
$$= F_x(x) \qquad (2)$$

i.e., $$F_y\{g(x)\} = F_x(x)$$

i.e., $$g(x) = F_x(x), \text{ from (1)}$$

(ii) X is uniform in (0, 1)

$\therefore$ $$F_x(x) = x$$

By (2), $$F_y(y) = F_x(x)$$

$\therefore$ $$F_y[g(x)] = x$$

$\therefore$ $$g(x) = F_y^{-1}(x)$$

(iii) X is uniform in (–1, 1)

$\therefore$ $$f_x(x) = \frac{1}{2} \text{ and } F_x(x) = \frac{1}{2}(x+1)$$

$$f_y(y) = 2\,e^{-2y},\ y > 0$$

$\therefore$ $$F_y(y) = \int_0^y 2e^{-2y}\,dy = 1 - e^{-2y}$$

By (2), $$1 - e^{-2y} = \frac{1}{2}(x+1)$$

i.e., $$1 - e^{-2g(x)} = \frac{1}{2}(x+1)$$

$\therefore$ $$g(x) = \frac{1}{2}\log\left(\frac{2}{1-x}\right)$$

Example 15 *If X and Y are independent RVs having density functions.*

$$f_1(x) = \begin{cases} 2e^{-2x}, & x \ge 0 \\ 0, & x < 0 \end{cases} \quad \textit{and}$$

$$f_2(y) = \begin{cases} 3e^{-3y}, & y \ge 0 \\ 0, & y < 0 \end{cases} \quad \textit{and}$$

find the density function of their sum U = X + Y. (MKU — Apr. 96)

Solution By corollary under Theorem 1,

$$f_u(u) = \int_0^u f_x(u-y)\,f_y(y)\,dy, \quad u > 0$$

$$= \int_0^u 2e^{-2(u-y)} \times 3e^{-3y}\,dy$$

$$= 6e^{-2u} \int_0^u e^{-y}\, dy$$

$$= 6e^{-2u}(1 - e^{-u}), \quad u > 0$$

Example 16 *If X and Y are independent RVs and if Y is uniformly distributed in (0, 1), show that the density of Z = X + Y is given by $f_z(z) = F_x(z) - F_x(z - 1)$.*

Solution By Theorem 1,

$$f_z(z) = \int_{-\infty}^{\infty} f_x(z - y) f_y(y)\, dy$$

Now $\qquad f_y(y) = 1$, provided $0 < y < 1$

$$\therefore \quad f_z(z) = \int_0^1 f_x(z - y)\, dy$$

$$= \int_{z-1}^{z} f_x(t)\, dt, \text{ by putting } z - y = t$$

$$= F_x(z) - F_x(z - 1)$$

Example 17 *If $f_x(x) = ce^{-cx}\, U(x)$ and $f_z(z) = c^2 z\, e^{-cz}\, U(z)$, find $f_y(y)$, if Z = X + Y and X and Y are independent.*

Solution

Note *U(x) is the unit step function defined as*

$$U(x) = 1, \quad \text{if } x \ge 0$$
$$= 0, \quad \text{if } x < 0$$

$$f_x(x) = c\, e^{-cx},\ x \ge 0, \text{ and } f_z(z) = c^2\, z\, e^{-cz},\ z \ge 0$$

By Theorem 1,

$$c^2 z\, e^{-cz} = \int_0^z c\, e^{-c(z - y)} f_y(y)\, dy \tag{1}$$

Recalling that

$$\frac{\partial}{\partial z} \int_a^{b(z)} \phi(z, y)\, dy = \int_a^{b(z)} \frac{\partial \phi}{\partial z}\, dy + \phi[z, b(z)] \frac{\partial b(z)}{\partial z},$$ where a is a constant, and

differentiating (1) with respect to z partially, we get,

$$c^2 e^{-cz} - c^3 z\, e^{-cz} = \int_0^z -c^2 e^{-c(z - y)} f_y(y)\, dy + cf_y(z)$$

$$= -c\{c^2 z\, e^{-cz}\} + cf_y(z), \text{ by (1)}$$

$$\therefore \quad f_y(z) = ce^{-cz} \quad z > 0$$

$$\therefore \quad f_y(y) = ce^{-cy} \quad y > 0$$

Example 18 *The current I and the resistance R in a circuit are independent continuous RVs with the following density functions.*

$$f_i(i) = 2i, \quad 0 \le i \le 1$$
$$= 0 \quad \text{elsewhere}$$
$$f_r(r) = r^2/9 \quad 0 \le r \le 3$$
$$= 0 \quad \text{elsewhere}$$

Find the pdf of the voltage E in the circuit, where E = IR.

Solution By Theorem 2,

$$f_z(z) = \int_{-\infty}^{\infty} \frac{1}{|y|} f_x\left(\frac{z}{y}\right) f_y(y)\, dy$$

when $Z = XY$ and X and Y are independent RVs

Taking $x = i$, $y = r$ and $z = e$, we have

$$f_e(e) = \int_{-\infty}^{\infty} \frac{1}{|r|} f_i\left(\frac{e}{r}\right) \times f_r(r)\, dr$$

$$= \int_{e}^{3} \frac{1}{r} \times 2\frac{e}{r} \times \frac{r^2}{9}\, dr$$

Note $\quad f_i\left(\frac{e}{r}\right) = 2\,\frac{e}{r}$, if $0 \le \frac{e}{r} \le 1$,

i.e., if $r \ge e$.

$$f_r(r) = \frac{r^2}{9}, \text{ if } 0 \le r \le 3$$

Hence the limits for r are taken as e and 3.

$$\therefore \quad f_e(e) = \frac{2e}{9}(3 - e), \quad 0 \le e \le 3$$

Example 19 *If X and Y are independent RVs each following N (0, 2) prove that* $Z = \dfrac{X}{Y}$ *follows a Cauchy's distribution.*

Solution $f_x(x) = \dfrac{1}{2\sqrt{2\pi}} e^{-x^2/8}$ and $f_y(y) = \dfrac{1}{2\sqrt{2\pi}} e^{-y^2/8} \quad -\infty < x, \quad y < \infty$

By Theorem 3

$$f_z(z) = \int_{-\infty}^{\infty} |y|\, f_x(yz) f_y(y)\, dy$$

$$= \int_{-\infty}^{\infty} |y| \frac{1}{2\sqrt{2\pi}} e^{-y^2z^2/8} \times \frac{1}{2\sqrt{2\pi}} e^{-y^2/8}\, dy$$

$$= \frac{1}{4\pi} \int_0^\infty y\, e^{-(1+z^2)y^2/8}\, dy \quad \text{(since the integrand is an even function)}$$

$$= \frac{1}{\pi} \times \frac{1}{1+z^2} \quad -\infty < z < \infty$$

which is the pdf of Cauchy's distribution.

Example 20 *If X and Y are independent RVs each following N(0, 2), find the pdf of Z = 2X + 3Y.*

Solution Introduce the auxiliary RV $W = Y$.

$\therefore \quad z = 2x + 3y$ and $w = y$

solving, $x = \frac{1}{2}(z - 3w)$ and $y = w$

$$J = \frac{\partial(x, y)}{\partial(z, w)} = \begin{vmatrix} \frac{\partial x}{\partial z} & \frac{\partial x}{\partial w} \\ \frac{\partial y}{\partial z} & \frac{\partial y}{\partial w} \end{vmatrix} = \begin{vmatrix} \frac{1}{2} & \frac{-3}{2} \\ 0 & 1 \end{vmatrix} = \frac{1}{2}$$

Since X and Y are independent normal RVs $f_{xy}(x, y) = \frac{1}{8\pi} e^{-(x^2+y^2)/8}$, $-\infty < x, y < \infty$. The joint pdf of (Z, W) is given by

$$f_{zw}(z, w) = |J|\, f_{xy}(x, y)$$

$$= \frac{1}{2} \times \frac{1}{8\pi} e^{-\{(z-3w)^2 + 4w^2\}/32} \qquad -\infty < z,\ w < \infty$$

The pdf of Z is the marginal pdf, obtained by integrating $f_{zw}(z, w)$ with respect to w over the range of w.

$$\therefore \qquad f_z(z) = \frac{1}{16\pi} \int_{-\infty}^{\infty} e^{-(13w^2 - 6zw + z^2)/32}\, dw$$

$$= \frac{1}{16\pi} e^{-z^2/8 \times 13} \int_{-\infty}^{\infty} e^{-13\left(w - \frac{3z}{13}\right)^2/32}\, dw$$

$$= \frac{1}{(2\sqrt{13})\sqrt{2\pi}} e^{-z^2/2(2\sqrt{13})^2} \qquad -\infty < z < \infty$$

which is $N(0, 2\sqrt{13})$.

Example 21 *If X and Y each follow an exponential distribution with parameter 1 and are independent, find the pdf of U = X – Y.*

$$f_x(x) = e^{-x},\ x > 0, \text{ and } f_y(y) = e^{-y},\ y > 0$$

Solution Since X and Y are independent.

$$f_{xy}(x, y) = e^{-(x+y)}; \quad x, y > 0$$

Consider the auxiliary RV $V = Y$ along with

$$U = X - Y.$$

$$\therefore \quad x = u + v \quad \text{and} \quad y = v$$

$$J = \begin{vmatrix} \dfrac{\partial x}{\partial u} & \dfrac{\partial x}{\partial v} \\ \dfrac{\partial y}{\partial u} & \dfrac{\partial y}{\partial v} \end{vmatrix} = \begin{vmatrix} 1 & 1 \\ 0 & 1 \end{vmatrix} = 1$$

The joint pdf of (U, V) is given by

$$f_{uv}(u, v) = |J|\, f_{xy}(x, y)$$
$$= e^{-(x + y)} = e^{-(u + 2v)}$$

The range space of (U, V) is found out from the map of the range space of (X, Y) under the transformations $x = u + v$ and $y = v$.

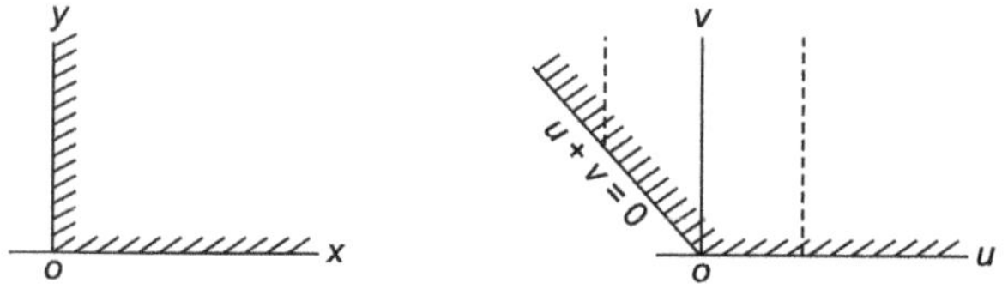

Therefore, the range space of (U, V) is given by $v > -u$, when $u < 0$ and $v > 0$, when $u > 0$.

Now the pdf of U is given by

$$f_u(u) = \int_{-u}^{\infty} e^{-(u + 2v)}\, dv, \quad \text{when } u < 0$$

and

$$= \int_{0}^{\infty} e^{-(u + 2v)}\, dv, \quad \text{when } u > 0$$

$$\therefore \quad f_u(u) = \frac{1}{2} e^{u}, \quad \text{when } u < 0.$$

and

$$= \frac{1}{2} e^{-u}, \quad \text{when } u > 0.$$

Example 22 *If the joint pdf of (X, Y) is given by $f_{xy}(x, y) = x + y$; $0 \le x, y \le 1$, find the pdf of $U = XY$.*

Solution Introduce the auxiliary RV $V = Y$.

$$\therefore \quad x = \frac{u}{v} \quad \text{and} \quad y = v$$

$$J = \begin{vmatrix} \dfrac{1}{v} & -\dfrac{u}{v^2} \\ 0 & 1 \end{vmatrix} = \frac{1}{v}$$

The joint pdf of (U, V) is given by

$$f_{uv}(u, v) = \frac{1}{|v|} f_{xy}(x, y) = \frac{1}{|v|}\left(\frac{u}{v} + v\right)$$

Range space of (X, Y) is given by $0 \le x \le 1$ and $0 \le y \le 1$

$\therefore$ Range space of (U, V) is given by

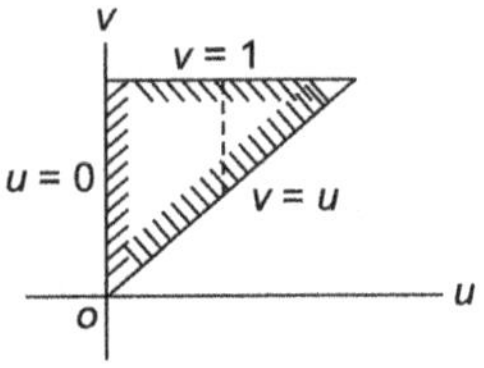

$$0 \le \frac{u}{v} \le 1 \text{ and } 0 \le v \le 1$$

i.e., $0 \le u \le v$ and $0 \le v \le 1$

The pdf of U is given by

$$f_u(u) = \int_u^1 f_{uv}(u, v)\, dv$$

$$= \int_u^1 \frac{1}{v}\left(\frac{u}{v} + v\right) dv$$

$$= 2(1 - u), \quad 0 < u < 1$$

Example 23 *If X and Y are independent RVs with $f_x(x) = e^{-x}\, U(x)$ and $f_y(y) = 3e^{-3y}\, U(y)$, find $f_z(z)$, if $Z = \frac{X}{Y}$.*

Solution Since X and Y are independent, $f_{xy}(x, y) = 3e^{-(x + 3y)}$, $x, y \ge 0$

Introduce the auxiliary RV $W = Y$.

$\therefore$ $x = zw$ and $y = w$

$$J = \begin{vmatrix} w & z \\ 0 & 1 \end{vmatrix} = w$$

The joint pdf of (Z, W) is given by

$$f_{zw}(z, w) = |J|\, f_{xy}(x, y)$$

$$= |w| \times 3e^{-(z+3)w}; \quad z, w \ge 0$$

The range space is obtained as follows:

Since $y \ge 0$, $w \ge 0$. Since $x \ge 0$, $zw \ge 0$.

As $w \ge 0$, $z \ge 0$.

The pdf of z is given by

$$f_z(z) = \int_0^{\infty} 3w\, e^{-(z+3)w}\, dw$$

$$= \frac{3}{(z+3)^2} \quad z \ge 0$$

Example 24 *If X and Y are independent RVs with pdf's e^{-x}, $x \geq 0$, and e^{-y}, $y \geq 0$, respectively, find the density functions of $U = \dfrac{X}{X+Y}$ and $V = X + Y$. Are U and V independent?*

Solution Since X and Y are independent, $f_{xy}(x, y) = e^{-(x+y)}$.

Solving the equations $u = \dfrac{x}{x+y}$ and $v = x + y$,

we get $x = uv$ and $y = v(1 - u)$.

$$J = \begin{vmatrix} v & u \\ -v & (1-u) \end{vmatrix} = v$$

The joint pdf of (U, V) is given by

$$f_{uv}(u, v) = |J|\, e^{-(x+y)}$$

$$= |v|\, e^{-v}$$

The range space of (U, V) is obtained as follows:

Since x and $y \geq 0$, $uv \geq 0$ and $v(1 - u) \geq 0$.

Therefore, either $u \geq 0$, $v \geq 0$ and $1 - u \geq 0$, i.e., $0 \leq u \leq 1$ and $v \geq 0$

or $u \leq 0$, $v \leq 0$ and $1 - u \leq 0$,

i.e., $u \leq 0$ and $u \geq 1$, which is absurd.

Therefore range space of (U, V) is given by $0 \leq u \leq 1$ and $v \geq 0$.

$\therefore\ f_{uv}(u, v) = v\, e^{-v};\quad 0 \leq u \leq 1$ and $v \geq 0$.

Pdf of U is given by $f_u(u) = \int_0^\infty v\, e^{-v}\, dv = 1, \quad 0 \leq u \leq 1$

i.e., U is uniformly distributed in $(0, 1)$.

Pdf of V is given by $f_v(v) = \int_0^1 v\, e^{-v}\, du$

$$= v\, e^{-v} \quad v \geq 0$$

Now $f_{uv}(u, v) = f_u(u) \times f_v(v)$

Therefore, U and V are independent RVs.

Example 25 *If X and Y are independent RVs each normally distributed with mean zero and variance σ^2, find the density functions of $R = \sqrt{X^2 + Y^2}$ and $\phi = \tan^{-1}\left(\dfrac{Y}{X}\right)$.*

Solution Since X and Y are independent $N(0, \sigma)$,

$$f_{xy}(x, y) = \frac{1}{2\pi\sigma^2}\, e^{-(x^2 + y^2)/2\sigma^2} \quad -\infty < x, \quad y < \infty$$

$r = \sqrt{x^2 + y^2}$ and $\theta = \tan^{-1}\left(\frac{y}{x}\right)$ are the transformations from cartesians to polars.

Therefore, the inverse transformations are given by $x = r\cos\theta$ and $y = r\sin\theta$.

$$J = \begin{vmatrix} \cos\theta & -r\sin\theta \\ \sin\theta & r\cos\theta \end{vmatrix} = r$$

The joint pdf of (R, ϕ) is given by

$$f_{r\theta}(r, \theta) = \frac{|r|}{2\pi\sigma^2} e^{-r^2/2\sigma^2} \quad r \geq 0 \quad 0 \leq \theta \leq 2\pi$$

Note *$(-\infty < x, y < \infty)$ and $(r \geq 0$ and $0 \leq \theta \leq 2\pi)$ both represent the entire xy-plane.*

The density function of R is given by

$$f_r(r) = \int_0^{2\pi} f_{r\theta}(r, \theta)\, d\theta = \frac{r}{\sigma^2} \times e^{-r^2/2\sigma^2} \quad r \geq 0$$

which is a Rayleigh distribution with parameter σ. The density function of ϕ is given by

$$f_\theta(\theta) = \int_0^{\infty} \frac{r}{2\pi\sigma^2} e^{-r^2/2\sigma^2}\, dr$$

$$= \frac{1}{2\pi}\int_0^{\infty} e^{-t}\, dt, \text{ on putting } t = \frac{r^2}{2\sigma^2}$$

$$= \frac{1}{2\pi}, \quad 0 \leq \theta \leq 2\pi,$$

which is a uniform distribution.

Exercise 3

Part A (Short answer questions)

1. If X and Y are two RVs where $Y = g(X)$, how are the density functions of X and Y related?
2. If the pdf of X is $f_x(x) = 2x$, $0 < x < 1$, find the pdf of $Y = 3X + 1$.
3. If the pdf of X is $f_x(x) = e^{-x}$, $x > 0$, find the pdf of $Y = 2X + 1$.
4. If the RV X is uniformly distributed in $(0, 2)$, find the pdf of $Y = X^3$.
5. If the RV X is uniformly distributed in $(1, 2)$ find the pdf of $Y = \frac{1}{X}$.
6. If X is uniformly distributed in $(0, 1)$, find the pdf of $Y = \frac{1}{2X+1}$.

7. If X is uniformly distributed in (1, 2) find the pdf of $Y = e^X$.
8. If the pdf of a RV X is $f_x(x) = 2x$, $0 < x < 1$, find the pdf of $Y = e^{-X}$.
9. If the cdf of a RV X is $F(x)$, show that the RV $Y = F(x)$ follows a uniform distribution.
10. A RV X assumes three values $-1, 0, 1$ with probabilities 1/3, 1/2, 1/6 respectively, find the probability distribution of $Y = 3X + 1$.
11. If X and Y are two RVs such that $Y = X^2$, how are the cdf's of X and Y related?
12. If X and Y are two RVs such that $Y = X^2$, how are the pdf's of X and Y related?
13. If the RV X is uniformly distributed in $(-3, 3)$, find the pdf of $Y = X^2$.
14. If the pdf of X is $f(x) = e^{-x}$, $x > 0$, find the pdf of $Y = X^2$.
15. If the RVs X and Y are related by $Y = |X|$, how are the cdf's of X and Y related?
16. If the RVs X and Y are related by $Y = |X|$, how are the pdf's of X and Y related?
17. If the RV X is uniformly distributed in $(-1, 1)$, find the pdf of $Y = |X|$.
18. If the RVs X and Y are related by $Y = \sqrt{X}$, how are their pdf's related?
19. If X is uniformly distributed in (0, 1), find the pdf of $Y = \sqrt{X}$.
20. If the pdf of a RV X is $f_X(x) = 2x$ in (0, 1), find the pdf of $Y = \sqrt{X}$.
21. If X is uniformly distributed in $\left(-\frac{\pi}{2}, \frac{\pi}{2}\right)$, find the pdf of $Y = \tan X$.
22. Write down the formula to find the pdf of $Z = X + Y$, if X and Y are independent RVs with pdf's $f_x(x)$ and $f_y(y)$ respectively.
23. Write down the formula to find the pdf of $Z = XY$ in terms of the pdf's of X, Y if they are independent.
24. Write down the formula for the pdf of $Z = \frac{X}{Y}$ in terms of the pdf's of X, Y if they are independent.
25. If $Z = g(X, Y)$ and $W = h(X, Y)$, how are the joint pdf's of (X, Y) and (Z, W) related?
26. If $Z = 2X + 3Y$ and $W = Y$, how are the joint pdf's of (X, Y) and (Z, W) related?
27. If $U = XY$ and $V = Y$, how are the joint pdf's of (X, Y) and (U, V) related?
28. If $U = \frac{X}{Y}$ and $V = Y$, how are the joint pdf's of (X, Y) and (U, V) related?
29. If $U = X + Y$ and $V = X - Y$, how are the joint pdf's of (X, Y) and (U, V) related?
30. If $X = R\cos\phi$ and $Y = R\sin\phi$, how are the joint pdf's of (X, Y) and (R, ϕ) related?

Part B

31. The RV X is of the continuous type with distribution function $F_x(x)$. If $g(x)$ is defined as

$$g(x) = \begin{cases} 1 & \text{if } x > 0 \\ -1 & \text{if } x \le 0 \end{cases}$$

what is the distribution function of $Y = g(X)$? (MSU — Apr. 96)

32. If $g(x)$ is defined as

$$g(x) = \begin{cases} -b & \text{if } x \le -b \\ x & \text{if } -b < x < b \\ b & \text{if } x \ge b \end{cases}$$

find the distribution function of $Y = g(X)$ in terms of that of X.

33. If X follows a normal distribution with mean zero and variance σ^2, find the density function of $Y = \frac{1}{2}(X + |X|)$.

34. If the density function of a continuous RV X is given by $f_x(x) = \frac{2}{9}(x + 1)$, for $-1 < x < 2$, and $= 0$, otherwise, find the density function of $Y = \frac{1}{2}(X + |X|)$.

35. If X is a continuous RV with density function $f_x(x) = e^{-x}$, $x > 0$, find the density function of $Y = 2X + 1$. Hence or otherwise find $P(Y \ge 5)$.

36. If the density function of a continuous RV X is given by $f_x(x) = 2x$, for $0 < x < 1$, and $= 0$, elsewhere, find the density and distribution functions of $Y = 3X + 1$.

37. If $Y = aX^2$ $(a > 0)$, where X is a Gaussian RV with zero mean and variance σ^2, find the pdf of the RV Y. (MSU — Nov. 96)

38. (i) If the continuous RV X is uniformly distributed in $(-3, 3)$, find the density function of $Y = 2X^2 - 3$.
    (ii) If the continuous RV X is uniformly distributed in $(-2, 2)$, find the density function of $Y = 6 - X^2$.

39. (i) A fluctuating electric current I may be considered as a uniformly distributed RV over the interval (9, 11). If this current flows through a 2-Ω resistor, find the density function of the power $P = 2I^2$.
    (ii) If the voltage E across a resistor is a RV uniformly distributed between 5 and 10 V, find the density function of the power $W = \frac{E^2}{r}$, when $r = 1000\ \Omega$.

40. (i) If the continuous RV X has density function $f_x(x) = 2\,e^{-2x}$, $x > 0$, find the density function of $Y = X^2$.
    (ii) If the density function of a continuous RV X is given by $f_x(x) = e^{-x}$, $x > 0$, find the density function of $Y = 3/(X + 1)^2$.

41. If the density function of a continuous RV X is given by $f_x(x) = e^{-x}$, $x > 0$, find the density function of $Y = X^3$.
42. If the RV X is uniform in $(-2\pi, 2\pi)$, find the density function of RV (i) $Y = X^3$ and (ii) $Y = X^4$.
43. If the radius R of a sphere is a continuous RV with pdf $f_r(r) = 6r(1 - r)$, $0 < r < 1$, find the pdf of (i) the surface area S of the sphere and (ii) the volume V of the sphere.
44. If the random variable X is uniformly distributed over $(-1, 1)$, find the density function of $Y = |X|$.
45. (i) If the resistance R follows a uniform distribution between 900 and 1000 Ω, find the density of the corresponding conductance $G = \dfrac{1}{R}$.
    (ii) If the RV X is uniformly distributed in (0, 1), find the pdf of $Y = \dfrac{1}{X+1}$.
46. (i) If the continuous RV X is uniformly distributed over (1, 3), obtain the pdf of the RV $Y = e^X$.
    (ii) If the density function of a random variable X is given by $f_x(x) = 2x$, for $0 < x < 1$, and $= 0$, elsewhere, find the pdf of the RV $Y = e^{-X}$.
47. If the RV X is uniformly distributed over $(-1, 1)$, find the density function of (i) $Y = \sin\left(\dfrac{\pi X}{2}\right)$ and (ii) $Y = \cos\left(\dfrac{\pi X}{2}\right)$.
48. If X is an arbitrary RV with continuous distribution function $F_x(x)$ and if $Y = F_x(X)$, show that Y is uniformly distributed in (0, 1).
49. If X and Y are independent RVs with density function $f_x(x) = a\, e^{-\alpha x} U(x)$ and $f_y(y) = \beta\, e^{-\beta y} U(y)$, find the density function of $Z = X + Y$.
50. If X and Y are independent RVs with identical uniform distributions in the interval $(-1, 1)$ find the density function of $Z = X + Y$.
51. If X and Y are independent RVs with density functions $f_x(x) = e^{-x}\, U(x)$ and $f_y(y) = 2e^{-2y} U(y)$, find the density function of $Z = X + Y$.
52. If X and Y are independent RVs with density function $f_x(x) = 1$, in $1 \le x \le 2$, and $f_y(y) = \dfrac{y}{6}$ in $2 \le y \le 4$, find the density function of $Z = XY$.
53. If X and Y are independent RVs such that $f_x(x) = \dfrac{x}{\alpha^2}\, e^{-x^2/2\alpha^2}\, U(x)$ and $f_y(y) = \dfrac{y}{\beta^2}\, e^{-y^2/2\beta^2}\, U(y)$ prove that the density function of $Z = \dfrac{X}{Y}$ is given by

$$f_z(z) = \frac{2\alpha^2}{\beta^2} \times \frac{z}{\left(z^2 + \dfrac{\alpha^2}{\beta^2}\right)} \times U(z)$$

54. If X and Y are independent RVs with identical uniform distributions in $(0, a)$, find the density function of $Z = |X - Y|$. (**Hint:** First find the cdf of Z)
55. If X and Y are independent RVs with identical uniform distributions in $(0, 1)$, find (i) the joint density function of (U, V), where $U = X + Y$ and $V = X - Y$, (ii) the density function of U and (iii) the density function of V.
56. Given the joint density function of X and Y as

$$f(x, y) = \frac{1}{2} x e^{-y}, \quad 0 < x < 2 \quad y > 0$$
$$= 0, \quad \text{elsewhere}$$

find the distribution function of $(X + Y)$. (BDU — Apr. 97)

Answers

Exercise 3

1. If $x = g^{-1}(y)$ is single valued, then $f_y(y) = \dfrac{f_x(x)}{|g'(x)|} = f_x(x)\left|\dfrac{dx}{dy}\right|$. If $x = g^{-1}(y)$ takes many values $x_1, x_2, \ldots, x_n$, then $f_y(y) = f_x(x_1)\left|\dfrac{dx_1}{dy}\right| + f_x(x_2)\left|\dfrac{dx_2}{dy}\right| + \ldots + f_x(x_n)\left|\dfrac{dx_n}{dy}\right|$.

2. $f_y(y) = \left|\dfrac{dx}{dy}\right| f_x(x) = \dfrac{2}{9}(y - 1)$ in $1 < y < 4$

3. $f_y(y) = \left|\dfrac{dx}{dy}\right| f_x(x) = \dfrac{1}{2} e^{-\frac{1}{2}(y-1)}, \quad y > 1$

4. $f_y(y) = \left|\dfrac{1}{3} y^{-2/3}\right| \dfrac{1}{2} = \dfrac{1}{6} y^{-2/3} \quad 0 < y < 8$

5. $f_y(y) = \left|-\dfrac{1}{y^2}\right| 1 = \dfrac{1}{y^2}$ in $\dfrac{1}{2} < y < 1$

6. $f_y(y) = \left|\dfrac{-(2x+1)^2}{2}\right| 1 = \dfrac{1}{2y^2}$ in $\dfrac{1}{3} < y < 1$

7. $f_y(y) = \left|\dfrac{1}{e^x}\right| 1 = \dfrac{1}{y}$ in $e < y < e^2$

8. $f_y(y) = |-e^x| 2x = -\dfrac{2}{y} \log y$ in $\dfrac{1}{e} < y < 1$

9. $f_y(y) = \left|\dfrac{1}{f_x(x)}\right| f_x(x) = 1$ in $0 < y < 1$

10. $Y \;:\; -2 \quad 1 \quad 4$

    $p_y \;:\; \dfrac{1}{3} \quad \dfrac{1}{2} \quad \dfrac{1}{6}$

11. $F_y(y) = F_x(\sqrt{y}) - F_x(-\sqrt{y})$ if $y \geq 0$

    $= 0$ if $y < 0.$

12. $f_y(y) = \dfrac{1}{2\sqrt{y}} \left\{f_y(\sqrt{y})\right\} + f_x(-\sqrt{y})$ if $y \geq 0$

    $= 0,$ if $y < 0$

13. $f_Y(y) = \dfrac{1}{2\sqrt{y}}\left\{\dfrac{1}{6} + \dfrac{1}{6}\right\} = \dfrac{1}{6\sqrt{y}}$ in $0 < y < 9$

14. Since $y = x^2$ is m.i. in $x > 0$, $f_y(y) = \left|\dfrac{1}{2x}\right| e^{-x} = \dfrac{1}{2\sqrt{y}} e^{-\sqrt{y}}$ in $y > 0$

15. $F_y(y) = F_x(y) - F_x(-y)$

16. $f_y(y) = f_x(y) + f_x(-y),\ y > 0$

17. $f_y(y) = \dfrac{1}{2} + \dfrac{1}{2} = 1$ in $0 < y < 1$

18. $f_y(y) = \left|\dfrac{dx}{dy}\right| f_x(x) = 2y f_x(y^2)$

19. $f_y(y) = 2y$ in $0 < y < 1$

20. $f_y(y) = 2y \times 2y^2 = 4y^3$ in $0 < y < 1$

21. $f_y(y) = \left|\dfrac{dx}{dy}\right| f_x(x) = \dfrac{1}{\pi}\,\dfrac{1}{1+y^2}$ in $-\infty < y < \infty$

22. $f_z(z) = \displaystyle\int_{-\infty}^{\infty} f_x(z-y)\, f_y(y)\, dy$

23. $f_z(z) = \displaystyle\int_{-\infty}^{\infty} \frac{1}{|y|}\, f_x\left(\frac{z}{y}\right) f_y(y)\, dy$

24. $f_z(z) = \displaystyle\int_{-\infty}^{\infty} |y|\, f_x(yz)\, f_y(y)\, dy$

25. $f_{zw}(z, w) = |J|\, f_{xy}(x, y)$, where $J = \begin{vmatrix} x_z & x_w \\ y_z & y_w \end{vmatrix}$

26. $f_{zw}(z, w) = \dfrac{1}{2} f_{xy}(x, y)$

27. $f_{uv}(u, v) = \dfrac{1}{|v|} f_{xy}(x, y)$

28. $f_{uv}(u, v) = |v| f_{xy}(x, y)$

29. $F_{uv}(u, v) = \dfrac{1}{2} f_{xy}(x, y)$

30. $f_{r\phi}(r, \theta) = |r| f_{xy}(x, y)$

31. $F_y(y) = 0$, if $y < -1$; $= F_x(0)$, if $-1 \le y < 1$ and $= 1$, if $1 \le y$

32. $F_y(y) = 0$, if $y < -b$; $= F_x(y)$, if $-b \le y < b$ and $= 1$, if $b \le y$

33. $f_y(y) = \dfrac{2}{\sigma\sqrt{2\pi}} e^{-y^2/2\sigma^2}$, $y \ge 0$ and 0, elsewhere

34. $f_y(y) = \dfrac{1}{4}(y + 1)$, $0 < y < 2$ and $= 0$, elsewhere

35. $f_y(y) = \dfrac{1}{2} e^{-(y-1)/2}$, $y > 1$; $1/e^2$

36. $f_y(y) = \dfrac{2}{9}(y - 1)$ for $1 < y < 4$ and $= 0$, elsewhere; $F_y(y) = (y - 1)^2/9$.

37. $f_y(y) = \dfrac{1}{\sigma\sqrt{2\pi ay}} e^{-y/2a\sigma^2}$, $y > 0$

38. (i) $f_y(y) = \dfrac{1}{6\sqrt{2}\sqrt{y+3}}$ for $-3 \le y \le 15$

(ii) $f_y(y) = \dfrac{1}{4\sqrt{6-y}}$ for $2 \le y \le 6$

39. (i) $f_p(p) = \dfrac{1}{8}\sqrt{\dfrac{2}{p}}$ for $162 \le p \le 242$

(ii) $f_w(w) = \sqrt{\dfrac{10}{w}}$ for $\dfrac{1}{40} \le w \le \dfrac{1}{10}$

40. (i) $f_y(y) = \dfrac{1}{\sqrt{y}} e^{-2\sqrt{y}}$ for $y > 0$

(ii) $f_y(y) = \dfrac{1}{6}\left(\dfrac{3}{y}\right)^{3/2} e^{1-\sqrt{3/y}}$ for $0 \le y \le 3$

41. $f_y(y) = \dfrac{1}{3} y^{-2/3} e^{-y^{1/3}}$ for $y > 0$

42. (i) $f_y(y) = \dfrac{1}{12\pi} y^{-2/3}$ for $-8\pi^3 \le y \le 8\pi^3$

(ii) $f_y(y) = \dfrac{1}{8\pi} y^{-3/4}$ for $0 \le y \le 16\pi^4$

43. (i) $f_s(s) = \frac{3}{4\pi}\left(1 - \sqrt{\frac{s}{4\pi}}\right)$ for $0 \le s \le 4\pi$

(ii) $f_v(v) = \frac{3}{2\pi}\left\{\left(\frac{3v}{4\pi}\right)^{-\frac{1}{3}} - 1\right\}$ for $0 \le v \le \frac{4\pi}{3}$

44. $f_y(y) = 1$ in $(0, 1)$

45. (i) $f_g(g) = \frac{1}{200g^2}$ for $\frac{1}{1100} \le g \le \frac{1}{900}$

(ii) $f_y(y) = \frac{1}{y^2}$ for $\frac{1}{2} \le y \le 1$

46. (i) $f_y(y) = \frac{1}{2y}$ for $e \le y \le e^3$

(ii) $f_y(y) = -\frac{2}{y}\log y$ for $\frac{1}{e} \le y \le 1$

47. (i) $f_y(y) = \frac{1}{\pi\sqrt{1 - y^2}}$ for $-1 \le y \le 1$

(ii) $f_y(y) = \frac{2}{\pi\sqrt{1 - y^2}}$ for $0 \le y \le 1$

49. $f_z(z) = \frac{\alpha\beta}{\beta - \alpha}(e^{-\alpha z} - e^{-\beta z})$ if $\beta \ne \alpha$

$= \alpha^2 z \quad e^{-\alpha z}$ if $\beta = \alpha$

50. $f_z(z) = \frac{1}{4}(2 + z)$ if $-2 < y < 0$

$= \frac{1}{4}(2 - z)$ if $2 > y > 0$

51. $f_z(z) = 2[e^{-z} - e^{-2z}]\,U(z)$

52. $f_z(z) = \begin{cases} \frac{1}{6}(z - 2) & \text{when } 2 \le z \le 4 \\ \frac{1}{12}(8 - z) & \text{when } 4 \le z \le 8 \end{cases}$

54. $f_z(z) = \frac{2}{a^2}(a - z)$, for $0 < z < a$

55. (i) $f_{uv}(u, v) = \frac{1}{2}$ for $0 \le u + v \le 2$ and $0 \le u - v \le 2$

(ii) $f_u(u) = \begin{cases} u & \text{if } 0 \le u \le 1 \\ 2 - u & \text{if } 1 \le u \le 2 \end{cases}$

(iii) $f_V(v) = \begin{cases} 1 + v & \text{if } -1 \le v \le 0 \\ 1 - v & \text{if } 0 \le v \le 1 \end{cases}$

56. If $Z = X + Y$, $F_z(z) = 0$, if $z < 0$

$$= \frac{1}{2}\left(\frac{z^2}{2} - e^{-z} - z + 1\right), \text{ if } 0 \le z < 2$$

$$= \frac{1}{2}\left[(1 - e^{-2} + (1 + e^2)(e^{-2} - e^{-z})\right], \text{ if } 2 \le z$$

4 Statistical Averages

Statistical Measures

A discrete random variable is no doubt completely described by its probability mass function or probability distribution. Similarly a continuous random variable is completely described by its probability density function. For many purposes, this description is often considered to consist too many details. It is sometimes simpler and more convenient to describe a random variable or to characterise its distribution by a few parameters, known as statistical measures, that are representative of the distribution.

The characteristics that are usually used to analyse the properties of a random variable or its probability distribution are the central tendency, dispersion, skewness and kurtosis. In this section, we shall discuss these characteristics and the measures used to study them.

Measures of Central Tendency

Central tendency of a random variable is the concentration of its values in the central part of the distribution.

A measure of central tendency is the value of the random variable which is representative of the entire distribution of the variable.

The measures of central tendency, also known as *measures of location* or *averages*, that are in common use are the following.

(i) Arithmetic mean or simply Mean
(ii) Median
(iii) Mode
(iv) Geometric Mean
(v) Harmonic Mean.

Definitions

1. If X is a discrete random variable with probability mass function $P(X = x_i) = p_i$; $i = 1, 2, 3, \ldots$, then *the mathematical expectation of* X or *the arithmetic mean of* X, denoted as $E(X)$, is defined by, $E(X) = \sum_i x_i p_i$.

If X is a continuous random variable with probability density function $f(x)$, then $E(X)$ is defined by

$$E(X) = \int_{R_X} x f(x)\,dx \quad \text{where } R_X \text{ is the range space of } X.$$

2. If X is a random variable, discrete or continuous, the value of $X = x$ for which the comulative distribution function $F(x) = \frac{1}{2}$ is called *the median* of X or the median of the distribution of X. In the case of a discrete random variable X, if there exists no x such that $F(x) = \frac{1}{2}$, then the median M is defined as

$$M = \frac{1}{2}\,(x_k + x_{k+1})$$

where, $F(x_k) < \frac{1}{2}$ and $F(x_{k+1}) > \frac{1}{2}$, where x_k and x_{k+1} are two consecutive values of X. In the case of a continuous random variable X, the median M is defined by

$$\int_{-\infty}^{M} f(x)\,dx = \int_{M}^{\infty} (x)\,dx = \frac{1}{2}$$

where $f(x)$ is the density function of X.

3. If X is a discrete random variable, then that value of X for which the probability is maximum is called *the mode* of the distribution of X (or *modal value* of X). If X is a continuous random variable with density function $f(x)$, then the value of $X = x$ for which $f(x)$ is maximum [i.e. for which $f'(x) = 0$ and $f''(x) < 0$] is the mode of X.
4. The *geometric mean* G of a random variable X is defined by $\log G = E(\log X)$.

   For a discrete random variable, $\log G = \sum_i (\log x_i)\, p_i$, where (x_i, p_i); $i = 1, 2, 3, \ldots$, is the probability distribution of X.

   For a continuous random variable $\log G = \int_{-\infty}^{\infty} (\log x)\, f(x)dx$.

5. *The harmonic mean* H of a random variable X is defined by $\frac{1}{H} = E\left(\frac{1}{X}\right)$.

   For a discrete random variable, $\frac{1}{H} = \sum_i \frac{1}{x_i}\, p_i$ and for a continuous random variable, $\frac{1}{H} = \sum_{-\infty}^{\infty} \frac{1}{x} f(x)dx$.

Mathematical Expectation and Moments

If $g(x)$ is a function of a random variable X, then the *expected value* of $g(X)$ is defined by

$$E\{g(X)\} = \begin{cases} \sum_i g(x_i)\, p_i, & \text{if } X \text{ is discrete} \\ \int_{-\infty}^{\infty} g(x)\, f(x)\, dx & \text{if } X \text{ is continuous} \end{cases}$$

Note *$E[ag(X) + bh(X)] = a\,E\{g(X)\} + bE\{h(X)\}$. In particular, $E(aX + b) = a\,E(X) + b$.*

The rth order *non-central moment* about the value 'a' of a random variable X, denoted by μ'_r is defined by

$$\mu'_r = E\{(X - a)^r\}$$
$$= \sum_i (x_i - a)^r\, p_i, \text{ when } X \text{ is discrete}$$
$$= \int_{-\infty}^{\infty} (x - a)^r f(x) dx, \text{ when } X \text{ is continuous.}$$

When $a = 0$, then $\mu'_r = E(X^r)$ is called the rth order *simple moment*. We note that $\mu'_1 = E(X)$ is the mean of X. Simple moments are also called *raw moments* or *ordinary moments*.

The moments about the mean value $\mu = E(X)$ are called *central moments* and denoted by μ_r

Thus $\mu_r = E\{(X - \mu)^r\}$, where $\mu = E(X)$.

We note the $\mu_1 = E(X - \mu) = E(X) - \mu = 0$.

The rth order non-central, simple and central *absolute moments* are defined by $E(|x - a|^r\}$, $E\{|X|^r\}$ and $E\{|X - \mu|^r\}$ respectively.

Relation Between Central and Non-central Moments

Let $\mu^1_n = E(X - a)^n$, $\therefore\ \mu'_1 = E(X - a) = \mu - a = \mathrm{d}$ (say)

Now $\mu_n = E(X - \mu)^n$, where $\mu = E(X)$

$$= E\{(X - a) - (\mu - a)\}^n$$
$$= E\{(X - a) - d\}^n$$
$$= E\left[\sum_{r=0}^{n} nC_r (X - a)^{n-r} (-d)^r\right]$$
$$= \sum_{r=0}^{n} nC_r E(X - a)^{n-r} (-1)^r d^r$$
$$= \sum_{r=0}^{n} nC_r\ \mu'_{n-r} (-1)^r d^r$$

i.e., $\mu_n = \mu'_n - nC_1 \cdot \mu'_{n-1}\, d + nC_2 \cdot \mu'_{n-2}\, d^2 - nC_3\, \mu'_{n-3}\, d^3 + \ldots (-1)^n d^n$.

In particular,

$$\mu_2 = \mu'_2 - d^2$$

$$\mu_3 = \mu'_3 - 3\mu'_2\, d + 2d^3; \text{ and}$$
$$\mu_4 = \mu'_4 - 4\mu'_3\, d + 6\mu'_2\, d^2 - 3d^4, \quad \text{where } d = \mu - a.$$

Dispersion

Though a measure of central tendency may be taken as a representative value of the random variable, it is important to know how the values are clustered around or scattered away from the measure of central tendency while studying the distribution of the random variable. The property of the random variable (or its distribution) by which its values are clustered around or scttering away from the central value is called *dispersion*, *spread*, *scatter* or *variability*.

Following are the measures of dispersion which are commonly used.

(i) Quartile Deviation
(ii) Mean Deviation
(iii) Standard Deviation

Definitions

1. The rth quartile of a continuous random variable (or its distribution), denoted by Q_r, is defined by

$$\int_{-\infty}^{Qr} f(x)\, dx = \frac{r}{4}\,;\; r = 1, 2, 3$$

Q_1 and Q_3 are called the first (lower) and the third (upper) *quartiles* respectively.

Q_2 is obviously the median (middle or second quartile).

Now $\dfrac{Q_3 - Q_1}{2}$, that is half of the difference between the upper and lower quartiles, is called *the semi-interquartile range* or *the quartile deviation* and denoted by Q

i.e. $$Q = \frac{Q_3 - Q_1}{2}$$

In the case of discrete random variable, the quartiles are defined in a manner similar to the definition of median.

2. Mean Deviation of a random variable X about the mean value μ is defined by

$$\text{M.D.} = E\{|X - \mu|\}$$

i.e. $$\text{M.D.} = \sum_i |x_i - \mu| p_i, \text{ when } X \text{ is discrete}$$

$$= \int_{-\infty}^{\infty} |x - \mu| f(x)dx, \text{ when } X \text{ is continuous}$$

Mean deviations about the averages median and mode may also be required in statistical analysis.

Property

Mean deviation about the median is the minimum.

Let $\phi(a)$ be the mean deviation about a.

i.e. $$\phi(a) = E\{|X - a|\} = \int_{-\infty}^{\infty} |x - a| f(x)dx$$

$$= \int_{-\infty}^{a} (a - x) f(x)dx + \int_{a}^{\infty} (x - a) f(x)dx$$

[$\because$ in $(-\infty, a)$, $x < a$ and in (a, ∞), $x > a$]

$\therefore$ $$\phi'(a) = \int_{-\infty}^{a} f(x)\, dx + (a - a) f(a) \frac{da}{da}$$

$$- \left[\int_{\infty}^{a} (-1) f(x)\, dx + (a - a) f(a) \frac{da}{da} \right]$$

$$\left[\because \frac{\partial}{\partial a} \int_{k}^{a} g(x, a)\, dx = \int_{k}^{a} \frac{\partial}{\partial a} g(x, a)\, dx + g(a, a) \frac{da}{da} \right]$$

i.e. $$\phi'(a) = \int_{-\infty}^{a} f(x)\, dx - \int_{a}^{\infty} f(x)\, dx$$

$\phi''(a) = f(a) + f(a) = 2f(a) > 0$ [$\because$ $f(x)$ is the density function]

$\phi'(a) = 0$, if $F(a) - \{1 - F(a)\} = 0$

i.e. $2F(a) = 1$

i.e., $F(a) = \frac{1}{2}$

i.e. a is the median.

Therefore, mean deviation about a is minimum, when a is the median.

3. Standard Deviation (S.D.) of a random variable X or its distribution is defined as the positive square root of the second order central moment μ_2, (called *the variance* of the random variable) denoted by σ_x.

i.e. $$\sigma_x = \sqrt{\mu_2} \text{ or } \sigma_x^2 = E\{(X - \mu)^2\}$$

The variance of random variable X itself is denoted by Var (X). When it is required to compare the dispersion or variability of two random variables (or their distributions), we cannot use their standard deviations which have the dimensions of the random variables. We therefore use dimensionless ratios, known as coefficients of variation.

The Coefficient of Variation

(C.V.) of a Random Variable X (or its distribution) is defined as

$$\text{C.V.} = \frac{\text{S.D.}}{\text{Mean}} \times 100$$

If $(\text{C.V.})_X > (\text{C.V.})_Y$, then the random variable X (or its distribution) is more variable (or less consistent) than the random variable Y (or its distribution).

Properties of variance

1. $\text{Var}(X) = E(X^2) - \{E(X)\}^2$

Proof From the relation between central and non-central moments, we have $\mu_2 = \mu_2' - d^2$, where $\mu_2' = E(X - a)^2$ and $d = \mu - a$ or $E(X) - a$

Taking $a = 0$, we get

$$\text{Var}(X) = \mu_2 = E(X^2) - \{E(X)\}^2$$

2. $\text{Var}\,(aX + b) = a^2\,\text{Var}(X)$

Proof $\text{Var}\,(aX + b) = E\{(aX + b) - E(aX + b)\}^2$, by definition

$$\begin{aligned} &= E\{(aX + b) - aE(X) - b\}^2 \\ &= E\{a[X - E(X)]\}^2 \\ &= a^2 E\{X - E(X)\}^2 \\ &= a^2\,\text{Var}\,(X) \end{aligned}$$

Skewness

Skewness, which means lack of symmetry, is the property of a random variable or its distribution by which we get an idea about the shape of the probability curve of the distribution. If the probability curve is not symmetrical but has a longer tail on one side than on the other, the distribution is said to be skewed. If a distribution is skewed, then the averages mean, median and mode will take different values and the quartiles will not be equidistant from the median.

The measures of skewness in common use are the following:

(i) Mean-Median,
(ii) Mean-Mode,
(iii) (Q_3 – Median) – (Median – Q_1) or $Q_3 + Q_1 - 2$ Median
(iv) the third order central moment $\mu_3 = E(X - \mu)^3$.

When the comparison of the skewness of the distributions of two random variables is to be made, we compute *coefficients of skewness* which are pure numbers, given below.

(i) Pearson's coefficient of skewness, S_k, defined by

$$S_k = \frac{\text{Mean} - \text{Median}}{\text{S.D.}} \text{ or equivalently}$$

$$S_k = \frac{3\,(\text{Mean} - \text{Mode})}{\text{S.D.}}$$

Note *For a moderately asymmetrical distribution, it is known that Mean – Median = 3(Mean – Mode) and so either of the above definitions may be used.*

(ii) Bowley's quartile coefficient of skewness S_k, defined by

$$S_k = \frac{Q_3 + Q_1 - 2\,\text{Median}}{Q_3 - Q_1}$$

(iii) Moment coefficient of skewness v_1, defined by

$$v_1 = \frac{\mu_3}{\mu_2^{3/2}} \text{ or } \frac{\mu_3}{\sigma^3}$$

If any of the above coefficients of skewness is positive, the distribution will be skewed towards the right, otherwise it will be skewed towards the left.

Kurtosis

Even if we know the measures of central tendency, dispersion and skewness of a radom variable (or its distribution), we cannot get a complete idea about the distribution. In order to analyse the distribution completely, another characteristic *kurtosis* is also required. Kurtosis means the convexity of the probability curve of the distribution. Using the measure of coefficient of kurtosis, we can have an idea about the flatness or peakedness of the probability curve near its top.

The only measure of kurtosis used is the fourth order central moment μ_4, given by

$$\mu_4 = E(X - \mu)^4$$

The coefficient of kurtosis, denoted by β_2 is defined as

$$\beta_2 = \frac{\mu_4}{\mu_2^2} \quad \text{or} \quad \frac{\mu_4}{\sigma^4}$$

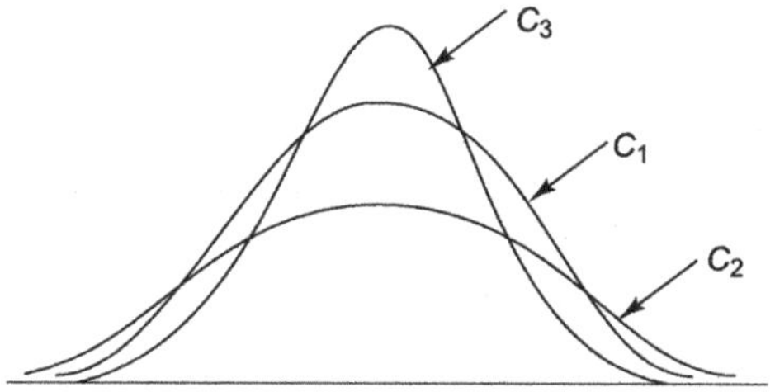

Curve C_1 which is neither flat nor peaked is called a *mesokurtic* curve, for which $\beta_2 = 3$.

Curve C_2 which is flatter than the curve C_1 is called *platykurtic* curve, for which $\beta_2 < 3$.

Curve C_3 which is more peaked than C_1 is called *leptokurtic* curve, for which $\beta_2 > 3$.

Pearson's Shape Coefficients

As $\beta_1 = v_1^2 = \frac{\mu_3^2}{\mu_2^3}$ and $\beta_2 = \frac{\mu_4}{\mu_2^2}$ determine the shape of the probability curve, they are called (Pearson's) shape coefficients.

Equivalently v_1 and $v_2 = \beta_2 - 3$ are also used as shape coefficients.

Expected Values of a Two-Dimensional RV

If (X, Y) is a two-dimensional discrete RV with joint probability mass function p_{ij}, then $E\{g(X, Y)\} = \sum_j \sum_i g(x_i, y_i\} p_{ij}$.

If (X, Y) is a two-dimensional continuous RV with joint pdf $f(x, y)$, then

$$E\{g(X, Y)\} = \int_{-\infty}^{\infty} \int_{-\infty}^{\infty} g(x, y) f(x, y)\, dx\, dy$$

Properties of Expected Values

We give below the proofs of the properties for continuous RVs. Students can prove the properties for discrete RVs.

(i) $E\{g(X)\} = \int_{-\infty}^{\infty} g(x) f_X(x)\, dx$, where $f_X(x)$ is the marginal density of X.

Proof
$$E\{g(X)\} = \int_{-\infty}^{\infty} \int_{-\infty}^{\infty} g(x) f(x, y)\, dx\, dy$$
$$= \int_{-\infty}^{\infty} g(x) \left[\int_{-\infty}^{\infty} f(x, y)\, dy \right] dx$$
$$= \int_{-\infty}^{\infty} g(x) f_X(x)\, dx$$

(ii) $E\{h(Y)\} = \int_{-\infty}^{\infty} h(y) f_Y(y)\, dy$

where $f_Y(y)$ is the marginal density of Y.
(Proof is left as an exercise to the student.)

(iii) $E(X + Y) = E(X) + E(Y)$

Proof
$$E(X + Y) = \int_{-\infty}^{\infty} \int_{-\infty}^{\infty} (x + y) f(x, y)\, dx\, dy$$
$$= \int_{-\infty}^{\infty} \int_{-\infty}^{\infty} x f(x, y) dx\, dy + \int_{-\infty}^{\infty} \int_{-\infty}^{\infty} y f(x, y) dx\, dy$$
$$= E(X) + E(Y)$$

(iv) In general, $E(XY) \neq E(X) \times E(Y)$, but if X and Y are independent RVs, $E(XY) = E(X) \times E(Y)$.

Proof
$$E(XY) = \int_{-\infty}^{\infty}\int_{-\infty}^{\infty} xyf(x, y)\mathrm{d}x\, \mathrm{d}y$$
$$= \int_{-\infty}^{\infty}\int_{-\infty}^{\infty} xyf_X(x)f_Y(y)\, \mathrm{d}x\, \mathrm{d}y$$
(since X and Y are independent)
$$= \int_{-\infty}^{\infty} xf_X(x)\mathrm{d}x \times \int_{-\infty}^{\infty} yf_Y(y)\mathrm{d}y$$
$$= E(X) \times E(Y)$$

In general, if X and Y are independent,
$$E\{g(X) \times h(Y)\} = E\{g(X)\} \times E\{h(Y)\}$$

Conditional Expected Values

If (X, Y) is a two-dimensional discrete RV with joint probability mass function p_{ij}, then the conditional expectations of $g(X, Y)$ are defined as follows:
$$E\{g(X, Y)/Y = Y_j\} = \sum_i g(x_i, y_j) \times P(X = x_i/Y = y_j)$$
$$= \sum_i g(x_i, y_j)\frac{p_{ij}}{p_{*j}}$$
and
$$E\{g(X, Y)/X = x_i\} = \sum_j g(x_i, y_j)\, p_{ij} / p_{i*}$$

If (X, Y) is a two-dimensional continuous RV with joint pdf $f(x, y)$, then
$$E\{g(X, Y)/Y\} = \int_{-\infty}^{\infty} g(x, y) \times f(x/y)\, \mathrm{d}x \text{ and}$$
$$E\{g(X, Y)/X\} = \int_{-\infty}^{\infty} g(x, y) \times f(y/x)\, \mathrm{d}y$$

In particular, the conditional means are defined as
$$\mu_{Y/X} = E(Y/X) = \int_{-\infty}^{\infty} yf(y/x)\, \mathrm{d}y \text{ and}$$
$$\mu_{X/Y} = E(X/Y) = \int_{-\infty}^{\infty} xf(x/y)\, \mathrm{d}x$$

The conditional variances are defined as
$$\sigma^2_{Y/X} = E\{(Y - \mu_{Y/X})^2\} = \int_{-\infty}^{\infty} (y - \mu_{Y/X})^2 f(y/x)\, \mathrm{d}y \text{ and}$$

$$\sigma^2_{X/Y} = E\{(X - \mu_{X/Y})^2\} = \int_{-\infty}^{\infty} (x - \mu_{X/Y})^2 f(x/y)\, dx$$

Properties

(i) If X and Y are independent RVs, then $E(Y/X) = E(Y)$ and $E(X/Y) = E(X)$.

Proof $$E(Y/X) = \int_{-\infty}^{\infty} y f(y/x) dy$$

$$= \int_{-\infty}^{\infty} y \frac{f(x, y)}{f_X(x)} dy$$

$$= \int_{-\infty}^{\infty} y \frac{f_X(x) \times f_Y(y)}{f_X(x)} dy$$

(since X and Y are independent)

$$= \int_{-\infty}^{\infty} y f_Y(y)\, dy = E(Y)$$

A similar proof can be given for the other result.

(ii) $E\,[E\{g(X, Y)/X\}] = E\{g(X, Y)\}$

Proof $$E\{g(X, Y)/X\} = \int_{-\infty}^{\infty} g(x, y) f(y/x)\, dy \quad (1)$$

Since $E\{g(X, Y)/X\}$ is a function of the RV X,

$$E[E\{g(X, Y)/X\}] = \int_{-\infty}^{\infty} E\{g(X, Y)/X\} f_X(x)\, dx$$

$$= \int_{-\infty}^{\infty}\int_{-\infty}^{\infty} g(x, y) f(y/x) f_X(x)\, dx\, dy \text{ [from (1)]}$$

$$= \int_{-\infty}^{\infty}\int_{-\infty}^{\infty} g(x, y) f(x, y)\, dx\, dy$$

$$= E\{g(X, Y)\}$$

In particular,

$$E\{E(Y/X)\} = E(Y) \text{ and similarly}$$
$$E\{E(X/Y)\} = E\,(X).$$

(iii) $E\{g_1(X) \times g_2(Y)\} = E[g_1(X) \times E\{g_2(Y)/X\}]$

Proof $E\{g_1(X) \times g_2(Y)\} = E[E\{g_1(X) \times g_2(Y)/X]$ (by Property (2))

$= E[g_1(X) \times E\{g_2(Y)/X\}]$ (since X is given)

In particular,

$$E(XY) = E[X \times E(Y/X)] \text{ and}$$
$$E(X^2Y^2) = E[X^2 \times E(Y^2/X)]$$

Worked Examples 4(A)

Example 1 *Find the mean, median Q.D., M.D., S.D., β_1 and β_2 of the following probability distribution.*

Solution

$X = x_i$	:	0	1	2	3	4	5	6	7	8
p_i	:	0.004	0.036	0.1	0.232	0.280	0.204	0.112	0.028	.004

$X = x_i$	p_i	$p_i x_i$	$p_i x_i^2$	$p_i x_i^3$	$p_i x_i^4$	$F(x);\ x \le x_i$
0	0.004	0.000	0.000	0.000	0.000	0.004
1	0.036	0.036	0.036	0.036	0.036	0.040
2	0.100	0.200	0.400	0.800	1.600	0.140
3	0.232	0.696	2.088	6.264	18.792	0.372
4	0.280	1.120	4.480	17.920	71.680	0.652
5	0.204	1.020	5.100	25.500	127.500	0.856
6	0.112	0.672	4.032	24.192	145.152	0.968
7	0.028	0.196	1.372	9.604	67.228	0.996
8	0.004	0.032	0.256	2.048	16.384	1.000
Total	1.000	3.972	17.764	86.364	448.372	–

Mean μ is given by

$$\mu = E(X) = \sum p_i x_i$$
$$= 3.972$$

$F(3) < 0.5$ and $F(4) > 0.5$

Therefore, median is given by med $= \frac{1}{2}(3 + 4) = 3.5$. Mode is the value of X for which $P(X = x)$ is maximum.

$P(X = 4) = 0.280$, that is the maximum probability.

$\therefore$ Mode $= 4.0$

$$\text{Var}(X) = E(X^2) - \{E(X)\}^2$$
$$= \Sigma p_i x_i^2 - \{E(X)\}^2$$
$$= 17.764 - (3.972)^2$$
$$= 1.987$$

$\therefore$ $\text{S.D.} = \sqrt{\text{Var}(X)} = 1.410$

$F(2) = 0.140 < 0.25$ and $F(3) = 0.372 > 0.25$

$\therefore$ $Q_1 = \frac{1}{2}(2 + 3) = 2.5$

$$F(4) = 0.652 < 0.75 \text{ and } F(5) = 0.856 > 0.75$$

$$\therefore \quad Q_3 = \frac{1}{2}(4 + 5) = 4.5$$

$$\text{Q.D.} = \frac{1}{2}(Q_3 - Q_1) = \frac{1}{2}(4.5 - 2.5) = 1.0$$

M.D. about the mean $= \Sigma\, p_i |x_i - \mu|$

$= 3.972 \times .004 + 2.972 \times .036 + 1.972 \times 0.100 + 0.972 \times 0.232 + 0.028 \times 0.280$
$+ 1.028 \times 0.204 + 2.028 \times 0.112 + 3.028 \times 0.028 + 4.028 \times 0.004 = 1.091$

$$\mu_3' = E(X^3) = \Sigma p_i x_i^3 = 86.364$$

Now $\quad \mu_3 = \mu_3' - 3\mu_2' d + 2d^3$ where $d = \mu - 0 = 3.972$

$$= 86.364 - 3 \times 17.764 \times 3.972 + 2 \times (3.972)^3$$

$$= 0.019 \qquad [(\because \mu_2' = E(x^2)]$$

$$\mu_4' = E(X^4) = \Sigma p_i x_i^4 = 448.372$$

Now $\quad \mu_4 = \mu_4' - 4\mu_3' d + 6\mu_4' d^2 - 3d^4$, where $d = 3.972$

$$= 448.372 - 4 \times 86.364 \times 3.972 + 6 \times 17.764 \times (3.972)^2 - 3 \times (3.972)^4$$

$$= 11.053$$

$$\beta_1 = \frac{\mu_3^2}{\mu_2^3} = \frac{(0.019)^2}{(1.987)^3} = 0.00005; \; v_1 = \sqrt{\beta_1} = 0.0068$$

$$\beta_2 = \frac{\mu^4}{\mu_2^2} = \frac{11.053}{(1.987)^2} = 2.800; \; v_2 = \beta_2 - 3 = -0.200$$

Example 2 *Find the Pearson's and Bowley's coefficients of skewness for the following probability distribution*

$X = x_i$:	1	2	3	4	5	6	7	8
p_i :	0.008	0.032	0.142	0.216	0.240	0.206	0.143	0.013

Solution

$X = x_i$	p_i	$p_i x_i$	$p_i x_i^2$	$F(x)\ [x \le x_i]$
1	0.008	0.008	0.008	0.008
2	0.032	0.064	0.128	0.040
3	0.142	0.426	1.278	0.182
4	0.216	0.864	3.456	0.398
5	0.240	1.200	6.000	0.638
6	0.206	1.236	7.416	0.844
7	0.143	1.001	7.007	0.987
8	0.013	0.104	0.832	1.000
Total	1	4.903	26.125	—

$$\text{Mean} = E(X) = \Sigma p_i x_i = 4.903$$

$$E(X^2) = \Sigma p_i x_i^2 = 26.125$$

$$\text{Var}(X) = E(X^2) - \{E(X)\}^2$$

$$= 26.125 - (4.903)^2$$

$$= 2.086$$

$$\therefore \quad (\text{S.D.})_X = \sqrt{2.086} = 1.444$$

$$Q_1 = \frac{1}{2}(3 + 4) = 3.5,\ Q_2 = \text{Median} = \frac{1}{2}(4 + 5) = 4.5$$

$$Q_3 = \frac{1}{2}(5 + 6) = 5.5.$$

Pearson's coefficient of skewness is given by

$$S_k = \frac{\text{Mean} - \text{Median}}{\text{S.D.}}$$

$$= \frac{4.903 - 4.5}{1.444}$$

$$= 0.279$$

Bowley's coefficient of skewness is given by

$$S_k = \frac{Q_3 + Q_1 - 2\text{Median}}{Q_3 - Q_1}$$

$$= \frac{5.5 + 3.5 - 2 \times 4.5}{5.5 - 3.5}$$

$$= 0$$

As Bowley's coefficient is zero, we should not conclude that the given solution is symmetrical, since the values of Q_1, Q_2 and Q_3 used are approximate. In this case, Pearson's coefficient is more satisfactory, as it indicates that the distributions positively skewed (or skewed to the right), which is true.

Example 3 *The first four moments of a distribution about the value 5 of the random variable X are 2, 20, 40 and 50. Compute a measure, each of central tendency, dispersion, skewness and kurtosis. Comment on the skewness and kurtosis of the distribution.*

Solution $\mu_1' = E(X - 5) = 2$

i.e. $E(X) - 5 = 2$

$\therefore$ Mean = a measure of central tendency = 7

$$\mu_2' = E(X - 5)^2 = 20$$

$$\mu_2 = \mu_2' - \mu_1'^2 = 20 - 4 = 16$$

$\therefore$ S.D. = a measure of dispersion = $\sqrt{\mu_2} = 4$

$$\mu_3' = E(X - 5)^3 = 40$$

$$\mu_3 = \mu_3' - 3\mu_2' \mu_1' + 2\mu_1'3$$

$$= 40 - 3 \times 20 \times 2 + 2 \times 2^3 = -64$$

A measure of skewness $= \mu_3 = -64$

$$\mu_4' = E(X-5)^4 = 50$$

$$\mu_4 = \mu_4' - 4\mu_3'\,\mu_1' + 6\mu_2' \cdot \mu_1'^2 - 3\mu_1'^4$$

$$= 50 - 4 \times 40 \times 2 + 6 \times 20 \times 4 - 3 \times 16 = 162$$

As $\mu_3 < 0$, the distribution is skewed to the left.

$$\beta_2 = \frac{\mu_4}{\mu_2^2} = \frac{162}{256} < 3.$$

Therefore the distribution is platykurtic.

Example 4 *The first three moments of a distribution about the value 2 of the random variable X are 1, 16 and – 40. Find the coefficient of variation and skewness of the distribution. Also find the first three simple moments.*

Solution $\mu_1' = E(X-2) = 1$

$\therefore$ $E(X) = \text{Mean} = 3$

$$\mu_2' = E(X-2)^2 = 16$$

$$\mu_2 = \mu_2' - \mu_1'^2 = 16 - 1 = 15$$

$$\mu_3' = E(X-2)^3 = -40$$

$$\mu_3 = \mu_3' - 4\mu_2'\,\mu_1' + 2\mu_1'^3$$

$$= -40 - 3 \times 16 \times 1 + 2 \times 1^3$$

$$= -86$$

$$\text{Coefficient of variation} = \frac{\text{S.D.}}{\text{Mean}} \times 100$$

$$= \frac{\sqrt{15}}{3} \times 100$$

$$= 129.1\%$$

$$\text{Coefficient of skewness} = \nu_1 = \frac{\mu_3}{\sigma^3}$$

$$= \frac{-86}{15\sqrt{15}} = -1.480$$

Now $E(X-0)^r = E\{(X-2)+2\}^r$

$$= E(X-2)^r + 2 \cdot rC_1\, E(X-2)^{r-1} + 4 \cdot rC_2\, E(X-2)^{r-2} + 8 \cdot rC_3\, E(X-2)^{r-3} + \ldots,$$

$\therefore$ $E(X) = E(X-2) + 2 = 1 + 2 = 3$

$$E(X^2) = E(X-2)^2 + 2 \times 2\, E(X-2) + 4 \times 1$$

$$= 16 + 4 + 4 = 24$$

$$E(X^3) = E(X-2)^3 + 2 \times 3 \cdot E(X-2)^2 + 4 \times 3 \cdot E(X-2) + 8 \times 1$$

$$= -40 + 96 + 12 + 8 = 76$$

Example 5 *Find the median, mode and quartile deviation of continuous random variable X, given that its density function is* $f(x) = \dfrac{k}{1+x^2}$, $-\infty < x < \infty$.

Solution $f(x)$ is the density function of X, $\int_{-\infty}^{\infty} f(x)\,dx = 1$

i.e. $$k \int_{-\infty}^{\infty} \frac{dx}{1+x^2} = 1$$

i.e. $$2k\,(\tan^{-1} x)_0^{\infty} = 1$$

i.e. $$2k \times \frac{\pi}{2} = 1 \text{ or } k = \frac{1}{\pi}$$

∴ $$f(x) = \frac{1}{\pi\,(1+x^2)}$$

which is the density function of Cauchy's distribution with parameter 1.

Note *The mean of this distribution does not exist, as the condition for the existence of the mean, i.e.* $\int_{-\infty}^{\infty} |x|\,f(x)\,dx < \infty$ *is not satisfied.*

The rth quartile Q_r is given by

$$\int_{-\infty}^{Q_r} f(x)\,dx = \frac{r}{4};\ r = 1, 2, 3$$

i.e. $$\int_{-\infty}^{Q_r} \frac{dx}{\pi(1+x^2)} = \frac{r}{4}$$

i.e. $$\frac{1}{\pi}\left[\tan^{-1} x\right]_{-\infty}^{Q_r} = \frac{r}{4}$$

i.e. $$\frac{1}{\pi}\left\{\tan^{-1} Q_r - \left(-\frac{\pi}{2}\right)\right\} = \frac{r}{4}$$

i.e. $$\tan^{-1} Q_r = \frac{\pi}{4}\,(r-2)$$

∴ $$Q_1 = \tan\left(-\frac{\pi}{4}\right) = -1$$

$$Q_2 = \text{Median} = \tan\,(0) = 0$$

and $$Q_3 = \tan\left(\frac{\pi}{4}\right) = 1$$

$$\therefore \qquad \text{Q.D.} = \frac{1}{2}(Q_3 - Q_1) = 1$$

Mode of X is the value of x for which $f(x)$ is maximum.

$$f(x) = \frac{1}{\pi(1+x^2)}$$

$$f'(x) = -\frac{2x}{\pi(1+x^2)^2}; f''(x) = \frac{2}{\pi}\left\{\frac{3x^2-1}{(1+x^2)^3}\right\}$$

Now $f'(x) = 0$ at $x = 0$ and $f''(0) < 0$

$\therefore$ $f(x)$ is maximum at $x = 0$.

$\therefore$ Mode of $X = 0$.

Example 6 *If the density function of a random variable X is given by $f(x) = kx(1 - x)$, $0 \le x \le 1$, find the A.M., H.M., median, mode, S.D., M.D. about the mean, β_1 and β_2.*

Solution Since $f(x)$ is a density function,

$$\int_0^1 f(x)\,dx = 1 \quad \text{or} \quad k\int_0^1 (x - x^2)\,dx = 1$$

$$\text{i.e.} \qquad k\left(\frac{x^2}{2} - \frac{x^3}{3}\right)_0^1 = 1 \quad \text{or} \quad \frac{k}{6} = 1 \quad \therefore \quad k = 6$$

$$\text{A.M.} = E(X) = \int_0^1 x \cdot 6x(1-x)\,dx$$

$$= 6\left(\frac{x^3}{3} - \frac{x^4}{4}\right)_0^1$$

$$= \frac{1}{2}$$

$$\frac{1}{\text{H.M.}} = \int_0^1 \frac{1}{x}.6x(1-x)\,dx$$

$$= (6x - 3x^2)_0^1 = 3$$

$$\therefore \qquad \text{H.M.} = \frac{1}{3}$$

Median M is given by $\int_0^M 6x(1-x)\,dx = \frac{1}{2}$

$$\text{i.e.} \qquad (3x^2 - 2x^3)_0^M = \frac{1}{2}$$

i.e. $$3M^2 - 2M^3 = \frac{1}{2} \quad \text{or} \quad 4M^3 - 6M^2 + 1 = 0$$

i.e. $$(2M - 1)(2M^2 - 2M - 1) = 0$$

$$\therefore \quad M = \frac{1}{2}, \frac{1}{2}(1 \pm \sqrt{3})$$

The values $\frac{1}{2}(1 \pm \sqrt{3})$ lie outside (0, 1) $\quad \therefore \quad M = \frac{1}{2}$.

Mode is the value of x at which

$$f(x) = 6x(1 - x) \text{ is maximum}$$

Now $$f'(x) = 6 - 12x \text{ and } f''(x) = -12$$

$$f'(x) = 0 \text{ at } x = \frac{1}{2},$$

where $f''(x) < 0$

$$\therefore \quad \text{Mode} = \frac{1}{2}$$

As the mean, median and mode are equal, the given distribution is symmetrical

$$E(X^2) = \int_0^1 6x^3(1 - x)\,dx$$

$$= \left(6\frac{x^4}{4} - 6\frac{x^5}{5}\right)_0^1 = \frac{3}{10}$$

$$\text{Var}(X) = E(X^2) - \{E(X)\}^2$$

$$= \frac{3}{10} - \frac{1}{4} = \frac{1}{20}$$

$$\therefore \quad \text{S.D.} = \frac{1}{2\sqrt{5}}$$

$$\text{M.D. about the mean} = E\left(\left|X - \frac{1}{2}\right|\right)$$

$$= \int_0^1 \left|x - \frac{1}{2}\right| 6x(1 - x)dx$$

$$= \int_0^{\frac{1}{2}} \left(\frac{1}{2} - x\right) 6x(1 - x)dx + \int_{\frac{1}{2}}^1 \left(x - \frac{1}{2}\right) 6x(1 - x)dx$$

$$= \int_0^{\frac{1}{2}} \left(3x - 9x^2 + 6x^3\right) dx + \int_{\frac{1}{2}}^1 \left(-3x + 9x^2 - 6x^3\right) dx$$

$$\left(3\frac{x^2}{2} - 3x^3 + \frac{3}{2}x^4\right)_0^{\frac{1}{2}} + \left(-\frac{3}{2}x^2 + 3x^2 - \frac{3}{2}x^4\right)_{\frac{1}{2}}^1$$

$$= \frac{3}{32} + \left(0 + \frac{3}{32}\right) = \frac{3}{16}$$

$$= \mu_3 = E\{X - E(X)\}^3 = E\left(X - \frac{1}{2}\right)^3$$

$$= \int_0^1 \left(x - \frac{1}{2}\right)^3 6x(1 - x)\,dx$$

$$= 6\int_{-\frac{1}{2}}^{\frac{1}{2}} t^3\left(\frac{1}{2} + t\right)\left(\frac{1}{2} - t\right) dt, \text{ on putting } t = x - \frac{1}{2}$$

$$= 6\int_{-\frac{1}{2}}^{\frac{1}{2}} t^3\left(\frac{1}{4} - t^2\right) dt$$

= 0, since the integrand is odd.

$$\mu_4 = E\{X - E(X)\}^4 = E\left(X - \frac{1}{2}\right)^4$$

$$= 6\int_0^1 \left(x - \frac{1}{2}\right)^4 x(1 - x)dx$$

$$= 6\int_{-\frac{1}{2}}^{\frac{1}{2}} t^4\left(\frac{1}{4} - t^2\right) dt, \text{ on putting } t = x - \frac{1}{2}$$

$$= 12\left(\frac{t^5}{20} - \frac{t^7}{7}\right)_0^{\frac{1}{2}}$$

$$= 12\left(\frac{1}{640} - \frac{1}{896}\right)$$

$$= 12 \times \frac{2}{128 \times 35} = \frac{3}{560}$$

Now $\beta_1 = \dfrac{\mu_3^2}{\mu_2^3} = 0$ and $\beta_2 = \dfrac{\mu_4}{\mu_2^2} = \dfrac{3}{560} \times 400 = \dfrac{15}{7}$

Example 7 *Find the mean and variance of the Pascal's (negative binomial distribution) distribution, given by* $P(X = k) = \binom{n+k-1}{k} p^n q^k$, $k = 0, 1, 2, \ldots$

Solution $$E(X) = \sum_{k=0}^{\infty} k \binom{n+k-1}{k} p^n q^k$$

$$= p^n[1 \times nC_1 q^1 + 2(n+1)C_2\, q^2 + 3(n+2)C_3 q^3 + \ldots]$$

$$= np^n q[1 + (n+1)C_1 q + (n+2)C_2 q^2 + \ldots]$$

$$= np^n q(1-q)^{-(n+1)} = \frac{nq}{p}$$

$$E(X^2) = \sum_{k=0}^{\infty} k^2 \binom{n+k-1}{k} p^n q^k$$

$$= p^n[1^2 nC_1 q^1 + 2^2(n+1)\, C_2 q^2 + 3^2\,(n+2)C_3 q^3 + \ldots]$$

$$= p^n\left[nq + \frac{1\times 2+2}{2!}(n+1)\, nq^2 + \frac{(2\times 3+3)}{3!}(n+2)\,(n+1)\, nq^3 + \cdots\right]$$

$$= np^n q\left[\left\{1 + \frac{(n+1)}{1!}q + \frac{(n+1)\,(n+2)}{2!}q^2 + \cdots\right\} + (n+1)q \times \left\{1 + \frac{(n+2)}{1!}q + \frac{(n+2)(n+3)}{2!}q^2 + \cdots\right\}\right]$$

$$= np^n q[(1-q)^{-(n+1)} + (n+1)q(1-q)^{-(n+2)}]$$

$$= np^n q\left[\frac{1}{p^{n+1}} + \frac{(n+1)q}{p^{n+2}}\right]$$

$$= \frac{nq}{p} + \frac{n(n+1)q^2}{p^2}$$

$\therefore$ $$\text{Var}(X) = E(X^2) - \{E(X)\}^2$$

$$= \frac{nq}{p} + \frac{n(n+1)q^2}{p^2} - \frac{n^2 q^2}{p^2}$$

$$= \frac{nq}{p}\left(1 + \frac{q}{p}\right) = \frac{nq}{p^2}$$

Example 8 *If the continuous RV X has Rayleight density* $f(x) = \frac{x}{\alpha^2} e^{-x^2/2\alpha^2} \times U(x)$, *find* $E(X^n)$ *and deduce the values of* $E(X)$ *and* $\text{var}(X)$.

Solution By definition,

$$E(X^n) = \int_0^{\infty} x^n \times \frac{x}{\alpha^2} e^{-x^2/2\alpha^2}\, dx$$

$$= \int_0^\infty (2\alpha^2 t)^{n/2} e^{-t}\, dt \qquad \left(\text{putting } \frac{x^2}{2\alpha^2} = t\right)$$

$$= 2^{n/2}\alpha^n \int_0^\infty t^{n/2} e^{-t}\, dt$$

$$= 2^{n/2}\, \alpha^n \,\Gamma(k+1 \text{ if } n = 2k$$

$$= 2^{n/2}\, \alpha^n \lfloor k = 2^{n/2}\, \alpha^n \lfloor n/2 \cdot \text{ if } n \text{ is even}$$

$$E(X^n) = 2^{n/2}\, \alpha^n \,\Gamma\left(\frac{2\pi + 3}{2}\right) \text{if } n = 2k + 1$$

$$= 2^{n/2}\, \alpha^n \, \frac{2k+1}{2} \times \frac{2k-1}{2} \cdots \frac{3}{2} \times \frac{1}{2} \times \Gamma\left(\frac{1}{2}\right)$$

$$= 2^{n/2}\, \alpha^n \frac{1 \times 3 \times 5 \times \cdots \times n}{2^{(n+1)/2}} \sqrt{\pi}$$

$$= 1 \times 3 \times 5 \times \ldots \times n\alpha^n \,\sqrt{\pi/2} \text{ if } n \text{ is odd}$$

$$\therefore \qquad E(X) = \alpha\sqrt{\pi/2}\,;\; E(X^2) = 2\alpha^2;\; \text{var}(X) = \left(2 - \frac{\pi}{2}\right)\alpha^2$$

Example 9 *A line of length a units is divided into two parts. If the first part is of length X, find E(X), var(X) and E{X(a – X)}.*

Solution Since the positions of the point of division are equally likely, X is uniformly distributed in $(0, a)$.

$$\therefore \qquad f(x) = \frac{1}{a}$$

$$E(X) = \int_0^a x f(x)\, dx = \frac{1}{a}\int_0^a x dx = \frac{a}{2}$$

$$E(X^2) = \int_0^a x^2 f(x)\, dx = \frac{a^2}{3}$$

$$\therefore \qquad \text{Var}(X) = E(X^2) - \{E(X)\}^2 = \frac{a^2}{3} - \frac{a^2}{4} = \frac{a^2}{12}$$

$$E\{X(a - X)\} = a\, E(X) - E(X^2) = \frac{a^2}{2} - \frac{a^2}{3} = \frac{a^2}{6}$$

Example 10 *If X is a continuous RV, prove that*

$$E(X) = \int_0^\infty \left[1 - F(x)\right] dx - \int_{-\infty}^0 \left[1 - F(x)\right] dx$$

Solution $E(X) = \int_{-\infty}^{\infty} xf(x)\,dx = \int_{-\infty}^{0} x\,dF(x) - \int_{0}^{\infty} x\,d\{1 - F(x)\}$

[since $F'(x) = f(x)$]

$$= [xF(x)]_{-\infty}^{0} - \int_{-\infty}^{0} F(x)\,dx - [x\{1 - F(x)\}]_0^{\infty} + \int_{0}^{\infty} \{1 - F(x)\}dx$$

$$= \int_{0}^{\infty} \{1 - F(x)\}\,dx - \int_{-\infty}^{0} F(x)\,dx$$

since $F(-\infty) = 0$ and $F(\infty) = 1$]

Example 11 *If the random variable X follows N(0, 2) and $Y = 3X^2$, find the mean and variance of Y.*

Solution Since X follows $N(0, 2)$, $E(X) = 0$ and $\text{var}(X) = 4$

$\therefore$ $E(X^2) = \text{var}(X) + \{E(X)\}^2 = 4$

Now $E(Y) = E(3X^2) = 3 \times 4 = 12$

$E(Y^2) = E(9X^4) = 9 \times 3 \times 2^4$

[since for the normal distribution $N(0, \sigma)$, $E(X^{2r}) = \dfrac{(2r)!\sigma^{2r}}{2^r r!}$]

$\text{Var}(Y) = E(Y^2) - \{E(Y)\}^2$

$= 27 \times 2^4 - 12^2 = 288$

Example 12 *Two cards are drawn at random with replacement from a box which contains 4 cards numbered 1, 1, 2 and 2. If X denotes the sum of the numbers shown on the two cards, find the mean and variance of X.*

Solution X can take the values 2, 3 and 4.

$P(X = 2) = P(\text{two 1's are drawn})$

$= \dfrac{2C_2}{4C_2} = \dfrac{1}{6}$

$P(X = 3) = P(\text{one 1 and one 2 are drawn})$

$= \dfrac{2C_1 \cdot 2C_1}{4C_2} = \dfrac{4}{6}$

$P(X = 4) = P(\text{two 2's are drawn})$

$= \dfrac{2C_2}{4C_2} = \dfrac{1}{6}$

The probability distrbution of X is as shown.

$X = x_i$	:	2	3	4
p_i	:	$\frac{1}{6}$	$\frac{4}{6}$	$\frac{1}{6}$

$$E(X) = \Sigma p_i x_i = \frac{2}{6} + \frac{12}{6} + \frac{4}{6} = 3$$

$$E(X^2) = \Sigma p_i x_i^2 = \frac{4}{6} + \frac{36}{6} + \frac{16}{6} = \frac{28}{3}$$

$$\text{Var }(X) = E(X^2) - \{E(X)\}^2$$

$$= \frac{28}{3} - 9 = \frac{1}{3}.$$

Example 13 *A lot is known to caontain 2 defectives and 8 non-defective items. If these items are inspected at random, one after another, what is the expected number of items that must be chosen in order to remove, both the defective ones?*

Solution Let the random variable X denote the number of items that must be drawn in order to remove both defective items.
clearly X takes the values 2, 3, 4, ..., 10.

$P(X = r) = P(r$ items are to be drawn to remove both defectives)

$= P\{$the first $(r - 1)$ items drawn should contain 1 defective and rth item drawn should be defective$\}$

$$= \frac{2C_1 \times 8C_{r-2}}{10C_{r-1}} \times \frac{1}{10 - (r-1)} = \frac{2 \times 8C_{r-2}}{10C_{r-1}(11 - r)}$$

$(r = 2, 3, \ldots, 10)$

The probability distribution of X will then be as follows:

$X = r$	2	3	4	5	6	7	8	9	10
p_r	1/45	2/45	3/45	4/45	5/45	6/45	7/45	8/45	9/45

$$E(X) = \sum_{r=2}^{10} r p_r = \frac{22}{3}$$

Example 14 *If a discrete random variable X takes the values 0, 1, 2, ... n, with probabilities* $\frac{1}{2^n}, \frac{nc_1}{2^n}, \frac{nc_2}{2^n}, \cdots, \frac{nc_n}{2^n}$ *respectively, show that the coefficient of variation of the distribution of X is* $\frac{100}{\sqrt{n}}$.

Solution $E(X) = \sum_{r=0}^{n} p_r x_r$

$$= \sum_{r=0}^{n} r \cdot nC_r / 2^n \qquad (1)$$

$$= \frac{1}{2^n} \sum_{r=0}^{n} r \frac{n!}{r!(n-r)!}$$

$$= \frac{n}{2^n} \sum_{r=1}^{n} r \frac{(n-1)!}{(r-1)!(n-r)!}$$

$$= \frac{n}{2^n} \sum_{r=1}^{n} (n-1)C_{(r-1)}$$

$$= \frac{n}{2^n} (1+1)^{n-1} \; [\because (1-1)^{n-1} = (n-1)\, C_0 + (n-1)\, C_1 + \ldots$$
$$+ (n-1)\, C_{n-1}]$$

$$= \frac{n}{2} \qquad (2)$$

$$E(X^2) = \sum_{r=0}^{n} p_r x_r^2$$

$$= \frac{1}{2^n} \sum_{r=0}^{n} r^2 \cdot nC_r$$

$$= \frac{1}{2^n} \sum_{r=0}^{n} \{r(r-1) + r\} \frac{n!}{r!(n-r)!}$$

$$= \frac{n(n-1)}{2^n} \sum_{r=2}^{n} \frac{(n-2)!}{(r-2)!(n-r)!} + \frac{1}{2^n} \sum_{r=0}^{n} r \cdot nC_r$$

$$= \frac{n(n-1)}{2^n} \sum_{r=2}^{n} (n-2)C_{r-2} + E(X), \text{ from (1)}$$

$$= \frac{n(n-1)}{2^n} (1+1)^{n-2} + \frac{n}{2}, \text{ using (2)}$$

$$= \frac{n(n-1)}{4} + \frac{n}{2} = \frac{n(n-1)}{4}$$

$$\therefore \quad \text{Var}(X) = E(X^2) - \{E(X\}^2$$

$$= \frac{n(n+1)}{4} - \frac{n^2}{4}$$

$$= \frac{n}{4}$$

$$(\text{C.V.})_X = \frac{\sigma_X}{E(X)} \times 100$$

$$= \frac{\sqrt{n}}{2} \times \frac{2}{n} \times 100$$

$$= \left(\frac{100}{\sqrt{n}}\right)\%$$

Example 15 *A Box contains 2^n tickets of which nC_r tickets bear the number r ($r = 0, 1, 2\ldots, n$). Two tickets are drawn from the box. Find the expectation of the sum of their numbers.*

Solution Total number of tickets in the box.

$$\sum_{r=0}^{n} nC_r = nC_0 + nC_1 + \ldots + nC_n$$

$$= (1 + 1)^n = 2^n, \text{ as given.}$$

Let the RVs X and Y represent the numbers on the first and second tickets respectively.

Then $\quad E(X + Y) = E(X) + E(Y)$

X can take the values 0, 1, 2, ..., n with probabilities $\frac{nC_0}{2^n}, \frac{nC_1}{2^n}, \cdots \frac{nC_n}{2^n}$ respectively.

$$\therefore \quad E(X) = 1 \times \frac{nC_1}{2^n} + 2 \times \frac{nC_2}{2^n} + \ldots + n \times \frac{nC_n}{2^n}$$

$$= \frac{n}{2^n}\{(n-1)C_0 + (n-1)C_1 + \ldots + (n-1)C_{n-1}\}$$

$$= \frac{n}{2^n}(1 + 1)^{n-1} = \frac{n}{2}$$

Similarly, $\quad E(Y) = \frac{n}{2}$

$\therefore \quad E(X + Y) = n$

Example 16 *If the joint pdf of (X, Y) is given by $f(x, y) = 24y(1 - x)$, $0 \le y \le x \le 1$, find $E(XY)$.*

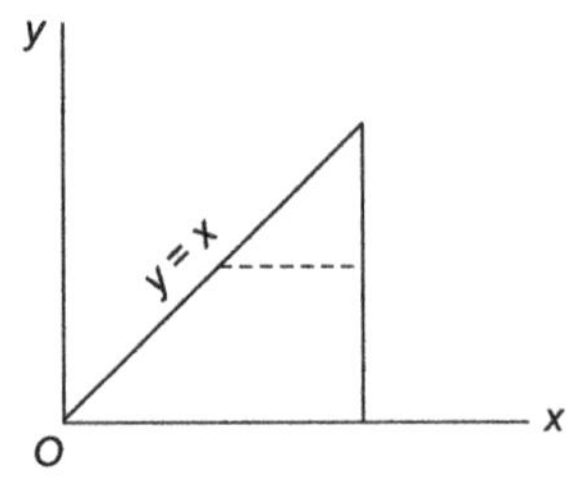

Solution $\quad E(XY) = \int_0^1 \int_y^1 xyf(x, y)\, dx\, dy$

$$= 24 \int_0^1 \int_y^1 xy^2(1 - x)\, dx\, dy$$

$$= 24 \int_0^1 y^2 \left(\frac{1}{6} - \frac{y^2}{2} + \frac{y^3}{3} \right) dy$$

$$= \frac{4}{15}$$

Example 17 *If X and Y are two independent RVs with $f_X(x) = e^{-x}\, U(x)$ and $f_Y(y) = e^{-y}\, U(y)$ and $Z = (X - Y)\, U(X - Y)$, prove that $E(Z) = 1/2$.*

Solution $U(X - Y) = \begin{cases} 1 \text{ if } X > Y \\ 0 \text{ if } X < Y \end{cases}$

$\therefore$ $$Z = \begin{cases} X - Y & \text{if } X > Y \\ 0 & \text{if } X < Y \end{cases}$$

$$E(Z) = \int_0^\infty \int_0^\infty z\, e^{-(x+y)}\, dx\, dy$$

[since X and Y are independent, $f(x, y) = f_X(x) \times f_Y(y)$]

$$= \int_0^\infty \int_y^\infty (x - y)\, e^{-(x+y)}\, dx\, dy$$

$$= \int_0^\infty e^{-y} [(x - y)(-e^{-x}) - e^{-x}]_y^\infty\, dy$$

$$= \int_0^\infty e^{-2y}\, dy = \frac{1}{2}$$

Example 18 *The joint pdf of (X, Y) is given by $f(x, y) = 24xy$; $x > 0$, $y > 0$, $x + y \le 1$, and $f(x, y) = 0$, elsewhere, find the conditional mean and variance of Y, given X.*

Solution

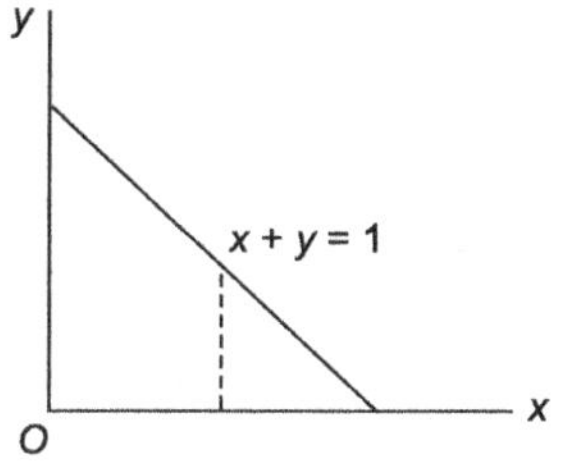

$$f_X(x) = \int_0^{1-x} 24xy\, dy$$

$$= 12x\,(1 - x)^2,\ 0 < x < 1$$

Now $$f(y/x) = \frac{f(x, y)}{f_X(x)} = \frac{2y}{(1-x)^2}, \ 0 < y < 1 - x$$

$$E(Y/X = x) = \int_0^{1-x} y f(y/x) \, dy$$

$$= \int_0^{1-x} \frac{2y^2}{(1-x)^2} \, dy = \frac{2}{3}(1 - x)$$

$$E(Y^2/x) = \int_0^{1-x} y^2 \times f(y/x) \, dy = \frac{1}{2}(1 - x)^2$$

$$\text{Var}(Y^2/x) = E(Y^2/x) - \{E(Y/x)\}^2$$

$$= \frac{1}{2}(1-x)^2 - \frac{4}{9}(1-x)^2$$

$$= \frac{1}{18}(1-x)^2$$

Example 19 *If (X, Y) is uniformly distributed over the semicircle bounded by* $y = \sqrt{1 - x^2}$ *and y = 0, find E(X/Y) and E(Y/X). Also verify the E{E(X/Y)} = E(X) and E{E(Y/X)} = E(Y).*

Solution

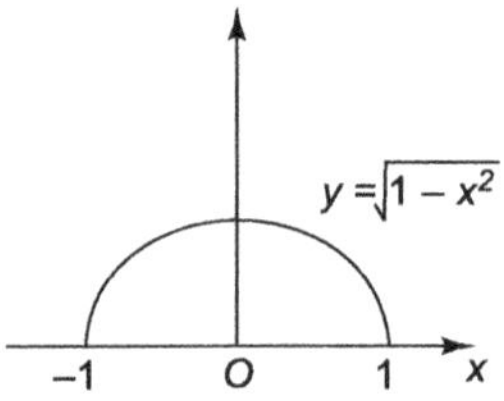

$$f(x, y) = k$$

$$\iint f(x, y) \, dy \, dx = 1$$

i.e., $$\int_{-1}^{1} \int_0^{\sqrt{1-x^2}} k \, dy \, dx = 1$$

i.e., $$2k \int_0^1 \sqrt{1 - x^2} \, dx = 1$$

$\therefore$ $$k = \frac{2}{\pi}$$

$$f_X(x) = \int_0^{\sqrt{1-x^2}} \frac{2}{\pi} \, dy = \frac{2}{\pi}\sqrt{1 - x^2}, \ -1 \le x \le 1$$

$$f_Y(y) = \int_{-\sqrt{1-y^2}}^{\sqrt{1-y^2}} \frac{2}{\pi}\,dx = \frac{4}{\pi}\sqrt{1-y^2}\,,\ 0 \le y \le 1$$

$$f(x/y) = \frac{f(x, y)}{f_Y(y)} = \frac{1}{2\sqrt{1-y^2}},\ -\sqrt{1-y^2} \le x \le \sqrt{1-y^2}$$

$$f(y/x) = \frac{1}{\sqrt{1-x^2}},\ 0 \le y \le \sqrt{1-x^2}$$

$$E(X) = \int_{-1}^{1} x f_X(x)\,dx = \frac{2}{\pi}\int_{-1}^{1} x\sqrt{1-x^2}\,dx = 0$$

(since the integrand is odd)

$$E(X/Y) = \int_{-\sqrt{1-y^2}}^{\sqrt{1-y^2}} x f(x/y)\,dx$$

$$= \frac{1}{2\sqrt{1-y^2}}\int_{-\sqrt{1-y^2}}^{\sqrt{1-y^2}} |x|\,dx = 0$$

(since the integrand is odd)

$$\therefore \quad E\{E(X/Y)\} = E\{0\} = 0 = E(X)$$

$$E(Y) = \int_0^1 y f_Y(y)\,dy = \frac{4}{\pi}\int_0^1 y\sqrt{1-y^2}\,dy = \frac{4}{3\pi}$$

$$E(Y/X) = \int_0^{\sqrt{1-x^2}} y f(y/x)\,dy = \frac{1}{\sqrt{1-x^2}}\cdot\left(\frac{y^2}{2}\right)_0^{\sqrt{1-x^2}} = \frac{1}{2}\sqrt{1-x^2}$$

$$\therefore \quad E\{E(Y/X)\} = E\left\{\frac{1}{2}\sqrt{1-X^2}\right\}$$

$$= \int_{-1}^{1} \frac{1}{2}\sqrt{1-x^2}\, f_X(x)\,dx$$

$$= \frac{2}{\pi}\int_0^1 (1-x^2)\,dx = \frac{4}{3\pi}$$

$$\therefore \quad E\{E(Y/X)\} = E(Y)$$

Example 20 *If (X, Y) follows a bivariate normal distribution $N(0, 0; \sigma_X, \sigma_Y; r)$, find $E(Y/X)$, $E(Y^2/X)$, $E(XY)$ and $E(X^2Y^2)$.*

Solution

$$f(x, y) = \frac{1}{2\pi\sigma_x\sigma_y\sqrt{1-r^2}}\exp\left\{-\frac{1}{2(1-r^2)}\left(\frac{x^2}{\sigma_x^2} - \frac{2rxy}{\sigma_x\sigma_y} + \frac{y^2}{\sigma_y^2}\right)\right\}$$

$$= \frac{1}{2\pi\sigma_x\sigma_y\sqrt{1-r^2}} \exp\left\{-\frac{1}{2\left(1-r^2\right)}\left(\frac{y}{\sigma_y}-\frac{rx}{\sigma_x}\right)^2 - \frac{x^2}{2\sigma_x^2}\right\}$$

$$f_X(x) = \frac{1}{\sigma_x\sqrt{2\pi}} \exp\left(-x^2/2\,\sigma^2_X\right)$$

[refer to the worked Example 12 in Chapter 2 on two-dimensional RVs]

$$\therefore \quad f(y/x) = \frac{f(x, y)}{f_X(x)} = \frac{1}{\sigma_y\sqrt{1-r^2}\sqrt{2\pi}}$$

$$\exp\left\{-\frac{1}{2(1-r^2)\sigma_y^2}\left(y - \frac{rx\sigma_y}{\sigma_x}\right)^2\right\}$$

which is a $N\left(\frac{rx\sigma_y}{\sigma_x}, \sigma_y\sqrt{1-r^2}\right)$

$$\therefore \quad E(Y/X = x) = \int_{-\infty}^{\infty} yf(y/x)\,dy = \frac{rx\sigma_y}{\sigma_y}$$

and $\quad \text{Var}(Y/X = x) = \sigma^2_y\,(1 - r^2)$

$$\text{Var}(Y/X) = E(Y^2/X) - \{E(Y/X)\}^2$$

$$\therefore \quad E(Y^2/X) = \sigma^2_y\,(1 - r^2) + r^2x^2\sigma_y^2/\sigma_x^2$$

By Property 3 of conditional expected values,

$$E(XY) = E\{XE(Y/X)\}$$

$$= E\left\{r\frac{\sigma_y}{\sigma_x}X^2\right\} = r\,\frac{\sigma_y}{\sigma_x} \times \sigma^2_x = r\sigma_x\sigma_y$$

Again, by the same property

$$E(X^2Y^2) = E\{X^2 \times E(Y^2/X)\}$$

$$= E\left[\sigma_y^2\left(1-r^2\right)X^2 + \frac{r^2\sigma_y^2}{\sigma_x^2}X^4\right]$$

$$= \sigma_y^2(1 - r^2)E(X^2) + \frac{r^2\sigma_y^2}{\sigma_x^2}E(X^4)$$

$$= \sigma_x^2\,\sigma_y^2\,(1 - r^2) + \frac{r^2\sigma_y^2}{\sigma_x^2} \times 3\sigma_x^4$$

$$= (1 + 2r^2)\sigma_x^2\sigma_y^2$$

Exercise 4(A)

Part A (Short answer questions)

1. What do you mean by central tendency and dispersion of a random variable?
2. What is meant by skewness and kurtosis with respect to the probability distribution of a RV?
3. Define β_1 and β_2 coefficients. What are their uses?
4. Define mathematical expectation of $g(X)$ when X is a (i) discrete RV, and (ii) continuous RV.
5. If X is and RV, prove that Var $(X) = E(X^2) - E^2(X)$ and hence prove that $E(X^2) \geq [E(X)]^2$.
6. If μ_X and σ_X are the mean and SD of the RVX, find μ_Y and σ_Y, where $Y = (X - \mu_X)/\sigma_X$.
7. Define the raw and central moments of an RV and state the relation between them.
8. For a RVX, $E(X) = 10$ and Var $(X) = 25$, find the positive values of a and b such that $E(X) = 0$ and Var $(Y) = 1$, where $Y = aX - b$.
9. A fair coin is tossed until a head appears. What is the expected value of the number of tosses required?
10. On the average, how many times must a die be thrown until one gets a 6?
11. The probability distribution of an RVX is given by

X	:	0	1	2	3
p_X	:	0.1	0.3	0.4	0.2

Find $E(Y)$ where $Y = X^2 + X$.

12. Find $E(X)$ and Var (X), when X is a discrete RV taking the values 1, 2, 3,… , n with equal probability.
13. Find the mean of the RVX, if its p.m.f. is given by $P(X = j) = (1 - a)\, a^j$; $j = 0, 1, 2, \ldots \infty$.
14. Find $E(X)$, if $P(X = x_k) = 2^{-k}$, where $x_k = \dfrac{(-1)^{k-1} 2^k}{k}$; $k = 1, 2, 3, \ldots \infty$.
15. Find $E(X)$, if X is a RV such that $P(X = r) = pq^r$; $r = 0, 1, 2, \ldots \infty$, where $p + q = 1$.
16. Find $E(X)$, if X is a discrete RV such that $P(X = r) = n\mathrm{C}_r p^r q^{n-r}$; $r = 0, 1, 2, \ldots, n$, where $p + q = 1$.
17. Find $E(X)$, if X is a discrete RV such that $P(X = r) = e^{-\lambda} \lambda^r / r!$; $r = 0, 1, 2, \ldots \infty$.
18. Find the mean and mode of the RVX, given that $P(X = r) = 2^{-r}$; $r = 1, 2, 3, \ldots \infty$.
19. Find the mean and variance of the RVX, whose pdf is given by $f(x) = me^{-mx}$; $x > 0$.
20. Find the mean and median of the RVX, whose pdf is given by $f(x) = 2x$; $0 \leq x \leq 1$.

21. Find the median and mode of the RVX, whose pdf is given by $f(x) = \sin x$; $0 \le x \le \dfrac{\pi}{2}$.
22. For a probability distribution, SD = 2, coefficient of variation = 5% and Pearson's coefficient of skewness = 0.5. Find the mean and mode of the distribution.
23. If the first three simple moments of an RVX are $\frac{1}{2}(n+1)$, $\frac{1}{6}(n+1)(2n+1)$ and $\frac{1}{4}n(n+1)^2$, find whether the distribution of X is symmetrical or skewed.
24. Show that for a discrete probability distribution, the β_2-coefficient is always greater than 1.
25. Define the expected value of $g(X, Y)$, where (X, Y) is a two-dimensional continuous RV with joint pdf $f(x, y)$.
26. If (X, Y) is a two-dimensional continuous RV, express $E\{g(X)\}$ and $E\{h(Y)\}$ in terms of the marginal densities of X and Y.
27. If X and Y are independent RVs, prove that $E(X\,Y) = E(X) \cdot E(Y)$.
28. If X and Y are independent RVs with means 2 and 3 and variances 1 and 2 respectively, find the mean and variance of $Z = 2X - 5Y$.
29. If (X, Y) is a two-dimensional RV, define $E\{g(X, Y)/X\}$ and $E\{g(X, Y)/Y\}$.
30. If (X, Y) is a two-dimensional continuous RV, define conditional mean and conditional variance of X, given Y.
31. If X and Y are independent RVs, prove that $E(Y/X) = E(Y)$.
32. If X and Y are independent RVs, prove that $E\{E(Y/X)\} = E(Y)$.
33. If X and Y are independent RVs, prove that $E(X\,Y) = E\,\{X \cdot E(Y/X)\}$
34. If the joint pdf of (X, Y) is given by $f(x, y) = 2 - x - y$, in $0 \le x, y \le 1$, find $E(X)$ and $E(Y)$.
35. If the joint pdf of (X, Y) is given by $f(x, y) = 2$, in $0 \le x < y \le 1$, find $E(X)$.

Part B

36. If t is any positive real number, show that the function defined by $p(x) = e^{-t}(1 - e^{-t})^{x-1}$; $x = 1, 2, 3, ...$ can represent the probability mass function of a RVx. Find the mean and variance of the distribution of X. [**Hind:** Put $e^{-t} = p$ and $1 - e^{-t} = q$]
37. A discrete RVX takes the values $0, \frac{1}{n}, \frac{2}{n}, ..., \frac{n-1}{n}, 1$ with equal probability. Find the mean and variance of the distribution of X.
38. A discrete RVX takes the values $a, a + 1, a + 2, ..., a + 2n$ with equal probability, find the SD and mean deviation of X about its mean.
39. Find the mean and variance of $Y = (2\,X + 1)^2$, if the probability distribution of X is given by

$X = i$	:	0	1	2	3
p_i	:	$\frac{1}{3}$	$\frac{1}{2}$	$\frac{1}{24}$	$\frac{1}{8}$

40. Find the mean, median, mode, QD, mean deviation about the mean and SD of the following probability distribution.

$X = x_i$	:	5	15	25	35	45	55	65	75
p_i	:	0.05	0.08	0.07	0.12	0.28	0.20	0.10	0.10

41. Compute β_1 and β_2 coefficients for the following probability distribution. Also comment on the skewness and kurtosis of the distribution.

$X = x_i$	:	1	2	3	4	5	6	7	8
p_i	:	0.008	0.032	0.142	0.216	0.240	0.206	0.143	0.013

42. The first four moments of a distribution about $X = 4$ are 1, 4, 10, 45 respectively. Find the mean, variance, β_1 and β_2.
43. The first three moments of a distribution about the value 3 of the RVX are 2, 10, –30 respectively. Show that the first three simple moments are 5, 31, 141. Find also μ_2 and μ_3.
44. The distribution of a RVX has mean 10, variance 16, $v_1 = 1$ and $\beta_2 = 4$. Obtain the first four simple moments of X.
45. What is the expectation of the number of failures preceding the first success in an infinite series of independent trials with constant probability p of success?
46. What is the expection of (i) the sum of the points on n dice, and (ii) the product of the points on n dice?
47. Three tickets are chosen at random without replacement from 100 tickets, numbered 1, 2, 3, ..., 100. Find the expectation of the sum of the numbers.
48. From an urn containing 3 red and 2 black balls, a man is to draw 2 balls at random, without replacement, being promised Rs 20/- for each red ball he draws and Rs. 10/- for each black ball. Find his expectation.
49. A person draws 3 balls from a bag containing 3 white, 4 red and 5 black balls. He is offered Rs 10/-, Rs 5/- and Rs. 2/-, if the draws 3 balls of the same colour, 2 balls of the same colour and one ball of different colour and 3 balls of different colours respectively. Find the expected value of the amount he gets.
50. Box A contains 5 cards numbered from 1 to 5 and Box B contains 4 cards numbered from 1 to 4. One card is drawn from each of these boxes. Find the mathematical expectation of the sum of the numbers on the cards drawn.
51. If the density function of a continuous RVX is given by $f(x) = \dfrac{kx}{(1+x)^3}$, $x \geq 0$, find the value of k, median and mode.
52. The density function of a continuous RVX is given by $f(x) = \dfrac{3}{4}x(2-x)$, $0 \leq x \leq 2$. Find the mean, median, mode, harmonic mean, MD about mean, SD, β_1 and β_2 of the distribution of X.

53. The density function of a continuous RVX is given by

$$f(x) = \begin{cases} \frac{1}{16}(3+x)^2, \text{ in } -3 \le x \le -1 \\ \frac{1}{16}(6-2x^2), \text{ in } -1 \le x \le 1 \\ \frac{1}{16}(3-x)^2, \text{ in } 1 \le x \le 3 \end{cases}$$

Find the mean, SD and MD about the mean.

54. If the density function of a continuous RVX is given by $f(x) = k\, e^{-|x|}$, $-\infty < x < \infty$, show that $k = \frac{1}{2}$, mean = 0, MD about the mean = 1 and SD = $\sqrt{2}$.

55. If the density function of a continuous RVX is given by $f(x) = \lambda e^{-\lambda(x-a)}$, $a \le x < \infty$, show that [(mean – SD) of the distribution of X] = a, $\beta_1 = 4$ and $\beta_2 = 9$.

56. If the continuous RVX has the density function $f(x) = 2x\, e^{-x^2}$, $x \ge 0$ and if $Y = X^2$, find the mean and variance of Y.

57. If X and Y are independent RVs with density functions $f_X(x) = \frac{8}{x^3}$, $x > 2$ and $f_Y(y) = 2y$, $0 < y < 1$ respectively and $Z = XY$, find $E(Z)$.

58. If the joint pdf of (X, Y) is given by $f(x, y) = 2$, $0 \le x < y \le 1$, find the conditional mean and conditional variance of X, given that $Y = y$.

59. If the joint pdf of (X, Y) is given by $f(x, y) = 21\, x^2 y^3$, $0 \le x < y \le 1$, find the conditional mean and conditional variance of X, given that $Y = y$, $0 < y < 1$.

60. If the joint pdf of (X, Y) is given by $f(x, y) = 3xy\,(x + y)$, $0 \le x, y \le 1$, verify that $E\{E(Y/X)\} = E(Y) = \frac{17}{24}$.

Linear Correlation

In many situations, the outcome of a random experiment will have two measurable characteristics, viz., will result in two random variables X and Y. Often we will be interested in finding whether the two different RV's are related to each other. If they are related, we will try to determine the nature of relationship and degree of relationship (correlation). Assuming that there is some correlation between X and Y, we will then try to find a formula expressing the relationship and use this formula to predict the most likely value of one RV corresponding to any given value of the other RV.

To examine whether the two RV's are inter-related, we collect n pairs of values of X and Y corresponding to n repetitions of the random experiment. Let them be $(x_1, y_1), (x_2, y_2), \ldots, (x_n, y_n)$. Then we plot the points with co-ordinates $(x_1, y_1), \ldots, (x_n, y_n)$ on a graph paper. The simple figure consisting of the plotted points is called a *scatter diagram*. From the scatter diagram, we can form a fairly good, though vague, idea of the relationship between X and Y. If the points are dense or closely packed, we may conclude that X and Y are correlated. On the other hand if the points are widely scattered throughout the graph paper, we may conclude that X and Y are either not correlated or poorly correlated.

Further if the points in the scatter diagram appear to lie near a straight line, we assume that the RV's have *linear correlation*. If they cluster round a well defined curve other than a straight line, the RV's are assumed to be *non-linear*. In this section we will assume linear correlation between the concerned RV's and discuss how to measure the degree of linear correlation.

Correlation Coefficient

As the variance $E\{X - E(X)\}^2$ measures the variations of the RV X from its mean value $E(X)$, the quantity $E\{[X - E(X)]\,[Y - E(Y)]\}$ measures the simultaneous variation of two RV's X and Y from their respective means and hence it is called *the covariance of* X, Y and denoted as Cov (X, Y).

Cov $(X, Y) = E\{[X - E(X)]\,[Y - E(Y)]\}$ is also called the *product moment* of X and Y and is also denoted as $p(X, Y)$.

Though $p(X, Y)$ is a useful measure of the degree of correlation between X and Y, it is to be expressed in mixed units of X and Y. To avoid this difficulty and to express the degree of correlation in absolute units, we divide $p(X, Y)$ by $\sigma_x \cdot \sigma_y$, so that $\dfrac{p(x, y)}{\sigma_x \sigma_y}$ is a mere number, free from the units of X and Y.

$\dfrac{p(x, y)}{\sigma_x \sigma_y}$ is a measure of intensity of linear relationship between X and Y and is called *Karl Pearson's Product Moment Correlation Coefficient* or simply *correlation coefficient* between X and Y. It is denoted by $r(X, Y)$ or r_{XY} or simply r.

Thus
$$r_{XY} = \frac{E\{[X - E(X)][Y - E(Y)]\}}{\sqrt{E\{X - E(X)\}^2 E\{Y - E(Y)\}^2}} \tag{1}$$

since σ_x, the standard deviation of X is the positive square root of the variance of X.

Now
$$\begin{aligned} & E[\{X - E(X)\}\ \{Y - E(Y)\}] \\ &= E[XY - E(Y) \cdot X - E(X) \cdot Y + E(X) \cdot E(Y)] \\ &= E(XY) - E(Y) \cdot E(X) - E(X) \cdot E(Y) + E(X) \cdot E(Y) \end{aligned}$$
[$\because$ $E(X)$ and $E(Y)$ are non-random constants]

$$= E\{XY\} - E(X) \times E(Y) \quad (2)$$

Also we know that $\quad E\{X - E(X)\}^2 = E(X^2) - \{E(X)\}^2 \quad (3)$

and $\quad E\{Y - E(Y)\}^2 = E\{Y^2\} - \{E(Y)\}^2 \quad (4)$

Using (2), (3) and (4) in (1), we get

$$r_{XY} = \frac{E(XY) - E(X)\cdot E(Y)}{\sqrt{\{E(X^2) - E^2(X)\}\{E(Y^2) - E^2(Y)\}}} \quad (5)$$

where $E^2(X)$ means $\{E(X)\}^2$.

We will mainly deal with linear correlation of discrete RV's X and Y. X will take the values $x_1, x_2, \ldots, x_n$ with frequency 1 each and Y will simultaneously take the values $y_1, y_2, \ldots, y_n$ with frequency 1 each. Hence $E(X) = \frac{1}{n}\sum x_i$; $E(X^2) = \frac{1}{n}\sum x_i^2$, $E(XY) = \frac{1}{n}\sum x_i\, y_i$ etc. Using these values in (5), the working formula for the compution of r_{XY} is got as

$$r_{XY} = \frac{\frac{1}{n}\Sigma x_i\, y_i - \frac{1}{n}\Sigma x_i . \frac{1}{n}\Sigma y_i}{\sqrt{\left\{\frac{1}{n}\Sigma x_i^2 - \left(\frac{1}{n}\Sigma x_i\right)^2\right\}\left\{\frac{1}{n}\Sigma y_i^2 - \left(\frac{1}{n}\Sigma y_i\right)^2\right\}}} \quad (6)$$

or

$$r_{XY} = \frac{n\,\Sigma xy - \Sigma x \cdot \Sigma y}{\sqrt{\{n\,\Sigma x^2 - (\Sigma x)^2\}\{n\,\Sigma y^2 - (\Sigma y)^2\}}} \quad (7)$$

Properties of Correlation Coefficient

1. $-1 \le r_{XY} \le 1$ or $|\text{Cov}(X, Y)| \le \sigma_X \cdot \sigma_Y$.

   Let us consider

   $E[a\{X - E(X)\} + \{Y - E(Y)\}]^2 = a^2\,\sigma_x^{\,2} + 2a\,C_{XY} + \sigma^2_Y \quad (1)$

   The R.H.S. expression is a quadratic expression in a, that is a real quantity. It is positive, as it is the expected value of a perfect square.

   Hence, by the property of quadratic expressions, the discriminant of the R.H.S. ≤ 0

   i.e., $\quad 4C^2_{XY} - 4\sigma^2_X\,\sigma^2_Y \le 0$

   i.e., $\quad C^2_{XY} \le \sigma_X^2 \cdot \sigma_Y^2 \quad (2)$

   i.e., $\quad \dfrac{C^2_{XY}}{\sigma^2_X \cdot \sigma^2_Y} \le 1$

   i.e., $\quad r^2_{XY} \le 1$

   $\therefore \quad |r_{XY}| \le 1$ or $-1 \le r_{XY} \le 1$

   From step (2), it is clear that $|C_{XY}| \le \sigma_X \cdot \sigma_Y$

Note *When $0 < r_{XY} \le 1$, the correlation between X and Y is said to be positive or direct.*

When $-1 \le r_{XY} \le 0$, the correlation is said to be negative or inverse.
When $-1 \le r_{XY} \le -0.5$ or $0.5 \le r_{XY} \le 1$, the correlation is assumed to be high, otherwise the correlation is assumed to be poor.

2. Correlation coefficient is independent of change of origin and scale.

i.e., If $U = \frac{X-a}{h}$ and $V = \frac{Y-b}{k}$, where $h, k > 0$, then $r_{XY} = r_{UV}$.

By the transformations, $X = a + hU$ and $Y = b + kV$

$\therefore \quad E(X) = a + hE(U)$ and $E(Y) = b + kE(V)$

$\therefore \quad X - E(X) = h\{U - E(U)\}$ and $Y - E(Y) = k\{V - E(V)\}$

Then $\quad C_{XY} = E[h\{U - E(U)\} \cdot k\{V - E(V)\} = hk\, C_{UV}$

$\sigma^2{}_X = E[h^2 \{U - E(U)\}^2] = h^2\, \sigma^2{}_U$

$\sigma^2{}_Y = E[k^2 \{V - E(V)\}^2] = k^2\, \sigma^2{}_V$

$$\therefore \quad r_{XY} = \frac{\text{Cov}(X, Y)}{\sqrt{\sigma_X^2 \cdot \sigma_Y^2}} = \frac{hk\, C_{UV}}{\sqrt{h^2 \cdot \sigma_U^2 \cdot k^2\, \sigma_V^2}}$$

$$= \frac{C_{UV}}{\sigma_U \cdot \sigma_V} = r_{UV}$$

Note *If X and Y take considerably large values, computation of r_{XY} will become difficult. In such problems, we may introduce change of origin and scale and compute r using the above property.]*

3. Two independent RV's X and Y are uncorrelated, but two uncorrelated RV's need not be independent.

When X and Y are independent, $E(XY) = E(X) \cdot E(Y)$.

$\therefore \quad C_{XY} = 0$ and hence $r_{XY} = 0$

viz., X and Y are uncorrelated.

The converse is not true, since $E(XY) = E(X) \cdot E(Y)$, when $r_{XY} = 0$.

This does not imply that X and Y are independent, as X and Y are independent only when $f(x, y) = f_X(x) \cdot f_Y(y)$.

Note *When $E(XY) = 0$, X and Y are said to be orthogonal RV's.*

4. $$r_{XY} = \frac{\sigma_X^2 + \sigma_Y^2 - \sigma_{(X-Y)}^2}{2\sigma_X\, \sigma_Y}$$

Let $\quad Z = X - Y$. Then $E(Z) = E(X) - E(Y)$

$\therefore \quad Z - E(Z) = [X - E(X)] - [Y - E(Y)]$

$$\therefore \quad \sigma^2{}_Z = E[Z - E(Z)]^2 = E[\{X - E(X)\} - \{Y - E(Y)\}]^2$$

$$= E\{X - E(X)\}^2 + E\{Y - E(Y)\}^2 - 2E\,[\{X - E(X)\}\,\{Y - E(Y)\}]$$

$$= \sigma^2{}_X + \sigma^2{}_Y - 2C_{XY}$$

$$\therefore \qquad C_{XY} = \frac{\sigma_X^2 + \sigma_Y^2 - \sigma_Z^2}{2}$$

$$\therefore \qquad r_{XY} = \frac{C_{XY}}{\sigma_X \sigma_Y} = \frac{\sigma_X^2 + \sigma_Y^2 - \sigma_{(X-Y)}^2}{2\sigma_X \sigma_Y}$$

Similarly we can prove that

$$\sigma^2_{(X+Y)} = \sigma^2_X + \sigma^2_Y + 2C_{XY}$$

and hence $r_{XY} = \dfrac{\sigma_{(X+Y)}^2 - \sigma_X^2 - \sigma_Y^2}{2\sigma_X \sigma_Y}$

Rank Correlation Coefficient

Sometimes the actual numerical values of X and Y may not be available, but the positions of the actual values arranged in order of merit (ranks) only may be available. The ranks of X and Y will in general, be different and hence may be considered as random variables. Let them be denoted by U and V. The correlation coefficient between U and V is called *the rank correlation coefficient* between (the ranks of) X, Y and denoted by ρ_{XY}.

Let us now derive a formula for ρ_{XY} or r_{UV}. Since U represents ranks of n values of X, U takes the values 1, 2, 3, $\cdots$, n.

Similarly V takes the same values 1, 2, 3, $\cdots$, n in a different order.

$$E(U) = E(V) = \frac{1}{n}(1 + 2 + \cdots + n) = \frac{n+1}{2}$$

$$E(U^2) = E(V^2) = \frac{1}{n}(1^2 + 2^2 + \cdots + n^2) = \frac{(n+1)(2n+1)}{6}$$

$$\therefore \qquad \sigma_U^2 = \sigma_V^2 = E(U^2) - E^2(U)$$

$$= \frac{(n+1)(2n+1)}{6} - \frac{(n+1)^2}{4}$$

$$= \frac{(n+1)}{12}\{2(2n+1) - 3(n+1)\}$$

$$= \frac{n^2 - 1}{12}$$

Let $\qquad D = U - V \quad \therefore E(D) = 0$

and $\qquad \sigma^2_D = E(D^2)$

By property (4) given above,

$$\rho_{XY} = r_{UV} = \frac{\sigma_U^2 + \sigma_V^2 - \sigma_D^2}{2\sigma_U \sigma_V}, \text{ where } D = U - V$$

$$= \frac{\left(\frac{n^2-1}{6}\right) - \sigma_D^2}{2\left(\frac{n^2-1}{12}\right)}$$

$$= 1 - \frac{6}{n^2-1}\sigma_D^2 \text{ or } 1 - \frac{6E\left(D^2\right)}{n^2-1}$$

$$= 1 - \frac{6\,\Sigma d^2}{n\left(n^2-1\right)} \quad \left[\because E\left(D^2\right) = \frac{1}{n}\Sigma d^2\right]$$

Note *The formula for the rank correlation coefficient is known as spearman's formula. The values of* r_{XY} *and* ρ_{XY} *(or* r_{UV}*) will be, in general, different.*

Worked Examples 4(B)

Example 1 *Compute the coefficient of correlation between X and Y, using the following data:*

X	:	1	3	5	7	8	10
Y	:	8	12	15	17	18	20

Solution

x_i	y_i	x_i^2	y_i^2	$x_i y_i$
1	8	1	64	8
3	12	9	144	36
5	15	25	225	75
7	17	49	289	119
8	18	64	324	144
10	20	100	400	200
34	90	248	1446	582

Thus $n = 6$

$\Sigma x_i = 34,\ \Sigma y_i = 90$

$\Sigma x_i^2 = 248,\ \Sigma y_i^2 = 1446$

$\Sigma x_i y_i = 582$

$$r_{XY} = \frac{n\,\Sigma xy - \Sigma x \cdot \Sigma y}{\sqrt{\left\{n\,\Sigma x^2 - (\Sigma x)^2\right\}\left\{n\,\Sigma y^2 - (\Sigma y)^2\right\}}}$$

$$= \frac{6 \times 582 - 34 \times 90}{\sqrt{\left\{6 \times 248 - (34)^2\right\}\left\{6 \times 1446 - (90)^2\right\}}}$$

$$= \frac{432}{\sqrt{332 \times 576}} = 0.9879$$

Example 2 *Compute the coefficients of correlation between X and Y using the following data:*

X	:	65	67	66	71	67	70	68	69
Y	:	67	68	68	70	64	67	72	70

Solution We effect change of origin in respect of both X and Y. The new origins are chosen at or near the average of extreme values. Thus we take $\frac{65+71}{2} = 68$ as the new origin for X and $\frac{64+72}{2} = 68$ as the new origin for Y. viz., we put $u_i = (x_i - 68)$ and $v_i = y_i - 68$ and find r_{UV}.

$X = x_i$	$Y = y_i$	$u_i = x_i - 68$	$v_i = y_i - 68$	u_i^2	v_i^2	$u_i v_i$
65	67	−3	−1	9	1	3
67	68	−1	0	1	0	0
66	68	−2	0	4	0	0
71	70	3	2	9	4	6
67	64	−1	−4	1	16	4
70	67	2	−1	4	1	−2
68	72	0	4	0	16	0
69	70	1	2	1	1	2
	Total	−1	2	29	39	13

$$r_{XY} = r_{UV} = \frac{n\Sigma uv - \Sigma u \cdot \Sigma v}{\sqrt{\left\{n\Sigma u^2 - (\Sigma u)^2\right\}\left\{n\Sigma v^2 - (\Sigma v)^2\right\}}}$$

$$= \frac{8 \times 13 - (-1) \times 2}{\sqrt{(8 \times 29 - 1)(8 \times 39 - 4)}} = \frac{106}{\sqrt{231 \times 308}} = 0.3974$$

Example 3 *Find the coefficient of correlation between X and Y using the following data:*

X	:	5	10	15	20	25
Y	:	16	19	23	26	30

Solution As the values of X are in arithmetic progression, we make the change of origin and scale, by choosing the middle most value 15 as the new origin and the common difference 5 as the new scale.

i.e., we put $U = \frac{X - 15}{5}$

As the values of Y are not in A.P., we are content with effecting a change of origin only i.e., we put $V = Y - \left(\frac{30+16}{2}\right) = Y - 23.$

x	y	$u = \frac{x-15}{5}$	$v = y - 23$	u^2	v^2	uv
5	16	− 2	− 7	4	49	14
10	19	− 1	− 4	1	16	4
15	23	0	0	0	0	0
20	26	1	3	1	9	3
25	30	2	7	4	49	14
	Total	0	− 1	10	123	35

$$r_{XY} = r_{UV} = \frac{n\Sigma uv - \Sigma u \cdot \Sigma v}{\sqrt{\left\{n\Sigma u^2 - (\Sigma u)^2\right\}\left\{n\Sigma v^2 - (\Sigma v)^2\right\}}}$$

$$= \frac{5 \times 35 - 0 \times (-1)}{\sqrt{(5 \times 10 - 0)(5 \times 125 - 1)}}$$

$$= \frac{175}{\sqrt{50 \times 624}} = 0.9907$$

Example 4 *The following table gives the bivariate frequency distribution of marks in an intelligence test obtained by 100 students according to their age:*

Age (x) in yrs / *Marks (y)*	*18*	*19*	*20*	*21*	*Total*
10–20	4	2	2	–	8
20–30	5	4	6	4	19
30–40	6	8	10	11	35
40–50	4	4	6	8	22
50–60	–	2	4	4	10
60–70	–	2	3	1	6
Total	19	22	31	28	100

Calculate the coefficient of correlation between age and intelligence.

Solution Since the frequencies of various values of x and y are not equal to 1 each the formula for the computation of r_{XY} is taken with a slight modification as given below:

$$r_{XY} = r_{UV} = \frac{N\Sigma f_{XY}\, uv - \Sigma f_X\, u \cdot \Sigma f_Y\, v}{\sqrt{\left\{N\Sigma f_X u^2 - (\Sigma f_X u)^2\right\}\left\{N\Sigma f_Y v^2 - (\Sigma f_Y\, v)^2\right\}}} \quad (1)$$

where $u = x - 20$, $v = \frac{y-35}{10}$, f_X represents frequencies of X-distribution, f_Y represents frequencies of Y-distribution and f_{XY} are the cell frequencies.

Mid y/mid x	18	19	20	21	f_Y	v	$f_Y v$	$f_Y v^2$	$\Sigma f_{XY} uv$
15	4	2	2	–	8	– 2	– 16	32	20
25	5	4	6	4	19	– 1	– 19	19	10
35	6	8	10	11	35	0	0	0	0
45	4	4	6	8	22	1	22	22	– 4
55	–	2	4	4	10	2	20	40	4
65	–	2	3	1	6	3	18	54	–3
f_X	19	22	31	28	100	Total	5	167	27
u	– 2	– 1	0	1	Total				
$f_X u$	– 38	– 22	0	28	– 32				
$f_X u^2$	76	22	0	28	126				
$\Sigma f_{XY} uv$	18	– 6	0	15	27				

Note $\Sigma f_{XY} uv$ *for the first row of the table is computed as follows.*

$$\begin{aligned}\Sigma f_{XY} uv &= f_{11} u_1 v_1 + f_{12} u_2 v_1 + f_{13} u_3 v_1 + f_{14} u_4 v_1 \\ &= 4(-2)(-2) + 2(-1)(-2) + 2(0)(-2) + 0(1)(-2) \\ &= 20\end{aligned}$$

Similarly, other $\Sigma f_{XY} uv$ values are computed. Value of $(\Sigma\Sigma f_{XY} uv)$ obtained as the total of the entries of the last column and as that of the last row must tally.

Using the relevant values obtained in the table in (1), we have

$$r_{XY} = \frac{100 \times 27 - (-32) \times 5}{\sqrt{\{100 \times 126 - (-32)^2\}\{100 \times 167 - 5^2\}}}$$

$$= \frac{2860}{\sqrt{13624 \times 16675}} = 0.1897$$

Example 5 *Calculate the correlation coefficient for the following ages of husbands (X) and wives (Y), using only standard deviations of X and Y :*

X:	*23*	*27*	*28*	*28*	*29*	*30*	*31*	*33*	*35*	*36*
Y:	*18*	*20*	*22*	*27*	*21*	*29*	*27*	*29*	*28*	*29*

Solution

x	y	$u = x - 30$	$v = y - 24$	u^2	v^2	$d = x - y$	d^2
23	18	– 7	– 6	49	36	5	25
27	20	– 3	– 4	9	16	7	49
28	22	– 2	– 2	4	4	6	36
28	27	– 2	3	4	9	1	1
29	21	– 1	– 3	1	9	8	64
30	29	0	5	0	25	1	1
31	27	1	3	1	9	4	16
33	29	3	5	9	25	4	16
35	28	5	4	25	16	7	49
36	29	6	5	36	25	7	49
	Total	0	10	138	174	50	306

$$\sigma_X^2 = \frac{1}{n}\Sigma u^2 - \left(\frac{1}{n}\Sigma u\right)^2 = \frac{1}{10}\times 138 = 13.8$$

$$\sigma_Y^2 = \frac{1}{n}\Sigma v^2 - \left(\frac{1}{n}\Sigma v\right)^2 = \frac{1}{10}\times 174 - \left(\frac{10}{10}\right)^2 = 16.4$$

$$\sigma^2_{(X-Y)} = \sigma^2_D = \frac{1}{n}\Sigma d^2 - \left(\frac{1}{n}\Sigma d\right)^2 = \frac{1}{10}\times 306 - \left(\frac{50}{10}\right)^2 = 5.6$$

$$r_{XY} = \frac{\sigma_X^2 + \sigma_Y^2 - \sigma_D^2}{2\sigma_X\,\sigma_Y} = \frac{13.8 + 16.4 - 5.6}{2\times\sqrt{13.8}\times\sqrt{16.4}}$$

$$= \frac{24.6}{30.0879}$$

$$= 0.8176$$

Example 6 *If the independent random variables X and Y have the variances 36 and 16 respectively, find the correlation coefficient between (X + Y) and (X – Y).*

Solution Let $U = X + Y$ and $V = X - Y$

$$E(U) = E(X) + E(Y);\ E(V) = E(X) - E(Y)$$

$$E(UV) = E(X^2 - Y^2) = E(X^2) - E(Y^2)$$

$$E(U^2) = E\{(X + Y)^2\} = E(X^2) + E(Y^2) + 2E(XY)$$

$$E(V^2) = E(X^2) + E(Y^2) - 2E(XY)$$

$$C_{UV} = E(UV) - E(U)\cdot E(V)$$

$$= E(X^2) - E(Y^2) - \{E^2(X) - E^2(Y)\}$$

$$= [E(X^2) - E^2(X)] - [E(Y^2) - E^2(Y)]$$

$$= \sigma^2_X - \sigma^2_Y = 36 - 16 = 20$$

$$\sigma^2_U = E(U^2) - E^2(U)$$

$$= \{E(X^2) + E(Y^2) + 2E(XY)\} - \{E^2(X) + E^2(Y) + 2E(X)\cdot E(Y)\}$$

$$= [E(X^2) - E^2(X)] + [E(Y^2) - E^2(Y)] + 2[E(XY) - E(X)\cdot E(Y)]$$

$$= 36 + 16 + 2\times 0$$

[$\because$ X and Y are independent and hence uncorrelated]

$$= 52$$

Similarly, $\sigma^2_V = 52$

Now $$r_{UV} = \frac{C_{UV}}{\sigma_U\cdot\sigma_V} = \frac{20}{52} = \frac{5}{13}$$

Example 7 *If X, Y and Z are uncorrelated RV's with zero means and standard deviations 5, 12 and 9 respectively and if U = X + Y and V = Y + Z, find the correlation coefficient between U and V.*

Solution $E(X) = E(Y) = E(Z) = 0$

$$\text{Var}(X) = E(X^2) - E^2(X) = 25 \quad \therefore \quad E(X^2) = 25$$

Similarly $E(Y^2) = 144$ and $E(Z^2) = 81$

X and Y are uncorrelated

$\therefore$ $r_{XY} = 0$. i.e., $E(XY) - E(X) \cdot E(Y) = 0$

$\therefore$ $E(XY) = 0$. Similarly $E(YZ) = 0$; $E(ZX) = 0$

Now $E(U) = E(X + Y) = 0$ and $E(V) = 0$

$$E(U^2) = E(X^2 + Y^2 + 2XY)$$
$$= 25 + 144 + 2 \times 0 = 169$$
$$E(V^2) = E(Y^2 + Z^2 + 2YZ)$$
$$= 144 + 81 + 2 \times 0 = 225$$

$\therefore$ $\sigma^2{}_U = E(U^2) - E^2(U) = 169$

and $\sigma^2{}_V = E(V^2) - E^2(V) = 225$

$$E(UV) = E\{(X + Y)(Y + Z)\}$$
$$= E(XY) + E(XZ) + E(YZ) + E(Y^2)$$
$$= 0 + 0 + 0 + 144 = 144$$

$$r_{UV} = \frac{E(UV) - E(U) \cdot E(V)}{\sigma_U \cdot \sigma_V} = \frac{144}{13 \times 15} = \frac{48}{65}$$

Example 8 *If X and Y are two RV's with variances* σ_X^2 *and* σ_Y^2 *respectively, find the value of k, if U = X + kY and V =* $X + \dfrac{\sigma_X}{\sigma_Y} \cdot Y$ *are uncorrelated.*

Solution U and V are uncorrelated.

$\therefore$ $\text{Cov}(U, V) = 0$

i.e., $E(UV) - E(U) \cdot E(V) = 0$

i.e.,
$$E\left\{(X + kY)\left(X + \frac{\sigma_X}{\sigma_Y} \cdot Y\right)\right\} - E(X + kY) \cdot E\left(X + \frac{\sigma_X}{\sigma_Y} Y\right) = 0$$

i.e.
$$E\left\{X^2 + k\frac{\sigma_X}{\sigma_Y} \cdot Y^2 + \left(k + \frac{\sigma_X}{\sigma_Y}\right) XY\right\}$$
$$-\left[\{E(X) + kE(Y)\}\left\{E(X) + \frac{\sigma_X}{\sigma_Y} E(Y)\right\}\right] = 0$$

i.e.,
$$E(X^2) + k\frac{\sigma_X}{\sigma_Y}E(Y^2) + \left(k + \frac{\sigma_X}{\sigma_Y}\right)E(XY)$$
$$-\left[E^2(X) + k\frac{\sigma_X}{\sigma_Y}E^2(Y) + \left(k + \frac{\sigma_X}{\sigma_Y}\right)E(X)\cdot E(Y)\right] = 0$$

i.e.,
$$\{E(X^2) - E^2(X)\} + \frac{k\sigma_X}{\sigma_Y}\{E(Y^2) - E^2(Y)\}$$
$$+\left(k + \frac{\sigma_X}{\sigma_Y}\right)\{E(XY) - E(X)\cdot E(Y)\} = 0$$

i.e.,
$$\sigma^2_X + k\frac{\sigma_X}{\sigma_Y}\cdot\sigma^2_Y + \left(k + \frac{\sigma_X}{\sigma_Y}\right)\text{Cov}(X, Y) = 0$$

Dividing throughout by σ_X,
$$(\sigma_X + k\sigma_Y) + (\sigma_X + k\sigma_Y)\cdot\frac{\text{cov}(X, Y)}{\sigma_X\sigma_Y} = 0$$

i.e., $(\sigma_X + k\sigma_Y)(1 + r_{XY}) = 0$

Assuming that $r_{XY} \neq -1$, we get
$$\sigma_X + k\sigma_Y = 0$$
$$\therefore \quad k = -\frac{\sigma_X}{\sigma_Y}$$

Example 9 *If (X, Y) is a two-dimensional RV uniformly distributed over the triangular region R bounded by y = 0, x = 3 and y = 4/3 x. Find $f_X(x)$, $f_Y(y)$, E(X), Var(X), E(Y), Var(Y) and ρ_{XY}.*

Solution

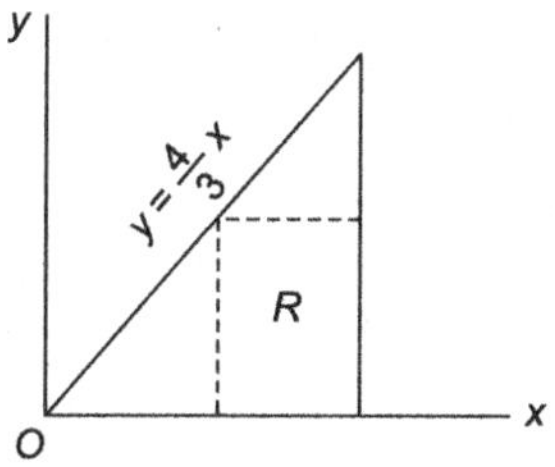

Since (X, Y) is uniformly distributed, $f(x, y)$ = a constant = k

Now
$$\iint f(x, y)\,dx\,dy = 1$$

i.e.,
$$\int_0^4 \int_{3y/4}^3 k\,dx\,dy = 1$$

i.e.,
$$k\int_0^4 \left(3 - \frac{3y}{4}\right)dy = 1$$

i.e., $6k = 1$

$\therefore$ $k = \frac{1}{6}$

$$f_Y(y) = \int_{3y/4}^{3} \frac{1}{6} dx = \frac{1}{8}(4 - y), 0 < y < 4$$

$$f_X(x) = \int_{3}^{4x/3} \frac{1}{6} dy = \frac{2}{9} x, 0 < x < 3$$

$$E(X) = \int x f_X(x)\, dx = \int_0^3 \frac{2}{9} x^2\, dx = 2$$

$$E(Y) = \int y f_Y(y)\, dy = \int_0^4 \frac{y}{8} \times (4 - y)\, dy = \frac{4}{3}$$

$$E(X^2) = \int_0^3 \frac{2}{9} \times x^3\, dx = \frac{9}{2}$$

$$E(Y^2) = \int_0^4 \frac{y^2}{8} \times (4 - y)\, dy = \frac{8}{3}$$

$$\text{Var}(X) = E(X^2) - \{E(X)\}^2 = \frac{9}{2} - 4 = \frac{1}{2}$$

$$\text{Var}(Y) = E(Y^2) - \{E(Y)\}^2 = \frac{8}{3} - \frac{16}{9} = \frac{8}{9}$$

$$E(XY) = \int_0^4 \int_{3y/4}^{3} \frac{1}{6} xy\, dx\, dy$$

$$= \frac{3}{64} \int_0^4 (16 - y^2) y\, dy = 3$$

$$\rho_{XY} = \frac{E(XY) - E(X) \times E(Y)}{\sigma_x \sigma_y} = \frac{3 - 2 \times \frac{4}{3}}{\frac{1}{\sqrt{2}} \times 2 \frac{\sqrt{2}}{3}} = \frac{1}{2}$$

Example 10 *Find the correlation co-efficient between X and Y, which are jointly normally distributed with*

$$f(x, y) = \frac{1}{2\pi \sigma_x \sigma_y \sqrt{1 - r^2}} \exp \left\{ -\frac{1}{2(1 - r^2)} \left(\frac{x^2}{\sigma_x^2} - \frac{2rxy}{\sigma_x \sigma_y} + \frac{y^2}{\sigma_y^2} \right) \right\}$$

Solution $\frac{x^2}{\sigma_x^2} - \frac{2rxy}{\sigma_x \sigma_y} + \frac{y^2}{\sigma_y^2} = \left(\frac{x}{\sigma_x} - \frac{ry}{\sigma_y} \right)^2 + (1 - r^2) \frac{y^2}{\sigma_y^2}$

$$= \frac{1}{\sigma_x^2}\left(x - \frac{ry\sigma x}{\sigma_y}\right)^2 + (1 - r^2)\,\frac{y^2}{\sigma_y^2}$$

$$E(XY) = \frac{1}{2\pi\sigma_x\,\sigma_y\sqrt{1-r^2}} \int_{-\infty}^{\infty}\int_{-\infty}^{\infty} \exp$$

$$\left\{-\frac{1}{2\left(1-r^2\right)}\left[\frac{1}{\sigma_x^2}\left(x - \frac{ry\sigma_x}{\sigma_y}\right)^2 + \frac{(1-r)^2\,y^2}{\sigma_y^2}\right]\right\} xy\,dx\,dy$$

$$= \frac{1}{\sigma_y\sqrt{2\pi}} \int_{-\infty}^{\infty} y \exp\left(\frac{-y^2}{2\sigma_y^2}\right) \int_{-\infty}^{\infty} \frac{x}{\left(\sigma_x\sqrt{1-r^2}\right)\sqrt{2\pi}}$$

$$\exp\left\{\frac{-\left(x - \dfrac{ry\sigma_x}{\sigma_y}\right)^2}{2\left(1-r^2\right)\sigma_x^2}\right\} dx\,dy \qquad (1)$$

The inner integral is the mean of the normal distribution with mean $\dfrac{ry\sigma_x}{\sigma_y}$ and variance $(1 - r^2)\,\sigma_X^2$.

$\therefore$ the inner integral $= \dfrac{ry\sigma_x}{\sigma_y}$

Using this value in (1),

$$E(XY) = \left(\frac{r\sigma_x}{\sigma_y}\right) \times \frac{1}{\sigma_y\sqrt{2\pi}} \int_{-\infty}^{\infty} y^2 \exp\left(-\frac{y^2}{2\sigma_y^2}\right) dy$$

$$= \frac{r\sigma_x}{\sigma_y}\, E(Y^2) \text{ for } N(0, \sigma_Y)$$

$$= \frac{r\sigma_x}{\sigma_y} \times \sigma_Y^2$$

$$= r\sigma_X\sigma_Y$$

$$\therefore \qquad \rho_{XY} = \frac{E(XY) - E(X)E(y)}{\sigma_x\,\sigma_y} = r$$

Example 11 *Ten students got the following percentage of marks in Mathematics and Physical sciences:*

Students:	*1*	*2*	*3*	*4*	*5*	*6*	*7*	*8*	*9*	*10*
Marks in Mathematics:	*78*	*36*	*98*	*25*	*75*	*82*	*90*	*62*	*65*	*39*
Marks in Phy. Sciences:	*84*	*51*	*91*	*60*	*68*	*62*	*86*	*58*	*63*	*47*

Calculate the rank correlation coefficient.

Solution Denoting the ranks in Mathematics and in Phy. Sciences by U and V, we have the following values of U and V:

U:	4	9	1	10	5	3	2	7	6	8	
V:	3	9	1	7	4	6	2	8	5	10	
D:	1	0	0	3	1	– 3	0	– 1	1	– 2	
D^2:	1	0	0	9	1	9	0	1	1	4	: $\Sigma d^2 = 26$

$$\rho_{XY} = r_{UV} = 1 - \frac{6\sum d^2}{n(n^2 - 1)}$$

$$= 1 - \frac{6 \times 26}{10 \times 99} = 0.8424$$

Example 12 *Ten competitors in a beauty contest were ranked by three judges as follows:*

	Competitors									
Judges	*1*	*2*	*3*	*4*	*5*	*6*	*7*	*8*	*9*	*10*
A:	*6*	*5*	*3*	*10*	*2*	*4*	*9*	*7*	*8*	*1*
B:	*5*	*8*	*4*	*7*	*10*	*2*	*1*	*6*	*9*	*3*
C:	*4*	*9*	*8*	*1*	*2*	*3*	*10*	*5*	*7*	*6*

Discuss which pair of judges have the nearest approach to common taste of beauty.

Solution

Rank by A (*U*)	*Rank by B* (*V*)	*Rank by C* (*W*)	$d_1 = U - V$	$d_2 = V - W$	$d_3 = U - W$	d_1^2	d_2^2	d_3^2
6	5	4	1	1	2	1	1	4
5	8	9	– 3	– 1	– 4	9	1	16
3	4	8	– 1	– 4	– 5	1	16	25
10	7	1	3	6	9	9	36	81
2	10	2	– 8	8	0	64	64	0
4	2	3	2	– 1	1	4	1	1
9	1	10	8	– 9	– 1	64	81	1
7	6	5	1	1	2	1	1	4
8	9	7	– 1	2	1	1	4	1
1	3	6	– 2	– 3	– 5	4	9	25
					Total:	157	214	158

$$r_{UV} = 1 - \frac{6\,\Sigma d_1^2}{n(n^2-1)} = 1 - \frac{6 \times 157}{10 \times 99} = 0.0485$$

$$r_{VW} = 1 - \frac{6\,\Sigma d_2^2}{n(n^2-1)} = 1 - \frac{6 \times 214}{10 \times 99} = -0.2970$$

$$r_{UW} = 1 - \frac{6\,\Sigma d_3^2}{n(n^2-1)} = 1 - \frac{6 \times 158}{10 \times 99} = 0.0424$$

Since r_{UV} is maximum, the judges A and B may be considered to have common taste of beauty to some extent compared to other pairs of judges.

Exercise 4(B)

Part A (Short answer questions)

1. What is a scatter diagram? What is its role in correlation analysis?
2. What do you mean by correlation between two random variables?
3. What is linear correlation? How will you find that two RV's are linearly correlated?
4. Define covariance of X, Y and coefficient of correlation between X and Y.
5. Why is r_{XY} preferred for measuring the degree of linear correlation to Cov (X, Y)?
6. State the properties of correlation coefficient.
7. State two different formulas used to compute r_{XY}.
8. Define rank correlation coefficient and write down the formula for computing it.
9. Prove that $-1 \le r_{XY} \le 1$.
10. Prove that $\sigma^2_{(X+Y)} - \sigma^2_{(X-Y)} = 4\,\text{Cov}(X, Y)$
11. If C_{XY} is the covariance of X and Y, prove that $C_{XY} = E(XY) - E(X) \cdot E(Y)$.
12. If X and Y are independent RV's prove that $r_{XY} = 0$. Is the converse true?
13. If X and Y are uncorrelated, prove that $\text{Var}(X + Y) = \text{Var}(X) + \text{Var}(Y)$.
14. When are two RVs said to be orthogonal?

Part B

15. Ten students got the following marks in Mathematics and Basic Engineering:

Marks in Mathematics	78	36	98	25	75	82	90	62	65	39
Marks in Basic Engg.	84	51	91	60	68	62	86	58	53	47

Calculate the coefficient of correlation.

16. Calculate the correlation coefficient between X and Y from the following data:

X:	65	66	67	67	68	69	70	72
Y:	67	68	65	68	72	72	69	71

17. Find the coefficient of correlation between X and Y using the following data:

X:	5.5	3.6	2.6	3.4	3.1	2.7	3.0	3.1	3.2	3.8
Y:	27	36	39	39	32	35	40	36	44	36

18. Compute the coefficient of correlation between X and Y from the following data:

X:	80	45	55	56	58	60	65	68	70	75	85
Y:	82	56	50	48	60	62	64	65	70	74	90

19. Find the coefficient of correlation between X and Y from the following data:

X:	10	14	18	22	26	30
Y:	18	12	24	6	30	36

20. Calculate the coefficient of correlation between X and Y, by finding variances only, from the following data:

X:	21	23	30	54	57	58	72	78	87	90
Y:	60	71	72	83	110	84	100	92	113	135

21. Calculate r_{XY} from the following data, where X represents production (in crore tons) and Y represents exports (in crore tons), using only the variances.

X:	55	56	58	59	60	60	62
Y:	35	38	38	39	44	43	44

22. The following table gives the frequency of scores obtained by 65 students in a general knowledge test according to age groups. Measure the degree of linear relationship between age and general knowledge:

Test scores	*Age in years* 19	20	21	22
225	4	4	2	1
275	3	5	4	2
325	2	6	8	5
375	1	4	6	8

23. Compute the value of r_{XY} between X, the ages of husbands and Y the ages of wives from the following data:

X / *Y*	*15–25*	*25–35*	*35–45*	*45–55*	*55–65*	*65–75*	*Total*
15–25	1	1	–	–	–	–	2
25–35	2	12	1	–	–	–	15
35–45	–	4	10	1	–	–	15
45–55	–	–	3	6	1	–	10
55–65	–	–	–	2	4	2	8
65–75	–	–	–	–	1	2	3
Total	3	17	14	9	6	4	53

24. Find the rank correlation coefficient between the ranks of the variable X and Y:

X:	10	15	12	17	13	16	24	14	22
Y:	30	42	45	46	33	34	40	35	39

25. The competitors in a musical contest were ranked by the three judges A, B, C in the following order:

Rank by A:	1	6	5	10	3	2	4	9	7	8
Rank by B:	3	5	8	4	7	10	2	1	6	9
Rank by C:	6	4	9	8	1	2	3	10	5	7

Using rank correlation technique, find which pair of judges have more or less the same taste in music.

26. If X, Y, Z are uncorrelated RV's having the same variance, find the correlation coefficient between $(X + Y)$ and $(Y + Z)$.
27. If X and Y are two uncorrelated RV's with zero means, prove that $U = X \cos \alpha + Y \sin \alpha$ and $V = X \sin \alpha - Y \cos \alpha$ are also uncorrelated.
28. X and Y are independent RV's with means 5 and 10 and variances 4 and 9 respectively. Obtain the correlation coefficient between U and V, where $U = 3X + 4Y$ and $V = 3X - Y$.
29. If X_1, X_2, X_3 are three uncorrelated RV's having variances v_1, v_2, v_3 respectively, obtain the coefficient of correlation between $(X_1 + X_2)$ and $(X_2 + X_3)$.
30. Show that (i) $E\{aX + bY\} = aE(X) + bE(Y)$ and (ii) var $(aX + bY) = a^2$ var $(X) + b^2$ var $(Y) + 2abC(X, Y)$, where $C(X, Y)$ is the covariance of (X, Y).
31. It two RV's are uncorrelated, prove that the variance of their sum is equal to the sum of their variances.
32. If the joint density fucntion of (X, Y) is given by $f(x, y) = 2 - x - y$, $0 \le x, y \le 1$, find $E(X)$, $E(Y)$, var (X), var (Y) and r_{XY}.
33. If the two dimensional RV (X, Y) is uniformly distributed in $0 \le x < y \le 1$, find $E(X)$, $E(Y)$, var (X), var (Y) and r_{XY}.
34. If the two dimensional RV (X, Y) is uniformly distributed over R, where R is defined by $\{(x, y)/x^2 + y^2 \le 1, y \ge 0\}$, find r_{XY}.
35. If the joint pdf of (X, Y) is given by $f(x, y) = x + y$, $0 \le x, y \le 1$, find r_{XY}.
36. Let X be a RV with mean value = 3 and variance = 2. Find the second moment of X about the origin. Another RV Y is defined by $Y = -6X + 22$. Find the mean value of Y and the correlation of X and Y.

Regression

When the random variables X and Y are linearly correlated, the points plotted on the scatter diagram, corresponding to n pairs of observed values of X and Y, will have a tendency to cluster round a straight line. This straight is called *the regression line*. The regression line can be taken as the best fitting straight line for the observed pairs of values of X and Y in the least square sense, with which the students are familiar.

When two RV's X and Y are linearly correlated, we may not know which variable takes independent values. If we treat X as the independent variable and hence assume that the values of Y depend on those of X, the regression line is called the *regression line of Y on X*. If we assume that the values of X depend on those of the independent variable Y, *the regression line of X on Y* is obtained. Thus in situations where the distinction cannot be made between the RV's X and Y as to which is the independent variable and which is the dependent variable, there will be two regression lines. However, when the value of $Y(X)$ is to be predicted corresponding to a specified value of $X(Y)$, we should make use of the regression line of $Y(X)$ on $X(Y)$.

Equation of the Regression Line of *Y* on X

The regression line of Y on X is the best-fitting straight line for the observed pairs of values $(x_1, y_1), (x_2, y_2), \cdots, (x_n, y_n)$, based on the assumption that x is the independent variable and y is the dependent variable. Hence, let the equation of the regression line of Y on X be assumed as $y = ax + b$. (1)

By the principle of least squares, the normal equations which give the values of a and b.

are $$\Sigma y_i = a\, \Sigma x_i + nb \tag{2}$$

and $$\Sigma x_i y_i = a\, \Sigma x_i^2 + b\, \Sigma x_i \tag{3}$$

Dividing equation (2) by n, we get

$$\bar{y} = a\,\bar{x} + b \tag{4}$$

where $\bar{x} = E(X)$ and $\bar{y} = E(Y)$. (1)–(4) gives the required equation as

$$y - \bar{y} = a(x - \bar{x}) \tag{5}$$

Eliminating b between equations (2) and (3)

we get $$a = \frac{n\Sigma x_i y_i - \Sigma x_i \cdot \Sigma y_i}{n\Sigma x_i^2 - (\Sigma x_i)^2}$$

$$= \frac{\frac{1}{n}\Sigma x_i y_i - \left(\frac{1}{n}\Sigma x_i\right)\cdot\left(\frac{1}{n}\Sigma y_i\right)}{\frac{1}{n}\Sigma x_i^2 - \left(\frac{1}{n}\Sigma x_i\right)^2}$$

or $$a = \frac{E(XY) - E(X)\cdot E(Y)}{E(X^2) - E^2(X)} = \frac{p_{XY}}{\sigma_X^2} \tag{6}$$

Using (6) in (5), we get the equation of the regression line of Y on X as

$$y - \bar{y} = \frac{p_{XY}}{\sigma_X^2}(x - \bar{x}) \tag{7}$$

or $$y - \bar{y} = \frac{r_{XY}\,\sigma_Y}{\sigma_X}(x - \bar{x}) \tag{8}$$

$$\left[\because r_{XY} = \frac{p_{XY}}{\sigma_X \sigma_Y}\right]$$

In a similar manner, assuming the equation of the regression line of X and Y as $x = ay + b$ and using the equations

$$\Sigma x_i = a\,\Sigma y_i + nb \text{ and } \Sigma x_i y_i = a\,\Sigma y^2{}_i + b\,\Sigma y_i,$$

we can get the equation of the regression line of X on Y as

$$x - \bar{x} = \frac{p_{XY}}{\sigma_Y^2}(y - \bar{y}) \tag{9}$$

or

$$x - \bar{x} = \frac{r_{XY}\,\sigma_X}{\sigma_Y}(y - \bar{y}) \tag{10}$$

Note *1.* $\frac{p_{XY}}{\sigma_X^2}$ *or* $\frac{r_{XY}\,\sigma_Y}{\sigma_X}$ *is called the regression coefficient of Y on X and denoted by* b_1 *or* b_{YX}*.* $\frac{p_{XY}}{\sigma_Y^2}$ *or* $\frac{r_{XY}\,\sigma_X}{\sigma_Y}$ *is called the regression coefficient of X on Y and denoted by* b_2 *or* b_{XY}*.*

2. *Clearly* $b_1 b_2 = r^2_{XY}$*, i.e.,* r_{XY} *is the geometric mean of* b_1 *and* b_2*.*

   $$\therefore \quad r_{XY} = \pm\sqrt{b_1 b_2}$$

   The sign of r_{XY} *is the same as that of* b_1 *or* b_2*, as* $b_1 = r_{xy}\frac{\sigma_Y}{\sigma_X}$ *and* $b_2 = r_{XY}\frac{\sigma_Y}{\sigma_X}$ *have the same sign as* r_{XY} *(*$\because$ σ_X *and* σ_Y *are positive).*

   Also $$\frac{b_1}{b_2} = \frac{\sigma_Y^2}{\sigma_X^2}$$

3. *When there is perfect linear correlation between X and Y, viz., when* $r_{XY} = \pm 1$*, the two regression lines coincide.*
4. *The point of intersection of the two regression lines is clearly the point whose co-ordinates are* $(\bar{x}, \bar{y})$*.*
5. *When there is no linear correlation between X and Y, viz., when* $r_{XY} = 0$*, the equations of the regression lines become* $y = \bar{y}$ *and* $x = \bar{x}$*, which are at right angles.*

Standard Error of Estimate of *Y*

Although we use the regression line of Y on X to predict the value of Y corresponding to a specified value of X we may also use it to estimate the value of Y corresponding to an observed value of $X = x_i$, say. The value of Y estimated in this manner need not, in general, be equal to the corresponding observed value of Y, namely, y_i. Hence the difference between Y and Y_E is called *the error of estimate of Y*. This error will vary from one observed value to the other and a random variable. The standard deviation of this RV $(Y - Y_E)$ is called *the standard error of estimate of Y* and denoted by S_Y.

Now $E\{Y - Y_E\} = E\left[Y - \left\{\bar{y} + \frac{r_{XY}\,\sigma_Y}{\sigma_X}(X - \bar{x})\right\}\right]$

$$= (\bar{y} - \bar{y}) - \frac{r_{XY}\,\sigma_Y}{\sigma_X}(\bar{x} - \bar{x})$$

$$= 0$$

$$\sigma^2_{(Y - Y_E)} = E\{(Y - Y_E)^2\} - E^2(Y - Y_E)$$

$$= E\left[Y - \left\{\bar{y} + \frac{r_{XY}\,\sigma_Y}{\sigma_X}(X - \bar{x})\right\}\right]^2$$

$$= E\left[(Y - \bar{y})^2 + \frac{r_{XY}^2\,\sigma_Y^2}{\sigma_X^2}(X - \bar{x})^2 - \frac{2r_{XY}\,\sigma_Y}{\sigma_X}(X - \bar{x})(Y - \bar{y})\right]$$

(i.e.) $$S^2{}_Y = \sigma_Y^2 + \frac{r_{XY}^2\,\sigma_Y^2}{\sigma_X^2}\sigma_X^2 - \frac{2r_{XY}\,\sigma_Y}{\sigma_X}\text{Cov}(X, Y)$$

$$= \sigma_Y^2 + r^2{}_{XY} \cdot \sigma_Y^2 - 2r_{XY}^2\,\sigma_Y^2$$

$$[\because \text{Cov}(X, Y) = r_{XY}\,\sigma_X\,\sigma_Y]$$

$$= (1 - r_{XY}^2)\,\sigma^2{}_Y \text{ or } S_Y = \sqrt{1 - r_{XY}^2}\,\sigma_Y \quad (1)$$

Similarly, the standard error of estimate of X, denoted by S_X is given by

$$S_X^2 = (1 - r_{XY}^2)\,\sigma^2{}_X \text{ or } S_X = \sqrt{1 - r_{XY}^2}\,\sigma_X \quad (2)$$

Note *We may use (1) or (2) to prove that* $|r_{XY}| \le 1$.

From (1), $S_Y = \sqrt{1 - r_{XY}^2}\,\sigma_Y$

Since S_Y *and* σ_Y *are positive,* $1 - r_{XY}^2 \ge 0$

$\therefore$ $r_{XY}^2 \le 1$

i.e., $|r_{XY}| \le 1$ or $-1 \le r_{XY} \le 1]$

Worked Examples 4(C)

Example 1 *Obtain the equations of the lines of regression from the following data:*

X:	1	2	3	4	5	6	7
Y:	9	8	10	12	11	13	14

Solution

X	Y	$U = X - 4$	$V = Y - 11$	U^2	V^2	UV
1	9	−3	−2	9	4	6
2	8	−2	−3	4	9	6
3	10	−1	−1	1	1	1
4	12	0	1	0	1	0
5	11	1	0	1	0	0
6	13	2	2	4	4	4
7	14	3	3	9	9	9
	Total	0	0	28	28	26

$$\bar{x} = E(X) = 4 + \frac{1}{n}\Sigma u = 4$$

$$\bar{y} = E(Y) = 11 + \frac{1}{n}\Sigma v = 11$$

$$\sigma^2_X = \frac{1}{n}\Sigma u^2 - \left(\frac{1}{n}\Sigma u\right)^2 = \frac{1}{7}\times 28 = 4$$

$$\sigma^2_Y = \frac{1}{n}\Sigma v^2 - \left(\frac{1}{n}\Sigma v\right)^2 = \frac{1}{7}\times 28 = 4$$

$$C_{XY} = \frac{1}{n}\Sigma uv - \left(\frac{1}{n}\Sigma u\right)\cdot\left(\frac{1}{n}\Sigma v\right) = \frac{1}{7}\times 26 = 3.7$$

The regression line of Y on X is

$$y - \bar{y} = \frac{p_{XY}}{\sigma_X^2}(x - \bar{x})$$

i.e., $$y - 11 = \frac{3.7}{4}(x - 4)$$

i.e., $3.7x - 4y + 29.2 = 0$

The regression line of X on Y is

$$x - \bar{x} = \frac{p_{XY}}{\sigma_X^2}(y - \bar{y})$$

i.e., $$x - 4 = \frac{3.7}{4}(y - 11)$$

i.e., $4x - 3.7y + 24.7 = 0$

Example 2 *Obtain the equations of the regression lines from the following data, using the method of least squares. Hence find the coefficient of correlation between X and Y. Also estimate the value of (i) Y, when X = 38 and (ii) X, when Y = 18.*

Solution

X:	22	26	29	30	31	31	34	35
Y:	20	20	21	29	27	24	27	31

Put $U = X - 29$ and $V = Y - 27$.

Let the equation of the regression line of Y on X be $y = Ax + B$ or equivalently $v = au + b$ (1)

The normal equations for finding a and b are

$$a\,\Sigma u + nb = \Sigma v \tag{2}$$

and $$a\,\Sigma u^2 + b\Sigma u = \Sigma uv \tag{3}$$

x	y	$u = x - 29$	$v = y - 27$	u^2	v^2	uv
22	20	– 7	– 7	49	49	49
26	20	– 3	– 7	9	49	21
29	21	0	– 6	0	36	0
30	29	1	2	1	4	2
31	27	2	0	4	0	0
31	24	2	– 3	4	9	– 6
34	27	5	0	25	0	0
35	31	6	4	36	16	24
	Total	6	– 17	128	163	90

Using the relevant values from the table in (2) and (3), we have

$$6a + 8b = -17 \tag{2$'$}$$

$$128a + 6b = 90 \tag{3$'$}$$

Solving (2)′ and (3)′, we get

$$a = 0.83;\ b = -2.75$$

Hence the regression line of Y on X is

$$y - 27 = 0.83\,(x - 29) - 2.75$$

i.e., $$y = 0.83x + 0.18 \tag{4}$$

Let the equation of the regression line of X on Y be $x = Cy + D$ or equivalently $u = cv + d$ (5)

The normal equations for finding c and d are

$$c\Sigma v + nd = \Sigma u \tag{6}$$

and $$c\Sigma v^2 + d\Sigma v = \Sigma uv \tag{7}$$

Using the relevant values from the table in (6) and (7), we have

$$-17c + 8d = 6 \tag{6$'$}$$

$$163c - 17d = 90 \tag{7$'$}$$

Solving (6)′ and (7)′, we get

$$c = 0.81;\ d = 2.47$$

Hence the regression line of X on Y is

$$x - 29 = 0.81(y - 27) + 2.47$$

i.e., $$x = 0.81y + 9.60 \tag{8}$$

Comparing equation (4) with

$$y - \bar{y} = r\frac{\sigma_Y}{\sigma_X}(x - \bar{x})$$

We get $$r\frac{\sigma_Y}{\sigma_X} = 0.83 \qquad (9)$$

Comparing equation (8) with

$$x - \bar{x} = \frac{r\sigma_X}{\sigma_Y}(y - \bar{y})$$

We get $$r\frac{\sigma_X}{\sigma_Y} = 0.81 \qquad (10)$$

From (9) and (10), we get $r^2 = 0.83 \times 0.81$

$\therefore$ $$r = 0.82 \left(\because b_1 = \frac{r\sigma_Y}{\sigma_X} \text{ and } b_2 = \frac{r\sigma_X}{\sigma_Y} \text{ are both positive} \right)$$

We use equation (4) to estimate the value of Y when $X = 38$.

$\therefore$ $$Y = 0.83 \times 38 + 0.18 = 31.72$$

Using equation (8) to estimate the value of X when $Y = 18$, we have

$$X = 0.81 \times 18 + 9.60 = 24.18$$

Example 3 *A study of prices of rice at Chennai and Madurai gave the following data:*

	Chennai	*Madurai*
Mean	*19.5*	*17.75*
S.D.	*1.75*	*2.5*

Also the coefficient of correlation between the two is 0.8. Estimate the most likely price of rice (i) at Chennai corresponding to the price of 18 at Madurai and (ii) at Madurai corresponding to the price of 17 at Chennai.

Solution Let the prices of rice at Chennai and Madurai be denoted by X and Y respectively. Then from the data,

$$\bar{x} = 19.5,\ \bar{y} = 17.75,\ \sigma_X = 1.75,\ \sigma_Y = 2.5 \text{ and } r_{XY} = 0.8.$$

Regression line of X on Y is

$$x - \bar{x} = \frac{r\sigma_X}{\sigma_Y}(y - \bar{y})$$

i.e., $$x - 19.5 = \frac{0.8 \times 1.75}{2.5}(y - 17.75)$$

$\therefore$ When $y = 18$,

$$x = 19.5 + \frac{0.8 \times 1.75}{2.5} \times 0.25$$

$$= 19.64$$

Regression line of Y on X is

$$y - \bar{y} = \frac{r\sigma_Y}{\sigma_X}(x - \bar{x})$$

i.e., $$y - 17.75 = \frac{0.8 \times 2.5}{1.75}(x - 19.5)$$

∴ When $x = 17$,

$$y = 17.75 + \frac{0.8 \times 2.5}{1.75} \times (-2.5)$$

$$= 14.89$$

Example 4 *In a partially destroyed laboratory record of an analysis of correlation data, the following results only are legible: Variance of X = 1. The regression equations are 3x + 2y = 26 and 6x + y = 31. What were (i) the mean values of X and Y? (ii) the standard deviation of Y? and (iii) the correlation coefficient between X and Y?*

Solution

(i) Since the lines of regression intersect at $(\bar{x}, \bar{y})$, we have $3\bar{x} + 2\bar{y} = 26$ and $6\bar{x} + \bar{y} = 31$

Solving these equations, we get $\bar{x} = 4$ and $\bar{y} = 7$.

(ii) Which of the two equations is the regression equation of Y on X and which one is the regression equation of X on Y are not known.

Let us tentatively assume that the first equation is the regression line of X on Y and the second equation is the regression line of Y on X. Based on this assumption, the first equation can be re-written as

$$x = -\frac{2}{3}y + \frac{26}{3} \tag{1}$$

and the other as $$y = -6x + 31 \tag{2}$$

Then $b_{XY} = -\frac{2}{3}$ and $b_{YX} = -6$

∴ $$r_{XY}^2 = b_{XY} \times b_{YX} = 4$$

∴ $$r_{XY} = -2, \text{ which is absurd.}$$

Hence our tentative assumption is wrong.

∴ The first equation is the regression line of Y on X and re-written as

$$y = -\frac{3}{2}x + 13 \tag{3}$$

The second equation is the regression line of X on Y and re-written as

$$x = -\frac{1}{6}y + \frac{31}{6} \tag{4}$$

Hence the correct $b_{YX} = -\dfrac{3}{2}$ and the correct $b_{XY} = -\dfrac{1}{6}$

$$\therefore \qquad r_{XY}^2 = b_{YX} \cdot b_{XY} = \frac{1}{4}$$

$$\therefore \qquad r_{XY} = -\frac{1}{2} \qquad (\because \text{ both } b_{YX} \text{ and } b_{XY} \text{ are negative})$$

(iii) Now
$$\frac{\sigma_Y^2}{\sigma_X^2} = \frac{b_{YX}}{b_{XY}} = \frac{-\frac{3}{2}}{-\frac{1}{6}} = 9$$

$$\therefore \qquad \sigma_Y^2 = 9 \times \sigma_x^2 = 9$$

$$\therefore \qquad \sigma_Y = 3$$

Example 5 *Given that $x = 4y + 5$ and $y = kx + 4$ are the regression lines of X on Y and of Y on X respectively, show that $0 \le k \le \frac{1}{4}$. If $k = \frac{1}{16}$, find the means of X and Y and r_{XY}.*

Solution From the given equations, we note that

$$b_{YX} = k \text{ and } b_{XY} = 4$$

$$r_{XY}^2 = b_{XY} \cdot b_{YX} = 4k$$

Since $0 \le r_{XY}^2 \le 1$, we get $0 \le 4k \le 1$

$$\therefore \qquad 0 \le k \le \frac{1}{4}.$$

When
$$k = \frac{1}{16},\ r_{XY}^2 = \frac{1}{4}$$

$$\therefore \qquad r_{XY} = \pm \frac{1}{2}$$

But both b_{YX} and b_{XY} are positive.

$$\therefore \qquad r_{XY} = \frac{1}{2}$$

When $k = \dfrac{1}{16}$, the regression equations become

$$x = 4y + 5 \qquad (1)$$

and
$$y = \frac{1}{16}x + 4 \qquad (2)$$

Solving equations (1) and (2), we get

$$x = 28 \text{ and } y = 5.75$$

$$\therefore \qquad \bar{x} = 28 \text{ and } \bar{y} = 5.75$$

Example 6 *Find the angle between the two lines of regression. Deduce the condition for the two lines to be (i) at right angles and (ii) coincident.*

Solution The equations of the regression lines

are $$y - \bar{y} = r\frac{\sigma_Y}{\sigma_X}(x - \bar{x}) \quad (1)$$

and $$x - \bar{x} = r\frac{\sigma_X}{\sigma_Y}(y - \bar{y}) \quad (2)$$

Slope of line (1) $= r\dfrac{\sigma_Y}{\sigma_X} = m_1$, say.

Slope of line (2) $= \dfrac{\sigma_Y}{r\sigma_X}$, m_2, say.

If θ is the acute angle between the two lines, then $\tan\theta = \dfrac{|m_1 - m_2|}{1 + m_1 m_2}$

$$= \frac{\left| r\dfrac{\sigma_Y}{\sigma_X} - \dfrac{\sigma_Y}{r\sigma_X} \right|}{1 + \dfrac{\sigma_Y^2}{\sigma_X^2}}$$

$$= \frac{\left| r - \dfrac{1}{r} \right| \sigma_X \sigma_Y}{\sigma_X^2 + \sigma_Y^2}$$

$$= \frac{(1 - r^2)}{|r|} \frac{\sigma_X \sigma_Y}{\sigma_X^2 + \sigma_Y^2}$$

The two regression lines are at angles when $\theta = \dfrac{\pi}{2}$, i.e., $\tan\theta = \infty$

i.e., $r = 0$

∴ When the linear correlation between X and Y is zero, the two lines of regression will be at right angles.

The two regression lines are coincident, when $\theta = 0$, i.e., when $\tan\theta = 0$

i.e., when $r = \pm 1$.

∴ When the correlation between X and Y is perfect, the two regression lines will coincide.

Example 7 *For two RV's X and Y with the same mean, the two regression equations are $y = ax + b$ and $x = cy + d$. Find the common mean, ratio of the standard deviations and also show that $\dfrac{b}{d} = \dfrac{1 - a}{1 - c}$.*

Solution If μ is the common mean, the point (μ, μ) lies on $y = ax + b$ and $x = cy + d$

[$\because$ They intersect at $(\bar{x}, \bar{y})$]

$$\therefore \quad \mu = a\mu + b \quad (1)$$

$$\mu = c\mu + d \quad (2)$$

From (1), $\mu = \dfrac{b}{1-a}$

From (2), $\mu = \dfrac{d}{1-c}$

$$\therefore \quad \frac{b}{1-a} = \frac{d}{1-c}$$

$$\therefore \quad \frac{b}{d} = \frac{1-a}{1-c}$$

Now $\dfrac{\sigma_Y^2}{\sigma_X^2} = \dfrac{b_{YX}}{b_{XY}} = \dfrac{a}{c}$;

$$\therefore \quad \frac{\sigma_Y}{\sigma_X} = \sqrt{\frac{a}{c}}$$

Example 8 *Find the standard error of estimate of Y on X and of X on Y from the following data:*

X:	1	2	3	4	5
Y:	2	5	9	13	14

Solution

x	y	x^2	y^2	xy
1	2	1	4	2
2	5	4	25	10
3	9	9	81	27
4	13	16	169	52
5	14	25	196	70
15	43	55	475	161

$$r_{XY} = \frac{n\Sigma xy - \Sigma x \cdot \Sigma y}{\sqrt{\left\{n\Sigma x^2 - (\Sigma x)^2\right\}\left\{n\Sigma y^2 - (\Sigma y)^2\right\}}}$$

$$= \frac{5 \times 161 - 15 \times 43}{\sqrt{\left\{5 \times 55 - (15)^2\right\}\left\{5 \times 475 - (43)^2\right\}}}$$

$$= \frac{160}{\sqrt{50 \times 526}} = 0.9866$$

$$\sigma^2_X = \frac{1}{n}\Sigma x^2 - \left(\frac{1}{n}\Sigma x\right)^2$$

$$= \frac{1}{5} \times 55 - \left(\frac{1}{5} \times 15\right)^2 = 2$$

$$\therefore \quad \sigma_X = 1.4142$$

$$\sigma^2_Y = \frac{1}{n} \Sigma y^2 - \left(\frac{1}{n} \Sigma y\right)^2$$

$$= \frac{1}{5} \times 475 - \left(\frac{1}{5} \times 43\right)^2$$

$$= 21.04$$

$$\therefore \quad \sigma_Y = 4.5869$$

$$S_Y = \sqrt{1 - r_{XY}^2} \cdot \sigma_Y = \sqrt{1 - (0.9866)^2} \times 4.5869$$

$$= 0.7484$$

$$S_X = \sqrt{1 - r_{XY}^2} \cdot \sigma_X = \sqrt{1 - (0.9866)^2} \times 1.4142$$

$$= 0.2307$$

Exercise 4(C)

Part A (Short answer questions)

1. What do you mean by regression line? What is its use?
2. For a given data of n pairs of values of X and Y, why should there be two regression lines?
3. Write down the analytic equations of the regression lines?
4. When will the two regression lines be (i) at right angles, (ii) coincident?
5. Define regression coefficients.
6. Prove that the correlation coefficient is the geometric mean of the regression coefficients.
7. Find the co-ordinates of the point of intersection of the regression lines.
8. What do you mean by standard error of estimate?
9. Write down the formulas for the standard errors of estimate of Y and X.
10. In the usual notation prove that
    (i) $r_{XY} \cdot S_X S_Y = (1 - r^2_{XY}) C_{XY}$ and
    (ii) $b_1 S^2_X = b_2 S^2_Y$.

Part B

11. Find the equations of the regression lines from the following data. Hence calculate the coefficient of correlation between X and Y.

X:	62	64	65	69	70	71	72	74
Y:	126	125	139	145	165	152	180	208

12. Find the equations of the regression lines from the following data. Also estimate the value of Y when $X = 71$ and the value of X when $Y = 70$.

X:	65	66	67	67	68	69	70	72
Y:	67	68	65	68	72	72	69	71

13. Find the equation of the regression line of Y on X using the method of least squares from the following data. Find the value of Y corresponding to $X = 18$.

X:	5	10	15	20	25
Y:	16	19	23	26	30

14. Obtain the line of regression of X on Y using the method of least squares from the following data. Find the value of X when $Y = 45$.

X:	4.7	8.2	12.4	15.8	20.7	24.9	31.9	35.0	39.1	38.8
Y:	4.0	8.0	12.5	16.0	20.0	25.0	31.0	36.0	40.0	40.0

15. Find the most likely price in Mumbai corresponding to the price of Rs. 70 at Chennai and that in Chennai corresponding to the price of Rs. 75 at Mumbai from the following:

	Chennai	Mumbai
Mean	65	67
S.D.	2.5	3.5

Coefficient of correlation between the prices in the two cities is 0.8.

16. In a partially destroyed laboratory record of an analysis of correlation data, the following results only are legible.
Variance of $X = 9$. Regression equations are $8x - 10y + 66 = 0$ and $40x - 18y = 214$. What were (i) the mean values of X and Y?
(ii) the correlation coefficient between X and Y and (iii) the standard deviation of Y?
17. The equations of two regression lines got in a correlation analysis are $3x + 12y = 19$ and $3y + 9x = 46$. Obtain (i) the correlation coefficient between X and Y, (ii) the mean values of X and Y and (iii) the ratio of the coefficient of variation of X to that of Y.
18. The equations of lines of regression are given by $x + 2y - 5 = 0$ and $2x + 3y - 8 = 0$ and variance of X is 12. Compute the values of $\bar{x}$, $\bar{y}$, $\sigma^2{}_Y$ and r_{XY}.
19. The regression lines of Y on X and of X on Y are respectively $y = a + bx$ and $x = c + dy$. Find the values of $\bar{x}$, $\bar{y}$ and r_{XY}. Can you find S_X snd S_Y from them ?
20. If the lines of regression of Y on X and X on Y are respectively $a_1x + b_1y + c_1 = 0$ and $a_2x + b_2y + c_2 = 0$, prove that $a_1b_2 \leq a_2b_1$. Find also the coefficient of correlation between X and Y and the ratio of the coefficient of variability of Y to that of X.

Characteristic Function

Although higher order moments of a RV X may be obtained directly by using the definition of $E(X^n)$, it will be easier in many problems to compute them through the characteristic function or equivalently through the moment generating function of the RV X. While the characteristic function always exists, the moment generating function need not.

Moment Generating Function (MGF) of a RV X (discrete or continuous) is defind as $E(e^{tX})$, where t is a real variable and denoted as $M(t)$.

If X is discrete, then $M(t) = \sum_r e^{tx_r} p_r$,

where X takes the values $x_1, x_2, x_3, \ldots$, with probabilities $p_1, p_2, p_3, \ldots$

If X is a continuous RV with density function $f(x)$, then

$$M(t) = \int_{-\infty}^{\infty} e^{tx} f(x)\, dx$$

Properties of MGF

(Proofs of the properties are omitted, as the proofs of the corresponding properties of characteristic function will be given later.)

1. $M(t) = \sum_{n=0}^{\infty} t^n E(X^n)/\underline{|n}$

   i.e., $E(X^n) = \mu'_n$ is the co-efficient of $\frac{t^n}{\underline{|n}}$ in the expansion of $M(t)$ in series of powers of t.

2. $\mu'_n = E(X^n) = \left[\frac{d^n}{dt^n} M(t)\right]_{t=0}$.

3. If the MGF of X is $M_X(t)$ and if $Y = aX + b$, then $M_Y(t) = e^{bt} M_X(at)$.

4. If X and Y are independent RVs and $Z = X + Y$, then $M_Z(t) = M_X(t) M_Y(t)$.

Characteristic function of a RV X (discrete or continuous) is defined as $E(e^{i\omega X})$ and denoted as $\phi(\omega)$.

If X is a discrete RV that can take the values $x_1, x_2, \ldots$, such that $P(X = x_r) = p_r$, then

$$\phi(\omega) = \sum_r e^{i\omega x_r} p_r$$

If X is a continuous RV with density function $f(x)$, then

$$\phi(\omega) = \int_{-\infty}^{\infty} e^{i\omega x} f(x)\, dx$$

Properties of Characteristic Function

1. $\mu'_n = E(X^n) =$ the coefficient of $\frac{i^n \omega^n}{n!}$ in the expansion of $\phi(\omega)$ in series of ascending powers of $i\omega$.

Proof $\phi(\omega) = E(e^{i\omega X})$

$$= E\left(1 + \frac{i\omega X}{\lfloor 1} + \frac{i^2\,\omega^2\,X^2}{\lfloor 2} + \cdots + \frac{i^n\,\omega^n\,X^n}{\lfloor n} \cdots\right)$$

$$= 1 + E(X) \times \frac{i\omega}{\lfloor 1} + E(X^2) \times \frac{i^2\,\omega^2}{\lfloor 2} + \cdots$$

$$= 1 + \mu'_1 \frac{i\omega}{\lfloor 1} + \mu'_2 \frac{i^2\,\omega^2}{\lfloor 2} + \cdots$$

i.e., $$\phi(\omega) = \sum_{n=0}^{\infty} \mu'_n \frac{i^n\,\omega^n}{\lfloor n} \tag{1}$$

Hence the result.

2. $\mu'_n = \dfrac{1}{i^n}\left[\dfrac{d^n}{d\omega^n}\phi(\omega)\right]_{\omega=0}$

Proof Differentiating both sides of (1) with respect to ω, n times and then putting $\omega = 0$,

$$[\phi^{(n)}(\omega)]_{\omega=0} = \mu'_n\, i^n$$

$$\therefore \qquad \mu'_n = \frac{1}{i^n}\,[\phi^{(n)}(\omega)]_{\omega=0}$$

3. If the characteristic function of a RV X is $\phi_X(\omega)$ and if $Y = aX + b$, then

$$\phi_Y(\omega) = e^{ib\omega}\,\phi_X(a\omega)$$ (BDU — Apr. 96)

Proof $\phi_Y(\omega) = E\{e^{i\omega Y}\}$

$$= E\{e^{i\omega\,(aX + b)}\}$$
$$= e^{ib\omega}\,E\{e^{i(a\omega)\,X}\}$$
$$= e^{ib\omega}\,\phi_X\,(a\omega)$$

4. If X and Y are independent RVs, then

$$\phi_{X+Y}(\omega) = \phi_X(\omega) \times \phi_Y(\omega)$$ (BDU — Apr. 96)

Proof $\phi_{X+Y}(\omega) = E\{e^{i\omega\,(X+Y)}\}$

$$= E\{e^{i\omega X} \times e^{i\omega Y}\}$$
$$= E(e^{i\omega X}) \times E(e^{i\omega Y})$$

[since X and Y are independent)

$$= \phi_X(\omega) \times \phi_Y(\omega)$$

5. If the characteristic function of a continuous RV X with density function

$f(x)$ is $\phi(\omega)$, then $f(x) = \dfrac{1}{2\pi}\displaystyle\int_{-\infty}^{\infty} \phi(\omega)\, e^{-ix\omega}\, d\omega$.

$$\phi(\omega) = \int_{-\infty}^{\infty} f(x)\, e^{i\omega x}\, dx$$

RS can be identified as the Fourier transform of $f(x)$.
Therefore, by Fourier inversion formula

$$f(x) = \frac{1}{2\pi}\int_{-\infty}^{\infty} \phi(\omega)\, e^{-i\omega x}\, d\omega$$

6. If the density function of X is known, the density function of $Y = g(X)$ can be found from the CF of Y, provided $Y = g(X)$ is one-to-one.

Proof Let the density function of X be $f_X(x)$.

Then $$\phi_{g(X)}(\omega) = \int_{-\infty}^{\infty} e^{ig(x)\omega} f_X(x)\, dx$$

Put $$g(X) = Y \text{ and hence } g(x) = y$$

Then $$\phi_Y(\omega) = \int_{-\infty}^{\infty} e^{iy\omega}\, h(y)\, dy, \text{ say}$$

Therefore, $h(y)$ is the density function of Y.

Cumulant Generating Function (CGF)

If $M(t)$ is the MGF of a RV X, then $\log_e M(t)$ is called the cumulant generating function of X and denoted by $K(t)$.

The coefficient of $\dfrac{t^r}{r!}$ in the expansion of $K(t)$ in ascending powers of t is called the *rth order cumulant of* X and denoted by λ_r.

i.e., $$K(t) = \sum_{r=1}^{\infty} \frac{\lambda_r t^r}{\underline{|r}}$$

Also $$\lambda_r = \left\{\frac{d^r}{dt^r} K(t)\right\}_{t=0}$$

If $\phi(\omega)$ is the characteristic function of a RV X, then $\log_e \phi(\omega)$ is called the *second characteristic function of* X and denoted by $\psi(\omega)$.

The coefficient of $\frac{i^r \omega^r}{\lfloor r}$ in the expansion of $\psi(\omega)$ in ascending powers of ω is the rth order cumulant of X and denoted by λ_r.

Thus $$\psi(\omega) = \sum_{r=1}^{\infty} \lambda_r i^r \omega^r/r!$$

Also $$\lambda_r = \frac{1}{i^r}\left\{\frac{d^r}{d\omega^r}\psi(\omega)\right\}_{\omega=0}$$

Joint Characteristic Function

If (X, Y) is a two-dimensional RV, then $E\left(e^{i\omega_1 X + i\omega_2 Y}\right)$ is called the joint characteristic function of (X, Y) and denoted by $\phi_{XY}(\omega_1, \omega_2)$.

It is easily verified that

(i) $\phi_{XY}(0, 0) = 1$

(ii) $E\{X^m Y^n\} = \frac{1}{i^{m+n}}\left[\frac{\partial^{m+n}}{\partial\omega_1^m \partial\omega_2^n}\phi_{XY}(\omega_1, \omega_2)\right]_{\omega_1=0, \omega_2=0}$

(iii) $\phi_X(\omega) = \phi_{XY}(\omega, 0)$ and $\phi_Y(\omega) = \phi_{XY}(0, \omega)$

(iv) If X and Y are independent

$\phi_{XY}(\omega_1, \omega_2) = \phi_X(\omega_1) \times \phi_Y(\omega_2)$ and conversely.

Worked Examples 4(D)

Example 1 *If X represents the outcome, when a fair die is tossed, find the MGF of X and hence find E(X) and Var(X).*

Solution The probability distribution of X is given by

$$p_i = p(X = i) = \frac{1}{6},\ i = 1, 2, \ldots, 6$$

$$M(t) = \sum_i e^{tx_i} p_i = \sum_{i=1}^{6} e^{ti} p_i$$

$$= \frac{1}{6}(e^t + e^{2t} + e^{3t} + e^{4t} + e^{5t} + e^{6t})$$

$$E(X) = [M'(t)]_{t=0} = \frac{7}{2}$$

$$E(X^2) = [M''(t)]_{t=0}$$

$$= \frac{1}{6}[e^t + 4e^{2t} + 9e^{3t} + 16e^{4t} + 25e^{5t} + 36e^{6t}]_{t=0} = \frac{91}{6}$$

$$\therefore \qquad \text{Var}(X) = \frac{91}{6} - \frac{49}{4} = \frac{35}{12}$$

Example 2 *If a RV X has the MGF* $M(t) = \dfrac{3}{3-t}$, *obtain the standard deviation of X.*

Solution $$M(t) = \frac{3}{3\left(1 - \frac{t}{3}\right)} = 1 + t/3 + t^2/9 + \cdots + \infty \qquad (1)$$

$$E(X) = \text{coefficient of } \frac{t}{\lfloor 1} \text{ in (1)} = \frac{1}{3}$$

$$E(X^2) = \text{coefficient of } \frac{t^2}{\lfloor 2} \text{ in (1)} = \frac{2}{9}$$

$$\text{Var}(X) = E(X^2) - \{E(X)\}^2 = \frac{1}{9}$$

$$\therefore \qquad \sigma_X = \frac{1}{3}$$

Example 3 *Find the MGF of the binomial distribution and hence find its mean and variance.*

Solution Binomial distribution is given by

$$p_r = p(X = r) = nC_r p^r q^{n-r}, \; r = 0, 1, 2, \ldots, n$$

$$\therefore \qquad M(t) = \sum_{r=0}^{n} e^{tr} p_r$$

$$= \sum_{r=0}^{n} e^{tr} nC_r p^r q^{n-r}$$

$$= \sum_{r=0}^{n} nC_r (p\, e^t)^r q^{n-r}$$

$$= (p\, e^t + q)^n$$

$$M'(t) = n(p\, e^t + q)^{n-1} \times p\, e^t$$

$$M''(t) = np[(p\, e^t + q)^{n-1} \times e^t + (n-1)(p\, e^t + q)^{n-2} p\, e^{2t}]$$

$$E(X) = M'(0) = np \qquad (\text{since } p + q = 1)$$

$$E(X^2) = M''(0) = np\,[1 + (n-1)p]$$

$$\text{Var}(X) = E(X^2) - \{E(X)\}^2$$

$$= np - np^2$$

$$= npq$$

Example 4 *Find the characteristic function of the Poisson distribution and hence find the values of the first four central moments.*

Solution Poisson distribution is given by

$$p_r = P(X = r) = e^{-\lambda}\lambda^r/\underline{|r} \quad r = 0, 1, 2, \cdots, \infty$$

$$\therefore \qquad \phi(\omega) = \sum_{r=0}^{\infty} e^{i\omega r} e^{-\lambda}\lambda^r/\underline{|r}$$

$$= \sum_{r=0}^{\infty} e^{-\lambda}(\lambda e^{i\omega})^r/\underline{|r}$$

$$= e^{-\lambda}\, e^{\lambda e^{i\omega}} = e^{-\lambda(1-e^{i\omega})}$$

$$\phi^{(1)}(\omega) = e^{-\lambda}\, e^{\lambda e^{i\omega}}\, i\lambda e^{i\omega}$$

$$\phi^{(2)}(\omega) = i\lambda\, e^{-\lambda}\, \{e^{\lambda e^{i\omega}} e^{i\omega} i + e^{\lambda e^{i\omega}}\, i\lambda\, e^{i2\omega}\}$$

$$= i^2\lambda\, e^{-\lambda}\, \{e^{i\omega} + \lambda\, e^{i2\omega}\}\, e^{\lambda e^{i\omega}}$$

$$\phi^{(3)}(\omega) = i^2\, \lambda\, e^{-\lambda}$$

$$[\{i\, e^{i\omega} + 2i\lambda\, e^{i2\omega}\}\, e^{\lambda e^{i\omega}} + \{e^{i\omega} + \lambda\, e^{i2\omega}\}\, e^{\lambda e^{i\omega}}\, i\lambda\, e^{i\omega}]$$

$$= i^3\lambda\, e^{-\lambda}\, e^{\lambda e^{i\omega}}\, \{e^{i\omega} + 3\lambda\, e^{i2\omega} + \lambda^2\, e^{i3\omega}\}$$

$$\phi^{(4)}(\omega) = i^3\, \lambda\, e^{-\lambda}\, [e^{\lambda e^{i\omega}}\, i\lambda\, e^{i\omega}\, \{e^{i\omega} + 3\lambda\, e^{i2\omega} + \lambda^2\, e^{i3\omega}\}$$

$$+ e^{\lambda e^{i\omega}}\, \{ie^{i\omega} + 6i\lambda\, e^{i2\omega} + 3\, i\lambda^2\, e^{i3\omega}\}$$

$$= i^4\, \lambda\, e^{-\lambda}\, e^{\lambda e^{i\omega}}\, \{e^{i\omega} + 7\lambda\, e^{i2\omega} + 6\lambda^2\, e^{i3\omega} + \lambda^3\, e^{i4\omega}\}$$

$$\therefore \qquad E(X) = \frac{1}{i}\, \phi^{(1)}(0) = \lambda$$

$$E(X^2) = \frac{1}{i^2}\, \phi^{(2)}(0) = \lambda\,(1 + \lambda)$$

$$E(X^3) = \frac{1}{i^3}\, \phi^{(3)}(0) = \lambda\,(1 + 3\lambda + \lambda^2)$$

$$E(X^4) = \frac{1}{i^4}\, \phi^{(4)}(0) = \lambda\,(1 + 7\lambda + 6\lambda^2 + \lambda^3)$$

The central moments are given by

$$\mu_k = E\{X - \mu_X)^k\}$$

$$\therefore \qquad \mu_1 = E\{(X - \lambda)\} = 0$$

$$\mu_2 = E\{(X - \lambda)^2\} = \lambda$$

$$\mu_3 = E\{(X - \lambda)^3\} = \lambda$$

$$\mu_4 = E\{(X - \lambda)^4\} = 3\lambda^2 + \lambda$$

Example 5 *Find the characteristic function of the geometric distribution given by* $P(X = r) = q^r\, p,\ r = 0, 1, 2, \ldots, \infty,\ p + q = 1$.

Solution Hence find the mean and variance.

$$\phi(\omega) = \sum_{r=0}^{\infty} e^{i\omega r} pq^r$$

$$= p \sum_{r=0}^{\infty} (qe^{i\omega})^r = p(1 - qe^{i\omega})^{-1}$$

$$\phi^{(1)}(\omega) = p(1 - qe^{i\omega})^{-2}\, i\, q\, e^{i\omega}$$

$$\phi^{(2)}(\omega) = i^2 pq\, [2(1 - q\, e^{i\omega})^{-3}\, qe^{i2\omega} + (1 - qe^{i\omega})^{-2}\, e^{i\omega}]$$

$\therefore$
$$E(X) = \frac{1}{i}\, \phi^1(0) = \frac{q}{p} \text{ and } E(X^2) = \frac{1}{i^2}\, \phi^{(2)}(0) = \frac{q}{p^2}\, (1 + q)$$

$\therefore$
$$\mu_X = \frac{q}{p} \text{ and } \sigma^2{}_X = \frac{q}{p^2}$$

Example 6 *Obtain the characteristic function of the normal distribution. Deduce the first four central moments.* (MKU — Apr. 96)

Solution Let X follow $N(\mu, \sigma)$

Then $Z = \dfrac{X - \mu}{\sigma}$ follows $N(0, 1)$, i.e., the standard normal distribution whose

density function is $\phi(z) = \dfrac{1}{\sqrt{2\pi}}\, e^{-z^2/2}$.

Now
$$\phi_Z(\omega) = \frac{1}{\sqrt{2\pi}} \int_{-\infty}^{\infty} e^{-z^2/2}\, e^{i\omega z}\, dz$$

$$= \frac{1}{\sqrt{2\pi}} \int_{-\infty}^{\infty} e^{-\frac{1}{2}(z - i\omega)^2 + \frac{i^2\omega^2}{2}}\, dz$$

$$= e^{-\omega^2/2}\, \frac{1}{\sqrt{2\pi}} \int_{-\infty}^{\infty} e^{-t^2} \cdot \sqrt{2}\, dt, \text{ on putting } \frac{z - i\omega}{\sqrt{2}} = t$$

$$= e^{-\omega^2/2} \left[\text{since} \int_{-\infty}^{\infty} e^{-t^2} dt = \Gamma\left(\frac{1}{2}\right) = \sqrt{\pi}\right]$$

$\therefore$
$$\phi_X(\omega) = \phi_{\sigma Z + \mu}(\omega)$$

$$= e^{i\mu\omega}\, \phi_Z(\sigma\omega) \quad \text{(by Property 3)}$$

$$= e^{i\mu\omega}\, e^{-\sigma^2\omega^2/2}$$

Now
$$\phi_X(\omega) = e^{i\omega(\mu + i\sigma^2\omega/2)}$$

$$= 1 + \frac{1}{\lfloor 1}\, i\omega\, (\mu + i\, \sigma^2\, \omega/2) + \frac{1}{\lfloor 2}\, i^2\, \omega^2\, (\mu + i\, \sigma^2\, \omega/2)^2 + \ldots$$

$$= 1 + \frac{i\omega\mu}{\lfloor 1} + \frac{i^2\,\omega^2}{\lfloor 2}(\sigma^2 + \mu^2) + \frac{i^3\,\omega^3}{\lfloor 3}\,(3\,\mu\,\sigma^2 + \mu^3)$$

$$+ \frac{i^4\,\omega^4}{\lfloor 4}\,(3\,\sigma^4 + 6\,\mu^2\,\sigma^2 + \mu^4) + \dots \qquad (1)$$

$$\therefore \qquad \mu'_1 = E(X) = \text{coefficient of } \frac{i\,\omega}{\lfloor 1} \text{ in (1)} = \mu$$

Similarly, $E(X^2) = \sigma^2 + \mu^2;$

$E(X^3) = 3\mu\sigma^2 + \mu^3$ and,

$E(X^4) = 3\sigma^4 + 6\mu^2\sigma^2 + \mu^4.$

Using the relation $\mu_k = E\{(X - \mu)^k\}$, we get

$$\mu_1 = 0,\ \mu_2 = \sigma^2,\ \mu_3 = 0 \text{ and } \mu_4 = 3\sigma^4$$

Example 7 *Find the characteristic function of the Erlang distribution given by* $f(x) = \dfrac{\lambda^n}{\lfloor n-1}\, x^{n-1}\, e^{-\lambda x}\, U(x)$ *and hence find its mean and variance.*

Solution $\phi(\omega) = \dfrac{\lambda^n}{\lfloor n-1} \displaystyle\int_0^\infty x^{n-1}\, e^{-(\lambda - i\omega)\,x}\, dx$

$$= \frac{\lambda^n}{\lfloor n-1 (\lambda - i\omega)^n} \int_0^\infty t^{n-1}\, e^{-t}\, dt, \text{ on putting } (\lambda - i\omega)\,x = t$$

$$= \left(\frac{\lambda}{\lambda - i\omega}\right)^n \frac{1}{\lfloor n-1} \Gamma(n) = \frac{1}{\left(1 - \dfrac{i\omega}{\lambda}\right)^n}$$

[since $\Gamma n = \lfloor n-1$ when n is a positive integer]

Now $\qquad \phi(\omega) = \left(1 - \dfrac{i\omega}{\lambda}\right)^{-n}$

$$= 1 + \frac{n}{\lfloor 1}\,\frac{i\omega}{\lambda} + \frac{n(n+1)}{\lfloor 2}\,\frac{i^2\,\omega^2}{\lambda^2} + \dots + \infty$$

$$\therefore \qquad E(X) = \frac{n}{\lambda} \text{ and } E(X^2) = \frac{n(n+1)}{\lambda^2}$$

$$\therefore \qquad \text{Var}(X) = \frac{n}{\lambda^2}$$

Note *If n = 1, Erlang's distribution becomes the exponential distribution. Hence the mean and variance of an exponential distribution with parameter λ are $1/\lambda$ and $1/\lambda^2$ respectively.*

Example 8 *Find the characteristic function of the Laplace distribution with pdf* $f(x) = \frac{\alpha}{2} e^{-\alpha |x|}$, $-\infty < x < \infty$. *Hence find its mean and variance*

Solution $\phi(\omega) = \frac{\alpha}{2} \int_{-\infty}^{\infty} e^{-\alpha |x|} e^{i\omega x} \, dx$

$$= \frac{\alpha}{2} \left[\int_{-\infty}^{0} e^{(\alpha + i\omega)x} dx + \int_{0}^{\infty} e^{-(\alpha - i\omega)x} \, dx \right]$$

$$= \frac{\alpha}{2} \left\{ \frac{1}{\alpha + i\omega} + \frac{1}{\alpha - i\omega} \right\} = \frac{\alpha^2}{\alpha^2 + \omega^2}$$

Now $\quad \phi(\omega) = \left(1 + \frac{\omega^2}{\alpha^2} \right)^{-1} = 1 - \frac{\omega^2}{\alpha^2} + \frac{\omega^4}{\alpha^4} + \cdots + \infty$

$$E(X) = 0 \text{ and } E(X^2) = \text{Var}(X) = \frac{2}{\alpha^2}$$

Example 9 *If X_1 and X_2 are two independent RVs that follow Poisson distribution with parameters λ_1 and λ_2, prove that $(X_1 + X_2)$ also follows a Poisson distribution with parameter $(\lambda_1 + \lambda_2)$.* (MKU — Apr. 97)

Solution (This property is called the reproductivity property of the Poisson distribution)

$$\phi_{X1}(t) = e^{\lambda_1 (e^{i\omega} - 1)}$$

$$\phi_{X2}(t) = e^{\lambda_2 (e^{i\omega} - 1)}$$

since X_1 and X_2 are independent RVs,

$\phi_{X1 + X2}(t) = e^{(\lambda_1 + \lambda_2)(e^{i\omega} - 1)}$, which is the characteristic function of a Poisson distribution with parameter $(\lambda_1 + \lambda_2)$. Hence the result.

Example 10 *Show that the distribution for which the characteristic function is $e^{-|\omega|}$ has the density function* $f(x) = \frac{1}{\pi} \times \frac{1}{1 + x^2}$ $-\infty < x < \infty$.

(BU – Apr. 96)

Solution By inversion Property 5,

$$f(x) = \frac{1}{2\pi} \int_{-\infty}^{\infty} \phi(\omega) \, e^{-ix\omega} \, d\omega$$

$$= \frac{1}{2\pi} \int_{-\infty}^{\infty} e^{-|\omega|} (\cos x\omega - i \sin x\omega)\, d\omega$$

$$= \frac{1}{\pi} \int_{0}^{\infty} e^{-\omega} \cos x\omega \, d\omega$$

(by properties of odd and even function)

$$= \frac{1}{\pi} \left[\frac{e^{-\omega}}{1 + x^2} (-\cos x\omega + x \sin x\omega) \right]_0^{\infty}$$

$$= \frac{1}{\pi} \times \frac{1}{1+x^2}, -\infty < x < \infty$$

Example 11 *Find the density function f(x) corresponding to the characteristic function defined as*

$$\phi(t) = \begin{cases} 1 - |t| & \text{for } |t| \le 1 \\ 0 & \text{for } |t| > 1 \end{cases}$$

Note *The letter t is used in the place of ω.*

Solution By inversion Property 5,

$$f(x) = \frac{1}{2\pi} \int_{-\infty}^{\infty} \phi(t)\, e^{-ixt}\, dt$$

$$= \frac{1}{2\pi} \int_{-1}^{1} \{1 - |t|\}\, e^{-ixt}\, dt$$

$$= \frac{1}{\pi} \int_{0}^{1} (1 - t) \cos xt \, dt$$

$$= \frac{1}{\pi} \left\{ (1-t) \frac{\sin xt}{x} - \frac{\cos xt}{x^2} \right\}_0^1$$

$$= \frac{1}{\pi} \frac{(1 - \cos x)}{x^2}$$

Example 12 *Express the first four cumulants in terms of central moments.*

Solution By definition, $K(t) = \log \{M(t)\}$

$$\therefore \quad \sum_{r=1}^{\infty} \frac{\lambda_r t^r}{\lfloor r} = \log \left\{ \sum_{r=0}^{\infty} \frac{\mu'_r t^r}{\lfloor r} \right\}$$

$$= \log \left\{ 1 + \mu_1' \, t + \frac{\mu_2'}{\lfloor 2} t^2 + \frac{\mu_3'}{\lfloor 3} t^3 + \cdots \right\}$$

$$= t \left(\mu_1' + \frac{\mu_2'}{2} t + \frac{\mu_3' \, t^2}{6} + \frac{\mu_4' \, t^3}{24} + \cdots \right)$$

$$- \frac{t^2}{2} \left(\mu_1' + \frac{\mu_2'}{2} t + \frac{\mu_3' \, t^2}{6} + \frac{\mu_4' \, t^3}{24} + \cdots \right)^2 + \ldots$$

Comparing like coefficients, we get,

$$\lambda_1 = \mu'_1; \; \lambda_2 = \mu'_2 - \mu'^2_1 = \mu_2$$

$$\lambda_3 = \mu'_3 - 3\mu'_2\mu'_1 + 2\mu'^3_1 = \mu_3$$

$$\lambda_4 = \mu'_4 - 4\mu'_3\mu'_1 - 3\mu'^2_2 + 12\mu'_2 \, \mu'^2_1 - 6 \, \mu'^4_1$$

$$= \mu_4 - 3\mu^2_2$$

Example 13 *If X and Y are two independent RVs, prove that the cumulant of (X + Y) of any order is the sum of the cumulants of X and Y of the same order.*

Solution By Property 4 of characteristic functions, $\phi_{X+Y}(\omega) = \phi_X(\omega) \times \phi_Y(\omega)$, when X and Y are independent.

$\therefore \quad \log\phi_{X+Y}(\omega) = \log\phi_X(\omega) + \log\phi_Y(\omega)$

i.e., $\quad \psi_{X+Y}(\omega) = \psi_X(\omega) + \psi_Y(\omega)$

i.e., $$\sum_{r=1}^{\infty} \lambda_r (X+Y) \frac{i^r \omega^r}{\lfloor r} = \sum_{r=1}^{\infty} \lambda_r (X) \frac{i^r \omega^r}{\lfloor r} + \sum_{r=1}^{\infty} \lambda_r (Y) \frac{i^r \omega^r}{\lfloor r}$$

$\therefore \quad \lambda_r(X + Y) = \lambda_r(X) + \lambda_r(Y)$

Example 14 *If the RV X follows N(0, σ), find the density function of $Y = aX^2$, using the characteristic function technique.*

Solution $$\phi_{aX^2}(\omega) = \int_{-\infty}^{\infty} e^{iax^2\omega} \times \frac{1}{\sigma\sqrt{2\pi}} e^{-x^2/2\sigma^2} \, dx \qquad (1)$$

Put $\quad y = ax^2$ in (1)

$$\therefore \quad dx = \frac{dy}{2ax} = \frac{dy}{2\sqrt{ay}}$$

$$\therefore \quad \phi_Y(\omega) = 2 \int_0^{\infty} e^{i\omega y} \times \frac{1}{2\sigma\sqrt{2\pi \, ay}} \times e^{-y/2a\sigma^2} \, dy \qquad (2)$$

But $$\phi_Y(\omega) = \int_{-\infty}^{\infty} e^{i\omega y} f_Y(y) \, dy \qquad (3)$$

Comparing (2) and (3), we get

$$f_Y(y) = \frac{1}{\sigma\sqrt{2\pi a y}}\, e^{-y/2a\sigma^2}\, U(y)$$

Example 15 *Two RVs X and Y have the joint characteristic function $\phi_{XY}(\omega_1, \omega_2) = \exp(-2\,\omega_1^2 - 8\,\omega_2^2)$. Show that X and Y are both zero mean RVs and also that they are uncorrelated.* (BDU — Apr. 97)

Solution By the property of joint CF

$$E(X) = \frac{1}{i}\left[\frac{\partial}{\partial\omega_1}\left(e^{-2\omega_1^2 - 8\omega_2^2}\right)\right]_{\omega_1=0,\,\omega_2=0}$$

$$= \left[e^{-8\omega_2^2}\, e^{-2\omega_1^2}\, 4i\omega_1\right]_{\omega_1=0,\,\omega_2=0}$$

$$= 0$$

$$E(Y) = \left[e^{-2\omega_1^2}\, e^{-8\omega_2^2}\, 16i\omega_2\right]_{\omega_1=0,\,\omega_2=0}$$

$$= 0$$

$$E(XY) = \frac{1}{i^2}\left[\frac{\partial^2}{\partial\omega_1\,\partial\omega_2}\left(e^{-2\omega_1^2 - 8\omega_2^2}\right)\right]_{\omega_1=0,\,\omega_2=0}$$

$$= \frac{\partial}{\partial\omega_1}\left\{e^{-2\omega_1^2}\, e^{-8\omega_2^2}\, 16\omega_2\right\}_{\omega_1=0,\,\omega_2=0}$$

$$= \left\{-64\,\omega_1\,\omega_2\, e^{-2\omega_1^2+8\omega_2^2}\right\}_{\omega_1=0,\,\omega_2=0}$$

$$= 0$$

$$C_{XY} = E(XY) - E(X)\times E(Y) = 0$$

Therefore, X and Y are uncorrected RVs.

Exercise 4(D)

Part A (Short answer questions)

1. Define the MGF of a RV X. Why is it called so?
2. State the properties of the MGF of a RV
3. Derive the relation between the MGFs of X and Y when $Y = aX + b$.
4. If X and Y are independent RVs and $Z = X + Y$, prove that $M_Z(t) = M_X(t) \times M_Y(t)$.
5. Define the characteristic function of a RV. How does it differ from the MGF?
6. State the properties of the characteristic function of a RV.

7. State the uniqueness theorem of characteristic functions.
8. If $Y = aX + b$, find the relation between the characteristic functions of X and Y.
9. If X and Y are 2 independent RVs prove that $\phi_{X+Y}(\omega) = \phi_X \times \phi_Y(\omega)$.
10. If the characteristic function of a continuous RV X is $\phi(\omega)$, express its density function $f(x)$ in terms of $\phi(\omega)$.
11. Define the cumulant generating function/the second characteristic function of a RV X and what is its use?
12. If the rth moment of a continuous RV X about the origin is $r!$, find the MGF of X.
13. If the MGF of a RV X is $\dfrac{2}{2-t}$ find the SD of X.
14. Find the MGF/CF of a uniform distribution in (a, b).
15. Find the MGF/CF of the binomial distribution.
16. Find the MGF/CF of the Poisson distribution.
17. State and prove the reproductive property of the Poisson distribution.
18. Find the CF/MGF of the geometric distribution.
19. Find the CF/MGF of the exponential distribution.
20. If the CF of the standard normal distribution $N(0, 1)$ is $e^{-\omega^2/2}$, find the CF of the general normal distribution $N(\mu, \sigma)$.
21. Find the CF of X whose pdf is given by $f(x) = \dfrac{1}{2} e^{-|x|}$, $-\infty < x < \infty$.

Part B

22. Find the MGF of a RV which is uniformly distributed over $(-1, 2)$ and hence find its mean and variance.
23. Find the characteristic function of the binomial distribution and hence find its mean and variance.
24. Obtain the MGF of the Poisson distribution; deduce the values of the first four central moments. (MKU — Apr. 96)
25. Find the characteristic function of the negative binomial distribution given by $P(X = r) = (n + r - 1)\, C_r\, q^r p^n$, $(r = 0, 1, 2, \ldots, \infty)$, $p + q = 1$, and hence find its mean and variance.
26. Find the MGF of the two-parameter exponential distribution whose density function is given by $f(x) = \lambda\, e^{-\lambda(x-a)}$, $x \geq a$. Hence find its mean and variance.
27. If the desnity function of a continuous RV X is given by $f(x) = \dfrac{1}{2} e^{-|x|}$, $-\infty < x < \infty$, find the MGF of X. Hence find its mean and variance.
28. Find the characteristic function of the Cauchy's distribution given by $f(x) = \dfrac{1}{\pi} \times \dfrac{1}{1+x^2}$ $-\infty < x < \infty$. Comment about the first two moments.

(**Hint**: Use contour integration.)

29. Find the characteristic function for the following probability density function:

$$f_X(x) = \frac{\lambda}{\pi\left(\lambda^2 + x^2\right)} \qquad \text{(BDU—Nov. 96)}$$

30. If X follows $N(\mu_X, \sigma_X)$ and Y follows $N(\mu_Y, \sigma_Y)$, prove, by using characteristic functions, that $(aX + bY)$ follows a normal distribution with mean $(a\mu_X + b\mu_Y)$ and variance $(a^2\,\sigma^2_X + b^2\,\sigma^2_Y)$.
31. Find the density function of the distribution for which the characteristic function is given by $\phi(t) = e^{-\sigma^2 t^2/2}$. [MU—Apr. 96]
32. If the raw moments of a continuous RV X are given by $E(X^n) = \lfloor n$, find the characteristic function of X and also the density function of X.
(**Hint**: Use contour integration to find pdf.)
33. Express the first 4 raw moments about the origin in terms of the cumulants.
34. Prove that the cumulants of all orders are equal for the Poisson distribution.
35. If X and Y are two jointly normal RVs whose joint pdf is $N(\mu_X, \mu_Y; \sigma_X, \sigma_Y; r)$, find the joint characteristic function of (X, Y).
36. If the random variable X is uniformly distributed over $(-\pi/2, \pi/2)$, find the pdf of $Y = \sin X$, using characteristic function technique.

Bounds on Probabilities

If we know the probability distribution of a random variable X (i.e., the pdf in the continuous case or the pmf in the discrete case), we may compute $E(X)$ and Var(X). Conversely, if $E(X)$ and Var(X) are known, we cannot construct the probability distribution of X and hence compute quantities such as $P\{|X - E(X)| \le k\}$. Although we cannot evaluate such probabilities from a knowledge of $E(X)$ and Var(X), several approximation techniques have been developed to yield upper and/or lower bounds to such probabilities. The most important of such techniques is Tchebycheff inequality.

Tchebycheff Inequality

If X is a RV with $E(X) = \mu$ and $\text{Var}(X) = \sigma^2$, then $P\{|X - \mu| \ge c\} \le \dfrac{\sigma^2}{c^2}$

where $c > 0$.

Proof Let X be a continuous RV with pdf $f(x)$.

Then
$$\sigma^2 = \int_{-\infty}^{\infty} (x-\mu)^2 f(x)\,dx$$

$$\ge \int_{-\infty}^{\mu - c} (x-\mu)^2 f(x)\,dx + \int_{\mu + c}^{\infty} (x-\mu)^2 f(x)\,dx$$

In the first integral, $x \le \mu - c$

$\therefore \qquad (x - \mu)^2 \ge c^2$

In the second integral, $x \ge \mu + c$

$\therefore \qquad (x - \mu)^2 \ge c^2$

$$\therefore \qquad \sigma^2 \ge c^2 \left\{ \int_{-\infty}^{\mu - c} f(x)\,dx + \int_{\mu + c}^{\infty} f(x)\,dx \right\} \qquad (1)$$

$$\begin{aligned} \text{RS of (1)} \quad &= c^2[1 - P\{\mu - c \le X \le \mu + c\}] \\ &= c^2[1 - P\{-c \le X - \mu \le c\}] \\ &= c^2[1 - P\{|X - \mu| \le c\}] \\ &= c^2 P\{|X - \mu| \ge c\} \qquad (2) \end{aligned}$$

Using (2) in (1), we get

$$P\{|X - \mu| \ge c\} \le \frac{\sigma^2}{c^2} \quad \text{or} \quad P\{|X - \mu| < c\} \ge 1 - \frac{\sigma^2}{c^2},\ c > 0$$

Alternative forms: If we put $c = k\sigma$, where $k > 0$, then Tchebycheff inequality takes the form

$$P\left\{\left|\frac{X - \mu}{k}\right| \ge \sigma\right\} \le \frac{1}{k^2}$$

$$\therefore \qquad P\left\{\left|\frac{X - \mu}{k}\right| \le \sigma\right\} \ge 1 - \frac{1}{k^2} \qquad \text{(BDU — Nov. 96)}$$

Note *Although we have proved the inequality for the continuous case, it holds good for the discrete RV also.*

Bienayme's Inequality

Let us first prove a basic result:

If X is a RV with $f(x) = 0$, when $x < 0$, and with $E(X) = \mu$, then for any $\alpha > 0$, $P(X \ge \alpha) \le \mu/\alpha$.

Proof $\quad \mu = E(X) = \int_0^{\infty} x f(x)\,dx$

$$\ge \int_{\alpha}^{\infty} x f(x)\,dx \qquad (\text{since } \alpha > 0)$$

i.e., $$\ge \alpha \int_{\alpha}^{\infty} f(x)\,dx \qquad (\text{since } x \ge \alpha)$$

i.e., $$\ge \alpha P(X \ge \alpha)$$

$$\therefore \qquad P(X \geq \alpha) \leq \frac{\mu}{\alpha} \tag{1}$$

[In particular, $P(X \geq \sqrt{\mu} \leq \sqrt{\mu}$].

In (1), replace X by $|X - a|^n$ and α by c^n, where X is an arbitrary RV and c, a, n are arbitrary numbers.

$|X - a|^n$ takes only positive values.

$$\therefore \qquad P\{|X - a|^n \geq c^n\} \leq \frac{E\{|X - a|^n\}}{c^n}$$

i.e.,
$$P(\{|X - a| \geq c\} \leq \frac{E\{|X - a|^n\}}{c^n} \tag{2}$$

This inequality is called the Bienayme's inequality.

If we take $n = 2$ and $a = \mu$ in (2), we get Tchebycheff's inequality.

Schwartz Inequality

If $f(x)$ and $g(x)$ are real functions of x, then

$$\left(\int_a^b f(x)g(x)\,dx\right)^2 \leq \int_a^b [f(x)]^2\,dx \times \int_a^b [g(x)]^2\,dx$$

Proof Consider $\int_a^b [f(x) - tg(x)]^2\,dx$, where t is real,

$$= t^2 \int_a^b g^2(x)\,dx - 2t \int_a^b f(x)g(x)\,dx + \int_a^b f^2(x)\,dx.$$

For every real t, LS ≥ 0.

Therefore, RS which is a quadratic expression in $t \geq 0$.

Therefore, discriminant ≤ 0.

i.e.,
$$\left[\int_a^b f(x)g(x)\,dx\right]^2 \leq \int_a^b f^2(x)\,dx \times \int_a^b g^2(x)\,dx$$

Note *(i) The equality holds, if $f(x) = kg(x)$.*

(ii) If f and g are complex, then

$$\left|\int f(z)\,g(z)\,dz\right|^2 \leq \int |f(z)|^2\,dz \int |g(z)|^2\,dz)$$

Cauchy-Schwartz Inequality

For any 2 RVs X and Y

$$\{E(XY)\}^2 \leq E(X^2)E(Y^2)$$

Proof Consider $E\{(X - tY)^2\} \geq 0$, where t is real.

i.e., $E(X^2) - 2tE(XY) + t^2E(Y^2) \geq 0$

Since the LS is a quadratic expression in t, discriminant of LS ≤ 0.

i.e., $\{E(XY)\}^2 \leq E(X^2) \times E(Y^2)$

Worked Examples 4(E)

Example 1 *A RV X has mean μ = 12 and variance σ^2 = 9 and an unknown probability distribution. Find $P(6 < X < 18)$.* (MU — Apr. 96, Nov. 96)

Solution Since the probability distribution of X is not known, we cannot find the value of the required probability. We can find only a lower bound for the probability using Tchebycheff's inequality.

$$P\{|X - \mu| \geq c\} \leq \frac{\sigma^2}{c^2}, c > 0$$

i.e., $$P\{|X - \mu| \leq c\} \geq 1 - \frac{\sigma^2}{c^2}$$

$\therefore$ $$P\{\mu - c \leq X \leq \mu + c\} \geq 1 - \frac{\sigma^2}{c^2}$$

Taking $\mu = 12$ and $\sigma^2 = 9$, we get,

$$P\{12 - c < X < 12 + c) \geq 1 - \frac{9}{c^2}$$

Putting $c = 6$, $P\{6 < X < 18\} \geq 1 - \frac{9}{36}$

i.e., $$P\{6 < X < 18\} \geq \frac{3}{4}$$

Example 2 *If the RV X is uniformly distributed over $\left(-\sqrt{3}, \sqrt{3}\right)$, compute P $\left\{|X - \mu| \geq \frac{3\sigma}{2}\right\}$ and compare it with the upper bound obtained by Tchebycheff's inequality.*

Solution $\mu = 0$, $\sigma^2 = 1$ and $f(x) = \frac{2}{2\sqrt{3}}$ for the given uniform distribution.

$$\therefore \quad P\left\{|X - \mu| \geq \frac{3\sigma}{2}\right\} = P\{|X| \geq 3/2\}$$

$$= 1 - P\{-3/2 < X < 3/2\}$$

$$= 1 - \int_{-3/2}^{3/2} \frac{1}{2\sqrt{3}}\, dx \quad \left(\text{since pdf} = \frac{1}{2\sqrt{3}}\right)$$

$$= 1 - \frac{\sqrt{3}}{2} = 0.134$$

By Tchebycheff's inequality

$$P\{|X - \mu| \geq k\sigma\} \leq \frac{1}{k^2}$$

$$\therefore \qquad P\{|X - \mu| \geq 3/2\ \sigma\} \leq \frac{4}{9} = 0.444,$$

which is a poor upper bound.

Example 3 *Can we find a RV X for which $P\{\mu - 2\sigma \leq X \leq \mu + 2\sigma\} = 0.6$?*

Solution $P\{\mu - 2\sigma \leq X \leq \mu + 2\sigma\} = P\{|X - \mu| \leq 2\sigma\}$

$$\geq 1 - \frac{1}{2^2} \text{ by Tchebycheff's inequality}$$

i.e., $$P\{\mu - 2\sigma \leq X \leq \mu + 2\sigma\} \geq 0.75$$

Therefore, there does not exist a RV X satistying the given condition.

Example 4 *A discrete RV X takes the values – 1, 0, 1 with probabilities $\frac{1}{8}$, $\frac{3}{4}$, $\frac{1}{8}$ respectively. Evaluate $P\{|X - \mu| \geq 2\sigma\}$ and compare it with the upper bound given by Tchebycheff's inequality.*

Solution $E(X) = -1 \times \frac{1}{8} + 0 \times \frac{3}{4} + 1 \times \frac{1}{8} = 0$

$$E(X^2) = 1 \times \frac{1}{8} + 0 \times \frac{3}{4} + 1 \times \frac{1}{8} = \frac{1}{4}; \ \text{Var}(X) = \frac{1}{4}$$

$$\therefore \qquad P\{|X - \mu| \geq 2\sigma\} = P\{|X| \geq 1\}$$

$$= P\{X = -1 \text{ or } X = 1\}$$

$$= \frac{1}{8} + \frac{1}{8} = \frac{1}{4} \tag{1}$$

By Tchebycheff's inequality

$$P\{|X - \mu| \geq 2\sigma\} \leq \frac{1}{2^2} = \frac{1}{4} \tag{2}$$

The two values coincide.

Example 5 *If X denotes the sum of the numbers obtained when 2 dice are thrown, obtain an upper bound for $P\{|X - 7| \geq 4\}$. Compare with the exact probability.*

Solution Let X_1, X_2 denote the outcomes of the first and second dice respectively.

$$E(X_1) = E(X_2) = \frac{1}{6}(1 + 2 + \cdots + 6) = \frac{7}{2}$$

$$E(X^2{}_1) = E(X^2{}_2) = \frac{1}{6}\,(1^2 + 2^2 + \cdots + 6^2) = \frac{91}{6}$$

$$\therefore \quad \text{Var}(X_1) = \text{Var}(X_2) = \frac{35}{12}$$

$$\therefore \quad E(X) = E(X_1 + X_2) = 7 \text{ and}$$

$$\text{Var}(X) = \text{Var}(X_1 + X_2) = 1^2 \times \frac{35}{12} + 1^2 \times \frac{35}{12} = \frac{35}{6}$$

BY Tchebycheff's inequality

$$P\{|X - \mu| \geq c\} \leq \frac{\sigma^2}{c^2}$$

$$\therefore \quad P\{|X - 7| \geq 4\} \leq \frac{35}{96} \tag{1}$$

$$\text{Now}\, P\{|X - 7| \geq 4\} = P\{X = 2, 3, 11 \text{ or } 12\}$$

$$= \frac{1}{36} + \frac{2}{36} + \frac{2}{36} + \frac{1}{36} = \frac{1}{6} \tag{2}$$

There is much difference between the two values.

Example 6 *A fair die is tossed 720 times. Use Tchebycheff's inequality to find a lower bound for the probability of getting 100 to 140 sixes.*

Solution Let X be the number of sixes obtained when a fair die is tossed 720 times.

$$p = P\{\text{getting '6' in a single toss}\} = \frac{1}{6}$$

$$q = \frac{5}{6} \text{ and } n = 720$$

X follows a binomial distribution with mean $np = 120$ and variance $npq = 100$, i.e., $\mu = 120$ and $\sigma = 10$.
By Tchebycheff's inequality,

$$P\{|X - \mu\} \leq k\sigma\} \geq 1 - \frac{1}{k^2}$$

i.e.,

$$P\{|X - 120| \leq 10k\} \geq 1 - \frac{1}{k^2}$$

i.e.,

$$P\{120 - 10k \leq X \leq 120 + 10k\} \geq 1 - \frac{1}{k^2}$$

Taking $k = 2$, we get,

$$P\{100 \leq X \leq 140\} \geq \frac{3}{4}$$

Therefore, required lower bound for the probability = 0.75.

Example 7 *Use Tchebycheff's inequality to find how many times a fair coin must be tossed in order that the probability that the ratio of the number of* *heads to the number of tosses will lie between 0.45 and 0.55 will be at least 0.95.*

Solution Let X be the number of heads obtained when a fair coin is tossed n times.
Then X follows a binomial distribution

$$B(np, \sqrt{npq}\,) \text{ where } p = q = \frac{1}{2}$$

$$\therefore \qquad \frac{X}{n} \text{ follows } B\left(\frac{1}{2}, \frac{1}{2\sqrt{n}}\right)$$

By Tchebycheff's inequality,

$$P\left\{\left|\frac{X}{n} - \frac{1}{2}\right| \le c\right\} \ge 1 - \frac{\sigma^2}{c^2}$$

i.e.,
$$P\{0.5 - c \le \frac{X}{n} \le 0.5 + c\} \ge 1 - \frac{1}{4nc^2}$$

Taking
$$c = 0.05,\ P\left\{0.45 \le \frac{X}{n} \le 0.55\right\} \ge 1 - \frac{100}{n}$$

Given that
$$P\left\{0.45 \le \frac{X}{n} \le 0.55\right\} \ge 0.95$$

$$\therefore \qquad 1 - \frac{100}{n} = 0.95$$

$$\therefore \qquad n = 2000$$

Example 8 *A RV X is exponentially distributed with parameter 1. Use Tchebycheff's inequality to show that P(–1 ≤ X ≤ 3) ≥ 3/4. Find the actual probability also.*

Solution For an exponential distribution with parameter λ, $E(X) = \frac{1}{\lambda}$ and

$\text{Var}(X) = \frac{1}{\lambda^2}$.

(refer to Worked Example 7 in the previous section of this chapter).

Here $\qquad \lambda = 1$

$\therefore \qquad \mu = 1$ and $\sigma = 1$

By Tchebycheff's inequality,

$$P\{|X - 1| \le 2\} \ge 1 - \frac{1}{4}$$

i.e.,
$$P\{-1 \le X \le 3\} \ge \frac{3}{4}$$

Density function of the exponential distribution with $\lambda = 1$ is given by $f(x) = e^{-x}$, $x > 0$.

$$\therefore \quad P\{-1 \le X \le 3\} = \int_{-1}^{3} e^{-x}\, dx$$

$$= \int_{0}^{3} e^{-x}\, dx = 1 - e^{-3} = 0.9502$$

Example 9 *Use Tchebycheff's inequality to prove that $P(X = \mu) = 1$, if $Var(X) = 0$*

Solution By Tchebycheff's inequality

$$P\{|X - \mu| \ge c\} \le \frac{\sigma^2}{c^2}$$

and $$P\{|X - \mu| \le c\} \ge 1 - \frac{\sigma^2}{c^2}$$

Putting $\sigma^2 = 0$, these inequalities become

$$P\{|X - \mu| \ge c\} \le 0$$

and $$P\{|X - \mu| \le c\} \ge 1 \text{ where } c > 0$$

The above results hold good even for arbitrarily chosen small values. Hence in the limit when $c \to 0$,

$$P\{|X - \mu| = 0\} = 1$$

i.e., $$P\{X = \mu\} = 1$$

Example 10 *A discrete RV X can assume the values $x = 1, 2, 3, \cdots$ with probability 2^{-x}. Show that $P\{|X - 2| \ge 2\} \le 1/2$, while the actual probability is 1/8.*

Solution $E(X) = \sum_{x=1}^{\infty} \frac{x}{2^x} = 2$ and $\text{Var}(X) = 2$

By Tchebycheff's inequality,

$$P\{|X - \mu| \ge c\} \le \frac{\sigma^2}{c^2}$$

Putting $\mu = 2$, $\sigma^2 = 2$, and $c = 2$, we get

$$P\{|X - 2| \ge 2\} \le \frac{1}{2}$$

Now $$P\{|X - 2| \ge 2\} = P\{X \ge 4\}$$

$$\frac{1}{2^4} + \frac{1}{2^5} + \frac{1}{2^6} + \cdots + \infty$$

$$= \frac{\frac{1}{2^4}}{1 - \frac{1}{2}} = \frac{1}{8}$$

Exercise 4(E)

Part A (Short answer questions)

1. State Tchebycheff's inequality.
2. A RV X has mean 10 and variance 16. Find a lower bound for $P\{5 < X < 15\}$.
3. A RV X has mean 10 and variance 16. Find an upper bound for $P\{|X - 10| \geq 5\}$.
4. If Tchebycheff's inequality for a RV X with mean 12 is $P\{6 < X < 18\} \geq \frac{3}{4}$, find the SD of X.
5. If Tchebycheff's inequality for a RV X with SD 3 is $P\{6 < X < 18\} \geq \frac{3}{4}$, find the mean of X.
6. If Tchebycheff's inequality for a RV X is $P\{-2 < X < 8\} \geq \frac{21}{25}$, find $E(X)$ and $\text{Var}(X)$.
7. If $E(X) = 8$ and $E(X^2) = 68$, find a lower bound for $P\{5 < X < 11\}$, using Tchebycheff's inequality.
8. State Beinayme's inequality. Deduce Tchebycheff's inequality.
9. State Schwarz inequality.
10. State Cauchy-Schwarz inequality.
11. Use Cauchy-Schwarz inequality to prove that $|\rho_{XY}| \leq 1$.
12. Does a RV X with mean μ and SD σ satisfying $P\{\mu - \sqrt{3}\ \sigma \leq X \leq \mu + \sqrt{3}\ \sigma\} = 0.5$ exist? Why?

Part B

13. If X is a RV with $E(X) = 3$ and $E(X^2) = 13$, find the lower bound for $P(-2 < X < 8)$, using Tchebycheff's inequality.
14. If the RV X is uniformly distributed over $\left(1 - \frac{1}{\sqrt{3}}, 1 + \frac{1}{\sqrt{3}}\right)$ compute $P\{|X - \mu| \geq 3/2\ \sigma\}$ and compare it with the upper bound obtained by Tchebycheff's inequality.
15. A discrete RV X can take the values $-a$, 0, a with probabilities 1/8, 3/4, 1/8 respectively. Compute $P\{|X| \geq 2\sigma\}$ and compare it with Tchebycheff's inequality bound.
16. Two dice are thrown once. If X is the sum of the numbers showing up, prove that $P\{|X - 7| \geq 3\} \leq \frac{35}{54}$ and compare this value with the exact probability.

17. A fair die is tossed 600 times. Use Tchebycheff's inequality to find a lower bound for the probability of getting 80 to 120 sixes.
18. An unbiased coin is tossed 100 times. Show that the probability that the number of heads will be between 30 and 70 is greater than 0.93.
19. If X is the number obtained in a throw of a fair die, show that the Tcheycheff's inequality gives $P\{|X - \mu| > 2.5\} < 0.47$, while the actual probability is zero.
20. Using Tchebycheff's inequality, find how many times a fair coin must be tossed in order that the probability that the ratio of the number of heads to the number of tosses will lie between 0.4 and 0.6 will be at least 0.9.
21. A random variable X has pdf $f(x) = e^{-x}$, $x \geq 0$. Use Tchebycheff's inequality to show that $P\{|X - 1| > 1\} < 1/4$ and show also that the actual probability is e^{-3}.
22. A random variable X has the pmf $P(X = 1) = 1/18$, $P(X = 2) = 16/18$, $P(X = 3) = 1/18$. Show that there is a value of c such that $P\{|X - \mu| \geq c\} = \frac{\sigma^2}{c^2}$, so that, in general, the bound given by Tchebycheff's inequality cannot be improved.

Convergence Concepts and Central Limit Theorem

Let us consider the sequence of RVs $X_1, X_2, X_3, \cdots X_n, \cdots$ or random sequence. The concept of convergence of random sequences is essential in the study of random signals. A few definitions and criteria that are used for determining the convergence of random sequences are given below.

(1) Convergence everywhere and almost everywhere

If $\{X_n\}$ is a sequence of RVs and X is a RV such that $\left[\lim\limits_{n \to \infty} (X_n) = X\right]$ i.e., $X_n \to X$ as $n \to \infty$, then the sequence $\{X_n\}$ is said to converge to X everywhere.

If $P\{X_n \to X\} = 1$ as $n \to \infty$, then the sequence $\{X_n\}$ is said to converge to X almost everywhere.

(2) Convergence in probability or stochastic convergence

If $P\{|X_n - X| > \varepsilon\} \to 0$ as $n \to \infty$, then the sequence $\{X_n\}$ is said to converge to X in probability or stochastically.

As a particular case of this kind of convergence we have the following result, known as *Bernoulli's law of large numbers*.

If X represents the number of successes out of n Bernoulli's trials with probability of success p (in each trial), then $\{X/n\}$ converges in probability to p.

i.e., $$P\left\{\left|\frac{X}{n} - p\right| > \varepsilon\right\} \to 0 \text{ as } n \to \infty$$

(3) Convergence in the mean square sense

If $E\{|X_n - X|^2\} \Rightarrow 0$ as $n \to \infty$, then the sequence $\{X_n\}$ is said to converge to X in the mean square sense.

(4) Convergence in distribution

If $F_n(x)$ and $F(x)$ are the distribution functions of X_n and X respectively such that $F_n(x) \to F(x)$ as $n \to \infty$ for every point of continuity of $F(x)$, then the sequence $\{X_n\}$ is said to converge to X in distribution.

Closely associated to the concept of convergence in distribution is a remarkable result known as central limit theorem, which is given below without proof.

Central Limit Theorem (Liapounoff's Form)

If $X_1, X_2, \cdots X_n, \cdots$, be a sequence of independent RVs with $E(X_i) = \mu_i$ and $\text{Var}(X_i) = \sigma^2_i$, $i = 1, 2, \cdots$, and if $S_n = X_1 + X_2 + \cdots X_n$, then under certain general conditions, S_n follows a normal distribution with mean $\mu = \sum_{i=1}^{n} \mu_i$ and variance $\sigma^2 = \sum_{i=1}^{n} \sigma_i^2$ as n tends to infinity.

Central Limit Theorem (Lindeberg–Levy's Form)

If $X_1, X_2, \cdots, X_n, \cdots$, be a sequence of independent identically distributed RVs with $E(X_i) = \mu$ and $\text{Var}(X_i) = \sigma^2$, $i = 1, 2, \cdots$, and if $S_n = X_1 + X_2 + \cdots + X_n$, then under certain general conditions, S_n follows a normal distribution with mean $n\mu$ and variance $n\sigma^2$ as n tends to infinity.

Corollary

If $\overline{X} = \dfrac{1}{n}(X_1 + X_2 + \cdots + X_n)$, then $E(\overline{X}) = \mu$ and $\text{Var}(\overline{X}) = \dfrac{1}{n^2}(n\,\sigma^2) = \dfrac{\sigma^2}{n}$

$$\therefore\ \overline{X} \text{ follows } N\left(\mu, \frac{\sigma}{\sqrt{n}}\right) \text{ as } n \to \infty$$

Worked Examples 4(F)

Example 1 *The lifetime of a certain brand of an electric bulb may be considered a RV with mean 1200 h and standard deviation 250 h. Find the probability, using central limit theorem, that the average lifetime of 60 bulbs exceeds 1250 h.*

Solution Let X_i represent the lifetime of the bulb.

$$E(X_i) = 1200 \text{ and Var } (X_i) = 250^2$$

Let $\overline{X}$ denote the mean lifetime of 60 bulbs.

By corollory of Lindeberg-Levy form of CLT

$$\overline{X} \text{ follows } N\left(1200, \frac{250}{\sqrt{60}}\right)$$

$$P(\overline{X} > 1250) = P\left(\frac{\overline{X}-1200}{\frac{250}{\sqrt{60}}} > \frac{1250-1200}{\frac{250}{\sqrt{60}}}\right)$$

$$= P\left(z > \frac{\sqrt{60}}{5}\right)$$

$$= P(z > 1.55),$$

where z is the standard normal variable

$$= 0.0606$$

(from the table of areas under normal curve)

Example 2 *A distribution with unknown mean μ has variance equal to 1.5. Use central limit theorem to find how large a sample should be taken from the distribution in order that the probability will be at least 0.95 that the sample mean will be within 0.5 of the population mean.* (MKU – Apr. 97)

Solution Let n be the size of the sample, a typical member of which is X_i.

Given: $E(X_i) = \mu$ and $\mathrm{Var}(X_i) = 1.5$.

Let $\overline{X}$ denote the sample mean.

By corollary under CLT,

$$\overline{X} \text{ follows } N\left(\mu, \frac{\sqrt{1.5}}{\sqrt{n}}\right)$$

We have to find n such that

$$P\{\mu - 0.5 < \overline{X} < \mu + 0.5\} \geq 0.95$$

i.e., $$P\{-0.5 < \overline{X} - \mu < 0.5\} \geq 0.95$$

i.e., $$P\{|\overline{X} - \mu| < 0.5\} \geq 0.95$$

i.e., $$P\left\{\frac{|\overline{X}-\mu|}{\sqrt{\frac{1.5}{n}}} < \frac{0.5}{\sqrt{\frac{1.5}{n}}}\right\} \geq 0.95$$

i.e., $$P\{|z| < 0.4082\sqrt{n}\} \geq 0.95$$

where z is the standard normal variable.

The least value of n is obtained from

$$P\{|z| < 0.4082\sqrt{n}\} = 0.95$$

From the table of areas under normal curve

$$P\{|z| < 1.96\} = 0.95$$

Therefore, least n is given by $0.4082\sqrt{n} = 1.96$, i.e., least $n = 24$.

Therefore, the size of the sample must be at least 24.

Example 3 *If $X_1, X_2, \cdots X_n$ are Poisson variates with parameter $\lambda = 2$, use the central limit theorem to estimate $P(120 \le S_n \le 160)$, where $S_n = X_1 + X_2 + \cdots X_n$ and $n = 75$.* (MU — Apr. 96)

Solution $E(X_i) = \lambda = 2$ and $\text{Var}(X_i) = \lambda = 2$

By CLT, S_n follows $N(n\mu, \sigma\sqrt{n})$

i.e., S_n follows $N(150, \sqrt{150})$

$$P\{120 \le S_n \le 160\} = P\left\{\frac{-30}{\sqrt{150}} \le \frac{S_n - 150}{\sqrt{150}} \le \frac{10}{\sqrt{150}}\right\}$$

$$= P\{-2.45 \le z \le 0.85\}$$

where z is the standard normal variable.

$$= 0.4927 + 0.2939, \text{ (from the normal tables)}$$

$$= 0.7866$$

Example 4 *Using the central limit theorem, show that, for large n,*

$$\frac{c^n}{\lfloor n-1} x^{n-1} e^{-cx} \cong \frac{c}{\sqrt{2\pi n}} e^{-(cx-n)^2/2n}, x > 0$$

(BDU — Apr. 96)

Solution Let $X_1, X_2, \cdots, X_n$ be independent RVs each of which is exponentially distri-buted with parameter c.

i.e., let the pdf of $X_i = c\, e^{-cx}, x > 0$

The characteristic function of X_i is given by

$$\phi_{Xi}(\omega) = (1 - i\omega/c)^{-1}$$

(refer to the Note under Worked Example 7 of the characteric function section).

By Property 4 of CFs, since $X_1, X_2, \cdots, X_n$ are independent RVs,

$$\phi_{(X1 + X2 + \cdots + Xn)}(\omega) = [\phi_{xi}(\omega)]^n$$

$$= \left(1 - \frac{i\omega}{c}\right)^{-n}$$

$$= \text{CF of Erlang distribution}$$

(refer to Worked Example 7 of Section 4(B))

Therefore, when n is finite, $(X_1 + X_2 + \cdots + X_n)$ follows the Erlang distribution whose pdf is given by

$$\frac{c^n}{\lfloor n-1} x^{n-1} e^{-cx}, x > 0 \tag{1}$$

When n tends to infinity, $(X_1 + X_2 + \cdots + X_n)$ follows a normal distribution with mean $nE(X_i) = \frac{n}{c}$ and variance $n\text{Var}(X_i) = \frac{n}{c^2}$ (Central limit theorem),

i.e., when $n \to \infty$, $(X_1 + X_2 + \cdots + X_n)$ follows a normal distribution whose pdf is given by

$$\frac{1}{\sqrt{\frac{n}{c^2}}\sqrt{2\pi}} \exp\left\{-\left(x-\frac{n}{c}\right)^2 \Big/ \frac{2n}{c^2}\right\}$$

$$= \frac{c}{\sqrt{2n\pi}} \exp\{-(cx-n)^2/2n\} \tag{2}$$

From (1) and (2), the required result follows.

Example 5 *Verify central limit theorem for the independent random variables X_k, where for each k, $P\{X_k = \pm 1\} = \frac{1}{2}$.* (MKU – Apr. 96)

Solution $E(X_k) = 1 \times 1/2 + (-1) \times 1/2 = 0$

$$\text{Var}(X_k) = 1^2 \times 1/2 + (-1)^2 \times 1/2 = 1$$

Consider $Y_n = \frac{1}{\sqrt{n}}(X_1 + X_2 + \cdots + X_n)$

$$E(Y_n) = 0 \text{ and } \text{Var}(Y_n) = \frac{1}{n} \times n = 1$$

Now $\phi_{Xk}(\omega) = E\{e^{i\omega xk}\}$

$$= e^{i\omega} \times 1/2 + e^{i\omega(-1)} \times \frac{1}{2}$$

$$= \cos\omega$$

$\therefore$ $\phi_{Yn}(\omega) = \phi_{\frac{1}{\sqrt{n}}(X_1 + X_2 + \ldots + X_n)}(\omega)$

$$= \left\{\phi_{X_k/\sqrt{n}}(\omega)\right\}^n \quad \text{(since } X_1, X_2, \cdots, X_n \text{ are independent)}$$

$$= \left[\cos\left(\frac{\omega}{\sqrt{n}}\right)\right]^n \quad [\text{since } \phi_{aX+b}(\omega) = e^{ib\omega}\phi_X(a\omega)]$$

$$= \left[1 - \frac{\omega^2}{2n} + \text{terms involving } \frac{1}{n^2} \text{ and higher powers of } \frac{1}{n}\right]^n$$

$$= \left(1 - \frac{\omega^2}{2n}\right)^n + \text{terms involving } \frac{1}{n} \text{ and higher powers of } \frac{1}{n}$$

$\therefore \qquad \lim_{n \to \infty} \phi_{yn}(\omega) = e^{\frac{-1}{2}\omega^2}$

which is the characteristic function of $N(0, 1)$.

$\therefore \qquad Y_n \to N(0, 1)$, as $n \to \infty$

Therefore, CLT holds good for the sequence $\{X_k\}$.

Example 6 *Show that the central limit theorem holds good for the sequence* $\{X_k\}$, *if*

$$P\{X_k = \pm k^{\alpha}\} = \frac{1}{2} \times k^{-2\alpha},\ P\{X_k = 0\} = 1 - k^{-2\alpha},\ \alpha < \frac{1}{2}. \quad \text{(BDU — Nov. 96)}$$

Note *Liapounoff's form of CLT holds good for a sequence* $\{X_k\}$, *if*

$$\lim_{n \to \infty} \frac{\sum_{k=1}^{n} E\left\{|X_k - \mu_k|^3\right\}}{\left\{\sum_{k=1}^{n} Var\left(X_k\right)\right\}^{\frac{3}{2}}} = 0 \quad \textit{Condition is assumed.}$$

Solution We have to verify whether this condition is satisfied by the given $\{X_k\}$.

Now $\qquad \mu_k = E(X_k) = \frac{1}{2} \times k^{-\alpha} - \frac{1}{2} \times k^{-\alpha} = 0$

$$E(X_k^2) = \frac{1}{2} \times k^{2\alpha} \times k^{-2\alpha} + \frac{1}{2} k^{2\alpha} \times k^{-2\alpha} = 1$$

$\therefore \qquad \text{Var}(X_k) = 1$

$$E\{|X_k - \mu_k|^3 = E\{|X_k|^3\}$$

$$= k^{3\alpha} \times \frac{1}{2} k^{-2\alpha} + k^{3\alpha} \times \frac{1}{2} k^{-2\alpha}$$

$$= k^{\alpha}$$

$\therefore \qquad \sum_{k=1}^{n} E\{|X_k - \mu_k|^3\} = 1^{\alpha} + 2^{\alpha} + \cdots + n^{\alpha} < n \times n^{\alpha} \qquad$ (since each term $\leq n\alpha$)

$$\sum_{k=1}^{n} \text{Var}(X_k) = n$$

$\therefore \qquad \lim_{n \to \infty} \dfrac{\sum_{k=1}^{n} E\left\{|X_k - \mu_k|^3\right\}}{\left\{\sum_{k=1}^{n} \text{Var}\left(X_k\right)\right\}^{\frac{3}{2}}}$

$$= \lim_{n \to \infty} \left(\frac{1^{\alpha} + 2^{\alpha} + \cdots + n^{\alpha}}{n^{3/2}} \right) < \lim_{n \to \infty} \frac{n \times n^{\alpha}}{n^{3/2}}$$

$$= \lim_{n \to \infty} \frac{1}{n^{\frac{1}{2} - \alpha}}$$

$$= 0 \text{ (since } (\alpha < 1/2)$$

i.e., the necessary condition is satisfied. Therefore, CLT holds good for the sequence $\{X_k\}$.

Exercise 4(F)

Part A (Short answer questions)

1. What is the difference between convergence everywhere and almost everywhere of a random sequence $\{X_n\}$?
2. Define stochastic convergence of a random sequence $\{X_n\}$.
3. State Bernoulli's law of large numbers.
4. Define convergence of a random sequence $\{X_n\}$ in the mean square sense.
5. Define convergence in distribution of a random sequence $\{X_n\}$.
6. State the Liapounoff's form of CLT.
7. State the Lindeberg–Levy's form of CLT.
8. What is the importance of CLT?

Part B

9. A random sample of size 100 is taken from a population whose mean is 60 and variance is 400. Using CLT, with what probability can we assert that the mean of the sample will not differ from $\mu = 60$ by more than 4? (MKU — Nov. 96)
10. The guaranteed average life of a certain type of electric light bulb is 1000 h with a standard deviation of 125 h. It is decided to sample the output so as to ensure that 90% of the bulbs do not fall short of the guaranteed average by more than 2.5%. Use CLT to find the minimum sample size.
11. If X_i, $i = 1, 2, \cdots, 50$, are independent RVs, each having a Poisson distribution with parameter $\lambda = 0.03$ and $S_n = X_1 + X_2 + \cdots X_n$, evaluate $P(S_n \geq 3)$, using CLT. Compare your answer with the exact value of the probability.
12. If V_i, $i = 1, 2, \cdots, 20$, are independent noise voltages received in an 'adder' and V is the sum of the voltages received, find the probability that the total incoming voltage V exceeds 105, using CLT. Assume that each of the random variables V_i is uniformly distributed over (0, 10).
13. 30 electronic devices $D_1, D_2, \cdots, D_{30}$ are used in the following manner. As soon as D_1 fails, D_2 becomes operative. When D_2 fails, D_3 becomes

operative and so on. If the time to failure of D_i is an exponentially distributed random variable with parameter $\lambda = 0.1$/h and T is the total time of operation of all the 30 devices, find the probability that T exceeds 350 h, using CLT.

14. Examine if the CLT holds good for the sequence $\{X_k\}$, if $P\{X_k = \pm 2^k\} = 2^{-(2k+1)}$, $P\{X_k = 0\} = 1 - 2^{-2k}$.
15. Show that the CLT does not hold if the RV's X_i have a Cauchy density.

Answers

Exercise 4(A)

6. $\mu_Y = 0$ and $\sigma_Y = 1$
8. $a = \frac{1}{5}, b = 2$
9. 2
10. 6
11. 5.4
12. $\frac{n+1}{2}; \frac{1}{12}(n^2-1)$
13. $\frac{a}{1-a}$
14. log 2
15. $\frac{q}{p}$
16. np
17. λ
18. 2 ; 1
19. $\frac{1}{m}; \frac{1}{m^2}$
20. $\frac{2}{3}; \frac{1}{\sqrt{2}}$
21. $\frac{\pi}{3}; \frac{\pi}{2}$
22. 40; 39
23. Symmetrical
28. –11; 54
34. $\frac{5}{12}; \frac{5}{12}$

35. $\frac{1}{3}$

36. e^t; $e^{2t}-e^t$

37. $\frac{1}{2}; \frac{n+2}{12n}$

38. $\sqrt{\frac{n(n+1)}{3}}; \frac{n(n+1)}{2n+1}$

39. 12; 223

40. 45; 40; 45; 10; 14; 18.49

41. (0.0193); (2.3888); skewed to the right; playtykurtic

42. 5; 3; 0; $\frac{26}{9}$

43. 6; – 74

44. 10, 116, 1544, 23184

45. $\frac{1-p}{p}$

46. (i) $\frac{7n}{2}$; (ii) $\left(\frac{7}{2}\right)^n$

47. (151.5)

48. Rs. 32/–

49. Rs. 4.52

50. $\frac{11}{2}$

51. 2; $1+\sqrt{2}; \frac{1}{2}$

52. 1; 1; 1; $\frac{2}{3}; \frac{3}{8}; \frac{1}{\sqrt{5}}; 0; \frac{15}{7}$

53. 0; 1; $\frac{13}{16}$

56. 1; 1

57. $\frac{8}{3}$

58. $\frac{y}{2}; \frac{y^2}{12}$

59. $\frac{3y}{4}; \frac{3y^2}{80}$

Exercise 4(B)

12. No
15. 0.7891
16. 0.6030
17. 0.0390
18. 0.8504
19. 0.6
20. 0.8762
21. 0.9173
22. 0.3875
23. 0.91
24. 0.4
25. A and C
26. $\frac{1}{2}$
28. 0
29. $\frac{v_2}{\sqrt{(v_1 + v_2)(v_2 + v_3)}}$
32. $\frac{5}{12}; \frac{5}{12}; \frac{1}{16}; \frac{1}{16}; -\frac{1}{9}$
33. $\frac{1}{3}; \frac{2}{3}; \frac{1}{18}; \frac{1}{18}; \frac{1}{2}$
34. 0
35. $-\frac{1}{11}$
36. 11; 4; 0

Exercise 4(C)

11. $y = 6x + 114$; $20x = 3y + 950$; 0.9
12. $y = 0.665x + 23.78$; $x = 0.54y + 30.74$; 70.995, 68.54
13. $y = 0.7x + 12.7$; 25.3
14. $x = 0.936y + 1.388$; 43.508
15. 72.6; 69.57
16. 13; 17; 0.6; 4
17. $-\frac{1}{2\sqrt{3}}; 5; \frac{1}{3}; \frac{2}{15\sqrt{3}}$
18. 1; 2; 4; –0.8.66

19. $\bar{x} = \dfrac{a+bc}{1-bd}$; $\bar{y} = \dfrac{c+ad}{1-bd}$; $r = \sqrt{bd}$; No.

20. $\sqrt{\dfrac{a_1 b_2}{a_2 b_1}}$; $\left(\dfrac{b_1 c_2 - b_2 c_1}{c_1 a_2 - c_2 a_1}\right)\sqrt{\dfrac{a_1 a_2}{b_1 b_2}}$

Exercise 4(D)

5. $M(t) = E(e^{tX})$, whereas $\phi(\omega) = E(e^{i\omega X})$
7. The characteristic function of a RV uniquely determines its pdf. (or) The necessary and sufficient condition for two distributions with pdf's $f_1(x)$ and $f_2(x)$ to be identical is that their characteristic functions $\phi_1(\omega)$ and $\phi_2(\omega)$ are identical.
11. It generates cumulants.

12. $M(t) = \sum\limits_{r=0}^{\infty} E(X^r)\, t^r/r! = \sum\limits_{r=0}^{\infty} t^r = \dfrac{1}{1-t}$

13. $\dfrac{2}{2-t} = \left(1 - \dfrac{t}{2}\right)^{-1} = 1 + \dfrac{t}{2} + \dfrac{t^2}{4} + \cdots$

$E(X) = \dfrac{1}{2}$; $E(X^2) = \dfrac{1}{2}$; $\text{Var}(X) = \dfrac{1}{4}$; $\sigma_x = \dfrac{1}{2}$

14. $$M(t) = E(e^{tX}) = \int_a^b \frac{1}{b-a}\, e^{tx}\, dx$$

$$= \frac{1}{b-a}\left(\frac{e^{bt} - e^{at}}{t}\right)$$

$$\phi(\omega) = \frac{1}{b-a}\left(\frac{e^{ib\omega} - e^{ia\omega}}{i\omega}\right)$$

15. $M(t) = \sum\limits_{r=0}^{n} nC_r\, p^r q^{n-r}\, e^{tr} = (q + pe^t)^n$

$\phi(\omega) = (q + pe^{i\omega})^n$

16. $M(t) = \sum\limits_{r=0}^{\infty} \dfrac{e^{-\lambda}\lambda^r}{r!}\, e^{tr} = e^{-\lambda(1-e^t)}$

$\phi(\omega) = e^{-\lambda(1-e^{i\omega})}$

18. $\text{CF} = \dfrac{p}{1-qe^{i\omega}}$

$$\text{MGF} = \frac{p}{1 - qe^t}$$

19. $\phi(\omega) = \int\limits_0^\infty \lambda e^{-\lambda x} e^{i\omega x}\, dx = \lambda \left\{ \frac{e^{-(\lambda - i\omega)x}}{-(\lambda - i\omega)} \right\}_0^\infty = \frac{\lambda}{\lambda - i\omega}$

$$M(t) = \frac{\lambda}{\lambda - t}$$

20. $\phi(\omega) = e^{i\mu\omega}\, e^{-\sigma^2\omega^2/2}$

21. $\phi(\omega) = \frac{1}{2} \int\limits_{-\infty}^{\infty} e^{-|x|} e^{i\omega x}\, dx$

$$= \frac{1}{2}\left[\int\limits_{-\infty}^{0} e^{(1 + i\omega)x}\, dx + \int\limits_0^\infty e^{-(1 - i\omega)x}\, dx \right]$$

$$= \frac{1}{2}\left[\frac{1}{1 + i\omega} + \frac{1}{1 - i\omega} \right] = \frac{1}{1 + \omega^2}$$

Part B

22. $\frac{e^{3t} - 1}{3te^t}; \frac{1}{2}; \frac{3}{4}$
23. $(pe^{i\omega} + q)^n$; np; npq
24. $e^{\lambda(e^t - 1)}$; 0; λ; λ; $3\lambda^2 + \lambda$
25. $p^n (1 - qe^{i\omega})^{-n}$; $\frac{nq}{p}$; $\frac{nq}{p^2}$
26. $\frac{\lambda e^{at}}{\lambda - t}; \frac{a\lambda + 1}{\lambda}; \frac{1}{\lambda^2}$
27. $\frac{1}{1 - t^2}$; 0; 2
28. $e^{-|\omega|}$; $\mu_X = 0$; σ_X does not exist.
29. $e^{-\lambda|\omega|}$
31. $f(x) = \frac{1}{\sigma\sqrt{2\pi}}, e^{-x^2/2\sigma^2} \quad -\infty < x < \infty$
32. $\phi(\omega) = \frac{1}{1 - i\omega}$; $f(x) = e^{-x}$, $0 < x < \infty$.
33. $\mu_1 = \lambda_1$; $\mu_2 = \lambda_2 + \lambda^2{}_1$; $\mu_3 = \lambda_3 + 3\lambda_2\, \lambda_1 + \lambda^3{}_1$ and

$$\mu_4 = \lambda_4 + 4\lambda_3\,\lambda_1 + 3\lambda^2{}_2 + 6\lambda_2\,\lambda^2{}_1 + \lambda^4{}_1$$

35. $\phi_{XY}(\omega_1, \omega_2) = e^{i(\mu_x\omega_1 + \mu_y\omega_2)^{-\frac{1}{2}}}$ etc. $-\dfrac{1}{2}(\sigma^2{}_x\,\omega_1^2 + 2r\sigma_x\,\sigma_y\,\omega_1\,\omega_2 + \sigma_y^2\,\omega_2^2)$

36. $f_Y(y) = \dfrac{1}{\pi}\cdot\dfrac{1}{\sqrt{1-y^2}}$ in $-1 \le y \le 1$

Exercise 4(E)

2. $P\{10 - c < X < 10 + c\} \ge 1 - \dfrac{16}{c^2}$; $c = 5$

$P\{5 < X < 15\} \ge \dfrac{9}{25}$

3. $P\{|X - \mu| \ge k\sigma\} \le \dfrac{1}{k^2}$

$k = \dfrac{5}{4}$

$P\{|X - 10| \ge 5\} \le \dfrac{16}{25}$

4. $P\{12 - c < X < 12 + c\} \ge 1 - \dfrac{\sigma^2}{c^2}$; $c = 6$

$1 - \dfrac{\sigma^2}{36} = \dfrac{3}{4}$; $\sigma = 3$

5. $P\{\mu - c < X < \mu + c\} \ge 1 - \dfrac{\sigma^2}{c^2}$; $1 - \dfrac{9}{c^2} = \dfrac{3}{4}$; $c = 6$; $\mu = 12$

6. $P\{\mu - c < X < \mu + c\} \ge 1 - \dfrac{\sigma^2}{c^2}$;

$\mu = \dfrac{-2+8}{2} = 3$; $c = 5$

$1 - \dfrac{\sigma^2}{25} = \dfrac{21}{25}$; $\sigma^2 = 4$

7. $\text{Var}(X) = 68 - 64 = 4$

$p\{8 - c < X < 8 + c\} \ge 1 - \dfrac{4}{c^2}$; $c = 3$

Reqiured Probability $\ge \dfrac{5}{9}$

11. $\{E(XY)^2 \le E(X^2)\,E(Y^2)$: Replace X by $X - \mu_x$ and Y by $Y - \mu_y$

$$\frac{\left[E(X-\mu_x)(Y-\mu_y)\right]^2}{E(X-\mu_x)^2\,E(Y-\mu_y)^2} \le 1;\ p_{xy}^2 \le 1;\ |p_{xy}| \le 1$$

12. By Tchebycheff's inequality, $P\{|X - \mu| \le \sqrt{3}\,\sigma\} \ge 1 - \frac{1}{3} = \left(\frac{2}{3}\right)$. Since 0.5 $< \frac{2}{3}$, such a RV does not exist.
13. 21/25
14. 4/9; 0.134
15. 1/4; 1/4
16. 1/3
17. 19/24
20. 250
22. $c = 1$

Exercise 4(F)

8. If $X_1, X_2, \cdots, X_n$ be a sequence of independent and identically distributed RVs with $E(X_i) = \mu$ and $\text{Var}(X_i) = \sigma^2$ and if $\overline{X} = \frac{1}{n}(X_1 + X_2 + \cdots + X_n)$, then $\overline{X}$ follows $N\left(\mu, \frac{\sigma}{\sqrt{n}}\right)$ as $n \to \infty$. This result is used in theory of sampling.
9. 0.9544
10. 41
11. 0.1112; 0.1915
12. 0.352
13. 0.1814
14. C.L.T does not hold good

5 Some Special Probability Distributions

Introduction

In Chapter 2, we have just stated the definitions of certain special probability distributions, both discrete and continuous. While constructing probabilistic models for observable phenomena, certain probability distributions arise more frequently than do others. We treat such distributions that play important roles in many engineering applications as special probability distributions. In this chapter we shall discuss a number of discrete as well as continuous (random variables) distributions in considerable detail.

Special Discrete Distributions

1. Binomial distribution

Definition: Let A be some event associated with a random experiment E, such that $P(A) = p$ and $P(\overline{A}) = 1 - p = q$. Assuming that p remains the same for all repetitions, if we consider n independent repetitions (or trials) of E and if the random variable (RV) X denotes the number of times the event A has occurred, then X is called a *binomial random variable* with parameters n and p or we say that X follows a *binomial distribution* with parameters n and p, or symbolically $B(n, p)$. Obviously the possible values that X can take, are 0, 1, 2, …, n.

By the theorem under Bernoulli's trials in chapter 1, the probability mass function of a binomial RV is given by

$$P(X = r) = nC_r\, p^r q^{n-r}; \quad r = 0, 1, 2, \ldots, n \text{ where } p + q = 1$$

Note *(1) Binomial distribution is a legitimate probability distribution since*

$$\sum_{r=0}^{n} P(X = r) = \sum_{r=0}^{n} nC_r\, q^{n-r} p^r$$

$$= (q + p)^n = 1$$

(2) *The name 'binomial distribution' is given since the probabilities $nC_r\, q^{n-r}\, p^r$ $(r = 0, 1, 2, \ldots, n)$ are the successive terms in the expansion of the binomial expression $(q + p)^n$.*

(3) *If we assume that n trials constitute a set and if we consider N sets, the frequency function of the binomial distribution is given by $f(r) = N\,p(r) = N \cdot nC_r\, q^{n-r}\, p^r$, $r = 0, 1, 2, \ldots, n$. In other words, the number of sets in which we get exactly r successes (the occurrences of the event A) $= N \cdot nC_r\, q^{n-r}\, p^r$; $r = 0, 1, 2, \ldots, n$.*

Mean and Variance of the Binomial Distribution

We have already found out $E(X)$ and Var(X) for the binomial distribution $B(n, p)$ using the moment generating function in Example 3 in Worked Example 4 (d). Here we shall find them directly using the definitions of $E(X)$ and Var(X).

$$E(X) = \sum_r x_r\, p_r$$

$$= \sum_{r=0}^{n} r \cdot nC_r\, p^r\, q^{n-r}$$

$$= \sum_{r=0}^{n} r \cdot \frac{n!}{r!(n-r)!}\, p^r\, q^{n-r} \tag{1}$$

$$= np \cdot \sum_{r=1}^{n} \frac{(n-1)!}{(r-1)!\{(n-1)-(r-1)\}!}\, p^{r-1}\, q^{(n-1)-(r-1)}$$

$$= np \sum_{r=1}^{n} (n-1)\, C_{r-1} \cdot p^{r-1} \cdot q^{(n-1)-(r-1)}$$

$$= np\,(q+p)^{n-1}$$

$$= np \tag{2}$$

$$E(X^2) = \sum_r x_r^2\, p_r = \sum_0^n r^2\, p_r$$

$$= \sum_{r=0}^{n} \{r(r-1) + r\}\, \frac{n!}{r!(n-r)!}\, p^r\, q^{n-r}$$

$$= n(n-1)p^2 \sum_{r=2}^{n} (n-2)C_{r-2}\, p^{r-2}\, q^{n-r} + np, \quad \text{[by (1) and(2)]}$$

$$= n(n-1)\, p^2\, (q+p)^{n-2} + np$$

$$= n(n-1)\, p^2 + np$$

$$\text{Var}(X) = E(X^2) - \{E(X)\}^2$$

$$= n(n-1)\, p^2 + np - n^2p^2$$

$$= np\,(1-p)$$

$$= npq$$

Recurrence Formula for the Central Moments of the Binomial Distribution

By definition, the kth order central moment μ_k is given by $\mu_k = E\{X - E(X)\}^k$.

For the binomial distribution $B(n, p)$,

$$\mu_k = \sum_{r=0}^{n} (r - np)^k \, nC_r \, p^r \, q^{n-r} \tag{1}$$

Differentiating (1) with respect to p, we get

$$\frac{d\mu_k}{dp} = \sum_{r=0}^{n} nC_r \,[-nk\,(r - np)^{k-1} \cdot p^r q^{n-r} + (r - np)^k \{rp^{r-1} q^{n-r} + (n - r)\, p^r q^{n-r-1} (-1\}]$$

$$= -\,nk\,\mu_{k-1} + \sum_{r=0}^{n} nC_r\,(r - np)^k p^{r-1} q^{n-r-1} \{rq - (n - r)p\}$$

$$= -\,nk\,\mu_{k-1} + \sum_{r=0}^{n} nC_r\,(r - np)^k p^{r-1} q^{n-r-1} (r - np)\ (\because p + q = 1)$$

$$= -\,nk\,\mu_{k-1} + \frac{1}{pq} \sum_{r=0}^{n} nC_r\, p^r q^{n-r} (r - np)^{k+1}$$

$$= -\,nk\,\mu_{k-1} + \frac{1}{pq}\,\mu_{k+1}$$

i.e.,
$$\mu_{k+1} = pq\left[\frac{d\mu_k}{dp} + nk\,\mu_{k-1}\right] \tag{2}$$

Using recurrence relation (2), we may compute moments of higher order, provided we know moments of lower order.

Putting $k = 1$ in (2), we get

$$\mu_2 = pq\left[\frac{d\mu_1}{dp} + n\mu_0\right]$$

$$= npq\ (\because \mu_0 = 1 \text{ and } \mu_1 = 0)$$

Putting $k = 2$ in (2), we get

$$\mu_3 = pq\left[\frac{d}{dp}\,\mu_2 + 2n\,\mu_1\right]$$

$$= pq\,\frac{d}{dp}\,[np\,(1 - p)]$$

$$= npq\,[1 - 2p] = npq\,(q - p)$$

Putting $k = 3$ in (2), we get

$$\mu_4 = pq\left[\frac{d}{dp}\,\mu_3 + 3n\,\mu_2\right]$$

$$= npq\left[\frac{d}{dp}\{p(1-p)(1-2p)\} + 3npq\right]$$
$$= npq\,[1 - 6p + 6p^2 + 3npq]$$
$$= npq\,[1 - 6pq + 3npq]$$
$$= npq\,[1 + 3\,pq\,(n-2)]$$

2. Poisson distribution

Definition: If X is a discrete RV that can assume the values 0, 1, 2, …, such that its probability mass function is given by

$$P(X = r) = \frac{e^{-\lambda}\lambda^r}{r!}; \quad r = 0, 1, 2, \ldots; \ \lambda > 0$$

then X is said to follow a *Poisson distribution* with parameter λ or symbolically X is said to follow $P(\lambda)$.
(**Note:** Poisson distribution is a legitimate probability distribution, since

$$\sum_{r=0}^{\infty} P(x = r) = \sum_{r=0}^{\infty} \frac{e^{-\lambda}\lambda^r}{r!}$$
$$= e^{-\lambda} e^{\lambda} = 1)$$

Poisson Distribution as Limiting Form of Binomial Distribution

Poisson distribution is a limiting case of binomial distribution under the following conditions:

(i) n, the number of trials is indefinitely large, i.e., $n \to \infty$.

(ii) p, the constant probability of success in each trial is very small, i.e., $p \to 0$.

(iii) $np(= \lambda)$ is finite or $p = \frac{\lambda}{n}$ and $q = 1 - \frac{\lambda}{n}$, where λ is a positive real number.

Proof If X is a binomially distributed RV with parameters n and p, then
$P(X = r) = nC_r\, p^r q^{n-r}$; $r = 0, 1, 2, \ldots, n$.

$$= \frac{n(n-1)(n-2)\cdots(n-r+1)}{r!}\, p^r (1-p)^{n-r}$$

$$= \frac{n(n-1)\cdots(n-r+1)}{r!}\left(\frac{\lambda}{n}\right)^r\left(1-\frac{\lambda}{n}\right)^{n-r}$$

(on putting $p = \frac{\lambda}{n}$)

$$= \frac{\lambda^r}{r!}\left[1\left(1-\frac{1}{n}\right)\left(1-\frac{2}{n}\right)\cdots\left(1-\frac{r-1}{n}\right)\right]\left(1-\frac{\lambda}{n}\right)^n \cdot \left(1-\frac{\lambda}{n}\right)^{-r}$$

$$\therefore \lim_{\substack{n\to\infty \\ np=\lambda=\text{finite}}} [P(X=r)] = \frac{\lambda^r}{r!}\lim_{n\to\infty}\left[\left(1-\frac{1}{n}\right)\left(1-\frac{2}{n}\right)\cdots\left(1-\frac{r-1}{n}\right)\right]$$

$$\lim_{n\to\infty}\left(1-\frac{\lambda}{n}\right)^{-r}\cdot\lim_{n\to\infty}\left(1-\frac{\lambda}{n}\right)^{n}$$

$$= \frac{\lambda^r}{r!}e^{-\lambda} \quad \left[\because \lim_{n\to\infty}\left(1-\frac{k}{n}\right) = 1, \quad \text{when } k \text{ is finite,}\right.$$

$$\left.\lim_{n\to\infty}\left(1-\frac{\lambda}{n}\right)^{-r} = 1 \text{ and } \lim_{n\to\infty}\left(1-\frac{\lambda}{n}\right)^{n} = e^{-\lambda}\right]$$

$$\text{But } P(X=r) = \frac{e^{-\lambda}\cdot\lambda^r}{r!}, \quad r = 0, 1, 2, \ldots$$

is the probability mass function of the Poisson random variable.

Thus the limit of the binomial RV (distribution) is the Poisson RV (distribution).

Note *(1) The above result means that we may compute binomial probabilities approximately by using the corresponding Poisson probabilities, whenever n is large and p is small.*

(2) When an event occurs rarely, the number of occurrences of such an event may be assumed to follow a Poisson distribution. The following are some of the examples, which may be analysed using Poisson distribution:

(i) the number of alpha particles emitted by a radioactive source in a given time interval

(ii) the number of telephone calls received at a telephone exchange in a given time interval

(iii) the number of defective articles in a packet of 100

(iv) the number of printing errors at each page of a book

(v) the number of road accidents reported in a city per day.

Mean and Variance of Poisson Distribution

We have already found out $E(X)$ and $\text{Var}(X)$ for the Poisson distribution $P(\lambda)$, using the characteristic function in Example 4 in Worked Example 4(d). Also, since the Poisson distribution is the limit of binomial distribution, the mean and variance of the Poisson distribution may be obtained as the limits of those of binomial distribution when $n \to \infty$,

i.e., if X is the Poisson RV

$$E(X) = \lim_{\substack{n\to\infty \\ np=\lambda}} (np) = \lambda$$

and $\quad \text{Var}(X) = \lim\limits_{\substack{p \to 0 \\ np = \lambda}} (npq) = \lim\limits_{p \to 0} [\lambda(1-p)] = \lambda.$

Now we shall find $E(X)$ and $\text{Var}(X)$ for the Poisson distribution directly using the definitions.

$$E(X) = \sum_r x_r p_r$$

$$= \sum_{r=0}^{\infty} r \frac{e^{-\lambda} \cdot \lambda^r}{r!} \tag{1}$$

$$= \lambda e^{-\lambda} \sum_{r=1}^{\infty} \frac{\lambda^{r-1}}{(r-1)!}$$

$$= \lambda e^{-\lambda} e^{\lambda} = \lambda \tag{2}$$

$$E(X^2) = \sum_r x_r^2 pr$$

$$= \sum_{r=0}^{\infty} \{r(r-1) + r\} e^{-\lambda} \frac{\lambda^r}{r!}$$

$$= \lambda^2 e^{-\lambda} \sum_{r=2}^{\infty} \frac{\lambda^{r-2}}{(r-2)!} + \lambda \qquad \text{[by (1) and (2)]}$$

$$= \lambda^2 e^{-\lambda} e^{\lambda} + \lambda = \lambda^2 + \lambda$$

$$\text{Var}(X) = E(X^2) - \{E(X)\}^2$$

$$= \lambda^2 + \lambda - \lambda^2 = \lambda$$

Recurrence Formula for the Central Moments of the Poisson Distribution

By definition, the kth order central moment μ_k is given by $\mu_k = E\{X - E(X)\}^k$.

For the Poisson distribution $P(\lambda)$,

$$\mu_k = \sum_{r=0}^{\infty} (r-\lambda)^k e^{-\lambda} \frac{\lambda^r}{r!} \tag{1}$$

Differentiating (1) with respect to λ, we get

$$\frac{d\mu_k}{d\lambda} = \sum_{r=0}^{\infty} \frac{1}{r!} [-k(r-\lambda)^{k-1} e^{-\lambda} \lambda^r + (r e^{-\lambda} \lambda^{r-1} - e^{-\lambda} \lambda^r)(r-\lambda)^k]$$

$$= -k \mu_{k-1} + \sum_{r=0}^{\infty} \frac{1}{r!} e^{-\lambda} \lambda^{r-1} (r-\lambda)^{k+1}$$

$$= -k\mu_{k-1} + \frac{1}{\lambda} \mu_{k+1}$$

i.e., $$\mu_{k+1} = \lambda \left(\frac{d\mu_k}{d\lambda} + k \mu_{k-1} \right) \tag{2}$$

Using recurrence relation (2), we may compute moments of higher order, provided we know moment of lower order.

Putting $k = 1$ in (2), we get

$$\mu_2 = \lambda\left(\frac{d\mu_1}{d\lambda} + \mu_0\right)$$

$$= \lambda \; (\because \mu_0 = 1 \text{ and } \mu_1 = 0)$$

Putting $k = 2$ in (2), we get

$$\mu_3 = \lambda\left(\frac{d\mu_2}{d\lambda} + 2\mu_1\right) = \lambda.$$

Putting $k = 3$ in (2), we get

$$\mu_4 = \lambda\left(\frac{d\mu_3}{d\lambda} + 3\mu_2\right) = \lambda(3\lambda + 1)$$

Note *The interesting property of the Poisson distribution is the equality of its mean, variance and third-order central moment.*

3. Geometric distribution

Definition: Let the RV X denote the number of trials of a random experiment required to obtain the first success (occurrence of an event A). Obviously X can assume the values 1, 2, 3, ...

Now $X = r$, if and only if the first $(r - 1)$ trials result in failure (occurrence of $\bar{A}$) and the rth trial results in success (occurrence of A). Hence

$$P(X = r) = q^{r-1}p; \quad r = 1, 2, 3, \ldots, \infty$$

where

$$P(A) = p \text{ and } P(\bar{A}) = q.$$

If X is a discrete RV that can assume the values 1, 2, 3, ..., ∞ such that its probability mass function is given by

$$P(X = r) = q^{r-1}\,p; \; r = 1, 2, \ldots, \infty \quad \text{where } p + q = 1$$

then X is said to follow *a geometric distribution*.

Note *Geometric distribution is a legitimate probability distribution, since*

$$\sum_{r=1}^{\infty} P(X = r) = \sum_{r=1}^{\infty} q^{r-1}p$$

$$= p(1 + q + q^2 + \ldots \infty)$$

$$= \frac{p}{1-q} = 1$$

Mean and Variance of Geometric Distribution

$$E(X) = \sum_r x_r p_r$$

$$= \sum_{r=1}^{\infty} rq^{r-1}p$$

$$= p[1 + 2q + 3q^2 + \dots + \infty]$$

$$= p(1-q)^{-2} = \frac{1}{p}$$

$$E(X^2) = \sum_r x_r^2 pr$$

$$= \sum_{r=1}^{\infty} r^2 q^{r-1} p$$

$$= p \sum_{r=1}^{\infty} \{r(r+1) - r\} q^{r-1}$$

$$= p[\{1 \times 2 + 2 \times 3q + 3 \times 4\, q^2 + \dots + \infty\} - \{1 + 2q + 3q^2 + \dots + \infty\}]$$

$$= p[2(1-q)^{-3} - (1-q)^{-2}]$$

$$= p\left(\frac{2}{p^3} - \frac{1}{p^2}\right) = \frac{1}{p^2}(2-p) = \frac{1}{p^2}(1+q)$$

$$\text{Var}(X) = E(X^2) - \{E(X)\}^2$$

$$= \frac{1}{p^2}(1+q) - \frac{1}{p^2} = \frac{q}{p^2}.$$

Note *Sometimes the probability mass function of a geometric RV X is taken as*

$$P(X = r) = q^r p;\ r = 0, 1, 2, \dots, \infty \text{ where } p + q = 1$$

It is this definition that was given in chapter II. If this definition is assumed, then

$$E(X) = \frac{q}{p} \text{ and } Var(X) = \frac{q}{p^2} \quad [\text{see example (5) in section 4(d)}]$$

4. Hypergeometric distribution

If X represents the number of defectives found, when n items are drawn without replacement from a lot of N items containing k defectives and $(N - k)$ non-defectives, clearly

$$P(X = r) = \frac{kC_r \cdot (N-k)C_{(n-r)}}{NC_n};\ r = 0, 1, 2, \dots, \min(n, k)$$

Note *If $n > k$, then the maximum value of X is k;*
If $n < k$, then the maximum value of X is n,
i.e., the maximum value of X is min (n, k),
i.e., r can take the values 0, 1, 2, ..., min (n, k).

Definition: If X is a discrete RV that can assume, non-negative values 0, 1, 2, ..., such that its probability mass function is given by

$$P(X = r) = \frac{k\,C_r \cdot (N-k)\,C_{(n-r)}}{N\,C_n}; \; r = 0, 1, 2, \ldots, \min(n, k)$$

then X is said to follow *a hypergeometric distribution* with the parameters N, k and n.

Note (1) *In the probability mass function of X, r can be assumed to take the values 0, 1, 2, ..., n, which is true when $n < k$. But when $n > k$, r can take the values 0, 1, 2, ... k. In other words,*
$P(X = r) = 0$, when $r = k + 1, k + 2, \ldots, n$. This value (namely zero) of the probability is provided by the probability mass function formula itself, since $kC_r = 0$, for $r = k + 1, k + 2, \ldots, n$.
Thus in the value of $P(X = r)$, min (n, k) can be replaced by n.

(2) *Hypergeometric distribution is a legitimatte probability distribution, since*

$$\sum_{r=0}^{n} P(X = r) = \sum_{r=0}^{n} \frac{kC_r\,(N-k)\,C_{(n-r)}}{NC_n} \qquad (1)$$

$$= \frac{1}{NC_n} NC_n = 1 \text{ since}$$

$$\sum_{r=0}^{n} kC_r \cdot (N-k)\,C_{n-r} = \text{coefficient of } x^n \text{ in } (1+x)^k (1+x)^{N-k}$$

$$= \text{coefficient of } x^n \text{ in } (1+x)^N$$

$$= NC_n.$$

Mean and Variance of Hypergeometric Distribution

$$E(X) = \sum_r x_r p_r$$

$$= \sum_{r=0}^{n} r\,kC_r \cdot (N-k)C_{(n-r)}/NC_n$$

$$= \frac{k}{N\,C_n} \sum_{r=1}^{n} (k-1)\,C_{(r-1)}\,(N-k)\,C_{(n-r)}$$

$$= \frac{k}{N\,C_n} \sum_{r'=0}^{n-1} k'C_{r'}\,(N-1-k')C_{(n-1-r')},$$

(on putting $k' = k - 1$ and $r' = r - 1$)

$$= \frac{k}{N C_n} \cdot (N-1)C_{n-1}, \quad \text{[by step (1) in note (2) given above]}$$

$$= \frac{nk}{N}.$$

$$E(X^2) = E\{X(X-1) + X\}$$

$$= E\{X(X-1)\} + \frac{nk}{N}$$

$$= \frac{nk}{N} + \sum_{r=0}^{n} r(r-1)\, kC_r \cdot (N-k)C_{(n-r)}/NC_n$$

$$= \frac{nk}{N} + \frac{k(k-1)}{NC_n} \sum_{r=2}^{n} (k-2)C_{(r-2)} \cdot (N-k)C_{(n-r)}$$

$$= \frac{nk}{N} + \frac{k(k-1)}{NC_n} \sum_{r'=0}^{n-2} k'C_{r'} \cdot (N-2-k')C_{(n-2-r')},$$

(on putting $k' = k-2$ and $r' = r-2$)

$$= \frac{nk}{N} + \frac{k(k-1)}{NC_n} \cdot (N-2)C_{n-2},$$

[by step (1) in note (2) given above]

$$= \frac{nk}{N} + \frac{k(k-1)\, n(n-1)}{N(N-1)}$$

$$\text{Var}(X) = E(X^2) - \{E(X)\}^2$$

$$= \frac{nk}{N} + \frac{k(k-1)\cdot n(n-1)}{N(N-1)} - \frac{n^2 k^2}{N^2}$$

$$= \frac{nk}{N^2(N-1)} [N(N-1) + N(k-1)(n-1) - (N-1)nk]$$

$$= \frac{nk}{N^2(N-1)} [N^2 - Nn - Nk + nk]$$

$$= \frac{nk(N-k)(N-n)}{N^2(N-1)}$$

Note *If we denote the proportion of defective items in the lot as p, i.e.,* $p = \frac{k}{N}$ *and* $q = 1 - p$, *then* $E(X) = np$ *and* $Var(x) = npq\left(\frac{N-n}{N-1}\right)$.

Binomial Distribution as Limiting Form of Hypergeometric Distribution

Hypergeometric distribution tends to binomial distribution as $N \to \infty$ and $\frac{k}{N} = p$.

Proof If X follows a hypergeometric distribution with parameters N, k and n, then

$$P\{X = r\} = \frac{kC_r \cdot (N-k)C_{n-r}}{NC_n}, \; r = 0, 1, 2, \ldots, n$$

$$= \frac{k(k-1)\ldots(k-r+1)}{r!} \cdot \frac{(N-k)(N-k-1)\ldots(N-k-n+r+1)}{(n-r)!} \times \frac{n!}{N(N-1)\ldots(N-n+1)}$$

$$= \frac{n!}{r!(n-r)!} \times$$

$$\frac{\left[\left(\frac{k}{N}\right)\left(\frac{k}{N}-\frac{1}{N}\right)\cdots\left\{\frac{k}{N}-\frac{(r-1)}{N}\right\}\right]\left[\left(1-\frac{k}{N}\right)\left\{1-\frac{k+1}{N}\right\}\cdots\left\{1-\frac{k+n-r-1}{N}\right\}\right]}{1\cdot\left(1-\frac{1}{N}\right)\left(1-\frac{2}{N}\right)\cdots\left\{1-\frac{(n-1)}{N}\right\}}$$

(by dividing each factor in the numerator and denominator by N.)

Putting $\frac{k}{N} = p$ and proceeding to the limit as $N \to \infty$, we get

$$\lim_{\substack{N \to \infty \\ \frac{k}{N} = p}} P\{X = r\} = nC_r \cdot p^r (1-p)^{n-r}$$

$$= nC_r p^r q^{n-r}; \; r = 0, 1, 2, \ldots, n$$

Thus the limit of a hypergeometric distribution is a binomial distribution.

Note *We know that the binomial distribution holds good when we draw samples with replacement (since the probability of getting a defective item has to remain constant), while the hypergeometric distribution holds good when we draw samples without replacement. If the lot size N is very large, there is not much difference in the proportions of defective items in the lot whether the item drawn is replaced or not. The previous result is simply a mathematical statement of this fact.*

Worked Examples 5(A)

Example 1 *Out of 800 families with 4 children each, how many families would be expected to have (i) 2 boys and 2 girls, (ii) at least 1 boy, (iii) at most 2 girls and (iv) children of both sexes. Assume equal probabilities for boys and girls.*

Considering each child as a trial, $n = 4$. Assuming that birth of a boy is a success, $p = \frac{1}{2}$ and $q = \frac{1}{2}$. Let X denote the number of successes (boys).

Solution (i) $P(\text{2 boys and 2 girls}) = P(X = 2)$

$$= 4C_2 \cdot \left(\frac{1}{2}\right)^2 \cdot \left(\frac{1}{2}\right)^{4-2}$$

$$= 6 \times \left(\frac{1}{2}\right)^4 = \frac{3}{8}$$

∴ No. of families having 2 boys and 2 girls

$= N \cdot (P(X = 2)$ (where N is the total no. of families considered)

$$= 800 \times \frac{3}{8}$$

$= 300.$

(ii) $P(\text{at least 1 boy}) = P(X \geq 1)$

$$= P(X = 1) + P(X = 2) + P(X = 3) + P(X = 4)$$

$$= 1 - P(X = 0)$$

$$= 1 - 4C_0 \cdot \left(\frac{1}{2}\right)^0 \cdot \left(\frac{1}{2}\right)^4$$

$$= 1 - \frac{1}{16} = \frac{15}{16}$$

∴ No. of families having at least 1 boy

$$= 800 \times \frac{15}{16} = 750.$$

(iii) $P(\text{at most 2 girls}) = P(\text{exactly 0 girl, 1 girl or 2 girls})$

$$= P(X = 4, X = 3 \text{ or } X = 2)$$

$$= 1 - \{P(X = 0) + P(X = 1)\}$$

$$= 1 - \left\{ 4C_0 \cdot \left(\frac{1}{2}\right)^4 + 4C_1 \cdot \left(\frac{1}{2}\right)^4 \right\}$$

$$= \frac{11}{16}$$

∴ No. of families having at most 2 girls

$$= 800 \times \frac{11}{16} = 550.$$

(iv) P(children of both sexes)

$$= 1 - P(\text{children of the same sex})$$
$$= 1 - \{P(\text{all are boys}) + P(\text{all are girls})\}$$
$$= 1 - \{P(X = 4) + P(X = 0)\}$$
$$= 1 - \left\{ 4C_4 \times \left(\frac{1}{2}\right)^4 + 4C_0 \times \left(\frac{1}{2}\right)^4 \right\}$$
$$= 1 - \frac{1}{8} = \frac{7}{8}$$

$\therefore$ No. of families having children of both sexes

$$= 800 \times \frac{7}{8} = 700.$$

Example 2 *An irregular 6-faced dice is such that the probability that it gives 3 even numbers in 5 throws is twice the probability that it gives 2 even numbers in 5 throws. How many sets of exactly 5 trials can be expected to give no even number out of 2500 sets?*

Solution Let the probability of getting an even number with the unfair dice be p.

Let X denote the number of even numbers obtained in 5 trials (throws).

Given: $P(X = 3) = 2 \times P(X = 2)$

i.e., $5C_3\, p^3q^2 = 2 \times 5C_2\, p^2q^3$

i.e., $p = 2q = 2(1 - p)$

$\therefore$ $3p = 2$ or $p = \frac{2}{3}$ and $q = \frac{1}{3}$

Now P(getting no even number)

$$= P(X = 0)$$
$$= 5C_0 \cdot p^0 \cdot q^5 = \left(\frac{1}{3}\right)^5 = \frac{1}{243}$$

$\therefore$ Number of sets having no success (even number) out of N sets $= N \times P(X = 0)$

$\therefore$ Required number of sets $= 2500 \times \frac{1}{243}$

$= 10$, nearly

Example 3 *The probability of a successful rocket launching is p. If launching attempts are made until 3 successful launchings have occurred, what is the probability that exactly 5 attempts will be necessary? What is the probability that fewer than 5 attempts will be necessary?*

If launching attempts are made until 3 consecutive successful launchings occur, what are the probabilities? [See example (1) in section 1 (c)]

Solution (i) Exactly 5 attempts will be required to get 3 successes (successful launchings of rockets), if 2 successes occur in the first 4 attempts and third success occurs in the fifth attempt.

$\therefore$ P(exactly 5 attempts are required)

$= P$(2 successes in 4 attempts)

$\times P$(success in the single 5th attempt)

$= 4C_2\, p^2q^2 \times p$

[$\because$ The no. of successes in the 4 independent attempts follow $B(4, p)$]

$= 6\, p^3q^2$

(ii) $P\{$fewer than 5 attemps are required$\}$

$= P\{$exactly 3 or 4 attempts are required$\}$

$= [P\{$2 successes in the first 2 attempts$\}$

$\times P$(success in the 3rd attempt)]

$+ [P\{$2 successes in the first 3 attempts$\}$

$\times P$(success in the 4th attempt)]

$= 2C_2\, p^2q^0 \times p + 3C_2\, p^2q^1 \times p$

$= p^3 + 3p^3q = p^3\,(1 + 3q)$

(iii) Five attempts will be required to get 3 consecutive successes, if the first 2 attempts result in failures and the last 3 attempts result in successes.

$\therefore$ Required probability $= q \cdot q \cdot p \cdot p \cdot p = p^3q^2$

(iv) Three attempts will be required to get 3 consecutive successes, if each attempt results in a success.

$\therefore$ Probability for this $= p^3$.

Four attempts will be required to get 3 consecutive successes, if the first attempt results in a failure and the remaining attempts result in a success each.

$\therefore$ Probability for this $= qp^3$

$\therefore$ $P\{$fewer than 5 attempts are required$\}$

$= p^3 + qp^3 = p^3\,(1 + q)$.

Example 4 *A communication system consists of n components, each of which will independently function with probability p. The total system will be able to operate effectively if at least one-half of its components function. For what values of p is a 5-component system more likely to operate effectively than a 3-component system?*

Solution Since the probability p of functioning of every component is a constant and the n components function independently, the no. of components X that function follow a binomial distribution with parameters n and p.

$\therefore$ $P(X = r) = nC_r\, p^rq^{n-r};\ r = 0, 1, 2, \ldots, n.$

P(5-component system functions effectively)

$= P(X = 3$ or 4 or $5)$

$$= \sum_{r=3}^{5} 5C_r\, p^rq^{5-r} \qquad (\because n = 5)$$

P(3-component system functions effectively)

$$= P(X = 2 \text{ or } 3)$$

$$= \sum_{r=2}^{3} 3C_r p^r q^{3-r} \qquad (\because n = 3)$$

5-component system will function more effectively than the 3-component system,

if $\quad \sum_{r=3}^{5} 5C_r p^r q^{5-r} \ge \sum_{r=2}^{3} 3C_r p^r q^{3-r}$

i.e., $\quad 10\, p^3 q^2 + 5p^4 q + p^5 \ge 3p^2 q + p^3$

i.e., $\quad 10p^3 (1 - 2p + p^2) + 5p^4 (1 - p) + p^5 \ge 3p^2(1 - p) + p^3$

i.e., $\quad 3p^2 (2p^3 - 5\, p^2 + 4p - 1) \ge 0$

i.e., $\quad 3p^2 (p - 1)^2 (2p - 1) \ge 0$

i.e., $\quad (2p - 1) \ge 0, \quad [\text{since } 3p^2 (p - 1)^2 \ge 0]$

i.e., $\quad p \ge \frac{1}{2}.$

Example 5 *If the probability that a child is a boy is p, where $0 < p < 1$, find the expected number of boys in a family with n children, given that there is at least one boy.*

Solution Let X be the no. of boys (successes) out of n children (trials)

Then X follows a $B(n, p)$.

Required to find $E\{X/X \ge 1\}$.

$$E\{X/X \ge 1\} = \sum_{r} r \cdot P(X = r/X \ge 1)$$

$$= \sum_{r=1}^{n} r \cdot \frac{P(X = r)}{P(X \ge 1)}$$

$$= \sum_{r=1}^{n} \frac{r \cdot nC_r\, p^r\, q^{n-r}}{1 - P(X = 0)}$$

$$= \frac{1}{1 - q^n} \sum_{r=0}^{n} r \cdot nC_r\, p^r q^{n-r}$$

$$= \frac{np}{1 - q^n}.$$

Example 6 *Two dice are thrown 120 times. Find the average number of times in which the number on the first die exceeds the number on the second die.*

Solution The number on the first die exceeds that on the second die, in the following combinations:

(2, 1); (3, 1), (3, 2); (4, 1), (4, 2), (4, 3); (5, 1), (5, 2), (5, 3); (5, 4); (6, 1), (6, 2), (6, 3), (6, 4), (6, 5),
where the numbers in the parentheses represent the numbers in the first and second dice respectively.

$\therefore$ P(success) = P(no. in the first dice exceeds the no. in the second dice)

$$= \frac{15}{36} = \frac{5}{12}$$

This probability remains the same in all the throws that are independent.

If X is the no. of successes, then X follows a binomial distribution with parameters $n\,(= 120)$ and $p\left(= \frac{5}{12}\right)$.

$$\therefore \qquad E(X) = np = 120 \times \frac{5}{12} = 50$$

Example 7 *Fit a binomial distribution for the following data:*

x:	*0*	*1*	*2*	*3*	*4*	*5*	*6*	*Total*
f:	*5*	*18*	*28*	*12*	*7*	*6*	*4*	*80*

Solution Fitting a binomial distribution means assuming that the given distribution is approximately binomial and hence finding the probability mass function and then finding the theoretical frequencies.

To find the binomial frequency distribution $N(q + p)^n$, which fits the given data, we require N, n and p. We assume N = total frequency = 80 and n = no. of trials = 6 from the given data.

To find p, we compute the mean of the given frequency distribution and equate it to np (mean of the binomial distribution).

x :	0	1	2	3	4	5	6	Total
f :	5	18	28	12	7	6	4	80
fx :	0	18	56	36	28	30	24	192

$$\bar{x} = \frac{\sum fx}{\sum f} = \frac{192}{80} = 2.4$$

i.e., $\quad np = 2.4$ or $6p = 2.4$

$\therefore \quad p = 0.4$ and $q = 0.6$

If the given distribution is nearly binomial, the theoretical frequencies are given by the successive terms in the expansion of $80(0.6 + 0.4)^6$. Thus we get,

	x:	0	1	2	3	4	5	6
Theoretical	f:	3.73	14.93	24.88	22.12	11.06	2.95	0.33

Converting these values into whole numbers consistent with the condition that the total frequency is 80, the corresponding binomial frequency distribution is as follows:

x:	0	1	2	3	4	5	6	Total
f:	4	15	25	22	11	3	0	80

Example 8 *The number of monthly breakdowns of a computer is a RV having a Poisson distribution with mean equal to 1.8. Find the probability that this computer will function for a month*

(a) without a breakdown,
(b) with only one breakdown and
(c) with atleast one breakdown.

Solution Let X denote the number of breakdowns of the computer in a month.

X follows a Poisson distribution with mean (parameter) $\lambda = 1.8$.

$$\therefore \quad P\{X = r\} = \frac{e^{-\lambda} . \lambda^r}{r!} = \frac{e^{-1.8} .(1.8)^r}{r!}$$

(a) $P(X = 0) = e^{-1.8} = 0.1653$
(b) $P(X = 1) = e^{-1.8}\ (1.8) = 0.2975$
(c) $P(X \geq 1) = 1 - P(X = 0) = 0.8347$

Example 9 *It is known that the probability of an item produced by a certain machine will be defective is 0.05. If the produced items are sent to the market in packets of 20, find the number of packets containing at least, exactly and at most 2 defective items in a consignment of 1000 packets using (i) binomial distribution and (ii) Poisson approximation to binomial distribution.*

Solution ***Use of binomial distribution***

p = probability that an item is defective = 0.05, $q = 0.95$ and n = No. of independent items (trials) considered = 20.

Let X denote the number of defectives in the n items considered.

$$P(X = r) = nC_r\, p^r\, q^{n-r}$$

(i) $\therefore \quad P(X = 2) = 20\ C_2\ (.05)^2\ (0.95)^{18} = 0.1887$

If N is the number of sets (packets), each set (packet) containing 20 trials (items), then the number of sets containing exactly 2 successes (defectives) is given by

$$N(X = 2) = N \times P(X = 2) = 1000 \times 0.1887 = 189, \text{ nearly}$$

(ii) $P(\text{at least 2 defectives}) = P(X \geq 2)$

$$= 1 - \{P(X = 0) + P(X = 1)\}$$
$$= 1 - [20C_0\ (0.05)^0\ (0.95)^{20} + 20C_1\ (0.05)^1\ (0.95)^{19}]$$
$$= 1 - [0.3585 + 0.3774]$$
$$= 0.2641$$

$$\therefore \quad N(X \geq 2) = N \times P\ (X \geq 2) = 1000 \times 0.2641 = 264, \text{ nearly}$$

(iii) $P(\text{at most 2 defectives}) = P(X \leq 2)$

$$= P(X = 0) + P(X = 1) + P(X = 2)$$

$$= \sum_{r=0}^{2} 20C_r\,(0.05)^r\,(0.95)^{20-r}$$
$$= 0.3585 + 0.3774 + 0.1887$$
$$= 0.9246$$

$\therefore \quad N(X \le 2) = N \times P(X \le 2)$

$$= 1000 \times 0.9246 = 925, \text{ nearly}$$

Use of Poisson distribution

Since $p = 0.05$ is very small and $n = 20$ is sufficiently large, binomial distribution may be approximated by Poisson distribution with parameter $\lambda = np = 1$.

$$\therefore \quad P(X = r) = \frac{e^{-\lambda} . \lambda^r}{r!} = \frac{e^{-1}}{r!}$$

(i) $\quad P(X = 2) = \dfrac{e^{-1}}{2!} = 0.1839$

$\therefore \quad N(X = 2) = 1000 \times 0.1839 = 184$, nearly

(ii) $\quad P(X \ge 2) = 1 - \{P(X = 0) + P(X = 1)\}$

$$= 1 - \{e^{-1} + e^{-1}\} = 0.2642$$

$\therefore \quad N(X \ge 2) = 1000 \times 0.2642 = 264$, nearly.

(iii) $\quad P(X \le 2) = \displaystyle\sum_{r=0}^{2} P(X = r) = \sum_{r=0}^{2} \frac{e^{-1}}{r!}$

$$= 0.9197$$

$\therefore \quad N(X \le 2) = 920$, nearly.

Example 10 *Prove the reproductive property of independent Poisson RVs. Hence find the probability of 5 or more telephone calls arriving in a 9 min period in a college switch-board, if the telephone calls that are received at the rate of 2 every 3 min follow a Poisson distribution.*

Solution Let X_1 and X_2 be independent RVs that follow Poisson distributions with parameters λ_1 and λ_2 respectively.

Let $\quad X = X_1 + X_2$

$$P\{X = n\} = P\{X_1 + X_2 = n\}$$
$$= \sum_{r=0}^{n} P\{X_1 = r] \cdot P\{X_2 = n-r\},$$

(since X_1 and X_2 are independent)

$$= \sum_{r=0}^{n} \frac{e^{-\lambda_1} \cdot \lambda_1^r}{r!} \, \frac{e^{-\lambda_2} \cdot \lambda_2^{n-r}}{(n-r)!}$$

$$= \frac{e^{-(\lambda_1 + \lambda_2)}}{n!} \sum_{r=0}^{n} \frac{n!}{r!(n-r)!} \lambda_1^r \cdot \lambda_2^{n-r}$$

$$= \frac{e^{-(\lambda_1+\lambda_2)}}{n!} \sum_{r=0}^{n} nC_r \cdot \lambda_1^r \times \lambda_2^{n-r}$$

$$= \frac{e^{-(\lambda_1+\lambda_2)} \cdot (\lambda_1+\lambda_2)^n}{n!}$$

Thus the sum of 2 independent Poisson variables with parameters λ_1 and λ_2 is also a Poisson variable with parameter $(\lambda_1 + \lambda_2)$.

This property, which can be extended to any finite number of independent Poisson variables is known as *the Reproductive Property* of Poisson RVs. [For an alternative proof, see example (9) in section 4(d)]

Let X_1, X_2, X_3 denote the number of telephone calls received in three consecutive 3-min periods.

Each of X_1, X_2, X_3 follows a Poisson distribution with parameter (mean) $\lambda = 2$.

$\therefore$ $X = X_1 + X_2 + X_3$ follow a Poisson distrubution with parameter 6.

Clearly X represents the number of calls received in a 9-min period.

Now $P(X \geq 5) = 1 - P\{X \leq 4)$

$$= 1 - \sum_{r=0}^{4} \frac{e^{-6} . 6^r}{r!}$$

$$= 1 - (0.0025 + 0.0149 + 0.0446 + 0.0892 + 0.1339)$$

$$= 1 - 0.2851 = 0.7149$$

Example 11 *If the number X of particles emitted during a 1-h period from a radioactive source has a Poisson distribution with parameter $\lambda = 4$ and that the probability that any emitted particle is recorded is $p = 0.9$, find the probability distribution of the number Y of the particles recorded in a 1-h period and hence the probability that no particle is recorded.*

Solution

$$P\{Y = n\} = \sum_{r=0}^{\infty} P[X = n + r \text{ and } n \text{ of them are recorded}]$$

$$= \sum_{r=0}^{\infty} \frac{e^{-\lambda} \cdot \lambda^{n+r}}{(n+r)!} (n+r)C_n \cdot p^n q^r$$

$$= \sum_{r=0}^{\infty} \frac{e^{-\lambda} \cdot (\lambda p)^n}{(n+r)!} \frac{(n+r)!}{n!r!} (\lambda q)^r$$

$$= \frac{e^{-\lambda} \cdot (\lambda p)^n}{n!} \sum_{r=0}^{\infty} \frac{1}{r!} (\lambda q)^r$$

$$= \frac{e^{-\lambda} \cdot (\lambda p)^n}{n!} e^{\lambda q}$$

$$= \frac{e^{-\lambda(1-q)} \cdot (\lambda p)^n}{n!} = \frac{e^{-\lambda p} \cdot (\lambda p)^n}{n!}, \quad n = 0, 1, 2, \ldots, \infty.$$

Therefore Y, the number of recorded particles, follows a Poisson distribution with parameter λp.

Hence $\quad P(Y = 0\} = e^{-\lambda p}$

$$= e^{-4 \times 0.9} = e^{-3.6}$$

$$= 0.0273.$$

Example 12 *If X and Y are independent Poisson RVs, show that the conditional distribution of X, given the value of X + Y, is a binomial distribution.*

Solution Let X and Y follow Poisson distributions with parameters λ_1 and λ_2 respectively.

Now $P\{X = r/(X + Y) = n\}$

$$= \frac{P\{X = r \text{ and } (X + Y) = n\}}{P\{(X + Y) = n\}} = \frac{P[X = r; Y = n - r]}{P[X + Y = n]}$$

(by independence of X and Y.)

$$= \frac{P\{X = r\} \cdot P\{Y = n - r\}}{P\{(X + Y) = n\}}$$

$$= \frac{\{e^{-\lambda_1} \cdot \lambda_1^r / r!\}\{e^{-\lambda_2} \cdot \lambda_2^{n-r} / (n-r)!\}}{e^{-(\lambda_1 + \lambda_2)} \cdot (\lambda_1 + \lambda_2)^n / n!}$$

(by the reproductive property)

$$= \frac{n!}{r!(n-r)!} \left(\frac{\lambda_1}{\lambda_1 + \lambda_2}\right)^r \cdot \left(\frac{\lambda_2}{\lambda_1 + \lambda_2}\right)^{n-r}$$

$$= nC_r\, p^r q^{n-r}, \text{ where } p = \frac{\lambda_1}{\lambda_1 + \lambda_2} \text{ and } q = \frac{\lambda_2}{\lambda_1 + \lambda_2}$$

Hence the required result.

Example 13 *Fit a Poisson distribution for the following distribution:*

x:	*0*	*1*	*2*	*3*	*4*	*5*	*Total*
f:	*142*	*156*	*69*	*27*	*5*	*1*	*400*

Solution Fitting a Poisson distribution for a given distribution means assuming that the given distribution is approximately Poisson and hence finding the probability mass function and then finding the theoretical frequencies.

To find the probability mass function

$$P\{X = r\} = \frac{e^{-\lambda} \cdot \lambda^r}{r!} \quad r = 0, 1, 2, \ldots, \infty$$

of the approximate Poisson distribution, we require λ, which is the mean of the Poisson distribution.

We find the mean of the given distribution and assume it as λ.

x :	0	1	2	3	4	5	Total
f:	142	156	69	27	5	1	400
fx :	0	156	138	81	20	5	400

$$\bar{x} = \frac{\sum f x}{\sum f} = \frac{400}{400} = 1 = \lambda$$

The theoretical frequencies are given by

$$\frac{N e^{-\lambda} \cdot \lambda^r}{r!} \quad \text{where } N = 400, \text{ obtained from the given distribution.}$$

$$= \frac{400\, e^{-1}}{r!}, \quad r = 0, 1, 2, \ldots, \infty$$

Thus, we get

	x:	0	1	2	3	4	5
Theoretical	f:	147.15	147.15	73.58	24.53	6.13	1.23

The theoretical frequencies for $x = 6, 7, 8, \ldots$ are very small and hence neglected.

Converting the theoretical frequencies into whole numbers consistent with the condition that the total frequency = 400, we get the following Poisson frequency distribution which fits the given distribution:

	x:	0	1	2	3	4	5
Theoretical	f:	147	147	74	25	6	1

Example 14 *If the probability that an applicant for a driver's licence will pass the road test on any given trial is 0.8, what is the probability that he will finally pass the test (a) on the fourth trial and (b) in fewer than 4 trials?*

Solution Let X denote the number of trials required to achieve the first success. Then X follows a geometric distribution given by

$$P(X = r) = q^{r-1} p; \quad r = 1, 2, 3, \ldots, \infty$$

Here $p = 0.8$ and $q = 0.2$

(a) $P(X = 4) = 0.8 \times (0.2)^{4-1}$

$= 0.8 \times 0.008 = 0.0064.$

(b) $P(X < 4) = \sum_{r=0}^{3} 0.8 \times (0.2)^{r-1}$

$= 0.8\, [(0.2)^0 + (0.2)^1 + (0.2)^2 + (0.2)^3]$

$= 0.9984.$

Example 15 *A and B shoot independently until each has hit his own target. The probabilities of their hitting the target at each shot are $\frac{3}{5}$ and $\frac{5}{7}$ respectively. Find the probability that B will require more shots than A.*

Solution Let X denote the number of trials required by A to get his first success. Then X follows a geometric distribution given by

$$P(X = r) = p_1 q_1^{r-1} = \frac{3}{5} \cdot \left(\frac{2}{5}\right)^{r-1}; \quad r = 1, 2, \ldots, \infty$$

Let Y denote the number of trials required by B to get his first success. Then Y follows a geometric distribution given by

$$P(Y = r) = p_2 \cdot q_2^{r-1} = \frac{5}{7} \cdot \left(\frac{2}{7}\right)^{r-1}; \quad r = 1, 2, \ldots, \infty$$

$P\{B$ requires more trials to get his first success than A requires to get his first success$\}$

$$= \sum_{r=1}^{\infty} P\{X = r \text{ and } Y = r + 1 \text{ or } r + 2, \ldots, \infty\}$$

$$= \sum_{r=1}^{\infty} [P\{X = r\} \cdot P\{Y = r + 1 \text{ or } r + 2, \ldots, \infty\}]$$

(by independence)

$$= \sum_{r=1}^{\infty} \frac{3}{5} \cdot \left(\frac{2}{5}\right)^{r-1} \cdot \sum_{k=1}^{\infty} \frac{5}{7} \cdot \left(\frac{2}{7}\right)^{r+k-1}$$

$$= \sum_{r=1}^{\infty} \frac{3}{7} \cdot \left(\frac{4}{35}\right)^{r-1} \sum_{k=1}^{\infty} \left(\frac{2}{7}\right)^{k}$$

$$= \sum_{r=1}^{\infty} \frac{3}{7} \cdot \left(\frac{4}{35}\right)^{r-1} \cdot \left(\frac{\frac{2}{7}}{1 - \frac{2}{7}}\right)$$

$$= \frac{6}{35} \sum_{r=1}^{\infty} \left(\frac{4}{35}\right)^{r-1} = \frac{6}{35} \cdot \frac{1}{1 - \frac{4}{35}} = \frac{6}{31}.$$

Example 16 *A coin is tossed until the first head occurs. Assuming that the tosses are independent and the probability of a head occurring is p, find the value of p so that the probability that an odd number of tosses is required is equal to 0.6.*

Can you find a value of p so that the probability is 0.5 that an odd number of tosses is required?

Solution Let X denote the number of tosses required to get the first head (success). Then X follows a geometric distribution given by

$$P(X = r) = pq^{r-1}; \quad r = 1, 2, \ldots,$$

$\therefore P(X = \text{an odd number}) = P(X = 1 \text{ or } 3 \text{ or } 5, \ldots)$

$$= \sum_{r=1}^{\infty} P(X = 2r - 1) = \sum_{r=1}^{\infty} p\, q^{2r-2}$$

$$= \frac{p}{q^2}(q^2 + q^4 + q^6 + \ldots)$$

$$= \frac{p}{q^2} \cdot \frac{q^2}{1 - q^2} = \frac{1}{1 + q} \quad \text{(since } p + q = 1\text{)} \tag{1}$$

Now $\dfrac{1}{1+q} = 0.6$, if $\dfrac{1}{2 - p} = 0.6$

i.e., $0.6\,p = 0.2$

i.e., $p = \dfrac{1}{3}$

Now $\dfrac{1}{1+q} = 0.5$, if $\dfrac{1}{2 - p} = 0.5$

i.e., $1 - \dfrac{p}{2} = 1$

i.e., $p = 0$

Though we get $p = 0$, it is meaningless, because

$$P(X = \text{an odd number})$$

$$= \sum_{r=1}^{\infty} pq^{2r-2}$$

$$= 0, \text{ when } p = 0.$$

Hence the value of p cannot be found out.

Example 17 *Establish the memoryless property of geometric distribution, that is, if X is a discrete RV following a geometric distribution, then* $P\{X > m + n/X > m\} = P\{X > n\}$, *where m and n are any two positive integers. Prove the converse also, if it is true.*

Solution Since X follows a geometric distribution,

$$P(X = r) = pq^{r-1};\ r = 1, 2, \ldots, \infty;\ p + q = 1$$

$$\therefore \quad P\{X > k = \sum_{r=k+1}^{\infty} pq^{r-1} = p(q^k + q^{k+1} + q^{k+2} + \ldots + \infty) = \frac{pq^k}{1 - q}$$

$$= q^k$$

$$\text{Now } P\{X > m + n/X > m\} = \frac{P\{X > m + n \text{ and } X > m\}}{P\{X > m\}}$$

$$= \frac{P\{X > m + n\}}{P\{X > m\}}$$

$$= \frac{q^{m+n}}{q^m} = q^n = P(X > n)$$

The converse of the above result is also true, i.e., if $P\{X > m + n/X > m\} = P\{X > n\}$, where m and n are any two positive integers, then X follows a geometric distribution.

Since X takes the values 1, 2, 3, …, $P(X \geq 1) = 1$.

Let $P\{X > 1\} = q$

$$\text{Now} \quad P\{X = (r + 1)\} = P\{X > r\} - P\{X > (r + 1)\} \tag{1}$$

$$\therefore \quad \frac{P\{X = (X + 1)\}}{P\{X > r\}} = 1 - \frac{P\{X > r + 1\}}{P\{X > r\}}$$

$$= 1 - P\{X > (r + 1)/X > r\}$$

$$= 1 - P(X > 1) \text{ (by the data)}$$

$$= 1 - q$$

$$\therefore \quad P\{X = r + 1)\} = (1 - q)\, P(X > r) \tag{2}$$

$$= (1 - q)\, [P\{X > (r - 1)\} - P\{X = r\}]$$

[from (1), on changing r to $(r - 1)$]

$$= (1 - q)\, [P\{X > (r - 1)\} - (1 - q)\, P(X > (r - 1)]$$

[from (2) on changing r to $(r - 1)$]

$$= (1 - q)\, q\, P\{X > (r - 1)\}$$

$$= (1 - q)\, q^2\, P\{X > (r - 2)\}$$

$$= (1 - q)\, q^{r-1}\, P(X > 1), \text{ proceeding similarly}$$

$$= (1 - q)q^r$$

$$\therefore \quad P\{X = r\} = pq^{r-1}, \text{ where } p = 1 - q \text{ and } r = 1, 2, \ldots$$

That is, X follows a geometric distribution.

Example 18 *If two independent RVs X and Y have the same geometric distribution, prove that the conditional distribution of X, given that X + Y = k is a discrete uniform distribution.*

Solution **Given:** $P(X = r) = P(Y = r) = pq^{r-1};\ r = 1, 2, \ldots$

$$\text{Now } P\{X = r/X + Y = k\} = \frac{P\{X = r \text{ and } X + Y = k\}}{P\{X + Y = k\}}$$

$$= \frac{P\{X = r \text{ and } Y = k - r\}}{P\{X + Y = k\}}$$

$$= \frac{P\{X = r\} \cdot P\{Y = k - r\}}{\sum_{r=1}^{k-1} P\{X = r\} \cdot P\{Y = k - r\}} \quad \text{(by independence)}$$

$$= \frac{pq^{r-1} \cdot pq^{k-r-1}}{\sum_{r=1}^{k-1} pq^{r-1} \cdot pq^{k-r-1}}$$

$$= \frac{q^{k-2}}{\sum_{r=1}^{k-1} q^{k-2}} = \frac{1}{k-1}; \quad r = 1, 2, 3, \ldots, (k-1)$$

Note *When $P\{X = r\}$ = constant, the discrete RV X is said to follow a discrete uniform distribution.*

Thus the conditional distribution of X, given that $X + Y = k$, is a discrete uniform distribution.

Example 19 *A taxi cab company has 12 Ambassadors and 8 Fiats. If 5 of these taxi cabs are in the workshop for repairs and an Ambassador is as likely to be in for repairs as a Fiat, what is the probability that*

(i) 3 of them are Ambassadors and 2 are Fiats,

(ii) at least 3 of them are Ambassadors and

(iii) all the 5 are of the same make?

Solution Let X denote the number of Ambassadors in the workshop out of the 5 taxi cabs.

We note that $N = 20$, $k = 12$, and $n = 5$ and X follows a hypergeometric distribution given by

$$P(X = r) = \frac{12\,C_r \cdot 8\,C_{5-r}}{20\,C_5} \quad r = 0, 1, \ldots, 5$$

(i) P(3 Ambassadors and 2 Fiats) $= \dfrac{12\,C_3 \cdot 8\,C_2}{20\,C_5}$

$$= \frac{220 \times 28}{15{,}504} = \frac{385}{969}$$

(ii) P(at least 3 Ambassaddors) $= P(X \geq 3)$

$$= P(X = 3) + P(X = 4) + P(X = 5)$$

$$= \sum_{r=3}^{5} \frac{12\,C_r \cdot 8\,C_{5-r}}{20\,C_5}$$

$$= \frac{1}{15{,}504}\{220 \times 28 + 495 \times 8 + 792 \times 1\} = \frac{682}{969}$$

(iii) P(all the 5 are of the same make)

$= P$(all are Ambassadors or all are Fiats)

$= P(X = 5 \text{ or } X = 0)$

$$= P(X = 5) + P(X = 0)$$

$$= \frac{12\,C_5 \cdot 8\,C_0}{20\,C_5} + \frac{12\,C_0 \cdot 8\,C_5}{20\,C_5}$$

$$= \frac{1}{15{,}504}\,(\,792 + 56\,) = \frac{53}{969}.$$

Example 20 *A panel of 7 judges is to decide which of the 2 final contestants A and B will be declared the winner; a simple majority of the judges will determine the winner. Assume that 4 of the judges will vote for A and the other 3 will vote for B. If we randomly select 3 of the judges and seek their verdict, what is the probability that a majority of them will favour A?*

Solution Let X denote the number of judges favouring A. We note that $N = 7$, $k = 4$ and $n = 3$ and X follows a hypegeometric distribution, given by

$$P(X = r) = \frac{4\,C_r \cdot 3\,C_{3-r}}{7\,C_3} \quad r = 0, 1, 2, 3$$

P(a majority of 3 chosen judges will favour A) $= P(X \geq 2)$

$$= P(X = 2) + P(X = 3)$$

$$= \frac{1}{7\,C_3}\,\{4C_2 \cdot 3C_1 + 4C_3 \cdot 3C_0\}$$

$$= \frac{1}{35}\,(\,6 \times 3 + 4 \times 1) = \frac{22}{35}$$

Exercise 5(A)

Part A (Short answer questions)

1. The mean and variance of a binomial distribution are 4 and $\frac{4}{3}$ respectively. Find $P(X \geq 1)$, if $n = 6$
2. If the recurrence relation for the central moments of the binomial distribution is $\mu_{r+1} = pq\left(nr\,\mu_{r-1} + \frac{d\mu_r}{dp}\right)$, find the value of β_1.
3. In 256 sets of 8 tosses of a coin, in how many sets one may expect heads and tails in equal numbers?
4. An experiment succeeds twice as often as it fails. Find the chance that in the next 4 trials, there shall be at least one success.
5. In a family of 4 children, what is the probability that there will be at least 1 boy and at least 1 girl, assuming equal probability for boys and girls.
6. If X has the distribution $B\left(25, \frac{1}{5}\right)$, find $P(X < \mu - 2\sigma)$ where μ and σ^2 are the mean and variance of the distribution.

7. Show that the largest value of the variance of a binomial distribution is $\frac{n}{4}$.
8. Find the mean and SD of the distribution whose moment generating functions is $(0.4\, e^t + 0.6\,)$.
9. When will the sum of 2 binomial variates having distributions $B(n_1, p_1)$ and $B(n_2, p_2)$ be also a binomial variate?
10. If X follows $B\left(3, \frac{1}{3}\right)$ and Y follows $B\left(5, \frac{1}{3}\right)$, find $P(X + Y \geq 1)$.
11. Write down the pmf of the Poisson distribution which is approximately equivalent to $B(100, 0.02)$.
12. If X is a Poisson variate such that $2P(X = 0) + P(X = 2) = 2P(X = 1)$, find $E(X)$.
13. If X is a poisson variate such that $E(X^2) = 6$, find $E(X)$.
14. If X is a Poisson variate such that $P(X = 0) = 0.5$, find $\text{Var}(X)$.
15. If X is a Poisson variate with parameter $\lambda > 0$, prove that
$$E(X^2) = \lambda E(X + 1).$$
16. If X is a Poisson variate with parameter λ, prove that
$$E(X \text{ is even}) = \frac{1}{2}\,(1 + e^{-2\lambda}).$$
17. If the MGF of a discrete RV X is $e^{4(et-1)}$, find $P(X = \mu + \sigma)$, where μ and σ are the mean and SD of X.
18. If X and Y are independent identical Poisson variates with mean 1, find $P(X + Y = 2)$.
19. Find the mean and variance of the discrete probability distribution given by
$P(X = r) = e^{-t}\,(1 - e^{-t})^{r-1} \quad r = 1, 2, 3, ..., \infty.$
20. If X is a geometric variate, taking values 1, 2, 3, …, ∞ find $P(X \text{ is odd})$.
21. Find the mean and variance of the distribution given by
$$P\,(X = r) = \frac{2}{3^r},\ r = 1, 2, ..., \infty.$$
22. For the geometric distribution of X, which represents the number of Bernoulli's trials required to get the first success $\text{Var}(X) = 2E(X)$. Find the pmf of the distribution.
23. Find the MGF of the geometric distribution, given by $P(X = r) = q^{r-1}\,p$, $r = 1, 2, \ldots, \infty$.
24. If the MGF of a discrete RV X, taking values 1, 2, …, ∞, is $e^t\,(5 - 4e^t)^{-1}$, find the mean and variance of X.
25. Define hypergeometric distribution and give an example for the situation where it arises.
26. Write down the mean and variance of the hypergeometric distribution given by
$$P(X = r) = kC_r(N - k)\,C_{n-r}/NC_n,\ r = 0, 1, 2, \ldots.$$

27. State the conditions under which the hypergeometric distribution tends to the binomial distribution. Hence deduce the mean and variance of the binomial distribution from those of the hypergeometric distribution.

Part B

28. It is known that diskettes produced by a certain company are defective with a probability 0.01 independently of each other. The company markets diskettes in packages of 10 and offers a money-back guarantee that atmost 1 of the 10 diskettes is defective. What proportion of diskettes are returned? If someone buys 3 diskettes, what is the probability that he will return exactly one of them?
29. Assuming that half the population is vegetarian and that 100 investigators each take 10 individuals to see whether they are vegetarians, how many would you expect to report that 3 people or less were vegetarians?
30. Show that, if 2 symmetric binomial distributions of degree n are so superposed that the r th term of the one coincides with the $(r + 1)$th term of the other, the distribution formed by adding superposed terms is a binomial distribution of degree $(n + 1)$.
[**Hint:** $nC_{r-1} + nC_r = (n + 1)C_r$]
31. A factory produces 10 articles daily. It may be assumed that there is a constant probability $p = 0.1$ of producing a defective article. Before these articles are stored, they are inspected and the defective ones are set aside. Suppose that there is a constant probability $r = 0.1$ that a defective article is misclassified. If X denotes the number of articles classified as defective at the end of a production day, find (a) $P(X = 3)$ and $P(X > 3)$.
[**Hint:** P(a defective article is classified as defective) = P(an article produced is defective) $\times$ P(it is classified as defective) = $0.1 \times 0.9 = 0.09$]
32. A fair coin is tossed 4 times. If X denotes the number of heads obtained and Y denotes the excess of the number of heads over the number of tails, obtain the probability mass function of Y.
33. An irregular 6-faced die is thrown and the expectation that in 10 throws it will give 5 even numbers is twice the expectation that it will give 4 even numbers. How many times in 10,000 sets of 10 throws would you expect to give no even number?
34. If m things are distributed among a men and b women, show that the probability that the number of things received by men is odd is
$$\frac{1}{2}\left[\frac{(b+a)^m - (b-a)^m}{(b+a)^m}\right].$$
[**Hint:** P(a thing is received by men) $= p = \dfrac{a}{a+b}$ and $q = \dfrac{b}{a+b}$].
35. At least one half of an airplane's engines are required to function in order for it to operate. If each engine independently functions with probability

p, for what values of p is a 4-engine plane to be preferred for operation to a 2-engine plane?

36. At least one half of an airplane's engines are required to function in order for it to operate. If each engine functions independently with probability of failure q, for what values of q is a 2-engine plane to be preferred for operation to a 4-engine plane?
37. If a fair coin is flipped an even number $2n$ times, show that the probability of getting more heads than tails is $\frac{1}{2}\left[1 - 2nC_n\left(\frac{1}{2}\right)^{2n}\right]$.

    $\left[\right.$**Hint:** P(more heads than tails) = P(less heads than tails = $\frac{1}{2}$ [1 – P(equal number of heads and tails)$\left.\right]$.
38. If a fair coin is tossed at random 5 independent times, find the conditional probability of 5 heads relative to the hypothesis that there are at least 4 heads.
39. A factory has 10 machines which may need adjustment from time to time during the day. Three of these machines are old, each having a probability of $\frac{1}{11}$ of needing adjustment during the day and 7 are new, having the corresponding probability of $\frac{1}{21}$. Assuming that no machine needs adjustment twice on the same day, find the probabilities that on a particular day
    (i) just 2 old and no new machine need adjustment and
    (ii) just 2 machines that need adjustment are of the same type.
40. The probability of a man hitting a target is $\frac{1}{4}$. (i) If he fires 7 times, what is the probability of his hitting the target at least twice? and (ii) How many times must he fire so that the probability of his hitting the target at least once is greater than $\frac{2}{3}$?
41. A set of 6 similar coins are tossed 640 times with the following results:

Number of heads :	0	1	2	3	4	5	6
Frequency :	7	64	140	210	132	75	12

    Calculate the binomial frequencies on the assumption that the coins are symmetrical.
42. Fit a binomial distribution for the following data and hence find the theoretical frequencies:

x:	0	1	2	3	4
f:	5	29	36	25	5

43. A car hire firm has 2 cars which it hires out day by day. The number of demands for a car on each day follows a Poisson distribution with mean 1.5. Calculate the proportion of days on which (i) neither car is used and (ii) some demand is not fulfilled.
44. The proofs of a 500-page book contains 500 misprints. Find the probability that there are at least 4 misprints in a randomly chosen page.
45. If the average number of claims handled daily by an insurance company is 5, what proportion of days will have less than 3 claims? What is the probability that there will be 4 claims in exactly 3 of the next 5 days. Assume that the number of claims on different days are independent.
46. In a certain factory producing razor blades, there is a small chance $\frac{1}{500}$ for any blade to be defective. The blades are supplied in packets of 10. Use Poisson distribution to calculate the approximate number of packets containing (i) no defective blade, (ii) at least 1 defective blade and (iii) at most 1 defective blade in a consignment of 10,000 packets.
47. An insurance company has discovered that only about 0.1% of the population is involved in a certain type of accident each year. If its 10,000 policy holders were randomly selected from the population, what is the probability that not more than 5 of its clients are involved in such an accident next year?
48. In a given city, 4% of all licenced drivers will be involved in at least 1 road accident in any given year. Determine the probability that among 150 licenced drivers randomly chosen in this city
    (i) only 5 will be involved in at least 1 accident in any given year and
    (ii) at most 3 will be involved in at least 1 accident in any given year.
49. A radioactive source emits on the average 2.5 particles per second. Find the probability that 3 or more particles will be emitted in an interval of 4s.
50. It has been established that the number of defective stereos produced daily at a certain plant is Poisson distributed with mean 4. Over a 2-day span, what is the probability that the number of defective stereos does not exceed 3?
51. In an industrial complex, the average number of fatal accidents per month is one-half. The number of accidents per month is adequately described by a Poisson distribution. What is the probability that 6 months will pass without a fatal accident?
52. If the numbers of telephone calls coming into a telephone exchange between 9 A.M. and 10 A.M. and between 10 A.M. and 11 A.M. are independent and follow Poisson distributions with parameters 2 and 6 respectively, what is the probability that more than 5 calls come between 9 A.M. and 11 A.M.?
53. Patients arrive randomly and independently at a doctor's consulting room from 5 P.M. at an average rate of one in 5 min. The waiting room can hold

12 persons. What is the probability that the room will be full, when the doctor arrives at 6 P.M.?

54. The number of blackflies on a broad bean leaf follows a Poisson distribution with mean 2. A plant inspector, however, records the number of flies on a leaf only if at least 1 fly is persent. What is the probability that he records 1 or 2 flies on a randomly chosen leaf? What is the expected number of flies recorded per leaf?
[**Hint:** If X is the number of flies on a leaf, we have to find $P\{X = r/X \geq 1\}$, $r = 1, 2$, and add them.]
55. A radioactive source emits particles at a rate of 10 per minute in accordance with Poisson law. Each particle emitted has a probability of $\frac{2}{5}$ of being recorded. Find the probability that at least 4 particles are recorded in a 2-min period.
56. Fit a Poisson distribution for the following distribution and hence find the expected frequencies.

x:	0	1	2	3	4	5	6
f:	314	335	204	86	29	9	3

57. If the probability that a certain test yields a positive reaction equals 0.4, what is the probability that fewer than 5 negative reactions occur before the first positive one?
58. In a test a light switch is turned on and off until it fails. If the probability that the switch will fail any time it is turned on or off is 0.001, what is the probability that the switch will not fail during the first 800 times it is turned on or off?
59. An item is inspected at the end of each day to see whether it is still functioning properly. If it is found to fail at the 10th inspection and not earlier, what is the maximum value of the probability of its failure on any day?
60. If X and Y are 2 independent RVs, each representing the number of failures preceding the first success in a sequence of Bernoulli's trials with p as the probability of success in a single trial, show that $P(X = Y) = \frac{p}{1+q}$, where $p + q = 1$.
61. A throws 2 dice until he gets 6 and B throws independently 2 other dice until he gets? Find the probability that B will require more throws than A.
62. If 2 independent RVs X and Y have identical geometric distributions with parameter p, find the probability mass function of $(X + Y)$ and hence the expected value of $(X + Y)$.
63. As part of an air-pollution survey, an inspector decides to examine the exhaust of 6 of a company's 24 trucks. If four of the company's trucks emit excessive amounts of pollutants, what is the probability that none of them will be included in the inspector's sample?

64. Among the 120 applicants for a job, only 80 are actually qualified. If 5 of the applicants are randomly selected for an in-depth interview, find the probability that only 2 of them will be qualified for the job using (i) hypergeometric probability and (ii) binomial approximation to hypergeometric probability.
65. If an auditor selects 5 returns from among 15 returns of which 9 contain illegitimate deductions, what is the probability that a majority of the selected returns contains illegitimate deductions?
66. In sampling a lot of 100 items, the sampling plan calls for inspection of 20 pieces. Find the probability of accepting a lot with 5 defectives, if we allow 1 defective in the sample.
67. If X and Y are independent binomial random variables having respective parameters (n, p) and (m, p), prove that the conditional probability mass function of X, given that $X + Y = k$, is that of a hypergeometric RV.
[**Hint:** The joint probability mass function of X and Y is that of a binomial distribution with parameters $(m + n, p)$.]

Special Continuous Distributions

1. Uniform or rectangular distribution

Definition: A continuous RV X is said to follow *a uniform* or *rectangular distribution* in any finite interval, if its probability density function is a constant in that interval.

If X follows a uniform distribution in $a < x < b$, then $f(x) = \frac{1}{b-a}$ in $a < x < b$, as explained below:

When X follows a uniform distribution in (a, b),

$$f(x) = k.$$

By the basic property of a probability density function,

$$\int_{R_X} f(x)\,dx = 1$$

$$\therefore \quad \int_a^b k\,dx = 1$$

$$\therefore \quad k = \frac{1}{b-a}$$

Thus for a uniform distribution in (a, b),

$$f(x) = \frac{1}{b-a}.$$

When X follows a uniform distribution in (a, b), it is symbolically written as: X follows $U(a, b)$.

Moments of the Uniform Distribution $U(a, b)$

Raw moments μ_r of the uniform distribution $U(a, b)$ about the origin are given by

$$\mu'_r = E\{X^r\}, \text{ where } X \text{ follows } U(a, b)$$

$$= \int_a^b x^r \frac{1}{b-a}\,dx$$

$$= \frac{b^{r+1} - a^{r+1}}{(r+1)(b-a)} \quad (1)$$

$$\therefore \quad E(X) = \text{Mean of } U(a, b) = \mu_1' = \frac{1}{2}(b+a) \quad (2)$$

Central moments μ_r of the uniform distribution $U(a, b)$ are given by

$$\mu_r = E[\{X - E(X)\}^r]$$

$$= E\{(X - \frac{1}{2}(b+a)\}^r]$$

$$= \int_a^b \frac{\left\{x - \frac{1}{2}(b+a)\right\}^r}{b-a}\,dx$$

$$= \frac{1}{b-a}\int_{-c}^{c} t^r\,dt, \text{ on putting } t = x - \frac{1}{2}(b+a) \text{ and } c = \frac{1}{2}(b-a)$$

$$= \begin{cases} 0 & \text{if } r \text{ is odd} \\ \dfrac{1}{r+1}\cdot\left(\dfrac{b-a}{2}\right)^r & \text{if } r \text{ is even} \end{cases}$$

Thus $\mu_{2n-1} = 0$ and $\mu_{2n} = \dfrac{1}{2n+1}\cdot\left(\dfrac{b-a}{2}\right)^{2n}$ for $n = 1, 2, 3, \ldots$ (3)

In particular, μ_2 = variance of $U(a, b) = \dfrac{1}{12}(b-a)^2$ (4)

$$\mu_3 = 0 \text{ and } \mu_4 = \frac{1}{80}(b-a)^4$$

The absolute central moments v_r of the uniform distribution $U(a, b)$ are given by

$$v_r = E\{|X - E(X)|^r\}$$

$$= \int_a^b \frac{\left|x - \frac{1}{2}(b+a)\right|^r}{b-a}\,dx$$

$$= \frac{1}{b-a}\int_{-c}^{c} |t|^r\,dt, \text{ on putting } t = x - \frac{1}{2}(b+a) \text{ and } c = \frac{1}{2}(b-a)$$

$$= \frac{2}{b-a} \int_0^c t^r \, dt, \quad \text{(since the integrand is an even function of } t\text{)}$$

$$= \frac{1}{r+1} \cdot \left(\frac{b-a}{2} \right)^r \tag{5}$$

Definition: $E\{|X - E(X)|\}$ is called the *mean deviation* (MD) about the mean of the RV X or of the corresponding distribution.

Thus the MD about the mean of the distribution $U(a, b)$ is given by

$$\nu_1 = \frac{1}{4}(b - a)$$

2. Exponential distribution

Definitions: A continuous RV X is said to follow an *exponential distribution* or *negative exponential distribution* with parameter $\lambda > 0$, if its probability density function is given by

$$f(x) = \begin{cases} \lambda e^{-\lambda x} & x \geq 0 \\ 0 & \text{otherwise} \end{cases}$$

We note that $\int_0^\infty f(x)\, dx = \int_0^\infty \lambda e^{-\lambda x} dx = 1$ and hence $f(x)$ is a legitimate density function.

Mean and Variance of the Exponential Distribution

Raw moments μ'_r about the origin of the exponential distribution are given by

$$\mu'_r = E(X^r)$$

$$= \int_0^\infty x^r \cdot \lambda e^{-\lambda x} dx$$

$$= \frac{1}{\lambda^r} \int_0^\infty y^r e^{-y} dy, \text{ (on putting } y = \lambda x)$$

$$= \frac{1}{\lambda^r} \Gamma(r+1)$$

$$= \frac{r!}{\lambda^r} \tag{1}$$

$\therefore$ $E(X)$ = Mean of the exponential distribution

$$= \mu_1' = \frac{1}{\lambda}, \text{ [from (1)]}$$

Putting $r = 2$ in (1), we get

$$\mu_2' = \frac{2}{\lambda^2}$$

$$\therefore \quad \text{Var}(X) = E(X^2) - \{E(X)\}^2$$

$$= \frac{2}{\lambda^2} - \frac{1}{\lambda^2} = \frac{1}{\lambda^2}$$

Memoryless Property of the Exponential Distribution

If X is exponentially distributed, then

$$P(X > s + t/X > s) = P(X > t), \text{ for any } s, t > 0$$

$$P(X > k) = \int_k^\infty \lambda\, e^{-\lambda x}\, dx$$

$$= \left(-e^{-\lambda x}\right)_k^\infty = e^{-\lambda k} \tag{1}$$

$$\text{Now } P(X > s + t/X > s) = \frac{P\{X > s + t \text{ and } X > s\}}{P\{X > s\}}$$

$$= \frac{P\{X > s + t\}}{P\{X > s\}}$$

$$= \frac{e^{-\lambda(s+t)}}{e^{-\lambda s}}, \text{ [by (1)]}$$

$$= e^{-\lambda t} = P(X > t).$$

Note *The converse of this result is also true. That is, if $P(X > s + t/X > s) = P(X > t)$, then X follows an exponential distribution. See Example (8) in Worked Example 5(b).]*

3. Erlang distribution or General Gamma distribution

Definition: A continuous RV X is said to follow an *Erlang distribution* or *General Gamma distribution* with parameters $\lambda > 0$ and $k > 0$, if its probability density function is given by

$$f(x) = \begin{cases} \dfrac{\lambda^k\, x^{k-1}\, e^{-\lambda x}}{\Gamma(k)}, & \text{for } x \geq 0 \\ 0, & \text{otherwise} \end{cases}$$

$$\text{We note that } \int_0^\infty f(x)\, dx = \frac{\lambda^k}{\Gamma(k)} \int_0^\infty x^{k-1}\, e^{-\lambda x}\, dx$$

$$= \frac{1}{\Gamma(k)} \int_0^\infty t^{k-1}\, e^{-t}\, dt, \text{ [on putting } \lambda x = t]$$

$$= 1$$

Hence $f(x)$ is a legitimate density function.

Note *1. When $\lambda = 1$, the Erlang distribution is called Gamma distribution or simple Gamma distribution with parameter k whose density function is $f(x) = \frac{1}{\Gamma(k)} x^{k-1} e^{-x}$, $x \geq 0$; $k > 0$.*

2. When $k = 1$, the Erlang distribution reduces to the exponential distribution with parameter $\lambda > 0$.

3. Sometimes, the Erlang distribution itself is called Gamma distribution.

Mean and Variance of Erlang Distribution

The raw moments μ_r' about the origin of the Erlang distribution are given by

$$\mu_r' = E(X^r)$$

$$= \int_0^\infty \frac{\lambda^k}{\Gamma(k)} x^{k+r-1} e^{-\lambda x} \, dx$$

$$= \frac{\lambda^k}{\Gamma(k)} \cdot \frac{1}{\lambda^{k+r}} \int_0^\infty t^{k+r-1} e^{-t} \, dt, \text{ (on putting } \lambda x = t)$$

$$= \frac{1}{\lambda^r} \frac{\Gamma(k+r)}{\Gamma(k)}$$

$$\therefore \quad \text{Mean} = E(X) = \frac{1}{\lambda} \cdot \frac{\Gamma(k+1)}{\Gamma(k)} = \frac{k}{\lambda}$$

$$\text{Var}(X) = E(X^2) - [E(X)]^2$$

$$= \frac{1}{\lambda^2} \cdot \frac{\Gamma(k+2)}{\Gamma(k)} - \left(\frac{k}{\lambda}\right)^2$$

$$= \frac{1}{\lambda^2} \{k(k+1) - k^2\} = \frac{k}{\lambda^2}$$

Reproductive Property of Gamma Distribution

The sum of a finite number of independent Erlang variables is also an Erlang variable. That is, if $X_1, X_2, \ldots, X_n$ are independent Erlang variables with parameters $(\lambda, k_1), (\lambda, k_2), \ldots, (\lambda, k_n)$, then $X_1 + X_2 + \ldots + X_n$ is also an Erlang variable with parameter $(\lambda, k_1 + k_2 + \ldots + k_n)$.

Let us first find the moment generating function of the Erlang variable X with parameters λ and k and use it to prove this property. MGF of X is given by

$$M_X(t) = E\{e^{tx}\}$$

$$= \int_0^\infty \frac{\lambda^k}{\Gamma(k)} x^{k-1} e^{-\lambda x} e^{tx} \, dx$$

$$= \frac{\lambda^k}{\Gamma(k)} \int_0^\infty x^{k-1} e^{-(\lambda - t)x} \, dx$$

$$= \frac{\lambda^k}{\Gamma(k)} \cdot \frac{1}{(\lambda - t)^k} \int_0^\infty y^{k-1} e^{-y}\,dy, \quad \text{(on putting } \lambda - t = y\text{)}.$$

$$= \left(\frac{\lambda}{\lambda - t}\right)^k \quad [\because \text{ the integral} = \Gamma(k)]$$

$$= \left(1 - \frac{t}{\lambda}\right)^{-k}$$

Now $M_{(X_1 + X_2 + \ldots + K_n)}(t) = M_{X_1}(t)\, M_{X_2}(t) \ldots M_{X_n}(t)$ (since $X_1, X_2, \ldots, X_n$ are independent)

[Refer to property (4) of MGF given in section 4(d) of chapter (4)]

$$= \left(1 - \frac{t}{\lambda}\right)^{-k_1} \left(1 - \frac{t}{\lambda}\right)^{-k_2} \ldots \left(1 - \frac{t}{\lambda}\right)^{-k_n}$$

$$= \left(1 - \frac{t}{\lambda}\right)^{-(k_1 + k_2 + \ldots + k_n)}$$

which is the MGF of an Erlang variable with parameters $(\lambda, k_1 + k_2 + \ldots + k_n)$. Hence the reproductive property.

Relation Between the Distribution Functions (cdf) of the Erlang Distribution With $\lambda = 1$ (or Simple Gamma Distribution) and (Poisson Distribution)

If X is a Poisson random variable with mean λ,

then $$P(X \le K) = \sum_{r=0}^{k} \frac{e^{-\lambda} \lambda^r}{r!} \tag{1}$$

Differentiating both sides with respect to λ, we get

$$\frac{d}{d\lambda} P(X \le k) = \sum_{r=0}^{k} \frac{1}{r!} \{e^{-\lambda} \cdot r\lambda^{r-1} - e^{-\lambda} \times \lambda^r\}$$

$$= e^{-\lambda} \cdot \sum_{r=0}^{k} \left[\frac{\lambda^{r-1}}{(r-1)!} - \frac{\lambda^r}{r!}\right]$$

$$= e^{-\lambda}\left[-1 + \left(1 - \frac{\lambda}{1!}\right) + \left(\frac{\lambda}{1!} - \frac{\lambda^2}{2!}\right) + \cdots + \left\{\frac{\lambda^{k-1}}{(k-1)!} - \frac{\lambda^k}{k!}\right\}\right]$$

$$= -\frac{e^{-\lambda}\lambda^k}{k!} \tag{2}$$

Integrating both sides of (2) with respect to λ from λ to ∞, we get

$$\left[\sum_{r=0}^{K} \frac{e^{-\lambda}\lambda^r}{r!}\right]^{\infty} = -\int_{\lambda}^{\infty} \frac{1}{k!} e^{-\lambda} \lambda^k\, d\lambda$$

i.e.,
$$\sum_{r=0}^{K} \frac{e^{-\lambda}\lambda^r}{r!} = \int_{\lambda}^{\infty} \frac{1}{k!} e^{-y} y^k \, dy$$

i.e.,
$$P(X \le k) = P(Y \ge \lambda),$$

[where Y is the Erlang variable with parameters 1 and $(k + 1)$

or
$$P(X \le k) = 1 - P(Y \le \lambda)$$

Note *The above relationship is valid only when the parameter k is a positive integer.*

4. Weibull Distribution

Definition: A continuous RV X is said to follow a *Weibull distribution* with parameters α, $\beta > 0$, if the RV $Y = \alpha X^\beta$ follows the exponential distribution with density function $f_y(y) = e^{-y}$, $y > 0$.

Density Function of the Weibull Distribution

Since $Y = \alpha \cdot X^\beta$, we have $y = \alpha \cdot x^\beta$.

By the transformation rule, derived in chapter 3, we have $f_X(x) = f_Y(y) \left| \frac{dy}{dx} \right|$, where $f_X(x)$ and $f_Y(y)$ are the density functions of X and Y respectively.

$$\therefore \quad f_X(x) = e^{-y} \alpha\beta x^{\beta - 1}$$
$$= \alpha\beta x^{\beta-1} e^{-\alpha x^\beta}; \; x > 0 \qquad [\because y > 0]$$

Note *When $\beta = 1$, Weibull distribution reduces to the exponential distribution with parameter α.*

Mean and Variance of the Weibull Distribution

The raw moments μ_r' about the origin of the Weibull distribution are given by

$$\mu_r' = E(X^r)$$
$$= \alpha\beta \int_0^\infty x^{r+\beta-1} e^{-\alpha x^\beta} dx$$
$$= \int_0^\infty \left(\frac{y}{\alpha}\right)^{\frac{r}{\beta}+1-\frac{1}{\beta}} e^{-y} \left(\frac{y}{\alpha}\right)^{\frac{1}{\beta}-1} dy,$$

on putting $y = \alpha x^\beta$ or $x = \left(\frac{y}{\alpha}\right)^{\frac{1}{\beta}}$

$$= \alpha^{-r/\beta} \int_0^\infty y^{r/\beta} e^{-y} dy$$

$$= \alpha^{-r/b} \Gamma\left(\frac{r}{\beta}+1\right)$$

$$\therefore \quad \text{Mean} = E(X) = \mu_1' = \alpha^{-\frac{1}{\beta}} \Gamma\left(\frac{1}{\beta}+1\right)$$

$$\text{Var}(X) = E(X^2) - \{E(X)\}^2$$

$$= \alpha^{-2/\beta}\left[\Gamma\left(\frac{2}{\beta}+1\right) - \left\{\Gamma\left(\frac{1}{\beta}+1\right)\right\}^2\right]$$

Note *Weibull distribution finds frequent applications in Reliability Theory. It is assumed as the probability distribution of the time to failure (or length of life) of a component in a system. Other distributions used to describe the failure law are the exponential and normal distributions. See Example (19) in Worked Example 5(b).*

5. Normal (or Gaussian) distribution

Definition: A continuous RV X is said to follow a *normal distribution* or *Gaussian distribution* with parameters μ and σ, if its probability density function is given by

$$f(x) = \frac{1}{\sigma\sqrt{2\pi}} e^{-(x-\mu)^2/2\sigma^2}; \quad -\infty < x < \infty$$

$$-\infty < \mu < \infty; \ \sigma > 0 \qquad (1)$$

Symbolically 'X follows $N(\mu, \sigma)$'. Sometimes it is also given as $N(\mu, \sigma^2)$. We shall use only the notation $N(\mu, \sigma)$ as in the earlier chapters.

$f(x)$ is a legitimate density function, as

$$\int_{-\infty}^{\infty} f(x)\,dx = \frac{1}{\sigma\sqrt{2\pi}} \int_{-\infty}^{\infty} e^{-(x-\mu)^2/2\sigma^2}\,dx$$

$$= \frac{1}{\sigma\sqrt{2\pi}} \int_{-\infty}^{\infty} e^{-t^2} \sigma\sqrt{2}\,dt, \quad \left(\text{on putting } t = \frac{x-\mu}{\sigma\sqrt{2}}\right)$$

$$= \frac{1}{\sqrt{\pi}} \int_{-\infty}^{\infty} e^{-t^2}\,dt$$

$$= \frac{1}{\sqrt{\pi}} 2\int_{0}^{\infty} e^{-t^2}\,dt = \frac{1}{\sqrt{\pi}} \cdot \Gamma\left(\frac{1}{2}\right) = \frac{1}{\sqrt{\pi}} \cdot \sqrt{\pi} = 1$$

Standard Normal Distribution

The normal distribution $N(0, 1)$ is called the standardised or simply the standard normal distribution, whose density function is given by

$$\phi(z) = \frac{1}{\sqrt{2\pi}} e^{-z^2/2} \quad -\infty < z < \infty$$

This is obtained by putting $\mu = 0$ and $\sigma = 1$ and by changing x and f respectively into z and ϕ. If X has distribution $N(\mu, \sigma)$ and if $Z = \frac{X-\mu}{\sigma}$, then we can prove that Z has distribution $N(0, 1)$.

[See the corollary under the property (6) of normal distribution]

The importance of $N(0, 1)$ is due to the fact that the values of $\phi(z)$ and $\int_0^z \phi(z)\, dz$ are tabulated.

Normal Probability Curve

The graph of $y = f(x)$, that is given above for $\sigma = \sigma_1$ and σ_2, is a well-known bell-shaped curve and is called the normal probability curve.

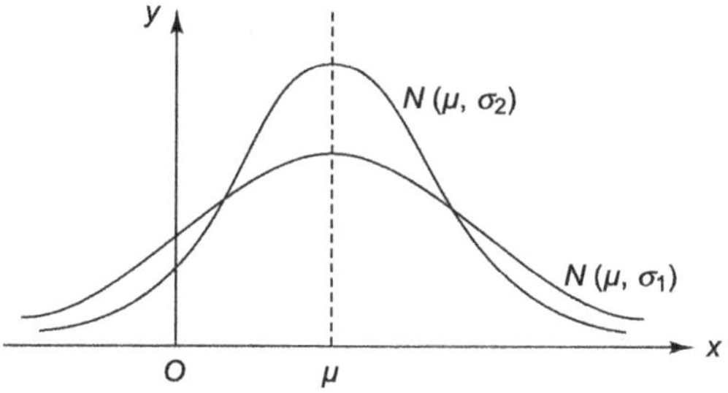

The curve is symmetrical about the ordinate at $x = \mu$. The ordinate $f(x)$ decreases rapidly as x increases numerically, the maximum (occurring at $x = \mu$) given by $\frac{1}{\sigma\sqrt{2\pi}}$. The curve extends upto infinity on either side of $x = \mu$ and the x-axis is an asymptote to the curve.

The graph is concave downward at $x = \mu$ and it is concave upward for large numerical values of x. The points at which the concavity changes are called the points of inflexion of the curve. They are found by solving the equation $y'' = 0$ [i.e., $f''(x) = 0$]. We can prove that the points of inflexion of the normal probability curve occur at $x = \mu \pm \sigma$, that is, at points which are at a distance of σ on either side of $x = \mu$. Thus if σ is relatively large, the curve tends to be flat, while if σ is small, the curve tends to be peaked. Hence the steepness of the curve is determined by σ. The two curves given in the figure relate to $N(\mu, \sigma_1)$ and $N(\mu, \sigma_2)$, where $\sigma_1 > \sigma_2$.

Properties of the Normal Distribution $N(\mu, \sigma)$

1. If X follows $N(\mu, \sigma)$, then $E(X) = \mu$ and $\text{Var}(X) = \sigma^2$

$$E(X) = \int_{-\infty}^{\infty} x f(x)\, dx$$

$$= \frac{1}{\sigma\sqrt{2\pi}} \int_{-\infty}^{\infty} x\, e^{-(x-\mu)^2/2\sigma^2}\, dx$$

$$= \frac{1}{\sqrt{\pi}} \int_{-\infty}^{\infty} (\mu + \sqrt{2}\,\sigma t)\, e^{-t^2}\, dt, \quad \left(\text{on putting } t = \frac{x-\mu}{\sigma\sqrt{2}}\right)$$

$$= \frac{\mu}{\sqrt{\pi}} \int_{-\infty}^{\infty} e^{-t^2}\, dt + \sqrt{\frac{2}{\pi}}\,\sigma \int_{-\infty}^{\infty} t\, e^{-t^2}\, dt$$

$$= \frac{\mu}{\sqrt{\pi}} \cdot \sqrt{\pi} = \mu.$$

(since the integrand in the second integral is an odd function of t)

$$E(X^2) = \frac{1}{\sigma\sqrt{2}} \int_{-\infty}^{\infty} x^2\, e^{-(x-\mu)^2/2\sigma^2}\, dx$$

$$= \frac{1}{\sqrt{\pi}} \int_{-\infty}^{\infty} (\mu + \sqrt{2}\,\sigma t)^2\, e^{-t^2}\, dt, \qquad \left(\text{on putting } t = \frac{x-\mu}{\sigma\sqrt{2}}\right)$$

$$= \frac{1}{\sqrt{\pi}} \left[\int_{-\infty}^{\infty} \mu^2 e^{-t^2}\, dt + 2\sqrt{2}\,\mu\sigma \int_{-\infty}^{\infty} t\, e^{-t^2}\, dt + 2\sigma^2 \int_{-\infty}^{\infty} t^2 e^{-t^2}\, dt \right]$$

$$= \mu^2 + 0 + \frac{2\sigma^2}{\sqrt{\pi}} \int_0^{\infty} t e^{-t^2}\, 2t\, dt, \qquad (\because t^2 e^{-t^2} \text{ is even})$$

$$= \mu^2 + \frac{2\sigma^2}{\sqrt{\pi}} \int_0^{\infty} u^{\frac{1}{2}}\, e^{-u}\, du, \qquad (\text{on putting } u = t^2)$$

$$= \mu^2 + \frac{2\sigma^2}{\sqrt{\pi}} \cdot \Gamma\left(\frac{3}{2}\right)$$

$$= \mu^2 + 2\,\frac{\sigma^2}{\sqrt{\pi}} \cdot \frac{1}{2}\,\Gamma\left(\frac{1}{2}\right)$$

$$= \mu^2 + \sigma^2 \qquad \left(\because \Gamma\left(\frac{1}{2}\right) = \sqrt{\pi}\right)$$

$$\therefore \quad \mathrm{Var}(X) = E(X^2) - \{E(X)\}^2$$

$$= \sigma^2$$

2. Median and mode of the normal distribution N(μ, σ)

For the normal distribution $N(\mu, \sigma)$, the median M is given by

$$\int_M^{\infty} \frac{1}{\sigma\sqrt{2\pi}} e^{-(x-\mu)^2/2\sigma^2}\,dx = \frac{1}{2}$$

i.e.,
$$\int_M^{\mu} \frac{1}{\sigma\sqrt{2\pi}} e^{-(x-\mu)^2/2\sigma^2}\,dx + \int_{\mu}^{\infty} \frac{1}{\sigma\sqrt{2\pi}} e^{-(x-\mu)^2/2\sigma^2} = \frac{1}{2}$$

i.e.,
$$\int_M^{\mu} \frac{1}{\sigma\sqrt{2\pi}} e^{-(x-\mu)^2/2\sigma^2}\,dx + \frac{1}{2} = \frac{1}{2},$$

$$\left\{\text{since } \int_{-\infty}^{\infty} \frac{1}{\sigma\sqrt{2\pi}} e^{-(x-\mu)^2/2\sigma^2}\,dx = 1 \text{ and the normal curve is symmetrical about } x = \mu\right\}$$

i.e.,
$$\int_M^{\mu} f(x)\,dx = 0$$

$$\therefore \quad M = \mu$$

For the normal distribution $N(\mu, \sigma)$,

$$f(x) = \frac{1}{\sigma\sqrt{2\pi}} e^{-(x-\mu)^2/2\sigma^2} \quad -\infty < x < \infty$$

$$\therefore \quad \log f(x) = k - \frac{1}{2\sigma^2}(x-\mu)^2$$

Differentiating with respect to x,

$$\frac{f'(x)}{f(x)} = -\frac{1}{\sigma^2}(x-\mu)$$

i.e.,
$$f'(x) = -\frac{1}{\sigma^2}(x-\mu)f(x)$$
$$= 0, \quad x = \mu$$
$$f''(x) = -\frac{1}{\sigma^2}\{(x-\mu)f'(x) + f(x)\}$$
$$\therefore [f''(x)]_{x=\mu} = -\frac{1}{\sigma^2} f(\mu) < 0$$

Therefore, $f(x)$ is maximum at $x = \mu$. That is, Mode of the distribution $N(\mu, \sigma) = \mu$.

Note *For the normal distribution, mean, median and mode are equal.*

3. Central moments of the normal distribution N(μ, σ)

Central moments μ_r of N(μ, σ) are given by $\mu_r = E(x-\mu)^r$

$$= \frac{1}{\sigma\sqrt{2\pi}} \int_{-\infty}^{\infty} (x-\mu)^r e^{-(x-\mu)^2/2\sigma^2}\,dx$$

$$= \frac{1}{\sqrt{\pi}} \int_{-\infty}^{\infty} (\sqrt{2}\ \sigma t)^r e^{-t^2} dt$$

$$= \frac{2^{r/2} \sigma^r}{\sqrt{\pi}} \int_{-\infty}^{\infty} t^r e^{-t^2} dt$$

Case (i): r is an odd integer, that is, $r = 2n + 1$.

$$\therefore \qquad \mu_{2n+1} = \frac{2^{(2n+1)/2} \sigma^{2n+1}}{\sqrt{\pi}} \int_{-\infty}^{\infty} t^{2n+1} e^{-t^2} dt$$

$= 0$, (since the integrand is an odd function of t)

Case (ii): r is an even integer, that is, $r = 2n$

$$\therefore \qquad \mu_{2n} = \frac{2^n \sigma^{2n}}{\sqrt{\pi}} \int_{-\infty}^{\infty} t^{2n} e^{-t^2} dt$$

$$= \frac{2^n \sigma^{2n}}{\sqrt{\pi}} \cdot 2 \int_0^{\infty} t^{2n} e^{-t^2} dt$$

{$\because$ the integrand is an even function of t)

$$= \frac{2^n \sigma^{2n}}{\sqrt{\pi}} \int_0^{\infty} u^{n-\frac{1}{2}} e^{-u} du, \qquad \text{[on putting } u = t^2)$$

$$= \frac{2^n \sigma^{2n}}{\sqrt{\pi}} \Gamma\left(n + \frac{1}{2}\right) \tag{1}$$

$$= \frac{2^n \sigma^{2n}}{\sqrt{\pi}} \left(\frac{2n-1}{2}\right) \Gamma\left(\frac{2n-1}{2}\right)$$

$$= \frac{2^n \sigma^{2n}}{\sqrt{\pi}} \cdot \left(\frac{2n-1}{2}\right) \cdot \left(\frac{2n-3}{2}\right) \Gamma\left(\frac{2n-3}{2}\right)$$

$$= \frac{2^n \sigma^{2n}}{\sqrt{\pi}} \cdot \frac{2n-1}{2} \cdot \frac{2n-3}{2} \cdots \frac{1}{2} \Gamma\left(\frac{1}{2}\right)$$

$= 1.3.5 \ldots (2n-1)\ \sigma^{2n}$

From (1), we get,

$$\mu_{2n-2} = \frac{2^{n-1} \sigma^{2n-2}}{\sqrt{\pi}} \cdot \Gamma\left(n - \frac{1}{2}\right) \tag{2}$$

From (1) and (2), we get

$$\frac{\mu_{2n}}{\mu_{2n-2}} = 2\sigma^2 (n - 1/2)$$

i.e., $\qquad \mu_{2n} = (2n-1)\sigma^2 \mu_{2n-2}$ (3)

(3) gives a recurrence relation for the even order central moments of the normal distribution $N(\mu, \sigma)$.]

4. Mean Deviation about the mean of the normal distribution $N(\mu, \sigma)$

For the normal distribution $N(\mu, \sigma)$,

$$\text{The MD about the mean} = \int_{-\infty}^{\infty} |x - \mu| \frac{1}{\sigma\sqrt{2\pi}} e^{-(x-\mu)^2/2\sigma^2} dx$$

$$= \frac{1}{\sigma\sqrt{2\pi}} \int_{-\infty}^{\infty} |\sqrt{2}\,\sigma t| e^{-t^2} \sqrt{2}\,\sigma\, dt$$

$$= \sqrt{\frac{2}{\pi}}\,\sigma \int_{-\infty}^{\infty} |t| e^{-t^2} dt$$

$$= 2\sqrt{\frac{2}{\pi}}\,\sigma \int_{0}^{\infty} t\, e^{-t^2} dt, \text{ (since the integrand is an even function of } t)$$

$$= \sqrt{\frac{2}{\pi}}\,\sigma \left(-e^{-t^2}\right)_0^{\infty}$$

$$= \sqrt{\frac{2}{\pi}}\,\sigma = \frac{4}{5}\,\sigma \text{ (approximately).}$$

5. Quartile deviation of the normal distribution $N(\mu, \sigma)$

For the normal distribution $N(\mu, \sigma)$, Q_1 is given by

$$\int_{Q_1}^{\mu} \frac{1}{\sigma\sqrt{2\pi}} e^{-(x-\mu)^2/2\sigma^2} dx = 0.25$$

i.e., $$\int_{(Q_1-\mu)/\sigma}^{0} \frac{1}{\sqrt{2\pi}} e^{-z^2/2} dz = 0.25, \quad \left(\text{on putting } z = \frac{x-\mu}{\sigma}\right)$$

i.e., $$\int_{0}^{(\mu-Q_1)/\sigma} \phi(z)\, dz = 0.25,$$

$$\left(\text{by symmetry of the normal curve and since } \frac{Q_1-\mu}{\sigma} < 0.\right)$$

From the table of normal areas (areas under standard normal curve), we get

$$\int_{0}^{0.674} \phi(z)\, dz = 0.25$$

$$\therefore \qquad \frac{\mu - Q_1}{\sigma} = 0.674$$

i.e., $\qquad Q_1 = \mu - 0.674\,\sigma$

By symmetry, $\qquad Q_3 = \mu + 0.674\,\sigma$

$$\therefore \quad QD = \frac{1}{2}(Q_3 - Q_1) = 0.674\,\sigma = \frac{2}{3}\,\sigma \text{ (approximately)}$$

6. Moment generating function of N(0, 1) and N(μ, σ)

The moment generating function of $N(0, 1)$ is given by

$$M_Z(t) = E(e^{tz})$$

$$= \int_{-\infty}^{\infty} e^{tz}\,\phi(z)\,dz,$$

[where $\phi(z)$ is the density function of $N(0, 1)$]

$$= \frac{1}{\sqrt{2\pi}} \int_{-\infty}^{\infty} e^{-z^2/2}\,e^{tz}\,dz$$

$$= \frac{1}{\sqrt{2\pi}} \int_{-\infty}^{\infty} e^{-(z^2 - 2tz)/2}\,dz$$

$$= \frac{1}{\sqrt{2\pi}} \int_{-\infty}^{\infty} e^{-\{(z-t)^2 - t^2\}/2}\,dz$$

$$= \frac{1}{\sqrt{2\pi}}\,e^{t^2/2}\sqrt{2} \int_{-\infty}^{\infty} e^{-u^2}\,du, \quad \left(\text{on putting } u = \frac{z-t}{\sqrt{2}}\right)$$

$$= \frac{1}{\sqrt{\pi}}\,e^{t^2/2}\,\Gamma\!\left(\frac{1}{2}\right) = e^{t^2/2} \quad \left(\because \Gamma\!\left(\frac{1}{2}\right) = \sqrt{\pi}\right)$$

The moment generating function of $N(\mu, \sigma)$ is given by

$$M_X(t) = M_{\sigma Z + \mu}(t), \quad \left(\text{since } Z = \frac{X - \mu}{\sigma}\right)$$

$$= e^{\mu t} \times \mu_Z(\sigma t), \quad \text{(by the property of MGF)}$$

$$= e^{\mu t}\,e^{\sigma^2 t^2/2}$$

$$= e^{t\left(u + \frac{\sigma^2 t}{2}\right)}$$

Now $$M_X(t) = 1 + \frac{t}{1!}\left(\mu + \frac{\sigma^2 t}{2}\right) + \frac{t^2}{2!}\left(\mu + \frac{\sigma^2 t}{2}\right)^2 + \frac{t^3}{3!}\left(\mu + \frac{\sigma^2 t}{2}\right)^3 + \frac{t^4}{4!}\left(\mu + \frac{\sigma^2 t}{2}\right)^4 + \ldots + \infty$$

$$\therefore \quad E(X) = \text{Coefficient of } \frac{t}{1!} = \mu$$

$$E(X^2) = \text{Coefficient of } \frac{t^2}{2!} = \sigma^2 + \mu^2$$

$$E(X^3) = \text{Coefficient of } \frac{t^3}{3!} = 3\mu\sigma^2 + \mu^3 \text{ and}$$

$$E(X^4) = \text{Coefficient of } \frac{t^4}{4!} = 3\sigma^4 + 6\mu^2\sigma^2 + \mu^4.$$

Using the relation μ_k = kth order central moment = $E\{(X - \mu)^k\}$, we get

$$\mu_1 = 0,\ \mu_2 = \sigma^2,\ \mu_3 = 0\ \mu_4 = 3\sigma^4$$

We could have got these values from the formulas $\mu_{2n+1} = 0$ and $\mu_{2n} = 1.3.5 \ldots (2n - 1)\, \sigma^{2n}$, which we have derived already.

Corollary

If X has the distribution $N(\mu, \sigma)$ then $Y = aX + b$ has the distribution $N(a\mu + b, a\sigma)$

$$M_X(t) = e^{\mu t + \sigma^2 t^2/2}$$

$$\begin{aligned} \therefore \quad M_Y(t) &= M_{aX+b}(t) \\ &= e^{bt} M_X(at) \\ &= e^{bt} e^{a\mu t + a^2\sigma^2 t^2/2} \\ &= e^{(a\mu + b)t + (a^2\sigma^2)t^2/2} \end{aligned}$$

which is the MGF of $N(a\mu + b, a\sigma)$.

In particular, if X has the distribution $N(\mu, \sigma)$, then $Z = \dfrac{X - \mu}{\sigma}$ has the distribution $N\left(\dfrac{1}{\sigma}\mu - \dfrac{\mu}{\sigma}, \dfrac{1}{\sigma}\cdot\sigma\right)$ that is, $N(0, 1)$.

7. Additive property of normal distribution

If $X_i (i = 1, 2, \ldots, n)$ be n independent normal RVs with mean μ_i and variance σ_i^2, then $\sum_{i=1}^{n} a_i X_i$ is also a normal RV with mean $\sum_{i=1}^{n} a_i \mu_i$ and variance $\sum_{i=1}^{n} a_i^2 \sigma_i^2$.

$$\begin{aligned} M_{\left(\sum_{i=1}^{n} a_i X_i\right)}(t) &= M_{a_1X_1}(t).\ M_{a_2X_2}(t) \ldots M_{a_nX_n}(t), \quad \text{(by independence)} \\ &= e^{a_1\mu_1 t + a_1^2\sigma_1^2 t^2/2} \times e^{a_2\mu_2 t + a_2^2\sigma_2^2 t^2/2} \times e^{a_n\mu_n t + a_n^2\sigma_n^2 t^2/2} \\ &= e^{(\Sigma a_i \mu_i)t + \Sigma a_i^2 \sigma_i^2 t^2/2} \end{aligned}$$

which is the MGF of a normal RV with mean $\sum a_i\mu_i$ and variance $\sum a_i^2 \sigma_i^2$. Hence the property.

Deductions:

1. Putting $a_1 = a_2 = 1$ and $a_3 = a_4 = \ldots = a_n = 0$, we get the following result, in particular:
   If X_1 is $N(\mu_1, \sigma_1)$ and X_2 is $N(\mu_2, \sigma_2)$, then $X_1 + X_2$ is $N(\mu_1 + \mu_2, \sqrt{\sigma_1^2 + \sigma_2^2}$. Similarly, $X_1 - X_2$ is $N(\mu_1 - \mu_2, \sqrt{\sigma_1^2 + \sigma_2^2})$.
2. Putting $a_1 = a_2 = \ldots = a_n = \dfrac{1}{n}$ and assuming that each X_i is $N(\mu, \sigma)$, then

$$\overline{X} = \frac{1}{n}\sum_{i=1}^{n} X_i \text{ has a normal distribution}$$

$$N\left\{\frac{1}{n}\sum_{i=1}^{n}\mu, \sqrt{\sum_{i=1}^{n}\frac{1}{n^2}\sigma^2}\right\}$$

   i.e., $N\{\mu, \sigma/\sqrt{n}\}$
   Thus if X_i $(i = 1, 2, \ldots, n)$ are independent and identically distributed normal variables with mean μ and standard deviation σ, then their mean $\overline{X}$ is $N\{\mu, \sigma/\sqrt{n}\}$.

8. Normal distribution as limiting form of binomial distribution

When n is very large and neither p nor q is very small, the standard normal distribution can be regarded as the limiting form of the standardised binomial distribution.

When X follows the binomial distribution $B(n, p)$, the standardised binomial variable Z is given by $Z = \dfrac{X - np}{\sqrt{npq}}$. As X varies from 0 to n with step size 1, Z varies from $\dfrac{-np}{\sqrt{npq}}$ to $\dfrac{np}{\sqrt{npq}}$ with step size $\dfrac{1}{\sqrt{npq}}$. When neither p nor q is very small and n is very large, Z varies from $-\infty$ to ∞ with infinitesimally small step size. Hence, in the limit, the distribution of Z may be expected to be a continuous distribution extending from $-\infty$ to ∞ and having mean 0 and standard deviation 1. In fact the limiting form of the distribution of Z is standard normal distribution as seen below:

If X follows $B(n, p)$, then the MGF of X is given by $M_X(t) = (q + p\, e^t)^n$.

If $\quad Z = \dfrac{X - np}{\sqrt{npq}}$, then

$$M_Z(t) = M_{\frac{1}{\sqrt{npq}}X - \frac{np}{\sqrt{npq}}}(t) = e^{\frac{-npt}{\sqrt{npq}}}\{q + p\, e^{t/\sqrt{npq}}\}^n$$

$$\therefore \quad \log M_Z(t) = -\frac{npt}{\sqrt{npq}} + n \log\{q + p\, e^{t/\sqrt{npq}}\}$$

$$= -\frac{npt}{\sqrt{npq}} + n\log\left[q + p\left\{1 + \frac{t}{\sqrt{npq}} + \frac{t^2}{2\,npq} + \frac{t^3}{6\,(npq)^{3/2}} + \cdots\right\}\right]$$

$$= -\frac{npt}{\sqrt{npq}} + n\log\left[1 + \left\{\frac{pt}{\sqrt{npq}} + \frac{pt^2}{2\,npq} + \frac{pt^3}{6\,(npq)^{3/2}} + \cdots\right\}\right]$$

$$= -\frac{npt}{\sqrt{npq}} + n\left[\frac{pt}{\sqrt{npq}}\left\{1 + \frac{t}{2\sqrt{npq}} + \frac{t^2}{6n^2p^2q^2} + \cdots\right\}\right.$$

$$\left. - \frac{1}{2}\cdot\frac{p^2t^2}{npq}\left\{1 + \frac{t}{2\sqrt{npq}} + \frac{t^2}{6n^2p^2q^2} + \cdots\right\}^2 + \cdots\right]$$

$$= \frac{t^2}{2} + \text{terms containing } \frac{1}{\sqrt{n}} \text{ and lower powers of } n$$

$$\therefore \lim_{n \to \infty} \log M_Z(t) = \frac{t^2}{2}$$

i.e., $$\log_e\left[\lim_{n \to \infty} M_z(t)\right] = \frac{t^2}{2}$$

$\therefore$ $$\lim_{n \to \infty} M_Z(t) = e^{t^2/2},$$

which is the MGF of the standard normal distribution. Hence the limit of the standardised binomial distribution, as n tends to ∞, is the standard normal distribution.

Note *We recall De Moivre–Laplace approximation for the sum of a large number of terms of the form $nC_r\, p^r q^{n-r}$ in terms of the integral of standard normal density function, which was discussed in section 1(c). It was stated that*

$$\sum_{r=r}^{r_2} nC_r\, p^r q^{n-r} = \int_{z_1}^{z_2} \phi(z)\, dz$$

where $z_1 = \dfrac{r_1 - np - \frac{1}{2}}{\sqrt{npq}}$ *and* $z_2 = \dfrac{r_2 - np + \frac{1}{2}}{\sqrt{npq}}$ *and $\phi(z)$ is the density function of the standard normal distribution.*

Importance of Normal Distribution

Normal distribution plays a very important role in statistical theory because of the following reasons:

(i) A large number of RVs, such as binomial and Poisson, occurring in many applications have a distribution closely resembling the normal distribution.

(ii) Many of the distributions of sample statistics, such as sample mean and sample variance, tend to normality for samples of large size. In particular, the sampling distributions like Student's t, Snedecor's F and Chi-square distributions tend to normality when the size of the sample is large.

(iii) Tests of significance for small samples are based on the assumption that samples have been drawn from normal populations.

(iv) Even if a variable is not normally distributed, it can sometimes be converted into a normal variable by simple transformation of the variable.

(v) Normal distribution is applied to a large extent in statistical Quality Control in industry.

Worked Examples 5(B)

Example 1 *If a string, 1 m long, is cut into 2 pieces at a random point along its length, what is the probability that the longer piece is at least twice the length of the shorter?*

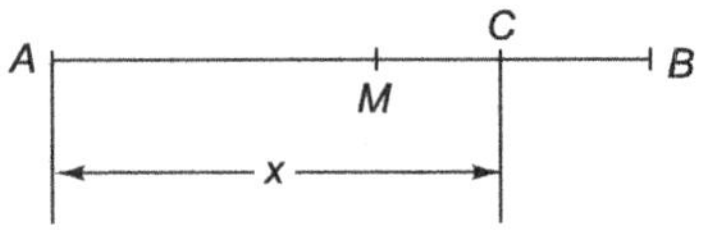

Let C be the point of cut on AB such that $AC = X$. Since all positions of C are equally likely, X is uniformly distributed over (0, 1) [$\because AB = 1$]

$\therefore \quad f(x) = 1.$

If X represents the length of the longer piece, then C lies in MB.

P\{longer piece length $\geq 2 \times$ shorter piece length\}

$$= P\{X \geq 2(1 - X)\} = P\left(X \geq \frac{2}{3}\right)$$

$$= \int_{\frac{2}{3}}^{1} 1\, dx = \frac{1}{3}$$

If X represents the length of the shorter piece, then C lies in AM.

In this case also, required probability

$$= P\{(1 - X) \geq 2X\} = P\left(X \geq \frac{1}{3}\right)$$

$$= \int_{0}^{\frac{1}{3}} 1\, dx = \frac{1}{3}$$

Hence, the required probability $= \frac{1}{3}$ [Also see Example (17) in Worked Example 2(A) and Problem (22) in Exercise 2(A)]

Example 2 *Buses arrive at a specified stop at 15 min intervals starting at 7 A.M., that is, they arrive at 7, 7:15, 7:30, 7:45 and so on. If a passenger arrives at the stop at a random time that is uniformly distributed between 7 and 7:30 A.M., find the probability that he waits*

(a) less than 5 min for a bus and

(b) at least 12 min for a bus.

Solution Let X denote the time in minutes past 7 A.M., when the passenger arrives at the stop.

Then X is uniformly distributed over (0, 30), i.e., $f(x) = \frac{1}{30}$, $0 < x < 30$

(a) The passenger will have to wait less than 5 min if he arrives at the stop between 7:10 and 7:15 or 7:25 and 7:30.

$\therefore$ Required probability $= P(10 < x < 15) + P(25 < x < 30)$

$$= \int_{10}^{15} \frac{1}{30}\, dx + \int_{25}^{30} \frac{1}{30}\, dx$$

$$= \frac{1}{3}$$

(b) The passenger will have to wait at least 12 min if he arrives at the stop between 7:00 and 7:03 or 7:15 and 7:18.

$\therefore$ Required probability $= P(0 < x < 3) + P(15 < x < 18)$

$$= \int_{0}^{3} \frac{1}{30}\, dx + \int_{15}^{18} \frac{1}{30}\, dx$$

$$= \frac{1}{5}$$

Example 3 *If the roots of the quadratic equation $x^2 - ax + b = 0$ are real and b is positive but otherwise unknown, what are the expected values of the roots of the equation.*

Solution Assume that b has a uniform distribution in the permissible range.

The roots of the equation $x^2 - ax + b = 0$ are given by

$$x = \frac{1}{2}\left(a \pm \sqrt{a^2 - 4b}\right)$$

Since the roots are real, $a^2 - 4b > 0$

i.e., $0 < b < \frac{a^2}{4}$ $(\because b > 0)$

Therefore, b is a random variable, uniformly distributed in $\left(0, \frac{a^2}{4}\right)$.

Therefore, its density function $f(b) = \frac{4}{a^2}$

$$E\{\text{the roots}\} = E\left\{\frac{1}{2}\left(a \pm \sqrt{a^2 - 4b}\right)\right\}$$

$$= \frac{1}{2}\int_0^{a^2/4} \left(a \pm \sqrt{a^2/4 - 4b}\right)\cdot\frac{4}{a^2}\, db$$

$$= \frac{2}{a^2}\left[ab \pm \frac{(a^2 - 4b)^{\frac{3}{2}}}{\frac{3}{2}(-4)}\right]_0^{a^2/4}$$

$$= \frac{2}{a^2}\left[\frac{a^3}{4} \mp \frac{1}{6}\left(0 - a^3\right)\right]$$

$$= a\left(\frac{1}{2} \pm \frac{1}{3}\right)$$

$$= \frac{5a}{6} \text{ and } \frac{a}{6}$$

Example 4 *Two points are taken at random on a given straight line of length 2 units. Prove that the probability of the distance between them exceeds 1 unit is $\frac{1}{4}$.*

[See Example (10) in Worked Example 2(B)]

Solution Let X and Y be the distances of the two points P, Q, taken on the line AB from A.

Each of X and Y follows a uniform distribution in (0, 2).

Therefore, the joint density function of $(X, Y) = f(x, y) = \frac{1}{4}$, $0 < x < 2$; $0 < y < 2$.

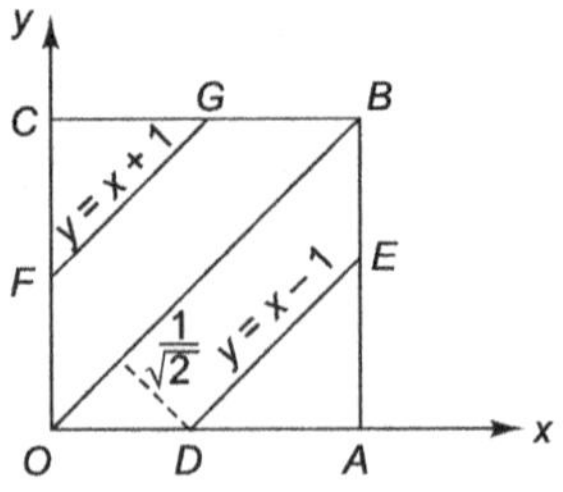

Now $\quad P\{PQ < 1\} = P\{|X - Y| < 1\}$

$$= P\{-1 < X - Y < 1\}$$

$$= \iint_{-1 < x - y < 1} f(x, y)\, dx\, dy$$

$$= \frac{1}{4} \iint_{-1 < x - y < 1} dx\, dy = \frac{1}{4}. \text{ Area of } ODEBGFO$$

$$= \frac{1}{2} \text{ Area of trapezium } ODEB, \text{ by symmetry}$$

$$= \frac{1}{2} \cdot \frac{1}{2} \cdot \frac{1}{\sqrt{2}} \left(\sqrt{2} + 2\sqrt{2}\right)$$

$$= \frac{3}{4}$$

$$\therefore \quad \{PQ > 1\} = \frac{1}{4}$$

Example 5 *If X is a RV with a continuous distribution function F(x), prove that Y = F(X) has a uniform distribution in (0, 1). Further,*

if
$$f(x) = \begin{cases} \frac{1}{2}(x - 1) & 1 \le x \le 3 \\ 0 & \text{otherwise} \end{cases}$$

find the range of Y corresponding to the range $1.1 \le X \le 2.9$.

Solution The distribution function of Y is given by

$$G_Y(y) = P(Y \le y)$$

$$= P\{F(X) \le y\}$$

$$= P\{X \le F^{-1}(y)\}$$ [The inverse exists, as $F(x)$ is non-decreasing and continuous]

$$= F[F^{-1}(y)] \quad [\because P\{X \le x\} = F(x)]$$

$$= y$$

Therefore, the density function of Y is given by

$$g_Y(y) = \frac{d}{dy}[G_Y(y)] = 1$$

Also the range of Y is $0 \le y \le 1$, since the range of $F(x)$ is (0, 1).

Therefore, Y follows a uniform distribution in (0, 1).

Note *The converse of this problem has been worked in Example (14) of Worked Example (3).*

When
$$f(x) = \begin{cases} \frac{1}{2}(x - 1) & 1 \le x \le 3 \\ 0 & \text{otherwise} \end{cases}$$

$$F(x) = \int_1^x \frac{1}{2}(x-1)\,dx = \frac{1}{4}(x-1)^2$$

Since $\quad Y = F(X),\ Y = \frac{1}{4}(X-1)^2$

$\therefore$ When $1.1 \le X \le 2.9,\ \frac{1}{4}(1.1-1)^2 \le Y \le \frac{1}{4}(2.9-1)^2$

i.e., the required range of Y is

$$0.0025 \le Y \le 0.9025$$

Example 6 *The mileage which car owners get with a certain kind of radial tire is a RV having an exponential distribution with mean 40,000 km. Find the probabilities that one of these tires will last (i) at least 20,000 km and (ii) at most 30,000 km.*

Solution Let X denote the mileage obtained with the tire

$$f(x) = \frac{1}{40{,}000} e^{-x/40{,}000} \quad x > 0$$

(i) $$P(X \ge 20{,}000) = \int_{20{,}000}^{\infty} \frac{1}{40{,}000} e^{-x/40{,}000}\,dx$$

$$= \left[-e^{-x/40{,}000}\right]_{20{,}000}^{\infty}$$

$$= e^{-0.5} = 0.6065$$

(ii) $$P(X \le 30{,}000) = \int_{0}^{30{,}000} \frac{1}{40{,}000} e^{-x/40{,}000}\,dx$$

$$= \left[-e^{-x/40{,}000}\right]_{0}^{30{,}000}$$

$$= 1 - e^{-0.75} = 0.5270$$

Example 7 *If the time T to failure of a component is exponentially distributed with parameter λ and if n such components are installed, what is the probability that one-half or more of these components are still functioning at the end of t hours?*

Solution The density function of T is given by

$$f(t) = \lambda e^{-\lambda t},\ t \ge 0$$

P(a component functions at the end of or after t hours)

$$= P(T \ge t) = \int_t^{\infty} \lambda e^{-\lambda t}\,dt = e^{-\lambda t}$$

If we consider a component functioning at the end or after t hours as a success in a single trial, we have $p = e^{-\lambda t}$ and $q = 1 - e^{-\lambda t}$.

Then the number X of successes in n independent trials follows a binomial distribution with parameters n and p.

$$\therefore \qquad P(X = r) = nC_r\, p^r q^{n-r};\ r = 0, 1, 2, \ldots, n$$

If n is even the required probability is given by

$$\sum_{r=\frac{n}{2}}^{n} P(X = r) = \sum_{r=\frac{n}{2}}^{n} nC_r\, e^{-\lambda rt}\,(1 - e^{-\lambda t})^{n-r}$$

If n is odd, the required probability is given by

$$\sum_{r=\frac{n+1}{2}}^{n} P(X = r) = \sum_{r=\frac{n+1}{2}}^{n} nC_r\, e^{-\lambda rt}\,(1 - e^{-\lambda t})^{n-r}$$

Example 8 *If a continuous RV X (> 0) possesses memoryless property, that is P(X > x + h) = P(X > x). P(X > h), then X follows an exponential distribution.*

Solution Let $\quad G(x) = P(X > x)$

$\therefore$ The given condition means that

$$G(x + h) = G(x)G(h)$$

$$\therefore \frac{G(x+h) - G(x)}{h} = G(x)\left\{\frac{G(h) - 1}{h}\right\}$$

$$= G(x)\cdot\frac{\{G(h) - G(0)\}}{h}$$

$$[\because G(0) = P(X > 0) = 1, \text{ as } X > 0]$$

Taking limits on both sides as $h \to 0$, we have

$$G'(x) = G(x) \cdot G'(0)$$

$$= -\lambda \cdot G(X), \quad [\text{on putting } \lambda = -G'(0)] \qquad (1)$$

Solving the differential equation (1), we get

$$\log G(x) = -\lambda x + \log C$$

i.e., $$G(x) = Ce^{-\lambda x} \qquad (2)$$

Using the fact that $G(0) = 1$ in (2), we get $C = 1$

Thus $\qquad G(x) = P(X > x) = e^{-\lambda x}$

Now the distribution function $F(x)$ of X is given by $F(x) = P(X \le x)$

$$= 1 - P(X > x) \quad \{= 1 - G(x)\}$$

$$= 1 - e^{-\lambda x}$$

Therefore, the density function $f(x)$ of X is given by

$$f(x) = F'(x) = \lambda e^{-\lambda x},\ x > 0$$

i.e., X follows an exponential distribution with parameter

$$\lambda = -G'(0) = F'(0) > 0$$

Example 9 *The time (in hours) required to repair a machine is exponentially distributed with parameter $\lambda = 1/2$.*

(a) What is the probability that the repair time exceeds 2 h?

(b) What is the conditional probability that a repair takes at least 10 h given that its duration exceeds 9 h?

Solution If X represents the time to repair the machine, the density function of X is given by

$$f(x) = \lambda e^{-\lambda x} = \frac{1}{2} e^{-\frac{x}{2}}, x > 0$$

(a) $$P(X > 2) = \int_2^\infty \frac{1}{2} e^{-\frac{x}{2}} dx$$

$$= \left(-e^{-\frac{x}{2}}\right)_2^\infty = e^{-1} = 0.3679$$

(b) $P\{X \geq 10/X > 9\} = P\{X > 1\}$, (by the memoryless property)

$$= \int_1^\infty \frac{1}{2} e^{-\frac{x}{2}} dx$$

$$= \left(-e^{-\frac{x}{2}}\right)_0^\infty = e^{-0.5} = 0.6065$$

Example 10 *The life length X of an electronic component follows an exponential distribution. There are 2 processes by which the component may be manufactured. The expected life length of the component is 100 h if process I is used to manufacture, while it is 150 h if process II is used. The cost of manufacturing a single component by process I is Rs 10, while it is Rs 20 for process II. Moreover if the component lasts less than the guaranteed life of 200 h, a loss of Rs. 50 is to be borne by the manufacturer. Which process is advantageous to the manufacturer?*

Solution If process I is used, the density function of X is given by

$$f(x) = \frac{1}{100} e^{-x/100}, x > 0.$$

$$\therefore P(X \geq 200) = \int_{200}^\infty \frac{1}{100} e^{-x/100} dx$$

$$= \left(-e^{-x/100}\right)_{200}^\infty = e^{-2}$$

$\therefore\ P(X < 200) = 1 - e^{-2}$

Similarly, if process II is used,

$$P(X \geq 200) = e^{-4/3} \text{ and } P(X < 200) = 1 - e^{-4/3}$$

Let C_1 and C_2 be the costs per component corresponding to the processes I and II respectively.

Then $C_1 = \begin{cases} 10, & X \geq 200 \\ 60, & X < 200 \end{cases}$

$$\therefore \quad E(C_1) = 10 \times P(X \geq 200) + 60 \cdot P(X < 200)$$
$$= 10\, e^{-2} + 60\,(1 - e^{-2})$$
$$= 60 - 50\, e^{-2} = 53.235$$

Now $C_2 = \begin{cases} 20, & X \geq 200 \\ 70, & X < 200 \end{cases}$

$$\therefore \quad E(C_2) = 20 \times P(X \geq 200) + 70 \times P(X < 200)$$
$$= 20\, e^{-4/3} + 70\,(1 - e^{-4/3})$$
$$= 70 - 50\, e^{-4/3} = 56.765$$

Since $E(C_1) < E(C_2)$, process I is advantageous to the manufacturer.

Example 11 *If the density function of a continuous RV X is $f(x) = ce^{-b(x-a)}$, $a \leq x$, where a, b, c are constants. Show that $b = c = \dfrac{1}{\sigma}$ and $a = \mu - \sigma$, where $\mu = E(X)$ and $\sigma^2 = Var(X)$.*

Solution Since $f(x)$ is a density function, $\int_a^\infty f(x)\, dx = 1$.

i.e., $$\int_a^\infty ce^{-b(x-a)}\, dx = 1$$

i.e., $$c\left\{\frac{e^{-b(x-a)}}{-b}\right\}_a^\infty = 1$$

i.e., $$\frac{c}{b} = 1 \text{ or } b = c \tag{1}$$

Now $$\mu = E(X) = \int_a^\infty bx\, e^{-b(x-a)}\, dx$$

$$= be^{ab}\left[x \cdot \left(\frac{e^{-bx}}{-b}\right) - \frac{e^{-bx}}{b^2}\right]_a^\infty$$

$$= be^{ab}\left[\frac{a}{b}e^{-ab} + \frac{1}{b^2}e^{-ab}\right]$$

$$= a + \frac{1}{b} \tag{2}$$

$$E(X^2) = \int_a^\infty bx^2 e^{-b(x-a)}\,dx$$

$$= be^{ab}\left[x^2\left(\frac{e^{-bx}}{-b}\right) - 2x\left(\frac{e^{-bx}}{-b^2}\right) + 2\left(\frac{e^{-bx}}{-b^3}\right)\right]_a^\infty$$

$$= b\left[\frac{a^2}{b} + \frac{2a}{b^2} + \frac{2}{b^3}\right]$$

$$= \frac{1}{b^2}(a^2b^2 + 2ab + 2)$$

$$\text{Var}(X) = E(X^2) - \{E(X)\}^2$$

$$= \frac{1}{b^2}(a^2b^2 + 2ab + 2) - \left(a^2 + \frac{2a}{b} + \frac{1}{b^2}\right)$$

i.e., $$\sigma^2 = \frac{1}{b^2} \text{ or } \sigma = \frac{1}{b} \tag{3}$$

From (1) and (3), we get $b = c = \dfrac{1}{\sigma}$.

From (2) and (3), $\mu - \sigma = a$.

Example 12 *In a certain city, the daily consumption of electric power in millions of kilowatt-hours can be treated as a RV having an Erlang distribution with parameters $\lambda = \dfrac{1}{2}$ and $k = 3$. If the power plant of this city has a daily capacity of 12 millions kilowatt-hours, what is the probability that this power supply will be inadequate on any given day.*

Solution Let X represent the daily consumption of electric power (in millions of kilowatt-hours). Then the density function of X is given as

$$f(x) = \frac{\left(\frac{1}{2}\right)^3}{\Gamma(3)} x^2 e^{-x/2}, x > 0$$

$P\{$the power supply is inadequate)

$$= P(X > 12) = \int_{12}^{\infty} f(x)\,dx \quad [\because \text{ The daily capacity is only 12}]$$

$$= \int_{12}^{\infty} \frac{1}{\Gamma(3)} \cdot \frac{1}{2^3} x^2 e^{-x/2}\,dx$$

$$= \frac{1}{16}\left[x^2\left(\frac{e^{-x/2}}{-\frac{1}{2}}\right) - 2x\left(\frac{e^{-x/2}}{\frac{1}{4}}\right) + 2\left(\frac{e^{-x/2}}{-\frac{1}{8}}\right)\right]_{12}^{\infty}$$

$$= \frac{1}{16} e^{-6} (288 + 96 + 16)$$

$$= 25\, e^{-6} = 0.0625$$

Example 13 *If a company employes n sales persons, its gross sales in thousands of rupees may be regarded as a RV having an Erlang distribution with $\lambda = \frac{1}{2}$ and $k = 80\sqrt{n}$. If the sales cost is Rs 8000 per salesperson, how many salespersons should the company employ to maximise the expected profit?*

Solution Let X represent the gross sales (in Rupees) by n salespersons.

X follows the Erlang distribution with parameters $\lambda = \frac{1}{2}$ and $k = 80{,}000\sqrt{n}$.

$$\therefore \quad E(X) = \frac{k}{\lambda} = 1{,}60{,}000\sqrt{n}$$

If y denotes the total expected profit of the company, then

$$y = \text{total expected sales} - \text{total sales cost}$$

$$= 1{,}60{,}000\sqrt{n} - 8000\,n$$

$$\frac{dy}{dn} = \frac{80{,}000}{\sqrt{n}} - 8000$$

$$= 0, \text{ when } \sqrt{n} = 10 \text{ or } n = 100$$

$$\frac{d^2y}{dn^2} = -\frac{40{,}000}{n^{3/2}} < 0, \text{ when } n = 100.$$

Therefore, y is maximum, when $n = 100$.

That is the company should employ 100 salespersons in order to maximise the total expected profit.

Example 14 *Consumer demand for milk in a certain locality, per month, is known to be a general Gamma (Erlang) RV. If the average demand is a litres and the most likely demand is b litres (b < a), what is the variance of the demand?*

Solution Let X represent the monthly consumer demand of milk.

Average demand is the value of $E(X)$.

Most likely demand is the value of the mode of X or the value of X for which its density function is maximum.

If $f(x)$ is the density function of X, then

$$f(x) = \frac{\lambda^k}{\Gamma(k)} x^{k-1} e^{-\lambda x} \quad x > 0$$

$$f'(x) = \frac{\lambda^k}{\Gamma(k)} [(k-1)\, x^{k-2} e^{-\lambda x} - \lambda x^{k-1} e^{-\lambda x}]$$

$$= \frac{\lambda^k}{\Gamma(k)} x^{k-2} e^{-\lambda x} \{(k-1) - \lambda x\}$$

$$= 0, \text{ when } x = 0,\ x = \frac{k-1}{\lambda}$$

$$f''(x) = \frac{\lambda^k}{\Gamma(k)} [-\lambda x^{k-2} e^{-\lambda x} + \{(k-1) - \lambda x\} \frac{d}{dx} \{x^{k-2} e^{-\lambda x}\}]$$

$$< 0, \text{ when } x = \frac{k-1}{\lambda}$$

Therefore $f(x)$ is maximum, when $x = \dfrac{k-1}{\lambda}$.

i.e., Most likely demand $= \dfrac{k-1}{\lambda} = b$ (1)

and $E(x) = \dfrac{k}{\lambda} = a$ (2)

Now $\text{Var}(X) = \dfrac{k}{\lambda^2} = \dfrac{k}{\lambda} \cdot \dfrac{1}{\lambda}$

$= a(a - b)$ [from (1) and (2)]

Example 15 *A random sample of size n is taken from a general Gamma (Erlang) distribution with parameters λ and k. Show that the mean $\overline{X}$ of the sample also follows a Gamma distribution with parameters $n\lambda$ and nk.*

Solution If X follows Erlang distribution with parameters λ and k, then the MGF of X is given by

$$M_X(t) = \left(1 - \frac{t}{\lambda}\right)^{-k}$$

If $X_1, X_2, \ldots, X_n$ are the members of the sample drawn, then each X_i follows Erlang distribution with MGF equal to $\left(1 - \dfrac{t}{\lambda}\right)^{-k}$ and also they are independent.

Therefore, by the reproductive property,

$$M_{X1 + X2 + \ldots + Xn}(t) = \left(1 - \frac{t}{\lambda}\right)^{-nk}$$

$$\therefore \quad M_{\overline{X}}(t) = M_{\frac{1}{n}(X_1 + X_2 + \ldots + X_n)}(t)$$

$$= M_{X1 + X_2 + \ldots + X_n}\left(\frac{t}{n}\right) \qquad [\because M_{aX}(t) = M_X(at)]$$

$$= \left(1 - \frac{t}{n\lambda}\right)^{-nk}$$

which is the MGF of an Erlang distribution with parameters $n\lambda$ and nk.

Therefore, $\overline{X}$ also follows an Erlang distribution with density function

$$\frac{(n\lambda)^{nk}}{\Gamma(nk)} \cdot x^{nk-1} \cdot e^{-n\lambda x}, x > 0.$$

Example 16 *If the conditional distribution of Y, given X = x, is an exponential distribution with parameter x and if the unconditional distribution of X is an Erlang distribution with parameters λ(> 0) and k(> 2), prove that the conditional distribution of X, given Y = y, is an Erlang distribution with parameters λ + y and k +1 .*

Solution **Given:** $F_{Y/X}(y) = x\, e^{-xy}, y > 0$ and $x > 0$

and $$f_X(x) = \frac{\lambda^k}{\Gamma(k)} x^{k-1} e^{-\lambda x}, x > 0$$

If $f(x, y)$ denotes the joint density function of (X, Y), then $f_{Y/X}(y) = \dfrac{f(x, y)}{f_X(x)}$

$$\therefore \quad f(x, y) = \frac{\lambda^k}{\Gamma(k)} x^k e^{-(\lambda + y)x}, x > 0, y > 0.$$

Now $f_Y(y)$ = the marginal density function of Y

$$= \int_0^\infty f(x, y)\, dx$$

$$= \frac{\lambda^k}{\Gamma(k)} \cdot \int_0^\infty x^k e^{-(\lambda + y)x} dx$$

$$= \frac{\lambda^k}{\Gamma(k)} \frac{1}{(\lambda + y)^{k+1}} \cdot \int_0^\infty t^k e^{-t} dt \text{ [on putting } (\lambda + y)x = t]$$

$$= \frac{\lambda^k}{\lceil(k)} \frac{1}{(\lambda + y)^{k+1}} \cdot \lceil(k+1) = \frac{k\,\lambda^k}{(\lambda + y)^{k+1}}, \quad y > 0$$

Now $f_{X/Y}(x) = \dfrac{f(x, y)}{f_Y(y)}$

$$= \frac{\dfrac{\lambda^k}{\lceil(k)} x^k e^{-(\lambda + y)x}}{\dfrac{k\,\lambda^k}{(\lambda + y)^{k+1}}} \quad x > 0 \text{ and } y > 0$$

$$= \frac{(\lambda + y)^{k+1}}{\lceil(k+1)} \cdot x^k e^{-(\lambda + y)x} \quad x > 0$$

This is the density function of an Erlang distribution with parameters $\lambda + y$ and $k + 1$.

Example 17 *Each of the 6 tubes of a radio set has a life length (in years) which may be considered as a RV that follows a Weibull distribution with parameters $\alpha = 25$ and $\beta = 2$. If these tubes function independently of one another, what is the probability that no tube will have to be replaced during the first 2 months of service?*

Solution If X represents the life length of each tube, then its density function $f(x)$ is given by

$$f(x) = \alpha\beta\, x^{\beta - 1} e^{-\alpha x^\beta} \quad x > 0$$

i.e., $f(x) = 50x\, e^{-25x^2} \quad x > 0$

Now P(a tube is not to be replaced during the first 2 months)

$$= P\left(X > \frac{1}{6}\right)$$

$$= \int_{\frac{1}{6}}^{\infty} 50x\, e^{-25x^2}\, dx$$

$$= \left(-e^{-25x^2}\right)_{\frac{1}{6}}^{\infty} = e^{-25/36}$$

$\therefore$ P(all the 6 tubes are not to be replaced during the first 2 months)

$= (e^{-25/36})^6$ (by independence)

$= e^{-25/6}$

$= 0.0155$

Example 18 *If the life X (in years) of a certain type of car has a Weibull distribution with the parameter $\beta = 2$, find the value of the parameter α, given that probability that the life of the car exceeds 5 years is $e^{-0.25}$. For these values of α and β, find the mean and variance of X.*

Solution The density function of X is given by

$$f(x) = 2\alpha\, x\, e^{-\alpha x^2},\ x > 0 \quad [\because\ \beta = 2]$$

Now $$P(X > 5) = \int_5^{\infty} 2\alpha\, x\, e^{-\alpha x^2}\, dx$$

$$= \left(-e^{-\alpha x^2}\right)_5^{\infty}$$

$$= e^{-25a}$$

Given that $P(X > 5) = e^{-0.25}$

$$\therefore \quad e^{-25\alpha} = e^{-0.25}$$

$$\therefore \quad \alpha = \frac{1}{100}$$

For the Weibull distribution with parameters α and β, $E(X) = \alpha^{-1/\beta}\,\Gamma\left(\frac{1}{\beta} + 1\right)$

$$\therefore \quad \text{Required mean} = \left(\frac{1}{100}\right)^{-\frac{1}{2}} \cdot \Gamma\left(\frac{3}{2}\right)$$

$$= 10 \times \frac{1}{2}\,\Gamma\left(\frac{1}{2}\right)$$

$$= 5\sqrt{\pi}.$$

$$\text{Var}(X) = a^{-\frac{2}{\beta}}\left[\Gamma\left(\frac{2}{\beta} + 1\right) - \left\{\Gamma\left(\frac{1}{\beta} + 1\right)\right\}^2\right]$$

$$= \left(\frac{1}{100}\right)^{-1}\left[\Gamma(2) - \left\{\Gamma\left(\frac{3}{2}\right)\right\}^2\right]$$

$$= 100\left[1 - \left(\frac{1}{2}\sqrt{\pi}\right)^2\right]$$

$$= 100\left(1 - \frac{\pi}{4}\right)$$

Example 19 *If the time T to failure of a component follows a Weibull distribution with parameters α and β, find the hazard rate or conditional failure rate at time t of the component.*

Solution Refer to Example (19) in Worked Example 2(A).

If $f(t)$ is the density function of T and $h(t)$ is the hazard rate at time t, then

$$h(t) = \frac{f(t)}{1 - F(t)}$$

where $F(t)$ is the distribution function of T.

Now $f(t) = a\beta \times t^{\beta - 1}\, e^{-\alpha t^{\beta}} \quad t > 0$

$\therefore$ $F(t) = P(T \le t)$

$$= \int_0^t \alpha\beta t^{\beta - 1}, e^{-\alpha t^{\beta}}\, dt$$

$$= \left[-e^{-\alpha t^{\beta}}\right]_0^t$$

$$= 1 - e^{-\alpha t \beta}$$

$\therefore$ $$h(t) = \frac{\alpha \beta t^{\beta - 1}\, e^{-\alpha t^{\beta}}}{e^{-\alpha t^{\beta}}}$$

$$= \alpha b t^{\beta - 1}$$

Example 20 *If Y is the smallest item of 3 independent observations X_1, X_2, X_3 from a Weibull distribution with parameters α and β, show that Y also has a Weibull distribution. What are its parameters?*

Solution Each of X_1, X_2, X_3 follows the Weibull distribution whose density function is given by

$$f(x) = \alpha\beta \times x^{\beta - 1}\, e^{-\alpha^{x\beta}}, x > 0$$

Now $P(Y > y) = P[\min (X_1, X_2, X_3) > y]$

$$= P(X_1 > y) \times P(X_2 > y) \times P(X_3 > y),$$

(since X_1, X_2, X_3 are independent.)

$$= \{P(X_i > y)\}^3 \quad (1)$$

Now $P(X_i > y) = \int_y^{\infty} \alpha\beta x^{\beta - 1}\, e^{-\alpha x^{\beta}}\, dx$

$$= \left(-e^{-\alpha x^{\beta}}\right)_y^{\infty}$$

$$= e^{-\alpha y \beta} \quad (2)$$

Using (2) in (1), we have

$$P(Y > y) = (e^{-\alpha y \beta})^3 = e^{-3\alpha y \beta}$$

Therefore, Y also has a Weibull distribution with parameters 3α and β.

Note *The result can be extended to n independent observations.*

Example 21 *There are 400 students in the first year class of an engineering college. The probability that any student requires a copy of a particular Mathematics book from the college library on any day is 0.1. How many copies of the book should be kept in the library so that the probability may be greater than 0.95 that none of the students requiring a copy from the library has to come back disappointed? (Use normal approximation to the binomial distribution).*

Solution $p = P(\text{a student requires the book}) = 0.1$ and $q = 0.9$

n = number of students = 400

If X represents the number of students requiring the book, then X follows a binomial distribution with mean = $np = 40$ and SD = $\sqrt{npq} = 6$.

As given in the problem, we may assume that X follows the distribution $N(40, 6)$.

Let m be the required number of books, satisfying the given condition.

i.e., $P(X < m) > 0.95$

i.e., $$P\left(-\infty < \frac{X - 40}{6} < \frac{m - 40}{6}\right) > 0.95$$

i.e., $$P\left(0 < Z < \frac{m - 40}{6}\right) > 0.45$$

where Z is the standard normal variate.

From the table of areas under normal curve, we find that

$$P\{0 < Z < 1.65\} > 0.45$$

$$\therefore \quad \frac{m - 40}{6} = 1.65$$

i.e., $\mu = 49.9$

Therefore, at least 50 copies of the book should be kept in the library.

Example 22 *The marks obtained by a number of students in a certain subject are approximately normally distributed with mean 65 and standard deviation 5. If 3 students are selected at random from this group, what is the probability that at least 1 of them would have scored above 75?*

Solution If X represents the marks obtained by the students, X follows the distribution $N(65, 5)$.

$P(\text{a student scores above 75})$

$$= P(X > 75) = P\left(\frac{75 - 65}{5} < \frac{X - 65}{5} < \infty\right)$$

$= P(2 < Z < \infty)$, (where Z is the standard normal variate)

$= 0.5 - P(0 < Z < 2)$

$= 0.5 - 0.4772$, (from the table of areas)

$= 0.0228$

Let $p = P(\text{a student scores above 75}) = 0.0228$ then $q = 0.9772$ and $n = 3$.

Since p is the same for all the students, the number Y, of (successes) students scoring above 75, follows a binomial distribution.

$P(\text{at least 1 student scores above 75})$

$= P(\text{at least 1 success})$

$= P(Y \geq 1) = 1 - P(Y = 0)$

$= 1 - nC_0 \times P^0 q^n$

$$= 1 - 3C_0\,(0.9772)^3$$
$$= 1 - 0.9333$$
$$= 0.0667$$

Example 23 *If the actual amount of instant coffee which a filling machine puts into '6-ounce' jars is a RV having a normal distribution with SD = 0.05 ounce and if only 3% of the jars are to contain less than 6 ounces of coffee, what must be the mean fill of these jars?*

Solution Let X be the actual amount of coffee put into the jars.

Then X follows $N(\mu, 0.05)$

Given: $P(X < 6) = 0.03$

$$\therefore \quad P\left\{-\infty < \frac{X-\mu}{0.05} < \frac{6-\mu}{0.05}\right\} = 0.03$$

$$\text{i.e.,} \quad P\left\{-\infty < Z < \frac{6-\mu}{0.05}\right\} = 0.03$$

$$\therefore \quad P\left\{0 < Z < \frac{\mu-6}{0.05}\right\} = 0.47, \quad \text{(by symmetry)}$$

From the table of areas, we have

$$P\{0 < Z < 1.808\} = 0.47$$

$$\therefore \quad \frac{\mu-6}{0.05} = 1.808$$

$$\therefore \quad \mu = 6.094 \text{ ounces.}$$

Example 24 *In an engineering examination, a student is considered to have failed, secured second class, first class and distinction, according as he scores less than 45%, between 45% and 60%, between 60% and 75% and above 75% respectively. In a particular year 10% of the students failed in the examination and 5% of the students got distinction. Find the percentages of students who have got first class and second class. (Assume normal distribution of marks).*

Solution Let X represent the percentage of marks scored by the students in the examination.

Let X follow the distribution $N(\mu, \sigma)$.

Given: $P(X < 45) = 0.10$ and $P(X > 75) = 0.05$

$$\text{i.e.,} \quad P\left(-\infty < \frac{x-\mu}{\sigma} < \frac{45-u}{\sigma}\right) = 0.10 \text{ and}$$

$$P\left(\frac{75-\mu}{\sigma} < \frac{X-\mu}{\sigma} < \infty\right) = 0.05$$

$$\text{i.e.,} \quad P\left(-\infty < Z < \frac{45-\mu}{\sigma}\right) = 0.10 \text{ and}$$

$$P\left(\frac{75-\mu}{\sigma} < Z < \infty\right) = 0.05$$

$$\therefore \quad P\left(0 < Z < \frac{\mu - 45}{\sigma}\right) = 0.40 \text{ and}$$

$$P\left(0 < Z < \frac{75 - \mu}{\sigma}\right) = 0.45$$

From the table of areas, we get

$$\frac{\mu - 45}{\sigma} = 1.28 \text{ and } \frac{75-\mu}{\sigma} = 1.64$$

i.e., $\mu - 1.28\,\sigma = 45$ (1)

and $\mu + 1.64\,\sigma = 75$ (2)

Solving equations (1) and (2), we get

$$\mu = 58.15 \text{ and } \sigma = 10.28$$

Now P(a student gets first class)

$$= P(60 < X < 75)$$

$$= P\left\{\frac{60 - 58.15}{10.28} < Z < \frac{75 - 58.15}{10.28}\right\}$$

$$= P\{0.18 < Z < 1.64\}$$

$$= P\{0 < Z < 1.64\} - P\{0 < Z < 0.18\}$$

$$= 0.4495 - 0.0714 = 0.3781$$

$\therefore$ Percentage of students getting first class = 38 (approximately)

Now percentage of students getting second class

= 100 – (sum of the percentages of students who have failed, got first class and got distinction)

= 100 – (10 + 38 + 5), approximately.

= 47 (approximately)

Example 25 *The percentage X of a particular compound contained in a rocket fuel follows the distribution N(33, 3), though the specification for X is that it should lie between 30 and 35. The manufacturer will get a net profit (per unit of the fuel) of Rs 100, if $30 < X < 35$, Rs 50, if $25 < X \le 30$ or $35 \le X < 40$ and incur a loss of Rs 60 per unit of the fuel otherwise. Find the expected profit of the manufacturer. If he wants to increase his expected profit by 50% by increasing the net profit on that category of the fuel that meets the specification, what should be the new net profit per unit of the fuel of this category?*

Solution $P(30 < X < 35) = P\left\{\frac{30-33}{3} < \frac{X-33}{3} < \frac{35-33}{3}\right\}$

$$= P\{-1 < Z < 0.67\}$$

$$= P\{0 < Z < 1\} + P\{0, < Z < 0.67\}$$

$= 0.3413 + 0.2486,$ (using the table of areas)

$= 0.5899$

$$P(25 < X \le 30) = P\left\{\frac{25-33}{3} < Z < \frac{30-33}{3}\right\}$$
$$= P(-2.67 < Z < -1)$$
$$= P(1 < Z < 2\ 67), \text{ (by symmetry)}$$
$$= P(0 < Z < 2.67) - P(0 < Z < 1)$$
$$= 0.4962 - 0.3413$$
$$= 0.1549$$

$$P(35 \le X < 40) = P\left\{\frac{35-33}{3} < Z < \frac{40-33}{3}\right\}$$
$$= P\{0.67 < Z < 2.33\}$$
$$= P\{0 < Z < 2.33\} - P\{0 < Z < 0.67\}$$
$$= 0.4901 - 0.2486$$
$$= 0.2415$$

$\therefore$ $P\{(25 < X \le 30)) \text{ or } (35 \le X < 40)\}$

$= P(25 < X \le 30) + P(35 \le X < 40),$

(since the 2 ranges are mutually exclusive)

$= 0.1549 + 0.2415 = 0.3964$

$P\{X < 25 \text{ or } X > 40\} = 1 - (0.5899 + 0.3964)$

$= 0.0137$

Profit/unit	*Probability*
Rs 100	0.5899
Rs 50	0.3964
Rs –60	0.0137

$\therefore$ E(Profit per unit) = Rs $(100 \times 0.5899 + 50 \times 0.3964 - 60 \times 0.0137)$

= Rs 79, approximately

Let the revised net profit per unit of the first category fuel be k.

Then E(Revised profit per unit)

= Rs $(k \times 0.5899 + 50 \times 0.3964 - 60 \times 0.0137)$

= Rs $(0.5899\,k + 18.998)$

E(Revised profit per unit) = Rs 79 + Rs 39.5, as per the manufacturer's wish

$\therefore$ $0.5899\,k + 18.998 = 118.5$

$$\therefore \quad k = \frac{118.5 - 18.998}{0.5899}$$

$= 168.68 \simeq$ Rs 169 nearly.

Example 26 *The marks obtained by the students in Mathematics, Physics and Chemistry in an examination are normally distributed with the means 52, 50 and 48 and with standard deviations 10, 8 and 6 respectively. Find the probability that a student selected at random has secured a total of (i) 180 or above and (ii) 135 or less.*

Solution Let X, Y, Z denote the marks obtained by students in Mathematics, Physics and Chemistry respectively.

Given: X follows $N(52, 10)$, Y follows $N(50, 8)$ and Z follows $N(48, 6)$.

By the additive property of normal distribution, $T = X + Y + Z$ follows the distribution

$$N\{52 + 50 + 48, \sqrt{10^2 + 8^2 + 6^2}\}$$

i.e., $N(150, 14.14)$

(i)
$$\begin{aligned} P(T \geq 180) &= P\left\{\frac{180 - 150}{14.14} < \frac{T - 150}{14.14} < \infty\right\} \\ &= P\{2.12 < Z < \infty\} \\ &= 0.5 - P\{0 < Z < 2.12\} \\ &= 0.5 - 0.4830, \quad \text{(from the table of areas)} \\ &= 0.0170 \end{aligned}$$

(ii)
$$\begin{aligned} P(T \leq 135) &= P\left\{\frac{T - 150}{14.14} < \frac{135 - 150}{14.14}\right\} \\ &= P(-\infty < Z < -1.06\} \\ &= P\{1.06 < Z < \infty\}, \quad \text{(by symmetry)} \\ &= 0.5 - P\{0 < Z < 1.06\} \\ &= 0.5 - 0.3554 \\ &= 0.1446 \end{aligned}$$

Example 27 *The independent RVs X and Y have distributions N(45, 2) and N(44, 1.5) respectively. What is the probability that randomly chosen values of X and Y differ by 1.5 or more?*

Solution X is $N(45, 2)$ and Y is $N(44, 1.5)$

$\therefore$ By the additive property,

$U = X - Y$ follows the distribution $N(1, \sqrt{4 + 2.25})$

i.e., $N(1, 2.5)$

Now $P\{X$ and Y differ by 1.5 or more$\}$

$$\begin{aligned} &= P\{|X - Y| \geq 1.5\} \\ &= P\{|U| \geq 1.5\} \\ &= 1 - P(|U| \leq 1.5) \\ &= 1 - P\{-1.5 \leq U \leq 1.5\} \\ &= 1 - P\left\{\frac{-1.5 - 1}{2.5} \leq \frac{U - 1}{2.5} \leq \frac{1.5 - 1}{2.5}\right\} \\ &= 1 - P\{-1 \leq Z \leq 0.2\} \\ &= 1 - \{P(0 \leq z \leq 1) + P(0 \leq z \leq 0.2)\} \end{aligned}$$

$= 1 - \{0.3413 + 0.0793\}$, (from the table of areas)
$= 0.5794.$

Example 28 *If X and Y are independent RVs, each following N(0, 3), what is the probability that the point (X, Y) lies between the lines 3X + 4Y = 5 and 3X + 4Y = 10?*

Solution X follows $N(0, 3)$ and Y follows $N(0, 3)$.

Therefore, by the additive property of normal distribution,

$U = 3X + 4Y$ follows $N[3 \times 0 + 4 \times 0, \sqrt{9 \times 9 + 16 \times 9}]$

i.e., $N(0, 15)$

Now $P\{$the point (X, Y) lies between the lines $3X + 4Y = 5$ and $3X + 4Y = 10\}$

$$= P\{5 < 3X + 4Y < 10\}$$
$$= P\{5 < U < 10\}$$
$$= P\left\{\frac{5-0}{15} < \frac{U-0}{15} < \frac{10-0}{15}\right\}$$
$$= P\{0.33 < Z < 0.67\},$$

where Z is the standard normal variable

$$= P(0 < Z < 0.67) - P(0 < Z < 0.33)$$
$$= 0.2486 - 0.1293, \text{ (from the table of areas)}$$
$$= 0.1193.$$

Example 29 *If X and Y are independent RVs following N(8, 2) and N(12, $4\sqrt{3}$) respectively, find the value of λ such that*

$P(2X - Y \le 2\lambda) = P(X + 2Y \ge \lambda)$

Solution By the additive property of normal distribution

$U = 2X - Y$ follows $N\{2 \times 8 - 12, \sqrt{4 \times 4 + 1 \times 48}\}$

i.e., $N(4, 8)$

and $V = X + 2Y$ follows $N\{8 + 2 \times 12, \sqrt{4 + 4 \times 48}\}$

i.e., $N(32, 14)$

Now $P(2X - Y \le 2\lambda) = P(X + 2Y \ge \lambda)$

i.e., $P(U \le 2\lambda) = P(V \ge \lambda)$

i.e., $$P\left(\frac{U-4}{8} \le \frac{2\lambda - 4}{8}\right)$$

$$= P\left(\frac{V-32}{14} \ge \frac{\lambda - 32}{14}\right)$$

i.e., $P\left(Z \le \frac{2\lambda - 4}{8}\right) = P\left(Z \ge \frac{\lambda - 32}{14}\right)$, where Z is the standard normal variable.

$$\therefore \quad \frac{2\lambda - 4}{8} = -\left(\frac{\lambda - 32}{14}\right)$$

i.e., $28\lambda - 56 = 256 - 8\lambda$

$\therefore$ $\lambda = \frac{26}{3}$

Example 30 Fit a normal distribution to the following frequency distribution and hence find the theoretical frequencies:

x:	125,	135,	145,	155,	165,	175,	185,	195,	205	Total
f:	1,	1,	14,	22,	25,	19,	13,	3,	2,	100

Solution To fit a normal distribution for the given data, we require the density function of the normal distribution which involves the mean and SD. Let us now compute the mean $\bar{x}$ and SD s of the given distribution and assume them as μ and σ of the approximate normal distribution.

x	f	$d = \frac{x - 165}{10}$	fd	fd^2
125	1	−4	−4	16
135	1	−3	−3	9
145	14	−2	−28	56
155	22	−1	−22	22
165	25	0	0	0
175	19	1	19	19
185	13	2	26	52
195	3	3	9	27
205	2	4	8	32
Total:	100	–	5	233

$$\bar{x} = A + \frac{c}{N}\sum fd = 165 + \frac{10}{100} \times 5 = 165.5$$

$$\sigma^2 = c^2\left\{\frac{1}{N}\sum fd^2 - \left(\frac{1}{N}\sum fd\right)^2\right\}$$

$$= 10^2\,(2.33 - 0.0025)$$

$$= 232.75$$

$\therefore$ $s = 15.26$

Therefore, the density function of the approximate normal distribution that fits the given distribution is

$$f(x) = \frac{1}{15.26\sqrt{2\pi}}\, e^{-(x - 165.5)^2/465.5} \quad -\infty < x < \infty$$

To find the theoretical frequency of the class $120 \le X \le 130$, whose mid-value is 125, we first get $P(120 \le X \le 130) = P\left\{\frac{120 - 165.5}{15.26} \le Z \le \frac{130 - 165.5}{15.26}\right\}$ and multiply this probability by the total frequency. Proceeding likewise, we get all the theoretical frequencies. The computations are shown in the table given below:

Class mid-value (x)	*Class* $(X_L \le X \le X_R)$	$Z_L \le Z \le Z_R$	$P(Z_L \le Z \le Z_R)$	*Theoretical frequency*	*Corrected frequency*
125	$120 \le X \le 130$	$-2.98 \le Z \le -2.33$	0.0085	0.85	1
135	$130 \le X \le 140$	$-2.33 \le Z \le -1.67$	0.0376	3.76	4
145	$140 \le X \le 150$	$-1.67 \le Z \le -1.02$	0.1064	10.64	11
155	$150 \le X \le 160$	$-1.02 \le Z \le -0.36$	0.2055	20.55	21
165	$160 \le X \le 170$	$-0.36 \le Z \le 0.29$	0.2547	25.47	25
175	$170 \le X \le 180$	$0.29 \le Z \le 0.95$	0.2148	21.48	21
185	$180 \le X \le 190$	$0.95 \le Z \le 1.61$	0.1174	11.74	12
195	$190 \le X \le 200$	$1.61 \le Z \le 2.26$	0.0418	4.18	4
205	$200 \le X \le 210$	$2.26 \le Z \le 2.92$	0.0102	1.02	1
					Total: 100

Exercise 5(B)

Part A (Short answer questions)

1. If X has uniform distribution in $(-3, 3)$, find $P(|X - 2| < 2)$.
2. If X has uniform distribution in $(-a, a)$, $a > 0$, find 'a' such that $P(|X| < 1) = P(|X| > 1)$.
3. If the MGF of a continuous RV X is $\frac{1}{t}(e^{5t} - e^{4t})$, $t \neq 0$, what is the distribution of X? What are its mean and variance?
4. A continuous RV X has the density function $c\, e^{-x/5}$; $x > 0$. Find c, $E(X)$ and $\text{Var}(X)$.
5. What do you mean by memoryless property of the exponential distribution?
6. If X and Y are independent identically distributed RVs, each with density function e^{-x}, $x > 0$, find the density function of $(X + Y)$.
7. Define Erlang distribution and also give its mean and variance.
8. Write down the MGF of simple Gamma distribution and hence find its mean and variance.
9. Give the values of β_1 and β_2 coefficients of the Erlang distribution with parameters $(1, k)$.
10. Find where the maximum occurs for the Erlang density function.
11. If X has uniform distribution in $(0, 2)$ and Y has exponential distribution with parameter λ, find λ such that $P(X < 1) = P(Y < 1)$.
12. If X has uniform distribution in $(-1, 3)$ and Y has exponential distribution with parameter λ, find λ such that $\text{Var}(X) = \text{Var}(Y)$.
13. Define Weibull distribution and also give its mean and variance.

14. Find the value of k, mean and variance of the normal distribution whose density function is $k \cdot 2^{-x^2}$ $-\infty < x < \infty$.
15. If X follows $N(30, 5)$ and Y follows $N(15, 10)$ show that $P(26 \le X \le 40) = P(7 \le Y \le 35)$.
16. If X follows $N(3, 2)$, find the value of k such that $P(|X - 3| > k) = 0.05$.
17. If $\log_{10}X$ follows $N(4, 2)$, find $P(1.202 < X < 83180000)$, given that $\log_{10}(1202) = 3.08$ and $\log_{10}(8318) = 3.92$.
18. For a certain normal distribution, the first moment about 10 is 40 and the fourth moment about 50 is 48. What are its mean and SD?
19. Show that, for a normal distribution, the quartile deviation, the mean deviation and the standard deviation are in the ratio 10 : 12 : 15.
20. If 2 normal universes A and B have the same total frequency, but the SD of A is k times the SD of B, prove that the maximum frequency of A is $\frac{1}{k}$ times that of B.
21. State the reproductive property of normal distribution.
22. If X and Y are independent RVs having $N(1, 2)$ and $N(2, 2)$ respectively find the density function of $(X + 2Y)$.
23. Why is normal distribution considered an important distribution?

Part B

24. X is uniformly distributed with mean 1 and variance $\frac{4}{3}$. If 3 independent observations of X are made, what is the probability that all of them are negative?
25. A point D is chosen on the line AB whose length is 1 and whose mid-point is C. If the distance X from D to A is a RV having a uniform distribution in (0, 1), what is the probability that AD, BD and AC will form a triangle?
26. A passenger arrives at a bus stop at 10 A.M., knowing that the bus will arrive at some time uniformly distributed between 10 A.M. and 10.30 A.M. What is the probability that he will have to wait longer than 10 min? If at 10.15 A.M. the bus has not yet arrived, what is the probability that he will have to wait at least 10 additional minutes?
27. A man and a woman agree to meet at a certain place between 10 A.M. and 11 A.M. They agree that the one arriving first will wait 15 min for the other to arrive. Assuming that the arrival times are independent and uniformly distributed, find the probability that they meet.
28. The RVs a and b are independently and uniformly distributed in the intervals (0, 6) and (0, 9) respectively. Find the probability that the roots of the exquation $x^2 - ax + b = 0$ are real.
29. If a, b, c are randomly chosen between 0 and 1, find the probability that the quadratic equation $ax^2 + bx + c = 0$ has real roots.

30. X, Y, Z are independent RVs, each following a uniform distribution in (0, 1). If U = maximum of X, Y, Z and V = minimum of X, Y, Z, find

(i) $P(U \leq \frac{1}{2})$, (ii) $P(V \geq \frac{1}{3})$ and (iii) $P\{U \leq \frac{1}{2}$ and $V \geq \frac{1}{3}\}$.

31. If the number of kilometres that a car can run before its battery wears out is exponentially distributed with an average value of 10,000 km and if the owner desires to take a 5000 km trip, what is the probability that he will be able to complete his trip without having to replace the car battery. Assume that the car has been used for some time. What is the probability, when the distribution is not exponential?
32. If X is exponentially distributed, prove that the probability that X exceeds its expected value is less than 0.5.
33. The amount of time that a watch will run without having to be reset is a RV having an exponential distribution with mean 120 days. Find the probability that such a watch will
    (i) have to be set in less than 24 days and
    (ii) not have to be reset in at least 180 days.
34. The daily consumption of milk in excess of 20,000 litres in a town is approximately exponentially distributed with parameter 1/3000. The town has a daily stock of 35,000L. What is the probability that of 2 days selected at random, the stock is insufficient for both days?
    [**Hint:** If Y denotes the daily consumption of milk, then $X = Y - 20{,}000$ follows the exponential distribution.]
35. The length of the shower on a tropical island during rainy season has an exponential distribution with parameter 2, time being measured in minutes. What is the probability that a shower will last more than 3 min? If a shower has already lasted for 2 min, what is the probability that it will last for at least one more minute?
36. If X is exponentially distributed with parameter λ, find the value of k such that

$$P(X > k)/P(X \leq K) = a.$$

37. If X is exponentially distributed with parameter, prove that the RV $Y = e^{-\lambda X}$ is uniformly distributed in (0, 1).
38. If X_1, X_2, X_3 are independent RVs having exponential distributions with parameters λ_1, λ_2, λ_3 respectively, prove that Y = minimum (X_1, X_2, X_3) follows an exponential distribution with parameter $(\lambda_1 + \lambda_2 + \lambda_3)$.
    [**Hint:** Find the distribution function of $Y = F(y) = 1 - P\{\min(X_1, X_2, X_3) > y\}$.]
39. The daily consumption of milk in a town in excess of 20,000L is approximately distributed as an Erlang variate with parameters

$\lambda = \dfrac{1}{10{,}000}$ and $k = 2$. The town has a daily stock of 30,000L. What is the probability that the stock is insufficient on a particular day?

40. Find the probabilities that the value of a RV will exceed 4, if it has an Erlang distribution with

(i) $\lambda = \frac{1}{3}$ and $k = 2$ and (ii) $\lambda = \frac{1}{4}$ and $k = 3$.

41. Show that for the Erlang distribution with parameters λ, k, (mean–mode)/SD $= \frac{1}{\sqrt{k}}$.

[**Hint:** If $f(x)$ is the Erlang density function, the mode is the value of x for which $f(x)$ is maximum.]

42. If X follows the Erlang distribution with parameters λ and k, prove that the expected value of the positive square root of X is $\Gamma\left(k + \frac{1}{2}\right) \Big/ \sqrt{\lambda}\,\Gamma k$.

43. If $X_1, X_2, \ldots, X_n$ are independent RVs, each following the same exponential distribution with parameter λ, prove that $X_1 + X_2 + \ldots + X_n$ follows an Erlang distribution with parameters λ and n.

[**Hint:** Use moment generating function. Also see example (7) in section 4(d).]

44. A random sample of size n is taken from a population which is exponentially distributed with parameter λ. If $\overline{X}$ is the sample mean, show that $n\lambda\overline{X}$ follows a simple Gamma distribution with parameter n.

[**Hint:** Use moment generating function.]

45. If the service life, in hours, of a semiconductor is a RV having a Weibull distribution with the parameters $\alpha = 0.025$ and $\beta = 0.5$,

(i) how long can such a semiconductor be expected to last? and

(ii) what is the probability that such a semiconductor will still be in operating condition after 4000 h?

46. Find the mode of the Weibull distribution with parameters α and β, when $\alpha > 1$.

47. If the hazard rate at time t of a system is given by $h(t) = \alpha\beta t^{\beta-1}$, prove that the time to failure of the system follows a Weibull distribution with parameters α and β.

48. If the RV X follows an exponential distribution with parameter 2, prove that $Y = X^3$ follows a Weibull distribution with parameters 2 and $\frac{1}{3}$.

49. Find the probability of failure-free performance of roller-bearings over a period of 10^4 h if the life expectancy of the bearings is defined by Weibull distribution with parameters $\alpha = 10^{-7}$ and $\beta = 1.5$.

[**Hint:** P(failure-free performance over a period t) = P(the component does not fail in $(0, t)$] = $P(T \geq t)$, where T is the life expectancy or time to failure of the component.]

50. The time when a country bus passes a certain point is distributed normally with a mean 9.25 A.M. and a SD of 3 min. What is the least time one could

arrive at this point and still have a probability of 0.99 of catching the bus? [**Hint:** If T is the time in minutes past 9 A.M., then T follows $N(25, 3)$.]

51. The marks obtained by a number of students in a certain subject are assumed to be approximately normally distributed with mean 55 and a SD of 5. If 5 students are taken at random from this set, what is the probability that 3 of them would have scored marks above 60?
52. The life lengths in hours of 2 electronic devices A and B have distributions $N(40, 6)$ and $N(45, 3)$ respectively. If the electronic device is to be used for a 45-h period, which device is to be preferred? If it is to be used for a 48-h period, which device is to be preferred?
53. The mean and SD of a certain group of 1000 high schoold grades, that are normally distributed are 78% and 11% respectively.
    (i) Find how many grades were above 90%?
    (ii) What was the highest grade of the lowest 10?
    (iii) What was the semi-interquartile range (Quartile deviation)?
    (iv) Within what limits did the middle 900 lie?
54. The local authorities in a certain city instal 10,000 electric lamps in the streets of the city. If these lamps have an average life of 1,000 burning hours with a standard deviation of 200 h, how many lamps might be expected to fail (i) in the first 800 burning hours? (ii) between 800 and 1200 burning hours? After how many burning hours would you expect (iii) 10% of the lamps to fail? (iv) 10% of the lamps to be still burning? Assume that the life of lamps is normally distributed.
55. In a normal population with mean 15 and SD 3.5, it is found that 647 observations exceed 16.25. What is the total number of observations in the population?
56. A RV has a normal distribution with SD 10. If the probability that the RV will take on a value less than 82.5 is 0.8212, what is the probability that it will take on a value greater than 58.3?
57. In a normal distribution, 7% of the items are under 35 and 89% are under 63. What are the mean and standard deviation of the distribution? What percentage of items are under 49?
58. A normal population has coefficient of variation equal to 2% and 8% of the population lies above 120 cm. What percentage of the population lies below 115 cm?
59. The breaking strength X of a certain kind of rope (in kg) has distribution $N(45, 1.8)$. Each 50 metre coil of rope brings a profit of Rs 1000, provided $X > 43$. If $X \leq 43$, the rope may be used for a different purpose and a profit of Rs 400 per coil is realised. Find the expected profit per coil.
60. The mean and standard deviation of marks in Mathematics are 45 and 10 respectively. The corresponding values for computer science are 50 and 15 respectively. Assuming that the marks in the two subjects are independent normal variates, find the probability that a student scores a total of marks lying between 100 and 120 in the 2 subjects.

61. If $\log_{10} X$ has the distribution $N(7, \sqrt{3})$ and $\log_{10} Y$ has the distribution $N(3, 1)$, find $P\{1.202 < \frac{X}{Y} < 8318 \times 10^4\}$, given that X and Y are independent.
[**Hint:** Find $P\{\log(1.202) < (\log X - \log Y) < \log (10^4 \times 8318\}$.]
62. If X and Y are independent RVs having normal distributions with a common mean μ, but with variances 4 and 48 respectively, such that $P(X + 2Y \le 3) = P(2X - Y \ge 4)$, determine μ.
63. Fit a normal distribution to the following distribution and hence find the theoretical frequencies:

Class	60–65	65–70	70–75	75–80	80–85	85–90	90–95	95–100	Total
Frequency	3	21	150	335	326	135	26	4	1000

Answers

Exercise 5(A)

1. $p = \frac{2}{3}, q = \frac{1}{3}; n = 6; P(X \ge 1) = 1 - nC_0\, p^0 q^n = \frac{728}{729}$

2. $\mu_3 = npq\,(q - p);\ \beta_1 = \frac{\mu_3^2}{\mu_2^3} = \frac{(q-p)^2}{npq}$

3. Reqd. no. $= 256 \times 8C_4 \left(\frac{1}{2}\right)^8 = 70$

4. $p = \frac{2}{3}, q = \frac{1}{3}; P(X \ge 1) = 1 - 4C_0\, p^0 q^4 = \frac{80}{81}$

5. Reqd. probability $= \sum_{r=1}^{3} 4C_r \left(\frac{1}{2}\right)^4 = \frac{7}{8}$

6. $\mu = 5, \sigma^2 = 4; P(X < \mu - 2\sigma) = P(X < 1) = \left(\frac{4}{5}\right)^{25}$

7. $V = np(1-p); \frac{dV}{dp} = n(1-2p) = 0$, when $p = \frac{1}{2}$ and $\frac{d^2V}{dp^2} < 0$

8. The MGF of $B(n, p)$ is $(q + p\,e^t)^n$. The given MGF is that of $B(6, 0.4)$. Hence mean = 2.4 and SD = 1.2

9. When $p_1 = p_2$, the sum is also a binomial variate

10. $(X + Y)$ follows $B\left(8, \frac{1}{3}\right)$; $P(X + Y \geq 1) = 1 - \left(\frac{2}{3}\right)^8 = \frac{6305}{6561}$

11. $P(X = r) = e^{-2}\, 2^r/r!$

12. $2e^{-\lambda} + e^{-\lambda}\frac{\lambda^2}{2} = 2\, e^{-\lambda}\,\lambda$. $\therefore$ $(\lambda - 2)^2 = 0$ or $\lambda = 2$

13. If λ is the parameter of the Poisson distribution, $\text{Var}(X) = E(X^2) - E^2(X)$.
i.e., $\lambda = 6 - \lambda^2$
$\therefore$ $\lambda^2 + \lambda - 6 = 0$ $\therefore$ $\lambda = E(X) = 2$, since $\lambda > 0$

14. If λ is the parameter, $e^{-\lambda} = 0.5$ $\therefore$ $\text{Var}(X) = \lambda = \log 2$

15. $E(X^2) = \lambda^2 + \lambda = \lambda(\lambda + 1) = \lambda E(X + 1)$

16. $P(X \text{ is even}) = e^{-\lambda}\left\{1 + \frac{\lambda^2}{2!} + \frac{\lambda^4}{4!} + \cdots\right\} = e^{-\lambda}\cosh\lambda = \frac{1}{2}\, e^{-\lambda}\,(e^{\lambda} + e^{-\lambda})$
$= \frac{1}{2}\,(1 + e^{-2\lambda})$

17. $e^{4(et - 1)}$ is the MGF of a Poisson distribution with parameter 4. $\therefore$ $\mu = 4$ and $\sigma = 2$.
$\therefore$ $P(X = 6) = e^{-4} \cdot 4^6/6!$

18. $(X + Y)$ is a Poisson variate with parameter 2.
$\therefore$ $P(X + Y = 2) = e^{-2}\, 2^2/2! = \frac{2}{e}$

19. The given distribution is a geometric distribution with $p = e^{-t}$
$\therefore$ $E(X) = \frac{q}{p} = e^t - 1$ and $\text{Var}(X) = \frac{q}{p^2} = e^t(e^t - 1)$.

20. $P(X = r) = pq^{\,r-1}$
$\therefore$ $P(X \text{ is odd}) = p + pq^2 + pq^4 + \ldots = \frac{p}{1 - q^2} = \frac{1}{1 + q}$.

21. The given distribution is a geometric distribution with $p = \frac{2}{3}$ and $q = \frac{1}{3}$.
$\therefore$ $E(X) = \frac{q}{p} = \frac{1}{2}$ and $\text{Var}(X) = \frac{q}{p^2} = \frac{3}{4}$

22. $\frac{q}{p^2} = 2\frac{q}{p}$ $\therefore$ $p = \frac{1}{2}$ $\therefore$ $P(X = r) = 2^{-r}$; $r = 1, 2, \ldots, \infty$

23. $M(t) = \sum_{r=1}^{\infty} e^{tr}\, q^{r-1}\, p = p\, e^t \sum_{r=1}^{\infty} (q\, e^t)^{r-1} = \frac{pe^t}{1 - qe^t}$

24. The given MGF $\frac{1}{5}e^t/(1 - \frac{4}{5}e^t)$ is that of a geometric distribution with $p = \frac{1}{5}$ and $q = \frac{4}{5}$.
$\therefore$ $E(X) = 4$ and $\text{Var}(X) = 20$.

26. Mean $= \dfrac{nk}{N}$ and Variance $= \dfrac{nk\,(N-k)\,(N-n)}{N^2\,(N-1)}$

27. Conditions: $N \to \infty$ and $\dfrac{k}{N} = p$; $\lim \left[n\left(\dfrac{k}{N}\right)\right] = np$;

$$\lim \left[n\left(\frac{k}{N}\right)\left(1-\frac{k}{N}\right)\left(\frac{1-\dfrac{n}{N}}{1-\dfrac{I}{N}}\right)\right] = npq$$

28. $P(X > 1) = .005$; 0.015
29. 17
31. $P(X = 3) = 0.0452$; $P(X > 3) = 0.0099$
32. $P(Y = k) = \dfrac{1}{16}.\ 4C_{\left(\frac{k}{2}+2\right)}$, $k = -4, -2, 0, 2, 4$
33. & 34. $\dfrac{2}{3} < p < 1$
35. $\dfrac{1}{3} < q < 1$
38. $\dfrac{1}{6}$
39. 0.016; 0.044
40. (i) 0.5551, (ii) 4
41. 10, 60, 150, 200, 150, 60, 10
42. 7, 26, 37, 34, 6
43. (i) 0.2231, (ii) 0.1913
44. 0.019
45. 0.1247; 0.0367
46. (i) 9802, (ii) 198, (iii) 9998
47. 0.067
48. (i) 0.1606, (ii) 0.1512
49. 0.0028
50. 0.0424
51. 0.0498
52. 0.8088
53. 0.1144
54. 0.6244, 2.3
55. 0.9577
56. 301, 362, 217, 87, 26, 6, 1
57. 0.92
58. 0.4529

59. $\dfrac{1}{10}$

61. $\dfrac{25}{61}$

62. $P(X + Y = k) = (k - 1)\, p^2 q^{k-2}, k = 2, 3, \ldots, \infty; \dfrac{2}{p}$

63. 0.2880
64. (i) 0.164, (ii) 0.165
65. 0.7134
66. 0.8121

Exercise 5(B)

1. $f(x) = \dfrac{1}{6}$; $P(|X - 2| < 2) = P(0 < X < 4) = P(0 < X < 3) = \dfrac{1}{2}$

2. $a > 1$; $P(|X| < 1) = \displaystyle\int_{-1}^{1} \frac{1}{2a}\, dx = \frac{1}{a}$; $P(|X| > 1) = 1 - \dfrac{1}{a}$ $\therefore a = 2$

3. Comparing the given MGF with that of uniform distribution, i.e., $\dfrac{1}{b-a}$.

   $\dfrac{e^{bt} - e^{at}}{a}$, X follows $U(4, 5)$ $E(X) = \dfrac{9}{2}$ and $\text{Var}(X) = \dfrac{1}{12}$.

4. $c\displaystyle\int_0^{\infty} e^{-x/5}\, dx$ $\therefore c = \dfrac{1}{5}$; Exponential distribution with parameter $\lambda = \dfrac{1}{5}$;

   $E(X) = \dfrac{1}{\lambda} = 5$; $\text{Var}(X) = \dfrac{1}{\lambda^2} = 25$

5. If X is exponentially distributed, then $P(X > s + t/X > s) = P(X > t)$, for any $s, t > 0$
6. X and Y follow exponential distribution with parameter 1. $\therefore M_X(t) = M_Y(t)$ $= \left(1 - \dfrac{t}{\lambda}\right)^{-1}$

   $\therefore M_{X+Y}(t) = M_X(t)\, M_Y(t) = (1 - \dfrac{t}{\lambda})^{-2}$ = MGF of Erlang distribution $(\lambda, 2)$. Its density function is $\lambda^2 x\, e^{-\lambda x}, x > 0$

8. $M(t) = (1 - t)^{-k} = 1 + kt + \dfrac{k(k+1)}{2!} t^2 + \ldots + \infty$

   $E(X) = k$; $E(X^2) = k(k + 1)$; $\text{Var}(X) = k$

9. Using $M(t) = (1 - t)^{-k}$, $\mu_3 = 2k$; $\mu_4 = 3k^2 + 6k$

$$\therefore\ \beta_1 = \mu^2{}_3/\mu_2{}^3 = \frac{4}{k} \text{ and } \beta_2 = \mu_4/\mu_2{}^2 = 3 + \frac{6}{k}$$

10. Maximum of $f(x) = \dfrac{\lambda k}{\Gamma(k)} |x|^{k-1} e^{-\lambda x}$ occurs at $x = \dfrac{\lambda - 1}{k}$

11. $P(X < 1) = \displaystyle\int_0^1 \frac{1}{2}\, dx = \frac{1}{2}$; $P(Y < 1) = \displaystyle\int_0^1 \lambda\ e^{-\lambda x}\, dx = 1 - e^{-\lambda} \cdot 1 - e^{-\lambda} = \frac{1}{2}$;

$$e^{-\lambda} = \frac{1}{2}\ \therefore\ \lambda = \log 2$$

12. $\text{Var}(X) = \dfrac{(3+1)^2}{12} = \dfrac{4}{3}$; $\text{Var}(Y) = \dfrac{1}{\lambda^2}$

$$\therefore\ \frac{1}{\lambda^2} = \frac{4}{3} \text{ gives } \lambda = \sqrt{3}/2$$

14. $f(x) = k\ e^{-(\log 2)\ x^2}$, Mean = 0; variance is given by $2\sigma^2 = \dfrac{1}{\log 2}$

$$\therefore\ \sigma = \frac{1}{\sqrt{\log 4}};\ k = \frac{1}{\sigma\sqrt{2\pi}} = \sqrt{\frac{\log 2}{\pi}}$$

15. $P(26 \le X \le 40) = P(-0.8 \le Z \le 2)$ and $P(7 \le Y \le 35) = P(-0.8 \le Z \le 2)$. Hence equality.

16. $P(|X-3| < k) = 0.95\ \therefore\ P(-k < X - 3 < k) = 0.95$

i.e., $P(-\frac{k}{2} < Z < \frac{k}{2}) = 0.95$; $\therefore\ \frac{k}{2} = 1.96$

$\therefore\ k = 3.92$

17. Reqd. probability $= P\{\log_{10}(1.202) < \log X < \log_{10} 8\ 3180000\} = P\{0.08 < \log X < 7.92\}$

$= P\{-1.96 < Z < 1.96\} = 0.95$

18. $E(X - 10) = 40$. $\therefore\ E(X) = \mu = 50$. $E(X - 50)^4 = 48$

$\therefore\ \mu_4 = 48$, i.e., $3\sigma^2 = 48\ \therefore\ \sigma = 2$

19. QD : MD : SD $= \frac{2}{3}\ \sigma : : \frac{4}{5}\ \sigma : \sigma = 10:12:15$

20. Max. ordinates of A and B are $\dfrac{1}{\sigma_A\sqrt{2\pi}}$ and $\dfrac{1}{\sigma_B\sqrt{2\pi}}$

$$\text{Max}\ (f_A) = C\ \frac{N}{\sigma_A\sqrt{2\pi}} \text{ and Max}\ (f_B) = C\ \frac{N}{\sigma_B\sqrt{2\pi}}$$

$$\therefore \quad \frac{\text{Max}(f_A)}{\text{Max}(f_B)} = \frac{\sigma_B}{\sigma_A} = \frac{1}{k}$$

21. See property (7) of normal distribution
22. By additive property, $(X + 2Y)$ follows $N(5, 2\sqrt{5})$. Its density function is

$$f(x) = \frac{1}{2\sqrt{10\pi}} e^{-(x-5)^2/40}, \; -\infty < x < \infty$$

24. $\frac{1}{64}$
25. $\frac{9}{16}$
26. $\frac{2}{3}; \frac{1}{3}$
27. $\frac{7}{16}$
28. $\frac{1}{3}$
29. $\frac{1}{9}$
30. (i) $\frac{1}{8}$, (ii) $\frac{8}{27}$, (iii) $\frac{1}{216}$
31. (i) 0.6065;

    (ii) $\frac{1 - F(t + 5000)}{1 - F(t)}$, where $F(t)$ is the distribution function of the life of the car.
33. (i) 0.1813, (ii) 0.2231
34. e^{-10}
35. (i) 0.0025, (ii) 0.1353
36. $\lambda^{-1} \log (1 + a^{-1})$
39. 0.7378
40. (i) 0.0879, (ii) 0.9197
45. (i) 3200h, (ii) 0.2057
46. $\beta(\alpha - 1)$
49. 0.9048
50. 9.18 A.M.
51. 0.0283
52. (i) B, (ii) B
53. (i) 138, (ii) 52%, (iii) 7.3%, (iv) 60% and 96.1%
54. (i) 1587, (ii) 6826, (iii) 744h, (iv) 1256h

55. 1800
56. 0.4332
57. 50.3, 10.33, 45%
58. 116.72, 2.33, 23%
59. Rs 920 nearly
60. 0.3074
71. 0.95
62. 2.1
63. 3, 31, 148, 322, 319, 144, 30, 3

6 Random Processes

In electrical systems voltage or current waveforms are used as signals for collecting, transmitting or processing information, as well as for controlling and providing power to a variety of devices. These signals (voltage or current waveforms) are functions of time and are of two classes—deterministic and random. Deterministic signals can be described by the usual mathematical functions with time t as the independent variable. But a random signal always has some element of uncertainty associated with it and hence it is not possible to determine its value exactly at any given point of time. However, we may be able to describe the random signal in terms of its average properties such as the average power in the random signal, its spectral distribution and the probability that the signal amplitude exceeds a given value. The probabilistic model used for characteristing a random signal is called a *random process* or *stochastic process*.

A random variable (RV) is a rule (or function) that assigns a real number to every outcome of a random experiment, while a random process is a rule (or function) that assigns a time function to every outcome of a random experiment. For example, consider the random experiment of tossing a dice at $t = 0$ and observing the number on the top face. The sample space of this experiment consists of the outcomes $\{1, 2, 3, \cdots, 6\}$. For each outcome of the experiment, let us arbitrarily assign a function of time t $(0 \le t < \infty)$ in the following manner.

Outcome:	1	2	3	4	5	6
Function of time:	$x_1(t) = -4$	$x_2(t) = -2$	$x_3(t) = 2$	$x_4(t) = 4$	$x_5(t) = -t/2$	$x_6(t) = t/2$

The set of functions $\{x_1(t), x_2(t), \cdots, x_6(t)\}$ represents a random process.

Definition: A random process is a collection (or ensemble) of RVs $\{X(s, t)\}$ that are functions of a real variable, namely time t where $s \in S$ (sample space) and $t \in T$ (parameter set or index set).

The set of possible values of any individual member of the random process is called *state space*. Any individual member itself is called a *sample function* or a realisation of the process.

Note (i) *If s and t are fixed, {X(s, t)} is a number.*
(ii) *If t is fixed {X(s, t)} is a RV.*
(iii) *If s is fixed, {X(s, t)} is a single time function.*
(iv) *If s and t are variables, {X(s, t)} is a collection of RVs that are time functions.*

Notation: As the dependence of a random process on s is obvious, s will be omitted hereafter in the notation of a random process. If the parameter set T is discrete, the random process will be noted by $\{X(n)\}$ or $\{X_n\}$.

If the parameter set T is continuous, the process will be denoted by $\{X(t)\}$.

Classification of Random Processes

Depending on the continuous or discrete nature of the state space S and parameter set T, a random process can be classified into four types:

(i) If both T and S are discrete, the random process is called a *discrete random sequence*. For example, if X_n represents the outcome of the nth toss of a fair dice, then $\{X_n, n \geq 1\}$ is a discrete random sequence, since $T = \{1, 2, 3, \cdots\}$ and $S = \{1, 2, 3, 4, 5, 6\}$.

(ii) If T is discrete and S is continuous, the random process is called a *continuous random sequence*.
For example, if X_n represents the temperature at the end of the nth hour of a day, then $\{X_n, 1 \leq n \leq 24\}$ is a continuous random sequence, since temperature can take any value in an interval and hence continuous.

(iii) If T is continuous and S is discrete, the random process is called a *discrete random process*.
For example, if $X(t)$ represents the number of telephone calls received in the interval $(0, t)$ then $\{X(t)\}$ is a discrete random process, since $S = \{0, 1, 2, 3, \cdots\}$.

(iv) If both T and S are continuous, the random process is called a *continuous Random process*. For example, if $X(t)$ represents the maximum temperature at a place in the interval $(0, t)$, $\{X(t)\}$ is a continuous random process. In the names given above, the word 'discrete' or 'continuous' is used to refer to the nature of S and the word 'sequence' or 'process' is used to refer to the nature of T.

Methods of Description of a Random Process

Since a random process is an indexed set of RVs, we can obviously use the joint probability distribution functions to describe a random process.

For a specific t, $X(t)$ is a RV as was observed earlier.

$F(x, t) = P\{X(t) \leq x\}$ is called the first-order distribution of the process $\{X(t)\}$

and $f(x, t) = \dfrac{\partial}{\partial x} F(x, t)$ is called the first-order density of $\{X(t)\}$.

$F(x_1, x_2, t_1, t_2) = P\{X(t_1) \leq x_1; X(t_2) \leq x_2\}$ is the joint distribution of the RVs $X(t_1)$ and $X(t_2)$ and is called the second-order distribution of the process $\{X(t)\}$

and $f(x_1, x_2, t_1, t_2) = \dfrac{\partial^2}{\partial x_1 \cdot \partial x_2} F(x_1, x_2, t_1, t_2)$ is called the second-order density of $\{X(t)\}$.

Similarly the nth order distribution $\{X(t)\}$ is the joint distribution $F\{x_1, x_2, \cdots, x_n; t_1, t_2, \cdots, t_n\}$ of the RVs $X(t_1), X(t_2), \cdots, X(t_n)$.

The first-order distribution function describes the instantaneous *amplitude* distribution of the process and the second-order distribution function tells us something about the structure of the signal in the time domain and hence the spectral content of the signal. Although the higher-order distributions describe the process in a more detailed manner, the first and second-order distribution functions are primarily used to describe the process.

Special Classes of Random Processes

The important feature of a random process is the relationship among the members of the family. Usually the nature of relationship is understood by the joint distribution function of the member RVs. A random process is said to be specified only when the parameter set, the state space and the nature of dependence relationship existing among the members of the family are specified.

Based on the dependence relationship among the members of the process, random processes are classified broadly into a few special types such as the ones explained below.

(i) Markov process

If, for $t_1 < t_2 < t_3 < \cdots < t_n < t$, $P\{X(t) \le x/X(t_1) = x_1, X(t_2) = x_2, \cdots, X(t_n) = x_n\} = P\{X(t) \le x/X(t_n) = x_n\}$, then the process $\{X(t)\}$ is called a Markov process.

In other words, if the future behaviour of a process depends only on the present state, but not on the past, the process is a Markov process.

A discrete parameter Markov process is called a Markov chain.

(ii) Process with independent increments

If, for all choices of $t_1, t_2, \cdots, t_n$ such that $t_1 < t_2 < t_3 < \cdots < t_n$, the random variables $X(t_2) - X(t_1), X(t_3) - X(t_2), \cdots, X(t_n) - X(t_{n-1})$ are independent, then the process $\{X(t)\}$ is said to be a random process with independent increments.

If $T = \{0, 1, 2, \cdots\}$ is the parameter set for $\{X_n\}$, then $\{Z_n\}$, where $Z_0 = X_0$ and $Z_n = X_n - X_{n-1}$, is a random sequence with independent increments if the RVs $Z_0, Z_1, Z_2, \cdots$, are independent.

Two processes with independent increments play an important role in the theory of random processes. One is the Poisson process that has a Poisson distribution for the increments and the other is the Wiener process with a Gaussian distribution for the increments. We will take up the study of Poisson and Gaussian processes in Chapter VIII.

(iii) Stationary processes

If certain probability distribution or averages do not depend on t, then the random process $\{X(t)\}$ is called stationary. A rigorous definition and detailed study of stationary processes will be taken up in the following articles.

Average Values of Random Processes

As in the case of RVs, random processes can be described in terms of averages or expected values, mostly derived from the first and second-order distributions of $\{X(t)\}$. *Mean* of the process $\{X(t)\}$ is the expected value of a typical member $X(t)$ of the process.

i.e., $\mu(t) = E\{X(t)\}$

Autocorrelation of the process $\{X(t)\}$, denoted by $R_{xx}(t_1, t_2)$ or $R_x(t_1, t_2)$ or $R(t_1, t_2)$, is the expected value of the product of any two members $X(t_1)$ and $X(t_2)$ of the process.

i.e., $R(t_1, t_2) = E\{X(t_1) \cdot X(t_2)\}$

Autocovariance of the process $\{X(t)\}$, denoted by $C_{xx}(t_1, t_2)$ or $C_x(t_1, t_2)$ or $C(t_1, t_2)$, is defined as

$$\begin{aligned} C(t_1, t_2) &= E[\{X(t_1) - \mu(t_1)\}\{X(t_2) - \mu(t_2)\}] \\ &= R(t_1, t_2) - \mu(t_1)\,\mu(t_2) \end{aligned}$$

Correlation co-efficient of the process $\{X(t)\}$, denoted by $\rho_{xx}(t_1, t_2)$ or $\rho(t_1, t_2)$, is defined as

$$\rho(t_1, t_2) = \frac{C(t_1, t_2)}{\sqrt{C(t_1, t_1) \times C(t_2, t_2)}}$$

where $C(t_1, t_1)$ is the variance of $X(t_1)$.

When we deal with 2 or more random processes, we can use joint distribution functions or averages to describe the relationship between them.

Cross-correlation of 2 processes $\{X(t)\}$ and $\{Y(t)\}$ is defined as

$$R_{xy}(t_1, t_2) = E\{X(t_1) \cdot Y(t_2)\}.$$

Cross-covariance of 2 processes $\{X(t)\}$ and $\{Y(t)\}$ is defined as

$$C_{xy}(t_1, t_2) = R_{xy}(t_1, t_2) - \mu_x(t_1) \times \mu_y(t_2)$$

Cross correlation co-efficient of 2 processes $\{X(t)\}$ and $\{Y(t)\}$ is defined as

$$\rho_{xy}(t_1, t_2) = \frac{C_{xy}(t_1, t_2)}{\sqrt{C_{xx}(t_1, t_1) \times C_{yy}(t_2, t_2)}}$$

Stationarity

A random process is called a *strongly stationary process* or *strict sense stationary process* (abbreviated as SSS process), if all its finite dimensional distributions are invariant under translation of time parameter.

That is, if the joint distribution (and hence the joint density) of $X(t_1), X(t_2), \cdots, X(t_n)$ is the same as that of $X(t_1 + h), X(t_2 + h), \cdots, X(t_n + h)$ for all $t_1, t_2, \cdots, t_n$ and $h (> 0)$ and for all $n \geq 1$, then the random process $\{X(t)\}$ is called a SSS process. If the definition given above holds good for $n = 1, 2, \cdots, k$ only and not for $n > k$, then the process is called *kth order stationary*.

Note *If a random process is a SSS process, as per the definition, its first-order densities must be invariant under translation of time, i.e., the densities of $X(t)$ and $X(t + h)$ are the same, i.e., $f(x, t) = f(x, t + h)$.*

This is possible only if $f(x, t)$ is independent of t.

Therefore, first-order densities (and hence distribution functions) of a SSS process are independent of time.

As a consequence, $E\{X(t)\}$ is also independent of t.

i.e., $E\{X(t)\} = \mu$ = *a constant*

Also the second-order densities must be invariant under translation of time, i.e., the joint pdf of $\{X(t_1), X(t_2)\}$ is the same as that of $\{X(t_1 + h), X(t_2 + h)\}$.

i.e., $f(x_1, x_2, t_1, t_2) = f(x_1, x_2, t_1 + h, t_2 + h)$

This is possible only if $f(x_1, x_2, t_1, t_2)$ is function of $t = t_1 - t_2$.

Therefore, second-order densities (and hence distribution functions) of a SSS process are functions of $\tau = t_1 - t_2$.

As a consequence, $R(t_1, t_2) = E\{X(t_1) \cdot X(t_2)\}$ is also a function of $\tau = t_1 - t_2$.

It is pointed out that if $E\{X(t)\}$ is a constant and $R(t_1, t_2)$ is a function of $(t_1 - t_2)$, the random process $\{X(t)\}$ need not be a SSS process.

The definition of strict sense stationarity can be extended as follows.

Two real-valued random processes $\{X(t)\}$ and $\{Y(t)\}$ are said to be *jointly stationary in the strict sense*, if the joint distribution of $X(t)$ and $Y(t)$ are invariant under translation of time.

The complex random process $\{Z(t)\}$, where $Z(t) = X(t) + iY(t)$, is said to be a SSS process if $\{X(t)\}$ and $\{Y(t)\}$ are jointly stationary in the strict sense.

Wide-sense stationarity

A random process $\{X(t)\}$ with finite first- and second-order moments is called a *weakly stationary process* or *covariance stationary process* or *wide-sense stationary process* (abbreviated as WSS process), if its mean is a constant and the autocorrelation depends only on the time difference.

i.e., if $E\{X(t)\} = \mu$ and

$$E\{X(t) \cdot X(t - \tau)\} = R(\tau)$$

Note: From the definitions given above, it is clear that a SSS process with finite first-and second-order moments is a WSS process, while a WSS process need not be a SSS process.)

A random process that is not stationary in any sense is called an evolutionary process.

Two random processes $\{X(t)\}$ and $\{Y(t)\}$ are said to be jointly stationary in the wide sense, if each process is individually a WSS process and $R_{xy}(t_1, t_2)$ is a function of $(t_1 - t_2)$ only.

Example of an SSS Process

Let X_n denote the presence or absence of a pulse at the nth time instant in a digital communication system or digital data processing system.

If $P\{X_n = 1\} = p$ and $P\{X_n = 0\} = 1 - p = q$, then the random process (sequence) $\{X_n, n \geq 1\}$, called the Bernoulli's process, is a SSS process, for, its first-order distribution is given by

$X_n = r$	1	0
$p(Xn = r)$	p	q

This distribution is the same for any X_n, i.e., for X_m and X_{m+p}.

Consider the second-order distribution of the process, i.e., the joint distribution of x_r and x_s.

	X_s	
X_r	1	0
1	p^2	pq
0	pq	q^2

This joint distribution is the same for the pair of members X_r and X_s and for the pair X_{r+p} and X_{s+p} of the process.

Consider the third-order distribution of the process, i.e., the joint distribution of X_r, X_s and X_t that is given below.

$$P\{X_r = 0, X_s = 0, X_t = 0\} = q^3$$
$$P\{X_r = 0, X_s = 0, X_t = 1\} = pq^2$$
$$P\{X_r = 0, X_s = 1, X_t = 0\} = pq^2$$
$$P\{X_r = 0, X_s = 1, X_t = 1\} = p^2q$$
$$P\{X_r = 1, X_s = 0, X_t = 0\} = pq^2$$
$$P\{X_r = 1, X_s = 0, X_t = 1\} = p^2q$$
$$P\{X_r = 1, X_s = 1, X_t = 0\} = p^2q$$
$$P\{X_r = 1, X_s = 1, X_t = 1\} = p^3$$

This joint distribution is the same for the triple of members X_r, X_s, X_t and for X_{r+p}, X_{s+p}, X_{t+p} of the process and so on, i.e., distributions of all orders are invariant under translation of time.

Note *If* $Y_n = \sum_{n=1}^{n} X_n$ = *the total number of pulses from time instant 1 through n, then the random process* $\{Yn, n \geq 1\}$, *called the Binomial process, is not a SSS process, for* $P\{Y_n = i\} = nC_i\, p^i\, q^{n-i}$ $(i = 0, 1, 2, \cdots, n)$ *depends on n, i.e., the distributions of* Y_m *and* Y_{m+p} *are not the same).*

Analytical Representation of a Random Process

Deterministic signals are usually expressed in simple analytical forms such as $X(t) = e^{-t^2}$ and $Y(t) = 20 \sin 10t$. It is sometimes possible to express a random process in an analytical form using one or more RVs. For example, consider an FM station that is broadcasting a 'tone', $X(t) = 100 \cos (10^8\, t)$, to a large number of receivers distributed randomly in a metropolitan area. The amplitude and phase of the waveform received by any receiver will depend on the distance between the transmitter and the receiver. Since there are a large number of receivers distributed randomly over an area, the distance can be considered as a continuous RV. Since the amplitude and the phase are functions of distance, they are also RVs. So we can represent the ensemble (collection) of received waveforms by a random process $\{X(t)\}$ of the form

$$X(t) = A \cos (10^8 t + \theta)$$

where A and θ are RVs representing the amplitude and phase of the received waveforms.

Such representation of a random process in terms of one or more RVs whose probability law is known is used in several applications in communication systems.

Worked Examples 6(A)

Example 1 *Examine whether the Poisson process $\{X(t)\}$, given by the probability law $P\{X(t) = r\} = e^{-\lambda t} (\lambda t)^r / r!$, $\{r = 0, 1, 2, \cdots\}$, is covariance stationary.*

Solution The probability distribution of $X(t)$ is a Poisson distribution with parameter λt.

$$\therefore \quad E\{X(t)\} = \lambda t \neq \text{a constant}$$

Therefore, the Poisson process is not covariance stationary.

Example 2 *The process $\{X(t)\}$ whose probability distribution under certain conditions is given by*

$$P\{X(t) = n\} = \frac{(at)^{n-1}}{(1+at)^{n+1}}, \; n = 1, 2, \cdots$$

$$= \frac{at}{1+at}, \; n = 0$$

Show that it is not stationary. (MU — Apr. 96)

Solution The probability distribution of $X(t)$ is

$X(t) = n:$	0	1	2	3	$\cdots$
$p_n:$	$\frac{at}{1+at}$	$\frac{1}{(1+at)^2}$	$\frac{at}{(1+at)^3}$	$\frac{(at)^2}{(1+at)^4}$	$\cdots$

$$E\{X(t)\} = \sum_{n=0}^{\infty} np_n$$

$$= \frac{1}{(1+at)^2} + \frac{2at}{(1+at)^3} + \frac{3(at)^2}{(1+at)^4} + \cdots$$

$$= \frac{1}{(1+at)^2} \{1 + 2\alpha + 3\alpha^2 + \cdots\}, \text{ where } \alpha = \frac{at}{1+at}$$

$$= \frac{1}{(1+at)^2}(1-\alpha)^{-2} = \frac{1}{(1+at)^2}(1+at)^2 = 1$$

$$E\{X^2(t)\} = \sum_{n=0}^{\infty} n^2 p_n = \sum_{n=1}^{\infty} n^2 \frac{(at)^{n-1}}{(1+at)^{n+1}}$$

$$= \frac{1}{(1+at)^2}\left[\sum_{n=1}^{\infty} n(n+1)\left(\frac{at}{1+at}\right)^{n-1} - \sum_{n=1}^{\infty} n\left(\frac{at}{1+at}\right)^{n-1}\right]$$

$$= \frac{1}{(1+at)^2}\left[\frac{2}{\left(1-\frac{at}{1+at}\right)^3} - \frac{1}{\left(1-\frac{at}{1+at}\right)^2}\right]$$

$$= 1 + 2at$$

$\therefore \quad \text{Var}\{X(t)\} = 2at$

If $\{X(t)\}$ is a stationary process, $E\{X(t)\}$ and $\text{Var}\{X(t)\}$ are constants. Since $\text{Var}\{X(t)\}$ is a function of t, the given process is not stationary.

Note *When $\{X(t)\}$ is a stationary process, $R\{t_1, t_2\} = E\{X(t_1)\, X(t_2)\}$ is function of $(t_1 - t_2)$.*

$\therefore$ *$E\{X^2(t)\}$ is a constant*

Also   *$E\{X(t)\}$ is a constant*

$\therefore$ *$Var\{X(t)\}$ is a constant*

Example 3 *Show that the random process $X(t) = A\cos(\omega_0 t + \theta)$ is wide-sense stationary, if A and ω_0 are constants and θ is a uniformly distributed RV in $(0, 2\pi)$.* (BDU — Apr. 97)

Solution Since θ is uniformly distributed in $(0, 2\pi)$

$$f_\theta(\theta) = \frac{1}{2\pi}, \; 0 < \theta < 2\pi$$

$$E\{X(t)\} = E\{A\cos(\omega_0 t + \theta)\}$$

$$= A\int_0^{2\pi} \frac{1}{2\pi}\cos(\omega_0 t + \theta)\,d\theta$$

$$\left[\text{since } E\{g(\theta)\} = \int g(\theta) f_\theta(\theta)\,d\theta\right]$$

$$= \frac{A}{2\pi}\{\sin(2\pi + \omega_0 t) - \sin \omega_0 t\}$$

$$= 0 = a \text{ constant}$$

$$E\{X(t_1)\,X(t_2)\} = E\{A^2\cos(\omega_0 t_1 + \theta) \times \cos(\omega_0 t_2 + \theta)\}$$

$$= \frac{A^2}{2} E\{\cos[(t_1 + t_2)\omega_0 + 2\theta] + \cos[\omega_0(t_1 - t_2)]\}$$

$$= \frac{A^2}{2} \int_0^{2\pi} \frac{1}{2\pi} [\cos \{(t_1 + t_2)\omega_0 + 2\theta\} + \cos\{(t_1 - t_2)\, \omega_0\}]\, d\theta$$

$$= \frac{A^2}{2} \cos \omega_0 (t_1 - t_2)$$

i.e., $R(t_1, t_2)$ = a function of $(t_1 - t_2)$

Therefore, $\{X(t)\}$ is a WSS process.

Example 4 *Given a RV Y with characteristic function*

$$\phi(\omega) = E\{e^{i\omega Y}\}$$
$$= E\{\cos \omega Y + i \sin \omega Y\}$$

and a random process defined by $X(t) = \cos(\lambda t + Y)$, show that $\{X(t)\}$ is stationary in the wide sense

if $\phi(1) = \phi(2) = 0$ (MSU — Apr. 96)

Solution $E\{X(t)\} = E\{\cos(\lambda t + Y)\}$

$$= \cos \lambda t \times E(\cos Y) - \sin \lambda t \times E(\sin Y) \quad (1)$$

Given $\phi(1) = 0$

i.e., $E\{\cos Y + i \sin Y\} = 0$

$\therefore$ $E(\cos Y) = 0 = E(\sin Y)$ (2)

Using (2) in (1), we get $E\{X(t)\} = 0$ (3)

$$E\{X(t_1) \times X(t_2)\} = E\{\cos(\lambda t_1 + Y) \times \cos(\lambda t_2 + Y)\}$$
$$= \cos \lambda t_1 \cos \lambda t_2\, E(\cos^2 Y) + \sin \lambda t_1 \sin \lambda t_2\, E(\sin^2 Y)$$
$$- \sin \lambda (t_1 + t_2)\, E(\sin Y \cos Y)$$
$$= \cos \lambda t_1 \cos \lambda t_2\, E\left(\frac{1}{2} + \frac{1}{2}\cos 2Y\right) + \sin \lambda t_1 \sin \lambda t_2$$
$$\cdot E\left(\frac{1}{2} - \frac{1}{2}\cos 2Y\right) - \frac{1}{2} \sin \lambda (t_1 + t_2)\, E(\sin 2Y) \quad (4)$$

Given: $\phi(2) = 0$

i.e., $E\{\cos 2Y + i \sin 2Y\} = 0$

$\therefore$ $E(\cos 2Y) = 0 = E(\sin 2Y)$ (5)

Using (5) in (4), we get

$$R(t_1, t_2) = E\{X(t_1) \times X(t_2)\} = \frac{1}{2}\{\cos \lambda t_1 \cos \lambda t_2 + \sin \lambda t_1 \sin \lambda t_2\}$$

$$= \frac{1}{2} \cos \lambda (t_1 - t_2) \quad (6)$$

From (3) and (6), it follows that $\{X(t)\}$ is a WSS process.

Example 5 *In the fair coin experiment, we define the process {X(t)} as follows.*

$X(t) = \sin \pi t$, if head shows, and

$= 2t$, if tail shows.

(a) Find E{X(t)} and (b) find F(x, t) for t = 0.25 (MKU — Apr. 96)

Solution (a) The probability distribution of $X(t)$ is given by

$$P\{X(t) = \sin \pi t\} = \frac{1}{2} \text{ and } P\{X(t) = 2t\} = \frac{1}{2}$$

$$\therefore \quad E\{X(t)\} = \frac{1}{2} \sin \pi t + t$$

(b) when $t = 0.25$, $P\left\{X(t) = \frac{1}{2}\right\} = \frac{1}{2}$ and $P\left\{X(t) = \frac{1}{\sqrt{2}}\right\} = \frac{1}{2}$

$\therefore$ $F(x, 0.25)$ is given by

$$F(x, 0.25) = 0, \text{ if } x < \frac{1}{2}$$

$$= \frac{1}{2}, \text{ if } \frac{1}{2} \le x < \frac{1}{\sqrt{2}}$$

$$= 1, \text{ if } \frac{1}{\sqrt{2}} \le x$$

Example 6 *If {X(t)} is a wide-sense stationary process with autocorrelation $R(\tau) = Ae^{-\alpha|\tau|}$, determine the second-order moment of the RV X(8) – X(5).*

(BDU — Nov. 96)

Solution Second moment of $X(8) - X(5)$ is given by

$$E[\{X(8) - X(5)\}^2] = E\{X^2(8)\} + E\{X^2(5)\} - 2E\{X(8)\, X(5)\} \quad (1)$$

Given: $R(\tau) = Ae^{-\alpha|\tau|}$

i.e., $R(t_1, t_2) = Ae^{-\alpha|t_1 - t_2|}$

$\therefore$ $E\{X^2(t)\} = R(t, t) = A$

$\therefore$ $E\{X^2(8)\} = E\{X^2(5)\} = A$ (2)

Also $E\{X(8) \times X(5)\} = R(8, 5) = Ae^{-3\alpha}$ (3)

Using (2) and (3) in (1), we get

$$E[\{X(8) - X(5)\}^2] = 2A(1 - e^{-3\alpha})$$

Example 7 *Show that the process $X(t) = A \cos \lambda t + B \sin \lambda t$ (where A and B are RVs) is wide-sense stationary, if*

(i) E(A) = E(B) = 0,

(ii) $E(A^2) = E(B^2)$ and (iii) E(AB) = 0

(MSU — Apr. 96)

Solution $E\{X(t)\} = \cos \lambda t \times E(A) + \sin \lambda t \times E(B)$ (1)

If $\{X(t)\}$ is to be a WSS process, $E\{X(t)\}$ must be a constant (i.e., independent of t).

In (1), if $E(A)$ and $E(B)$ are any constants other than zero, $E\{X(t)\}$ will be a function of t.

$$\therefore \quad E(A) = E(B) = 0$$

$$R(t_1, t_2) = E\{X(t_1) \times X(t_2)\}$$

$$= E\{(A \cos \lambda t_1 + B \sin \lambda t_1)(A \cos \lambda t_2 + B \sin \lambda t_2)\}$$

$$= E(A^2) \cos \lambda t_1 \cos \lambda t_2 + E(B^2) \sin \lambda t_1 \sin \lambda t_2 + E(AB)$$

$$X \sin \lambda(t_1 + t_2) \qquad (2)$$

If $\{X(t)\}$ is to be a WSS process, $R(t_1, t_2)$ must be a function of $(t_1 - t_2)$.

$\therefore$ In (2), $E(AB) = 0$ and $E(A^2) = E(B^2) = k$

Then $\quad R(t_1, t_2) = k \cos \lambda(t_1 - t_2)$

Example 8 *If the 2n RVs A_r and B_r are uncorrelated with zero mean and $E(A_r^2)$ $= E(B_r^2) = \sigma_r^2$, show that the process $X(t) = \sum_{r=1}^{n} (A_r \cos \omega_r t + B_r \sin \omega_r t)$ is wide-sense stationary. What are the mean and autocorrelation of $X(t)$?*

(MSU — Nov. 96)

Solution $E\{X(t)\} = \sum_{r=1}^{n} E(A_r) \cos \omega_r t + E\{B_r\} \sin \omega_r t = 0$

$$E\{X(t_1) \times X(t_2)\}$$

$$= E\left\{\sum_{r=1}^{n}\sum_{s=1}^{n} (A_r \cos \omega_r t_1 + B_r \sin \omega_r t_1) \times (A_s \cos \omega_s t_2 + B_s \sin \omega_s t_2)\right\}$$

Since $E\{A_r A_s\}$, $E\{B_r B_s\}$, $E\{A_r B_r\}$ and $E\{A_r B_s\}$ are all zero, for $r \neq s$, we have

$$E\{X(t_1) \times X(t_2)\} = \sum_{r=1}^{n} [E(A_r^2) \cos \omega_r t_1 \cos \omega_r t_2 + E(B_r^2) \sin \omega_r t_1 \sin \omega_r t_2]$$

$$= \sum_{r=1}^{n} \sigma_r^2 \cos \omega_r (t_1 - t_2)$$

Therefore, $\{X(t)\}$ is a WSS process.

Example 9 *Given a RV Ω with density $f(\omega)$ and another RV ϕ uniformly distributed in $(-\pi, \pi)$ and independent of Ω and $X(t) = a \cos(\Omega t + \phi)$, prove that $\{X(t)\}$ is a WSS process.*

Solution Recall that $E\{g(X, Y)\} = E[E\{g(X, Y)/X\}]$

$$\therefore \quad E\{X(t)\} = E\{a \cos(\Omega t + \phi)\}$$

$$= aE[E\{\cos(\Omega t + \phi)/\Omega\}]$$

$$= aE[\cos \Omega t \times E(\cos \phi) - \sin \Omega t \times E(\sin \phi)]$$

$$= aE\left[\cos \Omega t \int_{-\pi}^{\pi} \frac{1}{2\pi} \cos\phi \, d\phi - \sin \Omega t \int_{-\pi}^{\pi} \frac{1}{2\pi} \sin\phi \, d\phi\right]$$

$$= 0 \qquad \text{[since } \phi \text{ is uniform in } (-\pi, \pi)]$$

$$R(t_1, t_2) = E\{X(t_1) \times X(t_2)\}$$

$$= E\{a^2 \cos(\Omega t_1 + \phi) \cos(\Omega t_2 + \phi)\}$$

$$= a^2E\,[E\{\cos \Omega\, t_1 \cos \Omega\, t_2 \cos^2 \phi + \sin \Omega\, t_1 \sin \Omega\, t_2 \sin^2\phi$$
$$- (\sin \Omega\, t_1 \cos \Omega\, t_2 + \cos \Omega\, t_1 \sin \Omega\, t_2) \sin \phi \cos \phi/\Omega)\}$$

$$= a^2E\left[\cos \Omega t_1 \cos \Omega t_2 \int_{-\pi}^{\pi} \cos^2 \phi \, d\phi + \sin \Omega t_1 \sin \Omega t_2 \times \int_{-\pi}^{\pi} \sin^2 \phi \, d\phi - \sin\Omega(t_1 + t_2) \int_{-\pi}^{\pi} \sin\phi \cos\phi \, d\phi\right]$$

$$= \frac{1}{2} a^2E\,\{\cos \Omega\, t_1 \cos \Omega\, t_2 + \sin \Omega\, t_1 \sin \Omega\, t_2\}$$

$$= \frac{1}{2} a^2E\,\{\cos \Omega\,(t_1 - t_2)\}$$

= a function of $(t_1 - t_2)$, whatever be the value of $f(\omega)$.

Therefore, $\{X(t)\}$ is a WSS process.

Example 10 *Verify whether the sine wave process $\{X(t)\}$, where $X(t) = Y \cos\omega t$, where Y is uniformly distributed in (0, 1) is a SSS process.*

Solution $P\{(X(t) \le x\} = P\{Y \cos \omega t \le x\}$

$$= \begin{cases} P\left\{Y \le \dfrac{x}{\cos \omega t}\right\} & \text{if } \cos \omega t > 0 \\ P\left\{Y \ge \dfrac{x}{\cos \omega t}\right\} & \text{if } \cos \omega t < 0 \end{cases}$$

i.e.,
$$F_{x(t)}(x) = \begin{cases} F_y\left(\dfrac{x}{\cos \omega t}\right) & \text{if } \cos \omega t > 0 \\ 1 - F_y\left(\dfrac{x}{\cos \omega t}\right) & \text{if } \cos \omega t < 0 \end{cases}$$

$\therefore$
$$f_{x(t)}(x) = \frac{1}{|\cos \omega t|} \times f_y\left(\frac{x}{\cos \omega t}\right)$$

= *a* function of *t*

If $\{X(t)\}$ is to be a SSS process, its first-order density must be independent of t.

Therefore, $\{X(t)\}$ is not a SSS process.

Example 11 *If $X(t) = Y \cos \omega t + Z \sin \omega t$, where Y and Z are two independent normal RVs with $E(Y) = E(Z) = 0$, $E(Y^2) = E(Z^2) = \sigma^2$ and ω is a constant, prove that $\{X(t)\}$ is a SSS process of order 2.*

Solution Since $X(t)$ is a linear combination of Y and Z, that are independent, $X(t)$ follows a normal distribution with

$$E\{X(t)\} = \cos \omega t E(Y) + \sin \omega t E(Z) = 0$$

$$\text{and } \operatorname{Var}\{X(t)\} = \cos^2 \omega t \times E(Y^2) + \sin^2 \omega t \times E(Z^2) = \sigma^2$$

Since $X(t_1)$ and $X(t_2)$ are each $N(0, \sigma)$, $X(t_1)$ and $X(t_2)$ are jointly normal with the joint pdf given by

$$f(x_1, x_2, t_1, t_2) = \frac{1}{2\pi\sigma^2 \sqrt{1-r^2}} \exp\{-(x_1^2 - 2rx_1 x_2 + x_2^2)/2(1-r^2)\sigma^2\}$$

$$-\infty < x_1, x_2 < \infty \qquad (1)$$

In (1), r = correlation co-efficient between $X(t_1)$ and $X(t_2)$

$$= \frac{C(t_1, t_2)}{\sqrt{\operatorname{Var}\{X(t_1)\} \times \operatorname{Var}\{X(t_2)\}}}$$

$$= \frac{1}{\sigma^2} E\{X(t_1) \times X(t_2)\}$$

$$= \frac{1}{\sigma^2} E[(Y \cos \omega t_1 + Z \sin \omega t_1)(Y \cos \omega t_2 + Z \sin \omega t_2)]$$

$$= \frac{1}{\sigma^2} [E(Y^2) \cos \omega t_1 \cos \omega t_2 + E(Z^2) \sin \omega t_1 \sin \omega t_2]$$

[since $E(YZ) = 0$, as Y and Z are independent]

$$= \cos \omega (t_1 - t_2)$$

Now the joint pdf of $X(t_1 + h)$ and $X(t_2 + h)$ is given by a similar expression as in (1), where

$$r = \cos\omega \{(t_1 + h) - (t_2 + h)\} = \cos\omega (t_1 - t_2)$$

Thus the joint pdf's of $\{X(t_1), X(t_2)\}$ and $\{X(t_1 + h), X(t_2 + h)\}$ are the same.

Therefore, $\{X(t)\}$ is a SSS process of order 2.

Example 12 *Two random processes $X(t)$ and $Y(t)$ are defined by $X(t) = A \cos\omega_0 t + B \sin\omega_0 t$ and $Y(t) = B \cos \omega_0 t - A \sin \omega_0 t$. Show that $X(t)$ and $Y(t)$ are jointly wide-sense stationary, if A and B are uncorrelated RVs with zero means and the same variances and ω_0 is a constant.* (BDU — Apr. 97)

Solution $E(A) = E(B) = 0$; $\operatorname{Var}(A) = \operatorname{Var}(B)$

$$\therefore \qquad E(A^2) = E(B^2)$$

Since A and B are uncorrelated, $E(AB) = 0$.

Therefore, by Example 7, $\{X(t)\}$ and $\{Y(t)\}$ are individually WSS processes.

Now $\quad R_{xy}(t_1, t_2) = E\{X(t_1) \times Y(t_2)\}$

$$= E\{(A \cos \omega_0 t_1 + B \sin \omega_0 t_1)(B \cos \omega_0 t_2 - A \sin \omega_0 t_2)\}$$

$$= E(B^2) \sin \omega_0 t_1 \cos \omega_0 t_2 - E(A^2) \cos \omega_0 t_1 \sin \omega_0 t_2$$

$$= \sigma^2 \sin \omega_0 (t_1 - t_2) \text{ [assuming } E(A^2) = E(B^2) = \sigma^2]$$

$$= \text{a function of } (t_1 - t_2)$$

Therefore, $\{X(t)\}$ and $\{Y(t)\}$ are jointly WSS processes.

Example 13 *(i) Define Random Walk and prove that the limiting form of random walk is the wiener process and (ii) find the covariance function of the Wiener process.*

Solution

(i) Definition of Random Walk

Suppose we toss a fair coin every T seconds and instantly after each toss we move a distance d to the right if heads show and to the left if tails show. If the process starts at $t = 0$, our position at time $t = nT$ is a random sequence $X(nT)$ that may be simply denoted as $X(n)$ also. The process $\{X(nT)\}$ is called a random walk.

Suppose that r heads and $(n - r)$ tails have occurred in the first n tosses of the coin. Then the random walk consists of r steps to the right and $(n - r)$ steps to the left.

$\therefore \quad X(nT) = rd - (n - r)d$

$$= (2r - n)d = md, \text{ say}$$

Note that $X(nT)$ is a RV, taking the values md, where $m = -n, -n + 2, \cdots, n - 2, n$.

Now $\quad P\{X(nT) = md\} = P\{\text{getting } r \text{ heads in } n \text{ tosses}\}$

$$= nC_r \left(\frac{1}{2}\right)^r, \text{ where } \frac{m+n}{2}, \text{ since } 2r - n = m$$

Now $X(nT)$ can also be expressed as a sum as given below,

$$X(nT) = X_1 + X_2 + \cdots + X_n$$

where X_i represents the distance moved in the ith step.

The RVs X_i are independent, taking the values $\pm d$ with equal probability.

$\therefore \quad E\{X(nT)\} = \sum_{i=1}^{n} E(X_i) = 0$

and $\quad E\{X^2(nT)\} = \sum_{i=1}^{n} E(X_i^2)$

$$= \sum_{i=1}^{n} \left(\frac{1}{2} \times d^2 + \frac{1}{2} \times d^2 \right) = nd^2$$

We know that the limiting form of the binomial distribution with mean np and variance npq as $n \to \infty$ is the normal distribution $N\left(np, \sqrt{npq}\right)$.

i.e. $$nC_r\, p^r\, q^{n-r} \cong \frac{1}{\sqrt{2\pi\, npq}}\, e^{-(r-np)^2/2npq}$$

$\therefore$ $$P\{X(nT) = md\}\ \frac{1}{\sqrt{2\pi \frac{n}{4}}}\, e^{-m^2/2n} \left(\text{since } r = \frac{m+n}{2}\right) \quad (1)$$

Wiener Process as Limiting form of Random Walk

In (1), put $nT = t$, $md = x$ and $d^2 = \alpha T$ and take limits as $T \to 0$ and $n \to \infty$.

In the limit, $\{X(t)\}$ becomes a continuous process.

Now $$\frac{m}{\sqrt{n}} = \frac{x/d}{\sqrt{t/T}} = \frac{x}{\sqrt{d^2\, t/T}} = \frac{x}{\sqrt{\alpha t}} \quad (2)$$

Also $$\frac{n}{4} = E\{X^2(nT)\}$$
$$= nd^2$$
$$= n\alpha T$$
$$= \alpha t \quad (3)$$

If we use (2) and (3) in (1), when we proceed to limits,

$$P\{x \le X(t) \le x + \mathrm{d}x\} = \frac{1}{\sqrt{2\pi\alpha t}}\, e^{-x^2/2\alpha t}\, \mathrm{d}x, \ -\infty < x < \infty$$

i.e., the pdf of Wiener process $\{X(t)\}$ is

$$f_{x(t)}(x) = \frac{1}{\sqrt{2\pi\alpha t}}\, e^{-x^2/2\alpha t}$$

which is $N(0, \sqrt{\alpha t}\,)$.

(ii) Obviously, the random walk $\{X(nT)\}$ is a process with independent increments.

i.e., $\{X(n_2 T) - X(n_1 T)\}$ and $\{X(n_1 T) - X(0)\}$ are independent.

Since Wiener process $\{X(t)\}$ is the limiting form of random walk, $\{X(t_2) - X(t_1)\}$ and $X(t_1)$ are independent.

Let $t_1 < t_2$

Then $$E[\{X(t_2) - X(t_1)\} \times X(t_1)]$$

$$= E\{X(t_2) - X(t_1)\} \times E\{X(t_1)\}$$

$$= 0 \qquad \text{[since } E\{X(t)\} = 0]$$

i.e., $E\{X(t_1) \times X(t_2)\} = E\{X^2(t_1)\}$

$$= \alpha t_1 \text{ [since Var } \{X(t)\} = \alpha t]$$

i.e., $R(t_1, t_2) = \alpha t_1$

Similarly, when $t_2 < t$, $R(t_1, t_2) = \alpha t_2$.

$\therefore$ $R(t_1, t_2) = \alpha \min (t_1, t_2)$

$$C(t_1, t_2) = R(t_1, t_2) - \mu(t_1) \times \mu(t_2)$$

$$= \alpha \min (t_1, t_2) \qquad \text{[since } \mu(t) = 0]$$

Example 14 *If $X(t)$ with $X(0) = 0$ and $\mu = 0$ is a Wiener process, show that $Y(t) = \sigma X\left(\frac{t}{\sigma^2}\right)$ is also a Wiener process. Find its covariance function.*

(MU — Apr. 96)

Solution The pdf of Wiener process $\{X(t)\}$ is given by

$$f(x) = \frac{1}{\sqrt{2\pi\alpha t}} e^{-x^2/2\alpha t}$$

The standard deviation of $X(t)$ is given by

$$\sigma = \sqrt{\alpha t}$$

Therefore, the pdf of the RV $\left\{X\left(\frac{t}{\sigma^2}\right)\right\} = \left\{X\left(\frac{1}{\alpha}\right)\right\}$ is

$$f_1(x) = \frac{1}{\sqrt{2\pi}} e^{-x^2/2}$$

To find the pdf of $Y(t) = \sigma X\left(\frac{t}{\sigma^2}\right)$, let us use the transformation rule.

$$f_{Y(t)}(y) = f_1(x) \left|\frac{dx}{dy}\right| \text{ where } y = \sigma x$$

$$= \frac{1}{\sigma} f_1\left(\frac{y}{\sigma}\right)$$

$$= \frac{1}{\sqrt{\alpha t}} f_1\left(\frac{y}{\sqrt{\alpha t}}\right)$$

$$= \frac{1}{\sqrt{2\pi\alpha t}} e^{-y^2/2\alpha t}$$

which is the pdf of Wiener process.

The convariance function of the Wiener process was found out in the previous example.

Example 15 *Define semi-random telegraph signal process and random telegraph signal process and prove also that the former is evolutionary and the latter is wide-sense stationary.*

Solution If $N(t)$ represents the number of occurrences of a specified event in $(0, t)$ and $X(t) = (-1)^{N(t)}$, then $\{X(t)\}$ is called a *semi-random telegraph signal process.*

If $\{X(t)\}$ is a semi-random telegraph signal process, α is a RV which is independent of $X(t)$ and which assumes the values $+1$ and -1 with equal probability and $Y(t) = \alpha X(t)$, then $\{Y(t)\}$ is called a *random telegraph signal process.*

It will be proved in chapter VII that the distribution of $N(t)$ is Poisson with mean λt, where the probability of exactly one occurrence in a small interval of length $h = \lambda h$.

In other words, the process $\{N(t)\}$ is a Poisson process with the probability law

$$P\{N(t) = r\} = e^{-\lambda t} (\lambda t)^r / \underline{|r}; \; r = 0, 1, 2, \cdots, \infty$$

If $\{X(t)\}$ is the semi-random telegraph signal process, then as per the definition given above, $X(t)$ can take the values $+1$ and -1 only.

$$P\{X(t) = 1\} = P\{N(t) \text{ is even}\}.$$

$$= P\{N(t) = 0\} + P\{N(t) = 2\} + P\{N(t) = 4\}$$

$$+ \cdots + \infty \text{ (since the events are mutually exclusive)}$$

$$= e^{-\lambda t} \left\{1 + \frac{(\lambda t)^2}{\underline{|2}} + \frac{(\lambda t)^4}{\underline{|4}} + \cdots + \infty\right\}$$

$$= e^{-\lambda t} \cosh \lambda t$$

$$P\{X(t) = -1\} = P\{N(t) \text{ is odd}\}$$

$$= P\{N(t) = 1\} + P\{N(t) = 3\} + \cdots + \infty$$

(since the events are mutually exclusive)

$$= e^{-\lambda t} \left\{\frac{\lambda t}{\underline{|1}} + \frac{(\lambda t)^3}{\underline{|3}} + \cdots + \infty\right\}$$

$$= e^{-\lambda t} \sinh \lambda t$$

$$\therefore \quad E\{X(t)\} = 1 \times e^{-\lambda t} \cosh \lambda t + (-1) \times e^{-\lambda t} \sinh t$$

$$= e^{-\lambda t} e^{-\lambda t} = e^{-2\lambda t}$$

To find $E\{X(t_1) \times X(t_2)\}$, we require the joint probability distribution of $\{X(t_1), X(t_2)\}$.

Now $\quad P\{X(t_1) = 1, X(t_2) = 1\}$

$$= P\{X(t_1) = 1 / X(t_2) = 1\} \times P\{X(t_2) = 1\}$$

$= P\{\text{an even number of occurrences of the event in } (t_1 - t_2)\} \times P\{X(t_2) = 1\}$

$= e^{-\lambda\tau} \cosh \lambda\tau \times e^{-\lambda t_2} \cosh \lambda t_2$; where $\tau = t_1 - t_2$

Similarly, $P\{X(t_1) = -1, X(t_2) = -1\}$

$= e^{-\lambda\tau} \cosh \lambda\tau \; e^{-\lambda t_2} \sinh \lambda t_2$

$P\{X(t_1) = 1, X(t_2) = -1\} = e^{-\lambda\tau} \sinh \lambda\tau \; e^{-\lambda t_2} \sinh \lambda t_2$

and $P\{X(t_1) = -1, X(t_2) = 1\} = e^{-\lambda\tau} \sinh \lambda\tau \; e^{-\lambda t_2} \cosh \lambda t_2$

Now $X(t_1) \times X(t_2) = 1$, if $\{X(t_1) = 1$ and $X(t_2) = 1\}$ or $\{X(t_1) = -1$ and $X(t_2) = -1\}$

$\therefore \quad P\{X(t_1) \times X(t_2) = 1\} = e^{-\lambda(\tau + t_2)} \cosh \lambda\tau (\cosh \lambda t_2 + \sinh \lambda t_2)$

$= e^{-\lambda\tau} \cosh \lambda\tau$

and $P\{X(t_1) \times X(t_2) = -1\}$

$= e^{-\lambda(\tau + t_2)} \sinh \lambda\tau (\cosh \lambda t_2 + \sinh \lambda t_2)$

$= e^{-\lambda\tau} \sinh \lambda\tau$

$\therefore \quad R(t_1, t_2) = E\{X(t_1)\, X(t_2)\}$

$= 1 \times e^{-\lambda\tau} \cosh \lambda\tau - 1 \times e^{-\lambda\tau} \sinh \lambda\tau$

$= e^{-2\lambda\tau}$

$= e^{-2\lambda(t_1 - t_2)}$

Although $R(t_1, t_2)$ is a function of $(t_1 - t_2)$, $E\{X(t)\}$ is not a constant.

Therefore, $\{X(t)\}$ is *evolutionary*.

Let us now consider the random telegraph signal process $\{Y(t)\}$, where $Y(t) = \alpha X(t)$.

By definition, $P\{\alpha = 1) = \dfrac{1}{2}$ and $P(\alpha = -1) = \dfrac{1}{2}$

$\therefore \quad E(\alpha) = 0$ and $E(\alpha^2) = 1$

Now $E\{Y(t) = E(\alpha) \times E\{X(t)\}$

$= 0$ [since α and $X(t)$ are independent]

$E\{Y(t_1) \times Y(t_2)\} = E\{\alpha^2 X(t_1) \times X(t_2)\}$

$= E(\alpha^2) \times E\{X(t_1) \times X(t_2)\}$ (by independence)

$= 1 \times e^{-2\lambda(t_1 - t_2)}$

i.e., $R_{yy}(t_1, t_2) =$ a function of $(t_1 - t_2)$

Therefore, $\{Y(t)\}$ is a wide-sense stationary process.

Exercise 6(A)

Part A (Short answer questions)

1. What is the difference between a RV and a random process?
2. Define a random process and give an example of a random process.
3. Explain the terms 'state space' and 'parameter set' associated with a random process.
4. If $\{X(s, t)\}$ is a random process, what is the nature of $X(s, t)$ when (i) s is fixed and (ii) t is fixed?
5. What is the difference between a random sequence and random process?
6. What is a discrete random sequence? Give an example.
7. What is a continuous random sequence? Given an example.
8. What is a discrete random process? Give an example.
9. What is a continuous random process? Given an example.
10. How is a random process described mathematically?
11. Name 3 classes of RP's into which RP's are generally divided.
12. Name 2 important RP's with independent increments.
13. What do you mean by the mean and variance of a random process?
14. Define the autocorrelation of a RP $\{X(t)\}$.
15. Define the autocovariance of a RP $\{X(t)\}$.
16. Define the correlation coefficient of a RP$\{X(t)\}$.
17. Is the autocorrelation of a RP the same as the correlation coefficient of the process? Why?
18. Define the cross-correlation of 2 random processes.
19. When are 2 random processes said to be orthogonal?
20. Define the cross-covariance of 2 random processes.
21. Define the cross-correlation coefficient of 2 random processes.
22. Define a strict-sense stationary process and give an example.
23. Define a kth-order stationary process. When will it become a SSS process?
24. Prove that the first-order density function of a SSS process $\{X(t)\}$ is independent of t.
25. If $\{X(t)\}$ is a SSS process, prove that $E\{X(t)\}$ is a constant.
26. If $\{X(t)\}$ is a SSS process, prove that the joint pdf of $X(t_1)$ and $X(t_2)$ is a function of $(t_1 - t_2)$.
27. Prove that the autocorrelation of a SSS process $\{X(t)\}$ is a function of $(t_1 - t_2)$.
28. When are $\{X(t)\}$ and $\{Y(t)\}$ said to be jointly stationary in the strict sense?
29. When is a complex random process $\{Z(t)\}$, where $Z(t) = X(t) + iY(t)$, said to be a SSS process?
30. Define wide-sense stationary process. Give an example.
31. What is the difference between a SSS process and a WSS process?

32. When is a random process said to be evolutionary? Give an example of an evolutionary process.
33. When are the processes $\{X(t)\}$ and $\{Y(t)\}$ said to be jointly stationary in the wide sense?
34. If $\{X(t)\}$ is a stationary process in any sense, prove that $\text{Var}\{X(t)\}$ is a constant.
35. Give the one-dimensional density function of Wiener process. What are its mean and variance?
36. Define a semirandom telegraph signal process. Is it stationary?
37. Define a random telegraph signal process. Is it stationary?

Part B

38. If $X(t) = P + Qt$, where P and Q are independent RVs with $E(P) = p$, $E(Q) = q$, $\text{Var}(P) = \sigma^2{}_1$ and $\text{Var}(Q) = \sigma^2{}_2$, find $E\{X(t)\}$, $R(t_1, t_2)$ and $C(t_1, t_2)$. Is the process $\{X(t)\}$ stationary?
39. If $X(t) = \sin(\omega t + Y)$, where Y is uniformly distributed in $(0, 2\pi)$, prove that $\{X(t)\}$ is a wide-sense stationary process.
40. If $X(t) = Y\cos t + Z\sin t$ for all t where Y and Z are independent binary RVs, each of which assumes the values -1 and $+2$ with probabilities 2/3 and 1/3 respectively, prove that $\{X(t)\}$ is wide-sense stationary.
41. Calculate the autocorrelation function of the process $X(t) = A\sin(\omega_0 t + \phi)$, where A and ω_0 are constants and ϕ is a uniformly distributed RV in $(0, 2\pi)$. (BDU — Apr. 97)
42. Consider the random process $V(t) = \cos(\omega t + \theta)$, where θ is a RV with probability density

$$P(\theta) = \begin{Bmatrix} \dfrac{1}{2\pi} & -\pi \le \theta \le \pi \\ 0 & \text{elsewhere} \end{Bmatrix}$$

   (a) Show that the first and second moments of $V(t)$ are independent of time.
   (b) If θ = constant, will the ensemble mean of $V(t)$ be time-independent? (BU — Apr. 96)
43. If $X(t) = R\cos(\omega t + \phi)$, where R and ϕ are independent RVs and ϕ is uniformly distributed in $(-\pi, \pi)$, prove that $R(t_1, t_2) = \dfrac{1}{2} E(R^2) \times \cos\omega(t_1 - t_2)$.
44. In the fair-coin experiment, we define the process $X(t)$ as follows:
   $X(t) = \sin\pi t$, if head shows and
   $X(t) = 2t$, if tail shows.
   Find (i) $E\{X(t)\}$ and (ii) $F(x, t)$ for $t = 0.25, 0.5, 1$. (BDU — Nov. 96)

45. A stochastic process is described by $X(t) = A \sin t + B \cos t$, where A and B are independent RVs with zero means and equal standard deviations. Show that the process is stationary of the second order.
(MU — Nov. 96)
[**Hint:** Without knowing the distributions of A and B, it is not possible to prove strict sense stationarity of the second order. We can prove that $\{X(t)\}$ is WSS.]
46. Consider a random process $Z(t) = X_1 \cos\omega_0 t - X_2 \sin\omega_0 t$, where X_1 and X_2 are independent Gaussian RVs with zero mean and variance σ^2. Find $E\{z\}$ and $E\{z^2\}$. (BU — Nov. 96)
47. Suppose that $X(t)$ is a process with mean $\mu(t) = 3$ and autocorrelation $R(t_1, t_2) = 9 + 4\, e^{-0.2|t_1 - t_2|}$. Determine the mean, variance and the convariance of the RVs $Z = X(5)$ and $W = X(8)$. (BU – Nov. 96)
48. If the RVs A_i are uncorrelated with zero mean and $E\{|A_i|^2\} = \sigma^2_i$, prove that the process $X(t) = \sum_{i=1}^{n} A_i e^{j\omega_i t}$ is wide-sense stationary with zero mean. Show also that for $X(t)$, $R(z) = \sum_{i=1}^{n} \sigma^2_i\, e^{j\omega_i z}$. (MSU — Nov. 96)
[**Hint:** For a complex-valued random process $\{X(t)\}$, the autocorrelation is defined as $R(t_1, t_2) = E\{X(t_1) \times X^*(t_2)\}$, where $x^*(t_2)$ is the complex conjugate of $x(t_2)$.]
49. If $U(t) = X \cos t + Y \sin t$ and $V(t) = Y \cos t + X \sin t$, where X and Y are independent RVs such that $E(X) = 0 = E(Y)$, $E(X^2) = E(Y^2) = 1$, show that $\{U(t)\}$ and $\{V(t)\}$ are individually stationary in the wide sense, but they are not jointly wide-sense stationary.
50. If $X(t) = 5 \cos(10t + \theta)$ and $Y(t) = 20 \sin(10t + \theta)$, where θ is a RV uniformly distributed in $(0, 2\pi)$, prove that the processes $\{X(t)\}$ and $\{Y(t)\}$ are jointly wide-sense stationary.
51. If $X(t) = A \sin(\omega t + \theta)$, where A and ω are constants and θ is RV, uniformly distributed over $(-\pi, \pi)$, find the autocorrelation of $\{Y(t)\}$, where $Y(t) = X^2(t)$.
52. Find the first-order characteristic function of a Wiener process.
(MKU — Apr. 96)
53. Consider the process $W(t) = X(t) \cos\omega t + Y(t) \sin\omega t$, where $X(t)$ and $Y(t)$ are 2 real jointly stationary processes. What are the conditions for $W(t)$ to be WSS? In case $W(t)$ is wide-sense stationary, what is its autocorrelation in terms of autocorrelations of $X(t)$ and $Y(t)$? (MSU — Nov. 96)
54. Show that, if the process $X(t) = a \cos\omega t + b \sin\omega t$ is SSS, where a and b are independent RVs, then they are normal. (BDU — Apr. 96)
55. The RVs A and B are independent $N(0, \sigma)$ and p is the probability that the process $\{X(t)\}$, where $X(t) = A - Bt$, crosses the t-axis in the interval $(0, T)$. Show that $\pi p = \tan^{-1} T$. (BDU — Apr. 96)

[**Hint:** $Z = A/B$ follows the Cauchy distribution with $f(z) = \frac{1}{\pi} \times \frac{1}{1+z^2}$ and $p = P\{0 \le A/B \le T\}$].

Autocorrelation Function and its Properties

Definition: If the process $\{X(t)\}$ is stationary either in the strict sense or in the wide sense, then $E\{X(t)\, X(t-\tau)\}$ is a function of τ, denoted by $R_{xx}(\tau)$ or $R_x(\tau)$ or $R(\tau)$. This function $R(\tau)$ is called the autocorrelation function of the process $\{X(t)\}$.

Properties of R(τ)

1. $R(t)$ is an even function of τ

Proof $R(\tau) = E\{X(t) \times X(t-\tau)\}$

$$\therefore \quad R(-\tau) = E\{X(t) \times X(t+\tau)\}$$
$$= E\{X(t+\tau) \times X(t)\}$$
$$= R(\tau)$$

Therefore, $R(\tau)$ is an even function of τ.

2. $R(\tau)$ is maximum at $\tau = 0$

i.e., $$|R(\tau)| \le R(0)$$

Proof Cauchy–Schwarz inequality is

$$\{E(XY)\}^2 \le E(X^2) \times E(Y^2)$$

Put $$X = X(t) \text{ and } Y = X(t-\tau)$$

Then $$[E(X(t) \times X(t-\tau)\}]^2 \le E\{X^2(t)\} \times E\{X^2(t-\tau)\}$$

i.e., $$\{R(\tau)\}^2 \le [E\{X^2(t)\}]^2$$

[since $E\{X(t)\}$ and Var $\{X(t)\}$ are constant for a stationary process]

i.e., $$\{R(\tau)\}^2 \le \{R(0)\}^2$$

Taking square-root on both sides

$$|R(\tau)| \le R(0) \qquad [\text{since } R(0) = E\{X^2(t)\} \text{ is positive}]$$

3. If the autocorrelation function $R(\tau)$ of a real stationary process $\{X(t)\}$ is continuous at $\tau = 0$, it is continuous at every other point.

Proof Consider $E[\{X(t) - X(t-\tau)\}^2]$

$$= E\{X^2(t)\} + E\{X^2(t-\tau)\} - 2E\{X(t) \times X(t-\tau)\}$$
$$= R(0 + R(0) - 2R(\tau)$$
$$= 2[R(0) - R(\tau)] \qquad (1)$$

Since $R(\tau)$ is continuous at $\tau = 0$, $\lim\limits_{\tau \to 0} R(\tau) = R(0)$

i.e, $$\lim_{\tau \to 0} \{\text{R.S. of (1)}\} = 0$$

$\therefore \qquad \lim_{\tau \to 0} \{\text{L.S. of (1)}\} = 0$

$\therefore \qquad \lim_{\tau \to 0} \{X(t-\tau)\} = X(t)$

i.e. $\qquad X(t)$ is continuous for all t (2)

Consider $\qquad R(\tau + h) - R(\tau)$

$$= E[X(t) \times X\{t - (\tau + h)\}] - E[X(t) \times X(t-\tau)]$$

$$= E[X(t)\{X(t-\tau-h) - X(t-\tau)\}] \qquad (3)$$

Now $\qquad \lim_{h \to 0} [X\{(t-\tau)-h\} - X(t-\tau)] = 0$, by (2)

$\therefore \qquad \lim_{h \to 0} \{\text{R.S. of (3)}\} = 0$

$\therefore \qquad \lim_{h \to 0} \{\text{L.S. of (3)} = 0$

i.e., $\qquad \lim_{h \to 0} \{R(\tau + h)\} = R(\tau)$

i.e., $\qquad R(\tau)$ is continuous for all τ

4. If $R(\tau)$ is the autocorrelation function of a stationary process $\{X(t)\}$ with no periodic component, then $\lim_{\tau \to \infty} R(\tau) = \mu^2_x$, provided the limit exists.

Proof $\quad R(\tau) = E\{X(t) \times X(t-\tau)\}$

When τ is very large, $X(t)$ and $X(t-\tau)$ are two sample functions (members) of the process $\{X(t)\}$ observed at a very long interval of time.

Therefore, $X(t)$ and $X(t-\tau)$ tend to become independent [$X(t)$ and $X(t-\tau)$ may be dependent, when $X(t)$ contains a periodic component, which is not true].

$$\therefore \quad \lim_{\tau \to \infty} \{R(\tau)\} = E\{X(t)\} \times E\{X(t-\tau)\}$$

$$= \mu_x^2 \qquad \text{[since } E\{X(t)\} \text{ is a constant]}$$

i.e., $\qquad \mu_x = \sqrt{\lim_{\tau \to \infty} R(\tau)}$

Cross-Correlation Function and Its Properties

Definition: If the processes $\{X(t)\}$ and $\{Y(t)\}$ are jointly wide-sense stationary, then $E\{X(t) \times Y(t-\tau)\}$ is a function of τ, denoted by $R_{xy}(\tau)$. This function $R_{xy}(\tau)$ is called the cross-correlation function of the processes $\{X(t)\}$ and $\{Y(t)\}$.

We give below the properties of $R_{xy}(\tau)$ without proof. Proofs of these properties are left as exercises to the reader.

Properties

1. $R_{yx}(\tau) = R_{xy}(-\tau)$
2. $|R_{xy}(\tau)| \le \sqrt{R_{xx}(0) \times R_{yy}(0)}$

This means that the maximum of $R_{xy}(\tau)$ can occur anywhere, but it cannot exceed $\sqrt{R_{xx}(0) \times R_{yy}(0)}$.

3. $|R_{xy}(\tau)| \leq \frac{1}{2} \{R_{xx}(0) + R_{yy}(0)\}$
4. If the processes $\{X(t)\}$ and $\{Y(t)\}$ are orthogonal, then $R_{xy}(\tau) = 0$
5. If the processes $\{X(t)\}$ and $\{Y(t)\}$ are independent, then $R_{xy}(\tau) = \mu_x \times \mu_y$

Ergodicity

When we wish to take a measurement of a variable quantity in the laboratory, we usually obtain multiple measurements of the variable and average them to reduce measurement errors. If the value of the variable being measured is constant and errors are due to disturbances (noise) or due to the instability of the measuring instrument, then averaging is, in fact, a valid and useful technique. 'Time averaging' is an extension of this concept, which is used in the estimation of various statistics of random processes.

We normally use ensemble averages (or statistical averages) such as the mean and autocorrelation function for characterising random processes. To estimate ensemble averages, one has to compute a weighted average over all the member functions of the random process.

For example, the ensemble mean of a discrete random process $\{X(t)\}$ is computed by the formula $\mu_x = \sum x_i p_i$. If we have access only to a single sample function of the process, then we use its *time-average* to estimate the ensemble averages of the process.

Definition: If $\{X(t)]$ is a random process, then $\frac{1}{2T} \int_{-T}^{T} X(t)\, dt$ is called the *time-average* of $\{X(t)\}$ over $(-T, T)$ and denoted by $\overline{X}_T$.

In general, ensemble averages and time averages are not equal except for a very special class of random processes called *ergodic processes*. The concept of ergodicity deals with the equality of time averages and ensemble averages.

Definition: A random process $\{X(t)\}$ is said to be ergodic, if its ensemble averages are equal to appropriate time averages.

This definition implies that, with probability 1, any ensemble average of $\{X(t)\}$ can be determined from a single sample function of $\{X(t)\}$.

Note *Ergodicity is a stronger condition than stationarity and hence all random processes that are stationary are not ergodic. Moreover, ergodicity is usually defined with respect to one or more ensemble averages (such as mean and autocorrelation function) as discussed below and a process may be ergodic with respect to one ensemble average but not others.*

Mean-Ergodic Process

If the random process $\{X(t)\}$ has a constant mean $E\{X(t)\} = \mu$ and if $\overline{X}_T = \frac{1}{2T}\int_{-T}^{T} X(t)\,dt \to \mu$ as $T \to \infty$, then $\{X(t)\}$ is said to be mean-ergodic.

Mean-Ergodic Theorem

If $\{X(t)\}$ is a random process with constant mean μ and if $\overline{X}_T = \frac{1}{2T}\int_{-T}^{T} X(t)\,dt$, then $\{X(t)\}$ is mean-ergodic (or ergodic in the mean), provided $\lim_{T\to\infty} \{\text{Var}\,\overline{X}_T\} = 0$.

Proof $\overline{X}_T = \frac{1}{2T}\int_{-T}^{T} X(t)\,dt$

$$\therefore \quad E(\overline{X}_T) = \frac{1}{2T}\int_{-T}^{T} E\{X(t)\}dt$$

$$= \mu \tag{1}$$

By Tchebycheff's inequality

$$P\{|\overline{X}_T - E(\overline{X}_T)| \le \epsilon\} \ge 1 - \frac{\text{Var}(\overline{X}_T)}{\epsilon^2} \tag{2}$$

Taking limits as $T \to \infty$ and using (1) we get

$$P\left\{\left|\lim_{T\to\infty}(\overline{X}_T) - \mu\right| \le \epsilon\right\} \ge 1 - \frac{\lim_{T\to\infty}\text{Var}(\overline{X}_T)}{\epsilon^2}$$

$\therefore$ When $\lim_{T\to\infty} \text{Var}\,(\overline{X}_T) = 0$, (2) becomes

$$P\left\{\left|\lim_{T\to\infty}(\overline{X}_T) - \mu\right| \le \epsilon\right\} \ge 1$$

i.e. $\lim_{T\to\infty}(\overline{X}_T) = E\{X(t)\}$ with probability 1.

Note *This theorem provides a sufficient condition for the mean-ergodicity of a random process. That is, to prove the mean-ergodicity of $\{X(t)\}$, it is enough to prove* $\lim_{T\to\infty} \mathit{Var}\,(\overline{X}_T) = 0$.

Correlation Ergodic Process

The stationary process $\{X(t)\}$ is said to be correlation ergodic (or ergodic in the correlation), if the process $\{Y(t)\}$ is mean-ergodic, where $Y(t) = X(t + \tau) \times X(t)$.

That is, the stationary process $\{X(t)\}$ is correlation ergodic, if $\overline{Y}_T = \frac{1}{2T}\int_{-T}^{T} X(t+\tau)\, X(t)\, dt$ tends to

$E\{X(t+\tau)X(t)\} = R(\tau)$ as $T \to \infty$.

Distribution Ergodic Process

If $\{X(t)\}$ is a stationary process and if $\{Y(t)\}$ is another process such that

$$Y(t) = \begin{cases} 1 & \text{if } X(t) \le x \\ 0 & \text{if } X(t) > x \end{cases}$$

then $\{X(t)\}$ is said to be distribution-ergodic, if $\{Y(t)\}$ is mean-ergodic. That is, the stationary process $\{X(t)\}$ is distribution ergodic, if $\overline{Y}_T = \frac{1}{2T}\int_{-T}^{T} Y(t)\, dt \to$ $E\{Y(t)\}$ as $T \to \infty$.

We note that

$$\begin{aligned} E\{Y(t)\} &= 1 \times P\{X(t) \le x\} + 0 \times P\{X(t) > x\} \\ &= F_X(x) \end{aligned}$$

Thus the stationary process $\{X(t)\}$ is distribution-ergodic,

$$\text{if } \frac{1}{2T}\int_{-T}^{T} Y(t)\, dt \to F_X(x) \text{ as } T \to \infty$$

Worked Examples 6(B)

Example 1 *Given that the autocorrelation function for a stationary ergodic process with no periodic components is*

$$R_{xx}(\tau) = 25 + \frac{4}{1 + 6\tau^2}$$

Find the mean value and variance of the process $\{X(t)\}$.

(BDU — Apr. 97)

Solution By the property of autocorrelation function,

$$\begin{aligned} \mu_x^2 &= \lim_{\tau \to \infty} R_{xx}(\tau) \\ &= 25 \\ \therefore \quad \mu_x &= 5 \\ E\{X^2(t)\} &= R_{xx}(0) \\ &= 25 + 4 = 29 \\ \therefore \quad \text{Var}\{X(t)\} &= E\{X^2(t)\} - E^2\{X(t)\} \\ &= 29 - 25 = 4 \end{aligned}$$

Example 2 *Express the autocorrelation function of the process {X′(t)} in terms of the autocorrelation function of the process {X(t)}.*

Solution Consider

$$R_{xx'}(t_1, t_2) = E\{X(t_1) \times X'(t_2)\}$$

$$= E\left[X(t_1)\left\{\frac{X(t_2+h)-X(t_2)}{h}\right\}\right] \text{ as } h \to 0$$

$$= \lim_{h\to 0}\left[\frac{R_{xx}(t_1,t_2+h)-R_{xx}(t_1,t_2)}{h}\right]$$

$$= \frac{\partial}{\partial t_2} R_{xx}(t_1, t_2) \tag{1}$$

Similarly,

$$R_{x'x'}(t_1, t_2) = \frac{\partial}{\partial t_1} R_{xx'}(t_1, t_2) \tag{2}$$

Using (1) in (2),

$$R_{x'x'}(t_1, t_2) = \frac{\partial^2}{\partial t_1\, \partial t_2} R_{xx}(t_1, t_2) \tag{3}$$

If $\{X(t)\}$ is a stationary process, we put $t_1 - t_2 = \tau$. From (1), (2) and (3), we then get

$$R_{xx'}(\tau) = -\frac{\partial}{\partial \tau} R_{xx}(\tau)$$

$$R_{x'x'}(\tau) = \frac{\partial}{\partial \tau} R_{xx'}(\tau) \text{ and}$$

$$R_{x'x'}(\tau) = -\frac{\partial^2}{\partial \tau^2} R_{xx}(\tau)$$

Example 3 *Prove that the random process {X(t)} with constant mean is mean-ergodic,* if $\lim_{T\to\infty}\left[\frac{1}{4T^2}\int_{-T}^{T}\int_{-T}^{T} C(t_1, t_2)\, dt_1\, dt_2\right] = 0$

Solution As per mean-ergodic theorem, the condition for the mean-ergodicity of the process $\{X(t)\}$ is

$$\lim_{T\to\infty}\{\text{Var}\left(\overline{X}_T\right)\} = 0, \text{ where}$$

$$\overline{X}_T = \frac{1}{2T}\int_{-T}^{T} X(t)dt \text{ and } E\left(\overline{X}_T\right) = E\{X(t)\}$$

Now
$$\overline{X}_T^2 = \frac{1}{4T^2}\int_{-T}^{T}\int_{-T}^{T} X(t_1)\, X(t_2) dt_1 dt_2$$

$\therefore$
$$E\left\{\overline{X}_T^2\right\} = \frac{1}{4T^2}\int_{-T}^{T}\int_{-T}^{T} R(t_1, t_2) dt_1 dt_2$$

$$\therefore \quad \text{Var}\left(\overline{X}_T\right) = E\left\{\overline{X}_T^2\right\} - E^2\left(\overline{X}_T\right)$$

$$= \frac{1}{4T^2}\int_{-T}^{T}\int_{-T}^{T} [R(t_1, t_2) - E\{X(t_1)\}\, E\{X(t_2)\}]\mathrm{d}t_1\mathrm{d}t_2$$

$$= \frac{1}{4T^2}\int_{-T}^{T}\int_{-T}^{T} C(t_1, t_2)\mathrm{d}t_1\mathrm{d}t_2 \qquad (1)$$

Therefore the condition $\lim\limits_{T\to\infty} \{\text{Var}\left(\overline{X}_T\right)\} = 0$ is equivalent to the condition

$$\lim_{T\to\infty}\left[\frac{1}{4T^2}\int_{-T}^{T}\int_{-T}^{T} C(t_1, t_2)\mathrm{d}t_1\,\mathrm{d}t_2\right] = 0$$

Hence the result.

Example 4 *If $\overline{X}_T$ is the time-average of a stationary random process $\{X(t)\}$ over $\{-T, T\}$, prove that* $\text{Var}\left(\overline{X}_T\right) = \frac{1}{T}\int_0^{2T} C(\tau)\left[1 - \frac{|\tau|}{2T}\right] d\tau$ *and hence prove that the sufficient condition for the mean-ergodicity of the process $\{X(t)\}$ is*

(i) $\lim\limits_{T\to\infty} \frac{1}{T}\int_0^{2T} C(\tau)\left[1 - \frac{|\tau|}{2T}\right] d\tau = 0$ *and (ii)* $\int_{-\infty}^{\infty} |C(\tau)| d\tau < \infty$.

Solution Step (1) of the previous example gives

$$\text{Var}\left(\overline{X}_T\right) = \frac{1}{4T^2}\int_{-T}^{T}\int_{-T}^{T} C(t_1, t_2)\mathrm{d}t_1\mathrm{d}t_2 \qquad (1)$$

We shall convert the double integral (1) into a single definite integral with respect to the variable $\tau = t_1 - t_2$ as explained below:

The double integral (1) is evaluated over the area of the square bounded by $t_1 = -T, T$ and $t_2 = -T, T$ as shown in the figure.

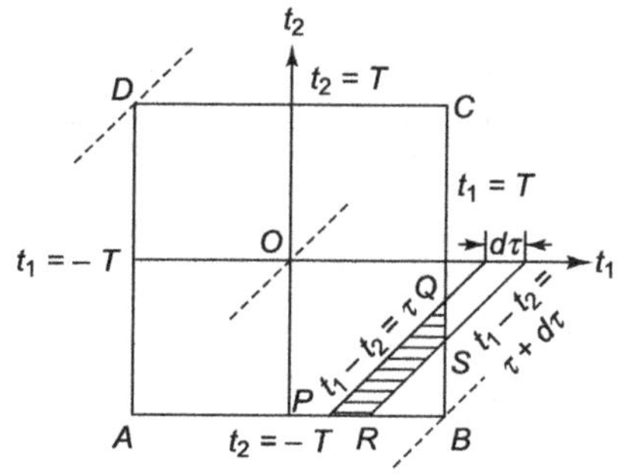

We divide the area of the square $ABCD$ into a number of strips parallel to the line $t_1 - t_2 = 0$. Let a typical strip be $PQRS$, where PQ is given by $t_1 - t_2 = \tau$ and RS is given by $t_1 - t_2 = \tau + \mathrm{d}\tau$.

When $PQRS$ is at the initial position D, $t_1 - t_2 = -2T$, i.e., the initial value of $\tau = -2T$.

When $PQRS$ is at the final position B, $t_1 - t_2 = 2T$, i.e. final value of $\tau = 2T$.

Hence to cover the given area $ABCD$, τ has to vary from $-2T$ to $2T$.

Since $d\tau$ is very small, $C(t_1 - t_2) = C(\tau)$ can be assumed to be a constant in the strip $PQRS$.

Now $$dt_1\, dt_2 = \text{elemental area in the } t_1t_2\text{-plane}$$
$$= \text{area of the small strip } PQRS \quad (2)$$

t_1 co-ordinate of P is obtained by solving the equations $t_1 - t_2 = \tau$ and $t_2 = -T$. Thus $(t_1)_P = \tau - T$.

$$\therefore \quad PB\,(= BQ) = T - (\tau - T) = 2T - \tau \text{ if } \tau > 0$$
$$= 2T + \tau \text{ if } \tau < 0$$

When $\tau > 0$

$$\text{Area of } PQRS = \text{Area of } \Delta\, PBQ - \text{Area of } \Delta\, RSB$$
$$= \frac{1}{2}(2T - \tau)^2 - \frac{1}{2}(2T - \tau - d\tau)^2$$
$$= (2T - \tau)\, d\tau, \text{ omitting } (d\tau)^2 \quad (3)$$

From (2) and (3),

$$dt_1 dt_2 = \{2T - |\tau|\} d\tau \quad (4)$$

Using (4) in (1),

$$\text{Var}(\overline{X}_T) = \frac{1}{2T} \int_{-2T}^{2T} C(\tau) \left\{1 - \frac{|\tau|}{2T}\right\} d\tau$$

i.e.,
$$\text{Var}(\overline{X}_T) = \frac{1}{T} \int_{0}^{2T} C(\tau) \left\{1 - \frac{|\tau|}{2T}\right\} d\tau \text{ (since the integrand is even)}$$

Therefore, sufficient condition for mean-ergodicity of a stationary process $\{X(t)\}$ can also be stated as

$$\lim_{T \to \infty} \left[\frac{1}{T} \int_{0}^{2T} C(\tau) \left\{1 - \frac{|\tau|}{2T}\right\} d\tau\right] = 0$$

(ii) The sufficient condition for mean ergodicity of $\{X(t)\}$ can also given as

$$\lim_{T \to \infty} \left[\frac{1}{2T} \int_{-2T}^{2T} C(\tau) \left\{1 - \frac{|\tau|}{2T}\right\} d\tau\right] = 0 \quad (5)$$

Since τ varies from $-2T$ to $2T$, $|\tau| \le 2T$.

$$\therefore \quad 1 - \frac{|\tau|}{2T} \le 1$$

$$\therefore \qquad \frac{1}{2T}\int_{-2T}^{2T} C(\tau)\left\{1-\frac{|\tau|}{2T}\right\}d\tau \le \frac{1}{2T}\int_{-2T}^{2T} |C(\tau)|d\tau$$

(5) will be true, only if $\lim\limits_{T\to\infty} \frac{1}{2T}\int_{-2T}^{2T} |C(\tau)|d\tau = 0$

i.e., if $\int_{-\infty}^{\infty} |C(\tau)|d\tau$ is finite

i.e., if $\int_{-\infty}^{\infty} |C(\tau)|d\tau < \infty$

Therefore, sufficient condition for mean ergodicity of the stationary process $\{X(t)\}$ can also be stated as $\int_{-\infty}^{\infty} |C(t)|dt < \infty$.

Example 5 *The random binary transmission process $\{X(t)\}$ is a WSS process with zero mean and autocorrelation function $R(\tau) = 1 - \frac{|\tau|}{T}$, where T is a constant. Find the mean and variance of the time average of $\{X(t)\}$ over $(0, T)$. Is $\{X(t)\}$ mean ergodic?*

Solution $\overline{X}_T = \frac{1}{T}\int_0^T X(t)dt$

$$E\left(\overline{X}_T\right) = E\{X(t)\} = 0$$

$$\text{Var}\left(\overline{X}_T\right) = \frac{1}{T}\int_{-T}^{T}\left\{1-\frac{|\tau|}{T}\right\} C(\tau)d\tau,$$

[see (5) in the previous problem]

$$= \frac{1}{T}\int_{-T}^{T}\left\{1-\frac{|\tau|}{T}\right\}^2 d\tau$$

$$= \frac{2}{T}\int_{0}^{T}\left\{1-\frac{\tau}{T}\right\}^2 d\tau = \frac{2}{3}$$

$$\lim_{T\to\infty}\{\text{Var}\left(\overline{X}_T\right)\} = \frac{2}{3} \neq 0$$

i.e., the condition for mean-ergodicity of $\{X(t)\}$ is not satisfied. Therefore, $\{X(t)\}$ is not mean-ergodic.

Example 6 *If $\{X(t)\}$ is a WSS process with mean μ and autocovariance function*

$$C_{xx}(\tau) = \begin{cases} \sigma_x^2\left(1 - \dfrac{|\tau|}{\tau_0}\right) & \text{for } 0 \le |\tau| \le \tau_0 \\ 0 & \text{for } |\tau| \ge \tau_0 \end{cases}$$

find the variance of the time average of $\{X(t)\}$ over $(0, T)$. Also examine if the process $\{X(t)\}$ is mean-ergodic.

Solution

$$\text{Var}(\overline{X}_T) = \frac{1}{T}\int_{-T}^{T}\left\{1 - \frac{|\tau|}{T}\right\} C(\tau)\, d\tau$$

$$= \frac{2\sigma_x^2}{T}\int_0^T \left(1 - \frac{\tau}{T}\right)\left(1 - \frac{\tau}{\tau_0}\right) d\tau, \text{ if } 0 < T \le \tau_0$$

$$= \frac{2\sigma_x^2}{T}\left(T - \frac{T}{2} - \frac{T^2}{2\tau_0} + \frac{T^2}{3\tau_0}\right)$$

$$= \sigma_x^2\left(1 - \frac{T}{3\tau_0}\right), \text{ if } 0 < T \le \tau_0 \tag{1}$$

and

$$\text{Var}(\overline{X}_T) = \frac{2\sigma_x^2}{T}\int_0^{\tau_0}\left(1 - \frac{\tau}{T}\right)\left(1 - \frac{\tau}{\tau_0}\right) d\tau, \text{ if } T \ge \tau_0$$

$$= \frac{\sigma_x^2 \tau_0}{T}\left(1 - \frac{\tau_0}{3T}\right), \text{ if } T \ge \tau_0 \tag{2}$$

When T is sufficiently large, (2) holds good.

$$\therefore \quad \lim_{T\to\infty} \text{Var}(\overline{X}_T) = \lim_{T\to\infty}\left\{\frac{\sigma_x^2\tau_0}{T}\left(1 - \frac{\tau_0}{3T}\right)\right\} = 0$$

Therefore, $\{X(t)\}$ is mean-ergodic.

Example 7 *If $X(t) = \mu + N(t)$, where $N(t)$ is white noise with $C(t_1, t_2) = \phi(t_1)\,\delta(t_1 - t_2)$, where $\phi(t)$ is a bounded function of t and δ is the unit impulse function, prove that $\{X(t)\}$ is a mean-ergodic process.*

Solution Let us consider the time-average of $\{X(t)\}$ over $(-T, T)$

$$\overline{X}_T = \frac{1}{2T}\int_{-T}^{T} X(t)\,dt$$

$$E\{\overline{X}_T\} = \frac{1}{2T}\int_{-T}^{T} E\{\mu + N(t)\}\,dt$$

$$= \mu \qquad \text{[since } E\{N(t)\} = 0, \text{ as } N(t) \text{ is white noise]}$$

Note: The unit impulse function $\delta(t-a)$ is defined as

$$\delta(t-a) = \begin{cases} \dfrac{1}{\in} & \text{if } a - \dfrac{\in}{2} \le t \le a + \dfrac{\in}{2} \\ 0 & \text{otherwise, where } \in \to 0 \end{cases}$$

$$\int_{-\infty}^{\infty} \phi(t)\, \delta(t-a) \mathrm{d}t = \int_{a-\frac{\in}{2}}^{a+\frac{\in}{2}} \phi(t) \frac{1}{\in} \mathrm{d}t = \phi(a)$$

[since $\phi(t)$ can be considered a constant in the small interval of length $\in$]

Now $$\operatorname{Var}(\overline{X}_T) = \frac{1}{4T^2} \int_{-T}^{T}\int_{-T}^{T} C(t_1\ t_2)\, \mathrm{d}t_1 \mathrm{d}t_2 \qquad \text{[refer to Example (3)]}$$

$$= \frac{1}{4T^2} \int_{-T}^{T}\int_{-T}^{T} \phi(t_1)\, \delta(t_1 - t_2)\, \mathrm{d}t_1 \mathrm{d}t_2$$

$$= \frac{1}{4T^2} \int_{-T}^{T} \phi(t_2)\, \mathrm{d}t_2 \qquad \text{(by the note above)}$$

$\therefore$ $\lim_{T\to\infty} \{\operatorname{Var}(\overline{X}_T) = 0$, since $\phi(\mathrm{t})$ is bounded

$\therefore$ $(\overline{X}_T) \to \mu$, as $T \to \infty$

i.e., $\{X(t)\}$ is a mean-ergodic process.

Example 8 *If the WSS process $\{X(t)\}$ is given by $X(t) = 10 \cos(100t + \theta)$, where θ is uniformly distributed over $(-\pi, \pi)$, prove that $\{X(t)\}$ is correlation ergodic.*

Solution $R(\tau) = E[10 \cos(100t + 100\tau + \theta) \times 10 \cos(100t + \theta)]$
$= 50 \cos(100\tau)$

Consider $$\overline{Z}_T = \frac{1}{2T} \int_{-T}^{T} X(t+\tau)\, X(t)\, \mathrm{d}t$$

$$= \frac{1}{2T} \int_{-T}^{T} 100 \cos(100\,t + 100\,\tau + \theta) \cos(100\,t + \theta)\, \mathrm{d}t$$

$$= \frac{25}{T} \int_{-T}^{T} \cos(100\,\tau) \mathrm{d}t + \frac{25}{T} \int_{-T}^{T} \cos(200\,t + 100\,\tau + 2\theta) \mathrm{d}t$$

$$= 50 \cos(100\,\tau) + \frac{25}{T} \int_{-T}^{T} \cos(200\,t + 100\,\tau + 2\theta)\, \mathrm{d}t$$

Now $\lim_{T \to \infty} (Z_T) = 50 \cos (100\, \tau)$

$= R(\tau)$

Therefore, $\{X(t)\}$ is correlation-ergodic

Exercise 6(B)

Part A (Short answer questions)

1. Define autocorrelation function (ACF) of a stationary process.
2. Prove that the ACF $R(\tau)$ of a real process is an even function of τ.
3. If $R(\tau)$ is the ACF of a complex process, prove that $R^*(\tau) = R(-\tau)$.
4. If $R(\tau)$ is the ACF of a stationary process, prove that $|R(\tau)| \le R(0)$.
5. If $R(\tau)$ is the ACF of a stationary process $\{X(t)\}$, prove that $\lim_{\tau \to \infty} [R(\tau)] = \mu_x^2$. Is it true for any stationary process?
6. Find the mean of the stationary process $\{X(t)\}$, whose ACF is given by
$$R(\tau) = \frac{25\tau^2 + 36}{6.25\tau^2 + 4}.$$
7. Find the variance of the stationary process $\{X(t)\}$, whose ACF is given by
$$R(\tau) = \frac{25\tau^2 + 36}{6.25\tau^2 + 4}.$$
8. Find the variance of the stationary process $\{X(t)\}$, whose ACF is given by $R(\tau) = 2 + 4e^{-2|\tau|}$.
9. Find the variance of the stationary process $\{X(t)\}$, whose ACF is given by
$$R(\tau) = 16 + \frac{9}{1 + 6\tau^2}.$$
10. Define the cross-correlation function and state any 2 of its properties.
11. Find the cross-correlation function of 2 stationary processes that are orthogonal.
12. When the jointly stationary processes $\{X(t)\}$ and $\{Y(t)\}$ are independent, prove that $R_{xy}(\tau) = \mu_x \mu_y$.
13. Define ensemble average and time average of a random process $\{X(t)\}$.
14. What is the difference between ensemble average and time average of a stochastic process $\{X(t)\}$?
15. When is a random process said to be ergodic? Give an example for an ergodic process.
16. Distinguish between stationarity and ergodicity.
17. What do you mean by mean-ergodicity of a RP?
18. State mean ergodic theorem.
19. State the sufficient conditions for the mean ergodicity of a RP $\{X(t)\}$.

20. State 2 different sufficient conditions for $\{X(t)\}$ with constant mean to be mean-ergodic.
21. Examine if the process $\{X(t)\}$, where $X(t) = X$, a random variable is mean-ergodic.
22. Give an example of a WSS process which is not mean-ergodic.
23. If $\overline{X}_T$ is the time-average of a stationary random process $\{X(t)\}$ over $(-T, T)$, express $\text{Var}(\overline{X}_T)$ in terms of the autocovariance function of $\{X(t)\}$ and hence state the sufficient condition for the mean-ergodicity of $\{X(t)\}$.
24. When is a random process said to be correlation ergodic?
25. When is a random process said to be distribution ergodic?

Part B

26. A stationary process has an autocorrelation function given by $R(\tau) = \dfrac{25\tau^2 + 36}{6.25\tau^2 + 4}$. Find the mean value, mean-square value and variance of the process.
27. If the autocorrelation of a process $\{X(t)\}$ is $R(t_1, t_2)$ and if $Y(t) = X(t + a) - X(a)$, where a is a constant, express $R_{yy}(t_1, t_2)$ in terms of R's.
28. If $\{X(t)\}$ is a WSS process with autocorrelation function $R_{xx}(\tau)$ and if $Y(t) = X(t + a) - X(t - a)$ show that $R_{yy}(\tau) = 2R_{xx}(\tau) - R_{xx}(\tau + 2a) - R_{xx}(\tau - 2a)$.
29. If $\{X(t)\}$ and $\{Y(t)\}$ are independent WSS processes with zero means, find the autocorrelation function of $\{Z(t)\}$, when
    (i) $Z(t) = a + bX(t) + cY(t)$
    (ii) $Z(t) = aX(t)Y(t)$
30. If $X(t) = A$, where A is a random variable, prove that $\{X(t)\}$ is not mean-ergodic.
31. If $S = \int_0^{10} X(t)\,dt$, show that $E(S^2) = \int_{-10}^{10} (10 - |\tau|)\, R_{xx}(\tau)\,d\tau$. Find also the mean and variance of S, if $E\{X(t)\} = 8$ and $R_{xx}(\tau) = 64 + 10e^{-2|\tau|}$.
32. A stationary zero mean random process $\{X(t)\}$ has the autocorrelation function $R_{xx}(\tau) = 10e^{-0.1\tau^2}$. Find the mean and variance of $\overline{X}_T = \dfrac{1}{5}\int_0^5 X(t)\,dt$.
33. If $\{X(t)\}$ is a WSS process with $E\{X(t)\} = 2$ and $R_{xx}(\tau) = 4 + e^{-|\tau|/10}$, find the mean and variance of $S = \int_0^1 X(t)\,dt$.

34. $\{X(t)\}$ is the random telegraph signal process with $E\{X(t)\} = 0$ and $R(\tau) = e^{-2\lambda|\tau|}$. Find the mean and variance of the time average of $\{X(t)\}$ over $(-T, T)$. Is it mean-ergodic?
35. The random process $\{X(t)\}$ is stationary with $E\{X(t)\} = 1$ and $R(\tau) = 1 + e^{-2|\tau|}$. Find the mean and variance of $S = \int_0^1 X(t)\,dt$.
36. If the autocovariance function of a stationary process $\{X(t)\}$ is given by $C(\tau) = Ae^{-\alpha|\tau|}$, prove that $\{X(t)\}$ is mean-ergodic. Also find $\text{Var}(\overline{X}_T)$, where $\overline{X}_T$ is the time average of $\{X(t)\}$ over $(-T, T)$
[**Hint:** $c(\tau) \to 0$ as $\tau \to \infty$].
37. If the autocorrelation function of a WSS process $\{X(t)\}$ is $R(\tau)$, show that $P\{|X(t+\tau) - X(t)| \geq a\} \leq 2\{R(0) - R(\tau)\}/a^2$.
[**Hint:** Use Tchebycheff's inequality].
38. Show that, if $\{X(t)\}$ is normal with $\mu_x = 0$ and $R_{xx}(\tau) = 0$ for $|\tau| > a$, then it is correlation-ergodic.
[**Hint:** $Z(t) = X(t+\lambda)X(t)$; $R_{zz}(\tau) = R^2(\lambda) + R^2(\tau) + R(\lambda+\tau)R(\lambda-\tau)$. Since $R_{zz}(\tau) \to 0$ as $\tau \to \infty$, $C_{zz}(\tau) \to 0$ as $\tau \to \infty$]
39. If the autocorrelation function of a stationary Gaussian process $\{X(t)\}$ is $R(\tau) = 10e^{-|\tau|}$, prove that $\{X(t)\}$ is ergodic both in mean and correlation.
40. Give an example of a WSS process which is not ergodic in mean.

Power Spectral Density Function

The autocorrelation function $R(\tau)$ tells us something about how rapidly we can expect the random signal $X(t)$ to change as a function of time. If the autocorrelation function decays rapidly (slowly), it indicates that the process can be expected to change rapidly (slowly). Morever if the autocorrelation function has periodic components, then the corresponding process also will have periodic components. Hence we may conclude that the autocorrelation function contains information about the expected frequency content of the random process.

For example, if we assume that $X(t)$ is a voltage waveform across a 1Ω resistance, then the ensemble average value of $X^2(t)$ is the average value of power delivered to the 1Ω resistance by $X(t)$.

i.e., Average power of $X(t) = E\{X^2(t)\}$

$$= R(0) \geq 0$$

Now if $R(0)$ can be expressed as

$$R(0) = \int_{-\infty}^{\infty} S(f)\,df, \text{ since } R(0) \text{ represents power,}$$

$S(f)$ will be expressed in units of power per Hertz. That is, $S(f)$ gives the distribution of power of $\{X(t)\}$ as a function of frequency and hence is called the *power spectral density function* or *simply spectral density* or *power spectrum* of the stationary process $\{X(t)\}$. We shall now give the mathematical definition of power spectral density function of a stationary process.

Definition: If $\{X(t)\}$ is a stationary process (either in the strict sense or wide sense) with autocorrelation function $R(\tau)$, then the Fourier transform of $R(\tau)$ is called the power spectral density function of $\{X(t)\}$ and denoted as $S_{xx}(\omega)$ or $S_x(\omega)$ or $S(\omega)$.

Thus $$S(\omega) = \int_{-\infty}^{\infty} R(\tau)e^{-i\omega\tau}\, d\tau \tag{1}$$

Sometimes ω is replaced by $2\pi f$, where f is the frequency variable, in which case the power spectral density function will be a function of f, denoted by $S(f)$.

Then $$S(f) = \int_{-\infty}^{\infty} R(\tau)\, e^{-i2\pi f\tau}\, d\tau \tag{2}$$

Note *Equation (1) or (2) is sometimes called the Wiener Khinchine relation. We shall mostly follow the definition (1) and denote the power spectral density as a function of ω only.*

Given the power spectral density function $S(\omega)$, the autocorrelation function $R(\tau)$ is given by the Fourier inverse transform of $S(\omega)$.

i.e, $$R(\tau) = \frac{1}{2\pi}\int_{-\infty}^{\infty} S(\omega)\, e^{i\tau\omega}\, d\omega \tag{3}$$

(or) $$R(\tau) = \int_{-\infty}^{\infty} S(f)\, e^{i2\pi\tau f}\, df \tag{4}$$

If $\{X(t)\}$ and $\{Y(t)\}$ are two jointly stationary random processes with cross-correlation function $R_{xy}(\tau)$, then the Fourier transform of $R_{xy}(\tau)$ is called the *cross-power spectral density* of $\{X(t)\}$ and $\{Y(t)\}$ and denoted as $S_{xy}(\omega)$.

i.e., $$S_{xy}(\omega) = \int_{-\infty}^{\infty} R_{xy}(\tau)e^{-i\omega\tau}\, d\tau$$

Properties of Power Spectral Density Function

1. The value of the spectral density function at zero frequency is equal to the total area under the graph of the autocorrelation function. By putting $\omega = 0$ in (1) or $f = 0$ in (2), we get

   $$S(0) = \int_{-\infty}^{\infty} R(\tau)\, d\tau,$$ which is the given property.

2. The mean square value of a wide-sense stationary process is equal to the total area under the graph of the spectral density. By Putting $f = 0$ in (4), we get

$$E\{X^2(t)\} = R(0) = \int_{-\infty}^{\infty} S(f)df, \text{ which is the given property.}$$

3. The spectral density function of a real random process is an even function.

Proof $S(\omega) = \int_{-\infty}^{\infty} R(\tau)\, e^{-i\omega\tau}\, d\tau$, by definition

$$\therefore \quad S(-\omega) = \int_{-\infty}^{\infty} R(\tau) e^{i\omega\tau}\, d\tau$$

Putting $\tau = -u$,

$$S(-\omega) = \int_{-\infty}^{\infty} R(-u)\, e^{-i\omega u}\, du$$

$$= \int_{-\infty}^{\infty} R(u)\, e^{-i\omega u}\, du$$

[since $R(\tau)$ is an even function of τ]

$$= S(\omega)$$

Therefore, $S(\omega)$ is an even function of ω.

4. The spectral density of a process $\{X(t)\}$, real or complex, is a real function of ω and non negative.

Proof $R(\tau) = E\{X(t)X^*(t-\tau)\}$

$$\therefore \quad R(-\tau) = E\{X(t)\, X^*(t+\tau)\}$$

$$\therefore \quad R^*(-\tau) = E\{X(t+\tau)\, X^*(t)\}$$

$$= R(\tau)$$

(or) $$R^*(\tau) = R(-\tau) \qquad (1)$$

Now $$S(\omega) = \int_{-\infty}^{\infty} R(\tau) e^{-i\omega\tau}\, d\tau$$

$$\therefore \quad S^*(\omega) = \int_{-\infty}^{\infty} R^*(\tau)\, e^{i\omega\tau}\, d\tau$$

$$= \int_{-\infty}^{\infty} R(-\tau) e^{i\omega\tau}\, d\tau \text{ by (1)}$$

$$= \int_{-\infty}^{\infty} R(u)\, e^{-i\omega u}\, du, \quad \text{by putting } u = -\tau$$

$$= S(\omega)$$

Hence $S(\omega)$ is a real function of ω.

Note: It will be proved that $S(\omega) \geq 0$, in Worked Example 14).

5. The spectral density and the autocorrelation function of a real WSS process form a Fourier cosine transform pair.

Proof $S(w) = \int_{-\infty}^{\infty} R(t)\,\{\cos wt - i \sin wt\} dt$

$$= 2\int_{-\infty}^{\infty} R(\tau)\cos\omega\tau\, d\tau$$

[since $R(\tau)$ is even]

= Fourier cosine transform of $[2R(\tau)]$

$$R(\tau) = \frac{1}{2\pi}\int_{-\infty}^{\infty} S(\omega)\,(\cos\tau\omega + i\sin\tau\omega) d\omega$$

$$= \frac{1}{\pi}\int_{0}^{\infty} S(\omega)\cos\tau\omega\, d\omega$$

[since $S(\omega)$ is even]

= Fourier inverse cosine transform of $\left[\frac{1}{2}S(\omega)\right]$

6. **Wienern–Khinchine Theorem**

If $X_T(\omega)$ is the Fourier transform of the truncated random process defined as

$$X_T(t) = \begin{cases} X(t) & \text{for } |t| \le T \\ 0 & \text{for } |t| > T \end{cases}$$

where $\{X(t)\}$ is a real WSS process with power spectral density function $S(\omega)$, then

$$S(\omega) = \lim_{T\to\infty}\left[\frac{1}{2T}E\left\{|X_T(\omega)|^2\right\}\right]$$

Proof Given: $X_T(\omega) = \int_{-\infty}^{\infty} X_T(t)\, e^{-i\omega t}\, dt$

$$= \int_{-T}^{T} X(t)\, e^{-i\omega t}\, dt$$

Since $\{X(t)\}$ is real

$$|X_T(\omega)|^2 = X_T(\omega)X_T(-\omega)$$

$$= \int_{-T}^{T} X(t_1)e^{-i\omega t_1}\, dt_1 \int_{-T}^{T} X(t_2)\, e^{i\omega t_2}\, dt_2$$

$$= \int_{-T}^{T} \int_{-T}^{T} X(t_1)\, X(t_2)\, e^{-i\omega(t_1 - t_2)} \mathrm{d}t_1\, \mathrm{d}t_2$$

$$\therefore \quad E|X_T(\omega)|^2 = \int_{-T}^{T} \int_{-T}^{T} E\{X(t_1)\, X(t_2)\}\, e^{-i\omega(t_1 - t_2)} \mathrm{d}t_1\, \mathrm{d}t_2$$

$$= \int_{-T}^{T} \int_{-T}^{T} R(t_1 - t_2)\, e^{-i\omega(t_1 - t_2)} \mathrm{d}t_1\, \mathrm{d}t_2$$

[since $\{X(t)\}$ is WSS]

$$= \int_{-T}^{T} \int_{-T}^{T} \phi(t_1 - t_2) \mathrm{d}t_1\, \mathrm{d}t_2, \text{ say} \qquad (1)$$

The double integral (1) is evaluated over the area of the square *ABCD* bounded by $t_1 = -T, T$ and $t_2 = -T, T$ as shown in the figure.

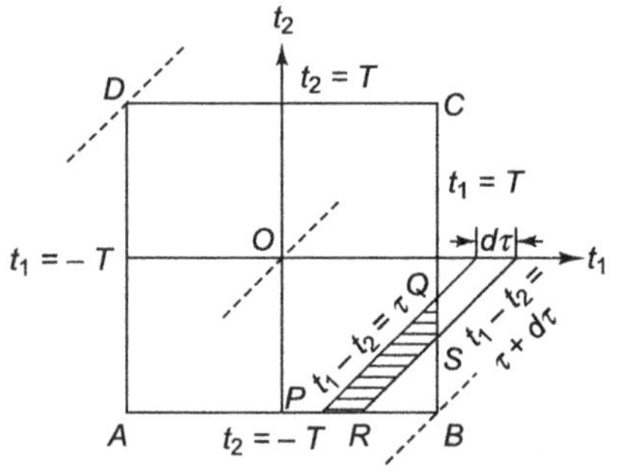

We divide the area of the square into a number of strips like *PQRS*, where *PQ* is given by $t_1 - t_2 = \tau$ and *RS* is given by $t_1 - t_2 = \tau + \mathrm{d}\tau$.

When *PQRS* is at the initial position *D*, $t_1 - t_2 = -2T$. When it is at the final position *B*, $t_1 - t_2 = 2T$. Hence when τ varies from $-2T$ to $2T$, the area *ABCD* is covered.

Now $\quad \mathrm{d}t_1\, \mathrm{d}t_2$ = elemental area of the $t_1 t_2$-plane

= area of *PQRS* $\qquad (2)$

$(t_1)_P = \tau - T$ and PC (= BQ) $= T - (\tau - T)$

$= 2T - \tau, \quad$ if $\tau > 0$

$= 2T + \tau, \quad$ if $\tau < 0$

When $\tau > 0$,

Area of $PQRS = \Delta\, PCQ - \Delta\, RCS$

$$= \frac{1}{2}(2T - \tau)^2 - \frac{1}{2}(2T - \tau - \mathrm{d}\tau)^2$$

$$= (2T - \tau)\, \mathrm{d}\tau, \text{ omitting } (\mathrm{d}\tau)^2 \qquad (3)$$

From (2) and (3),

$$dt_1\, dt_2 = (2T - |\tau|)\, d\tau \tag{4}$$

Using (4) in (1), we get,

$$E\{|X_T(\omega)|^2\} = \int_{-2T}^{2T} \phi(\tau)\,(2T - |\tau|)d\tau$$

$$\therefore \quad \frac{1}{2T} E\{|X_T(\omega)|^2\} = \int_{-2T}^{2T} \phi(\tau) \left\{1 - \frac{|\tau|}{2T}\right\} d\tau$$

$$\therefore \quad \lim_{T\to\infty} \frac{1}{2T} E\{|X_T(\omega)|^2\}$$

$$= \lim_{T\to\infty} \int_{-2T}^{2T} \phi(\tau)d\tau - \lim_{T\to\infty} \frac{1}{2T} \int_{-2T}^{2T} |\tau|\phi(\tau)d\tau$$

$$= \int_{-\infty}^{\infty} \phi(\tau)d\tau$$

[assuming that $\int_{-\infty}^{\infty} |\tau|\, \phi(\tau)d\tau$ is bounded]

$$= \int_{-\infty}^{\infty} R(\tau)\, e^{-i\omega\tau}\, d\tau$$

[provided $\int_{-\infty}^{\infty} |\tau|\, R(\tau)\, e^{-i}\omega\tau\, d\tau$ is bounded]

$= S(\omega)$, by definition of $S(\omega)$

Note *This theorem provides an alternative method for finding $S(\omega)$ for a WSS process.*

Linear systems with random inputs

Mathematically, a *system* is a functional relationship between the input $x(t)$ and the output $y(t)$. Usually this relationship is written as $y(t) = f[x(t)]$, $-\infty < t < \infty$.

If we assume that $x(t)$ represents a sample function of a random process $\{X(t)\}$, the system produces an output or response $y(t)$ and the ensemble of the output functions forms a random process $\{Y(t)\}$. The process $\{Y(t)\}$ can be considered as the output of the system or transformation 'f' with $\{X(t)\}$ as the input. The system is completely specified by the operator 'f'.

We recall that $X(t)$ actually means $X(s, t)$, where $s \in S$ (sample space). If the system operates only on the variable t treating s as a parameter, it is called a *deterministic system*. If the system operates on both t and s, it is called *stochastic*. We shall consider only deterministic systems in our study.

Definitions: If $f[a_1 X_1(t) \pm a_2 X_2(t)] = a_1 f[X_1(t)] \pm a_2 f[X_2(t)]$, then f is called a *linear system.*

If $Y(t + h) = f[X(t + h)]$, where $Y(t) = f[X(t)]$, f is called a *time-invariant system* or $X(t)$ and $Y(t)$ are said to form a time-invariant system.

If the output $Y(t_1)$ at a given time $t = t_1$ depends only on $X(t_1)$ and not on any other past or future values of $X(t)$, then the system f is called a *memoryless system.*

If the value of the output $Y(t)$ at $t = t_1$ depends only on the past values of the input $X(t)$, $t \le t_1$, i.e., $Y(t_1) = f[X(t); t \le t_1]$, then the system is called a *causal system.*

System in the Form of Convolution

Very often in electrical systems, the output $Y(t)$ is expressed as a convolution of the input $X(t)$ with a *system weighting function* $h(t)$, i.e., the input–output relationship will be of form

$$Y(t) = \int_{-\infty}^{\infty} h(u)\, X(t-u) \mathrm{d}u \qquad (1)$$

Unit Impulse Response of the System

Refer to the Note in Worked Example 7 of the previous section, in which we have proved that

$$\int_{-\infty}^{\infty} \phi(t)\, \delta(t-a) \mathrm{d}t = \phi(a)$$

where $\delta(t - a)$ is the unit impulse function at a.
If we take $a = 0$, we get

$$\int_{-\infty}^{\infty} \phi(t)\, \delta(t) \mathrm{d}t = \phi(0) \qquad (2)$$

Put $\qquad X(t) = \delta(t)$ in (1)

Then
$$Y(t) = \int_{-\infty}^{\infty} h(u)\, \delta(t-u) \mathrm{d}u$$

$$= \int_{-\infty}^{\infty} h(t-u)\, \delta(u)\, \mathrm{d}u$$

(by the property of the convolution)

$$= h(t-0), \text{ by (2)}$$
$$= h(t)$$

Thus if the input of the system is the unit impulse function, then the output or response is the system weighting function. Hence the system weighting function $h(t)$ will be hereafter called unit impulse response function.

Properties

1. If a system is such that its input $X(t)$ and its output $Y(t)$ are related by a convolution integral, i.e., if $Y(t) = \int_{-\infty}^{\infty} h(u)\, X(t-u)\, du$, then the system is a linear time-invariant system.

Proof Let $X(t) = a_1 X_1(t) + a_2 X_2(t)$

Then
$$Y(t) = \int_{-\infty}^{\infty} h(u)\, [a_1 X_1(t-u) + a_2 X_2(t-u)] du$$
$$= a_1 Y_1(t) + a_2 Y_2(t)$$

Therefore, the system is linear. If $X(t)$ is replaced by $X(t+h)$, then

$$\int_{-\infty}^{\infty} h(u)\, X\left(\overline{t+h}-u\right) du = Y(t+h)$$

Therefore, the system is time-invariant.

Note *If $h(t)$ is absolutely integrable, viz., $\int_{-\infty}^{\infty} |h(t)| dt < \infty$, then the system is said to be stable in the sense that every bounded input gives a bounded output.*

In addition, if $h(t) = 0$, when $t < 0$, the system is said to be *causal*.

2. If the input to a time-invariant, stable linear system is a WSS process, the output will also be a WSS process.

Proof $Y(t) = \int_{-\infty}^{\infty} h(u)\, X(t-u) du$

$$\therefore\ E\{Y(t)\} = \int_{-\infty}^{\infty} h(u)\, E\{X(t-u)\} du$$

$$= \mu_x \int_{-\infty}^{\infty} h(u)\, du \qquad \text{[since } \{X(t)\} \text{ is WSS]}$$

= a finite constant, independent of t (since the system is stable)

$$R_{yy}(t_1, t_2) = E\{Y(t_1)\, Y(t_2)\}$$
$$= E \int_{-\infty}^{\infty} \int_{-\infty}^{\infty} h(u_1) h(u_2)\, X(t_1 - u_1)\, X(t_2 - u_2) du_1 du_2$$
$$= \int_{-\infty}^{\infty} \int_{-\infty}^{\infty} h(u_1)\, h(u_2)\, R_{xx}(t_1 - u_1, t_2 - u_2) du_1 du_2$$

$$= \int_{-\infty}^{\infty} \int_{-\infty}^{\infty} h(u_1)\, h(u_2)\, R_{xx}[(t_1 - t_2) - (u_1 - u_2)]\mathrm{d}u_1 \mathrm{d}u_2$$

[since $\{X(t)\}$ is WSS]

Since the RS is a function of $(t_1 - t_2)$, so will be the LS. Therefore, $R_{yy}(t_1, t_2)$ will be a function of $(t_1 - t_2)$. Therefore, $\{Y(t)\}$ is a WSS process.

3. If $\{X(t)\}$ is a WSS process and if

$$Y(t) = \int_{-\infty}^{\infty} h(u)\, X(t - u)\mathrm{d}u, \text{ then}$$

(i) $R_{xy}(\tau) = R_{xx}(\tau) * h(-\tau)$ and

(ii) $R_{yy}(\tau) = R_{xy}(\tau) * h(\tau)$, where * denotes convolution. Also

(iii) $S_{xy}(\omega) = S_{xx}(\omega)H^*(\omega)$ and

(iv) $S_{yy}(\omega) = S_{xx}(\omega)|H(\omega)|^2$

Proof $Y(t) = \int_{-\infty}^{\infty} X(t - \alpha)h(\alpha)\mathrm{d}\alpha$

$$\therefore \quad Y(t).X(t - \tau) = \int_{-\infty}^{\infty} X(t - \alpha)X(t - \tau)h(\alpha)\mathrm{d}\alpha$$

Taking expections (E) on both sides,

$$R_{yx}\{\tau\} = \int_{-\infty}^{\infty} R_{xx}(\tau - \alpha)\, h(\alpha)\}\mathrm{d}\alpha \qquad [\because \{X(t)\} \text{ is WSS}]$$

i.e., $R_{yx}\{\tau\} = R_{xx}(\tau) * h(\tau)$

$\therefore$ $R_{yx}\{-\tau\} = R_{xx}(-\tau) * h(-\tau)$, changing τ to $-\tau$.

i.e., $R_{xy}(\tau) = R_{xx}(\tau) * h(-\tau)$, (1)

since $R_{yx}(-\tau) = R_{xy}(\tau)$

and $R_{xx}(-\tau)$ is even.

$$\text{Now} \quad Y(t)\, Y(t - \tau) = \int_{-\infty}^{\infty} X(t - \alpha)\, Y(t - \tau)h(\alpha)\mathrm{d}\alpha$$

$$\therefore\ E\{Y(t)\, Y(t - \tau)\} = \int_{-\infty}^{\infty} R_{xy}(\tau - \alpha)h(\alpha)\mathrm{d}\alpha$$

assuming that $\{X(t)\}$ and $\{Y(t)\}$ are jointly WSS

i.e., $R_{yy}(\tau) = R_{xy}(\tau) * h(\tau)$ (2)

Taking Fourier transforms of (1) and (2), we get

$S_{xy}(\omega) = S_{xx}(\omega)H^*(\omega)$ (3)

where $H^*(\omega)$ is the conjugate of $H(\omega)$ and $S_{yy}(\omega) = S_{xy}(\omega)H(\omega)$ (4)

Note *Inserting (1) in (2), $Ryy(\tau) = Rxx(\tau)*h(\tau)*h(-\tau)$* (5)

Inserting (3) in (4), $S_{yy}(\omega) = |H(\omega)|^2 S_{xx}(\omega)$ (6)

However alternative proofs for (5) and (6) are given in the following properties.

4. If $\{X(t)\}$ is a WSS process and if

$$Y(t) = \int_{-\infty}^{\infty} h(u)X(t-u)\mathrm{d}u, \text{ then}$$

$$R_{yy}(\tau) = R_{xx}(\tau)*K(\tau),$$

where $$K(t) = h(t)*h(-t) = \int_{-\infty}^{\infty} h(u)h(t+u)\mathrm{d}u$$

Proof $$Y(t) = \int_{-\infty}^{\infty} X(u)\, h(t-u)\mathrm{d}u$$

$$\therefore \quad Y(t)\,Y(t-\tau) = \int_{-\infty}^{\infty} X(u)h(t-u)\mathrm{d}u \int_{-\infty}^{\infty} X(v)\, h(t-\tau-v)\mathrm{d}v$$

$$= \int_{-\infty}^{\infty}\left[\int_{-\infty}^{\infty} X(u)\, X(v)\, h(t-u)\, \mathrm{d}u\right] h(t-\tau-v)\mathrm{d}v$$

$$\therefore \quad R_{yy}(\tau) = \int_{-\infty}^{\infty}\left[\int_{-\infty}^{\infty} R_{xx}(u-v)\, h(t-u)\, \mathrm{d}u\right] h(t-\tau-v)\mathrm{d}v$$

[taking expectations on both sides]

In the inner integral, put $u - v = -w$, treating v and t as parameters and also change the order of integration.

$$\therefore \quad R_{yy}(\tau) = \int_{-\infty}^{\infty} Rxx(-w)\left\{\int_{-\infty}^{\infty} h(t-v+w)\, h(t-\tau-v)\right\}\mathrm{d}w$$

In the inner integral, put $t - \tau - v = \alpha$, treating w, t and τ as parameters.

$$\therefore \quad R_{yy}(\tau) = \int_{-\infty}^{\infty} R_{xx}(-w)\left[\int_{-\infty}^{\infty} h(\alpha)\, h(\tau + w + \alpha)\, \mathrm{d}\alpha\right]\mathrm{d}w$$

$$= \int_{-\infty}^{\infty} R_{xx}(-w)K(\tau + w)\, \mathrm{d}w \qquad \text{[by definition of } K(t)\text{]}$$

$$= \int_{-\infty}^{\infty} R_{xx}(\beta)\, K(\tau - \beta)\, \mathrm{d}\beta \qquad (\text{putting } \beta = -w)$$

$$= R_{xx}(\tau)*K(\tau)$$

5. The power spectral densities of the input and output processes in the system are connected by the relation $S_{yy}(\omega) = |H(\omega)|^2 \, S_{xx}(\omega)$, where $H(\omega)$ is the Fourier transform of unit impuse response function $h(t)$.

Proof By the previous property

$$R_{yy}(\tau) = R_{xx}(\tau) * K(\tau), \text{ where } K(t) = h(t) * h(-t)$$

Taking Fourier transform on both sides,

$$S_{yy}(\omega) = F\{K(\tau)\} \, S_{xx}(\omega) \tag{1}$$

Let $\quad H(\omega) = F\{h(t)\}$

$$= \int_{-\infty}^{\infty} h(t)\, e^{-i\omega t}\, dt \tag{2}$$

$$\therefore \quad H^*(\omega) = \int_{-\infty}^{\infty} h(t)\, e^{i\omega t}\, dt$$

$$= \int_{-\infty}^{\infty} h(-s) e^{-i\omega s}\, ds$$

(putting $s = -t$)

$$= F\{h(-t)\} \tag{3}$$

Now $\quad K(t) = h(t) * h(-t)$

$\therefore \quad F\{K(t)\} = F\{h(t)\}\; F\{h(-t)$

(by convolution theorem)

$$= H(\omega)\, H^*(\omega)$$

[by (2) and (3)]

$$= |H(\omega)|^2 \tag{4}$$

Inserting (4) in (1), we get

$$S_{yy}(\omega) = |H(\omega)|^2 \, S_{xx}(\omega)$$

Definition: $H(\omega)$ is called the *system function* or the *power transfer function*. It is the Fourier transform of the unit impulse response function of the system.

Worked Examples 6(C)

Example 1 *The autocorrelation function of the random telegraph signal process is given by $R(\tau) = a^2 e^{-2\gamma|\tau|}$. Determine the power density spectrum of the random telegraph signal.* (BDU — Apr. 97)

Solution $\quad S(\omega) = \int_{-\infty}^{\infty} R(\tau)\, e^{-i\omega\tau}\, d\tau$

$$= a^2 \int_{-\infty}^{\infty} e^{-2\gamma|\tau|} (\cos \omega\tau - i \sin \omega\tau) d\tau$$

$$= 2a^2 \int_0^{\infty} e^{-2\gamma\tau} \cos\omega\tau \, d\tau$$

$$= \left[\frac{2a^2 e^{-2\gamma\tau}}{4\gamma^2 + \omega^2} (-2\gamma \cos \omega\tau + \omega \sin \omega\tau) \right]_0^{\infty}$$

$$= \frac{4a^2\gamma}{4\gamma^2 + \omega^2}$$

Example 2 *The autocorrelation function of the Poisson increment process is given by*

$$R(\tau) = \begin{cases} \lambda^2 & \text{for } |\tau| > \epsilon \\ \lambda^2 + \dfrac{\lambda}{\epsilon}\left(1 - \dfrac{|\tau|}{\epsilon}\right) & \text{for } |\tau| \le \epsilon \end{cases}$$

Prove that its spectral density is given by

$$S(\omega) = 2\pi\lambda^2 \delta(\omega) + \frac{4\lambda \sin^2(\omega\epsilon/2)}{\epsilon^2 \omega^2}$$

Solution $S(\omega) = \int_{-\epsilon}^{\epsilon} \left\{\lambda^2 + \frac{\lambda}{\epsilon}\left(1 - \frac{|\tau|}{\epsilon}\right)\right\} e^{-i\omega\tau} \, d\tau + \int_{-\infty}^{\epsilon} \lambda^2 e^{-i\omega\tau} \, d\tau$

$$+ \int_{\epsilon}^{\infty} \lambda^2 e^{-i\omega\tau} \, d\tau$$

$$= \frac{\lambda}{\epsilon} \int_{-\epsilon}^{\epsilon} \left(1 - \frac{|\tau|}{\epsilon}\right) e^{-i\omega\tau} \, d\tau + \int_{-\infty}^{\infty} \lambda^2 e^{-i\omega\tau} \, d\tau$$

$$= \frac{2\lambda}{\epsilon} \int_0^{\epsilon} \left(1 - \frac{\tau}{\epsilon}\right) \cos \omega\tau \, d\tau + F\{\lambda^2\}$$

where $F(\lambda^2)$ is the Fourier transform of λ^2.

$$= \frac{2\lambda}{\epsilon} \left[\left(1 - \frac{\tau}{\epsilon}\right) \frac{\sin \omega\tau}{\omega} + \frac{1}{\epsilon}\left(\frac{-\cos \omega\tau}{\omega^2}\right)\right]_0^{\epsilon} + F(\lambda^2)$$

$$= \frac{2\lambda}{\epsilon^2 \omega^2} (1 - \cos \omega\epsilon) + F(\lambda^2)$$

$$= \frac{4\lambda \sin^2(\omega\epsilon/2)}{\epsilon^2 \omega^2} + F(\lambda^2) \qquad (1)$$

The Fourier inverse transform of $S(\omega)$ is given by

$$R(\tau) = F^{-1}\{S(\omega)\}$$

$$= \frac{1}{2\pi} \int_{-\infty}^{\infty} S(\omega) e^{i\tau\omega} \, d\omega$$

Let us now find $R(\tau)$ corresponding to $S(\omega) = 2\pi\lambda^2 \, \delta(\omega)$, where $\delta(\omega)$ is the unit impulse function.

i.e., $R(\tau) = F^{-1} \{2\pi \, \lambda^2 \, \delta(\omega)\}$

$$= \frac{2\pi\lambda^2}{2\pi} \int_{-\infty}^{\infty} \delta(\omega) \, e^{i\tau\omega} \, d\omega$$

$$= \lambda^2 \left[\text{since} \int_{-\infty}^{\infty} \phi(t) \, \delta(t) \, dt = \phi(0) \right]$$

$\therefore \quad F(\lambda^2) = 2\pi \, \lambda^2 \, \delta(\omega)$ (2)

Inserting (2) in (1) the required result is obtained.

Example 3 *Find the power spectral density of a WSS process with autocorrelationfunction*

$$R(\tau) = e^{-\alpha\tau^2}$$

Solution $S(\omega) = \int_{-\infty}^{\infty} e^{-\alpha\tau^2} \, e^{-i\omega\tau} \, d\tau$

$$= e^{-\omega^2/4\alpha} \int_{-\infty}^{\infty} e^{-\alpha\left(\tau + \frac{i\omega}{2\alpha}\right)^2} \, d\tau$$

$$= \frac{1}{\sqrt{\alpha}} \, e^{-\omega^2/4\alpha} \int_{-\infty}^{\infty} e^{-x^2} \, dx, \text{ putting } \sqrt{\alpha}\left(\tau + \frac{i\omega}{2\alpha}\right) = x$$

$$= \sqrt{\frac{\pi}{\alpha}} \, e^{-\omega^2/4\alpha} \left[\text{since} \int_{-\infty}^{\infty} e^{-x^2} \, dx = \left|\overline{\left(\frac{1}{2}\right)}\right. = \sqrt{\pi} \right]$$

Example 4 *A random process {X(t)} is given by X(t) = A cos pt + B sin pt, where A and B are independent RVs such that E(A) = E(B) = 0 and E(A^2) = E(B^2) = σ^2. Find the power spectral density of the process [refer to Problem 7 in Exercise 6(a)].*

Solution The autocorrelation function of the given process can be found as

$$R(\tau) = \sigma^2 \cos p\tau$$

$$S(\omega) = \int_{-\infty}^{\infty} \sigma^2 \cos p\tau \, e^{-i\omega\tau} \, d\tau \qquad (1)$$

Consider $\quad F^{-1} \{\pi \, \sigma^2 \, [\delta(\omega + p) + \delta(\omega - p)]\}$

$$= \frac{1}{2\pi} \pi\sigma^2 \int_{-\infty}^{\infty} [\delta(\omega + p) + \delta(\omega - p)] \, e^{i\tau\omega} d\omega$$

$$= \frac{\sigma^2}{2} [e^{-i\tau p} + e^{i\tau p}] \qquad \left\{\text{since} \int_{-\infty}^{\infty} \phi(t)\, \delta(t-a)\, dt = \phi(a)\right\}$$

$$= \sigma^2 \cos p\tau$$

$$\therefore \quad F(\sigma^2 \cos p\tau) = \pi\sigma^2 [\delta(\omega + p) + \delta(\omega - p)] \qquad (2)$$

Using (2) in (1), we get,

$$S(\omega) = \pi\sigma^2 [\delta(\omega + p) + \delta(\omega - p)]$$

Example 5 *If $Y(t) = X(t + a) - X(t - a)$, prove that $R_{yy}(\tau) = 2 R_{xx}(\tau) - R_{xx}(\tau + 2a) - R_{xx}(\tau - 2a)$. Hence prove that $S_{yy}(\omega) = 4 \sin^2 a\omega \, S_{xx}(\omega)$*

[refer to Problem 28 in Exercise 6(b)].

Solution $R_{yy}(\tau) = 2R_{xx}(\tau) - R_{xx}(\tau + 2a) - R_{xx}(\tau - 2a)$

Taking Fourier transforms on both sides.

$$S_{yy}(\omega) = 2S_{xx}(\omega) - \int_{-\infty}^{\infty} R_{xx}(\tau + 2a)\, e^{-i\omega\tau} d\tau$$

$$- \int_{-\infty}^{\infty} R_{xx}(\tau - 2a)\, e^{-i\omega\tau} d\tau$$

$$= 2S_{xx}(\omega) - e^{i2a\omega} \int_{-\infty}^{\infty} R_{xx}(u)\, e^{-i\omega u} du$$

$$- e^{-i2a\omega} \int_{-\infty}^{\infty} R_{xx}(v)\, e^{-i\omega v} dv$$

(putting $\tau + 2a = u$ in the first integral and $\tau - 2a = v$ in the second integral)

i.e., $S_{yy}(\omega) = 2S_{xx}(\omega) - \{e^{i2a\omega} + e^{-i2a\omega}\} S_{xx}(\omega)$

$$= 2(1 - \cos 2a\omega)\, S_{xx}(\omega)$$

$$= 4 \sin^2 a\omega \, S_{xx}(\omega)$$

Example 6 *If the process $\{X(t)\}$ is defined as $X(t) = Y(t)\, Z(t)$, where $\{Y(t)\}$ and $\{Z(t)\}$ are independent WSS processes, prove that*

(i) $R_{xx}(\tau) = R_{yy}(\tau)\, R_{zz}(\tau)$ and

(ii) $S_{xx}(\omega) = \dfrac{1}{2\pi} \displaystyle\int_{-\infty}^{\infty} S_{yy}(\alpha)\, S_{zz}(\omega - \alpha) d\alpha$

[refer to problem 29(ii) in Exercise 6(b) for part (i)].

Solution $S_{xx}(\omega) = F\{R_{xx}(\tau)\} = F\{R_{yy}(\tau)\, R_{zz}(\tau)\}$ (1)

Consider $F^{-1}\left[\displaystyle\int_{-\infty}^{\infty} S_{yy}(\alpha)\, S_{zz}(\omega - \alpha)\, d\alpha\right]$

$$= \frac{1}{2\pi} \int_{-\infty}^{\infty} \int_{-\infty}^{\infty} S_{yy}(\alpha)\, S_{zz}(\omega - \alpha) e^{i\omega\tau}\, d\alpha\, d\omega$$

Putting $\alpha = y$ and $\omega - \alpha = z$, we get (from calculus)

$$d\alpha\, d\omega = \begin{vmatrix} \alpha_y & \alpha_z \\ \omega_y & \omega_z \end{vmatrix} dy\, dz = \begin{vmatrix} 1 & 0 \\ 1 & 1 \end{vmatrix} dy\, dz$$

$\therefore$

$$F^{-1}\left[\int_{-\infty}^{\infty} S_{yy}(\alpha)\, S_{zz}(\omega - \alpha)\, d\alpha \right]$$

$$= \frac{1}{2\pi} \int_{-\infty}^{\infty} \int_{-\infty}^{\infty} S_{yy}(y)\, S_{zz}(z)\, e^{i(y+z)\tau}\, dy\, dz$$

$$= \frac{1}{2\pi} \int_{-\infty}^{\infty} S_{yy}(y)\, e^{iy\tau} dy \int_{-\infty}^{\infty} S_{zz}(z)\, e^{iz\tau}\, dz$$

$$= F^{-1}\{S_{yy}(\omega)\} \times 2\pi\, F^{-1}\{S_{zz}(\omega)\}$$
$$= 2\pi\, R_{yy}(\tau)\, R_{zz}(\tau)$$

$$\therefore \quad F\{R_{yy}(\tau) R_{zz}(\tau)\} = \frac{1}{2\pi} \int_{-\infty}^{\infty} S_{yy}(\alpha)\, S_{zz}(\omega - \alpha)\, d\alpha \qquad (2)$$

Using (2) in (1), we get $S_{xx}(\omega)$ in the required form.

Example 7 *If the power spectral density of a WSS process is given by*

$$S(\omega) = \begin{cases} \frac{b}{a}(a - |\omega|), & |\omega| \le a \\ 0, & |\omega| > a \end{cases}$$

find the autocorrelation function of the process.

Solution The autocorrelation function $R(\tau)$ is given by

$$R(\tau) = F^{-1}\{S(\omega)\}$$

$$= \frac{1}{2\pi} \int_{-\infty}^{\infty} S(\omega)\, e^{i\tau\omega}\, d\omega$$

$$= \frac{1}{2\pi} \int_{-a}^{a} \frac{b}{a} (a - |\omega|)\, e^{i\tau\omega}\, d\omega$$

$$= \frac{1}{\pi} \int_{0}^{a} \frac{b}{a} (a - \omega) \cos\tau\omega\, d\omega$$

$$= \frac{b}{\pi a} \left\{ (a - \omega) \frac{\sin \tau\omega}{\tau} - \frac{\cos \tau\omega}{\tau^2} \right\}_0^a$$

$$= \frac{b}{\pi a \tau^2} (1 - \cos a\tau)$$

$$= \frac{ab}{2\pi} \left(\frac{\sin a \frac{\tau}{2}}{a \frac{\tau}{2}} \right)^2$$

Example 8 *The power spectrum of a WSS process {X(t)} is given by* $S(\omega) = \dfrac{1}{\left(1 + \omega^2\right)^2}$. *Find its autocorrelation function* $R(\tau)$ *and average power.*

Solution $$R(\tau) = \frac{1}{2\pi} \int_{-\infty}^{\infty} \frac{1}{\left(1+\omega^2\right)^2} e^{i\tau\omega} \, d\omega \qquad (1)$$

The integral in (1) is evaluated by contour integration technique as given below.

Consider $\int_C \frac{e^{iaz}}{\left(1 + z^2\right)^2} dz$, where C is the closed contour consisting of the real axis from $-R$ to $+R$ and the upper half of the circle $|z| = R$.

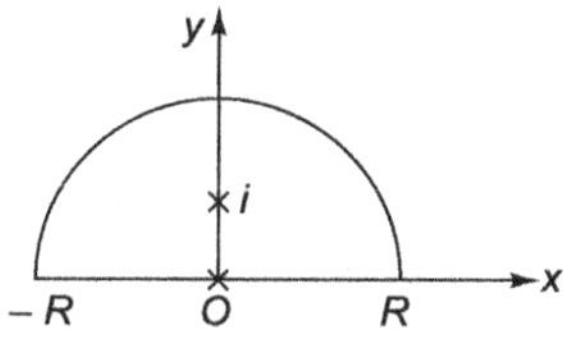

The only singularity of the integrand lying within C is the double pole $z = i$.

$$(\text{Residue})_{z=i} = \frac{1}{\underline{|1}} \left[\frac{d}{dz} \left\{ \frac{e^{iaz}}{(z+i)^2} \right\} \right]_{z=i} = \frac{-i}{4} (1 + a) e^{-a}$$

Using Cauchy's residue theorem, taking limits as $R \to \infty$ and using Jorden's lemma, we get

$$\int_{-\infty}^{\infty} \frac{e^{iax}}{\left(1+x^2\right)^2} dx = \frac{\pi}{2} (1 + a) e^{-a} \qquad (2)$$

Using (2) in (1),

$$R(\tau) = \frac{1}{4} (1 + \tau) e^{-\tau}$$

Average power of

$$\{X(\tau)\} = R(0) = 0.25.$$

Example 9 *The power spectral density function of a zero mean WSS process $\{X(t)\}$ is given by*

$$S(\omega) = \begin{cases} 1, & |\omega| < \omega_0 \\ 0, & \text{elsewhere} \end{cases}$$

Find $R(\tau)$ and show also that $X(t)$ and $X\left(t + \frac{\tau}{\omega_0}\right)$ are uncorrelated.

Solution $R(\tau) = \frac{1}{2\pi} \int_{-\infty}^{\infty} S(\omega)\, e^{i\tau\omega}\, d\omega$

$$= \frac{1}{2\pi} \int_{-\omega_0}^{\omega_0} e^{i\tau\omega}\, d\omega$$

$$= \frac{1}{2\pi} \left(\frac{e^{i\tau\omega}}{i\tau}\right)_{-\omega_0}^{\omega_0}$$

$$= \frac{1}{\pi\tau} \sin\omega_0\tau$$

$$\therefore \quad E\left\{X\left(t + \frac{\pi}{\omega_0}\right) X(t)\right\} = R\left(\frac{\pi}{\omega_0}\right) = \frac{1}{\pi\tau} \sin\pi = 0$$

Since the mean of the process is zero,

$$C\left\{X\left(t + \frac{\pi}{\omega_0}\right) X(t)\right\} = E\left\{X\left(t + \frac{\pi}{\omega_0}\right) X(t)\right\} = 0$$

Therefore, $X(t)$ and $X\left(t + \frac{\pi}{\omega_0}\right)$ are uncorrelated.

Example 10 *The short-time moving average of a process $\{X(t)\}$ is defined as $Y(t) = \frac{1}{T} \int_{t-T}^{t} X(s)ds$. Prove that $X(t)$ and $Y(t)$ are related by means of a convolution type integral. Find the unit impulse response of the system also.*

Solution $Y(t) = \frac{1}{T} \int_{t-T}^{t} X(s)\, ds$ (1)

Putting $s = t - u$ and treating t as a parameter, (1) becomes

$$Y(t) = \frac{1}{T} \int_{0}^{T} X(t-u)\, du \qquad (2)$$

Let us define the unit impulse response of the system as follows:

$$h(t) = \begin{cases} \dfrac{1}{T}, & \text{for } 0 \le t \le T \\ 0, & \text{otherwise} \end{cases}$$

Then (2) can be expressed as

$$Y(t) = \int_{-\infty}^{\infty} h(u)\, X(t-u)\, du$$

which is a convolution type integral.

Example 11 *If the input $x(t)$ and the output $y(t)$ are connected by the differential equation $T\dfrac{dy(t)}{dt} + y(t) = x(t)$, prove that they can be related by means of a convolution type integral. Assume that $x(t)$ and $y(t)$ are zero for $t \le 0$.*

Solution The given differential equation $y'(t) + \dfrac{1}{T} y(t) = \dfrac{1}{T} x(t)$ is a linear equation. Its solution is

$$y(t)\, e^{t/T} = \int \frac{1}{T} x(u)\, e^{u/T}\, du + c$$

i.e.,

$$y(t) = \frac{1}{T}\int x(u)\, e^{-(t-u)/T}\, du + c$$

Since $y(0) = 0$,

$$y(t) = \frac{1}{T}\int_0^t x(u)\, e^{-(t-u)/T}\, du$$

(or)

$$y(t) = \frac{1}{T}\int_0^t x(t-u) e^{-u/T}\, du \qquad (1)$$

Given:

$$x(t) = 0, \text{ for } t < 0$$

$\therefore$ $$x(t-u) = 0, \text{ for } t < u$$

$\therefore$ (1) can be written as

$$y(t) = \frac{1}{T}\int_0^{\infty} x(t-u)\, e^{-u/T}\, du \qquad (2)$$

Now if we define

$$h(t) = \begin{cases} \dfrac{1}{T} e^{-t/T}, & \text{for } t \ge 0 \\ 0, & \text{otherwise} \end{cases}$$

(2) can be rewritten as

$$y(t) = \int_{-\infty}^{\infty} h(u) x(t-u)\, du$$

Hence the result.

Example 12 *$X(t)$ is the input voltage to a circuit (system) and $Y(t)$ is the output voltage. $\{X(t)\}$ is a stationary random process with $\mu_x = 0$ and $R_{xx}(\tau) = e^{-\alpha|\tau|}$. Find μ_y, $S_{yy}(\omega)$ and $R_{yy}(\tau)$, if the power transfer function is*

$$H(\omega) = \frac{R}{R + iL\omega}.$$

Solution $$Y(t) = \int_{-\infty}^{\infty} h(\alpha) X(t-\alpha)\, d\alpha$$

$$\therefore \quad E\{Y(t)\} = \int_{-\infty}^{\infty} h(\alpha) E\{X(t-\alpha)\}\, d\alpha$$

$$= 0 \qquad [\text{since } E\{X(t-\alpha)\} = \mu_x = 0]$$

$$S_{xx}(\omega) = \int_{-\infty}^{\infty} R_{xx}(\tau)\, e^{-i\omega\tau}\, d\tau$$

$$= \int_{-\infty}^{0} e^{\alpha\tau} e^{-i\omega\tau}\, d\tau + \int_{0}^{\infty} e^{-\alpha\tau} e^{-i\omega\tau}\, d\tau$$

$$= \left\{\frac{e^{(\alpha - i\omega)\tau}}{\alpha - i\omega}\right\}_{-\infty}^{0} + \left\{\frac{e^{-(\alpha + i\omega)\tau}}{-(\alpha + i\omega)}\right\}_{0}^{\infty}$$

$$= \frac{1}{\alpha - i\omega} + \frac{1}{\alpha + i\omega} = \frac{2\alpha}{\alpha^2 + \omega^2}$$

Now $$S_{yy}(\omega) = S_{xx}(\omega)\, |H(\omega)|^2$$

$$= \frac{2\alpha}{\alpha^2 + \omega^2} \times \frac{R^2}{R^2 + L^2\omega^2}$$

$$= \frac{\{(2\alpha R^2/(R^2 - L^2\alpha^2))\}}{\alpha^2 + \omega^2} + \frac{\{2\alpha R^2/(\alpha^2 - R^2/L^2)\}}{R^2 + L^2\omega^2}$$

$$= \frac{2\alpha\left(\frac{R}{L}\right)^2}{\left(\frac{R}{L}\right)^2 - \alpha^2} \times \frac{1}{\alpha^2 + \omega^2} + \frac{2\alpha R^2/L^2}{\alpha^2 - \left(\frac{R}{L}\right)^2} \times \frac{1}{\left(\frac{R}{L}\right)^2 + \omega^2}$$

$$= \lambda \frac{1}{\alpha^2 + \omega^2} + \mu \frac{1}{\left(\frac{R}{L}\right)^2 + \omega^2}, \text{ say}$$

$$\therefore \quad R_{yy}(\tau) = \frac{\lambda}{2\pi} \int_{-\infty}^{\infty} \frac{e^{i\tau\omega}}{\alpha^2 + \omega^2}\, d\omega + \frac{\mu}{2\pi} \int_{-\infty}^{\infty} \frac{e^{i\tau\omega}}{\left(\frac{R}{L}\right)^2 + \omega^2}\, d\omega \qquad (1)$$

We can prove that, by contour integration technique,

$$\int_{-\infty}^{\infty} \frac{e^{iaz}}{z^2+b^2}\,dz = \frac{\pi}{b}\,e^{-ab};\ a>0 \tag{2}$$

Using (2) in (1)

$$R_{yy}(\tau) = \frac{\left(\frac{R}{L}\right)^2}{\left(\frac{R}{L}\right)^2-\alpha^2}\,e^{-\alpha|\tau|} + \frac{\left(\frac{R}{L}\right)^2\alpha}{\alpha^2-\left(\frac{R}{L}\right)^2}\,e^{-\left(\frac{R}{L}\right)|\tau|}$$

Example 13 *Given that* $Y(t) = \frac{1}{2\in}\int_{t-\in}^{t+\in} X(\alpha)\,d\alpha$, *where* $\{X(t)\}$ *is a WSS process, prove that* $S_{yy}(\omega) = \frac{\sin^2 \in\omega}{\in^2\omega^2}\,S_{xx}(\omega)$.

Hence find the relation between $R_{xx}(\tau)$ *and* $R_{yy}(\tau)$.

Solution Putting $\alpha = t-u$, we get $Y(t) = \frac{1}{2\in}\int_{-\in}^{\in} X(t-u)\,du$.

If we define $h(t)$ as follows

$$h(t) = \begin{cases} \frac{1}{2\in}, & \text{for } |t| \le \in \\ 0, & \text{for } |t| > \in \end{cases}$$

then $$Y(t) = \int_{-\infty}^{\infty} h(u)\,X(t-u)\,du$$

$\therefore$ $$S_{yy}(\omega) = |H(\omega)|^2\,S_{xx}(\omega), \text{ where}$$

$$H(\omega) = F\{h(t)\}$$

$$= \int_{-\in}^{\in} \frac{1}{2\in}\,e^{-i\omega t}\,dt = \frac{\sin\in\omega}{\in\omega}$$

i.e., $$S_{yy}(\omega) = \frac{\sin^2\in\omega}{(\in\omega)^2}\,S_{xx}(\omega)$$

$\therefore$ $$R_{yy}(\tau) = F^{-1}\left\{\frac{\sin^2\in\omega}{\in^2\omega^2}\,S_{xx}(\omega)\right\}$$

$$= F^{-1}\left\{\frac{\sin^2\in\omega}{\in^2\omega^2}\right\} * R_{xx}(\tau) \tag{1}$$

We can prove that

if $$R(\tau) = \begin{cases} 1 - \dfrac{|\tau|}{2\in} & \text{if } |\tau| \le 2\in \\ 0 & \text{if } |\tau| > 2\in \end{cases}$$

then $$S(\omega) = 2\in \times \frac{\sin^2 \in\omega}{\in^2 \omega^2}$$ [refer to Problem 35 in Exercise 6(c)]

$$\therefore\ F^{-1}\left(\frac{\sin^2 \in\omega}{\in^2 \omega^2}\right) = \begin{cases} \dfrac{1}{2\in}\left(1 - \dfrac{|\tau|}{2\in}\right) & \text{if } |\tau| \le 2\in \\ 0 & \text{if } |\tau| > 2\in \end{cases} \quad (2)$$

Using (2) in (1)

$$R_{yy}(\tau) = \frac{1}{2\in} \int_{-2\in}^{2\in} \left(1 - \frac{|u|}{2\in}\right) R_{xx}(\tau - u)\, du$$

Example 14 ***Property (4) of power spectral density***. *Prove that the spectral density of any WSS process is non-negative. i.e., $S(w) \ge 0$.*

Solution If possible, let $S(\omega) < 0$ at $\omega = \omega_0$. That is, let $S(\omega) < 0$ in $\omega_0 - \frac{\in}{2} < \omega < w_0 + \frac{\in}{2}$, where $\in$ is very small. Let us assume that the system function of the convolution type linear system is

$$H(\omega) = \begin{cases} 1, & \omega_0 < \dfrac{\in}{2} < \omega < \omega_0 + \dfrac{\in}{2} \\ 0, & \text{elsewhere} \end{cases}$$

Note *In this case, the system is called a narrow band filter.*

Now $$S_{yy}(\omega) = |H(\omega)|^2 S_{xx}(\omega)$$

$$= \begin{cases} S_{xx}(\omega), & \omega_0 - \dfrac{\in}{2} < \omega < \omega_0 + \dfrac{\in}{2} \\ 0\ , & \text{elsewhere} \end{cases}$$

$$E\{Y^2(t)\} = R_{yy}(0)$$

$$= \frac{1}{2\pi} \int_{-\infty}^{\infty} S_{yy}(\omega)\, d\omega$$

$$= \frac{1}{2\pi} \int_{\omega_0 - \frac{\in}{2}}^{\omega_0 + \frac{\in}{2}} S_{xx}(\omega)\, d\omega$$

$$= \frac{\in}{2\pi} S_{xx}(\omega_0)$$

[since $S_{xx}(\omega)$ can be considered a contant $S_{xx}(\omega_0)$, as the band is narrow]
Since $E\{Y^2(t)\} \geq 0$, $S_{xx}(\omega_0) \geq 0$, which is contrary to our initial assumption.
Therefore $S_{xx}(\omega) \geq 0$, everywhere, since $\omega = \omega_0$ is arbitrary.

Exercise 6(C)

Part A (Short answer questions)

1. Define the power spectral density (PSD) function of a stationary process.
2. Express each of ACF and PSD of a stationary process in terms of the other.
3. Write down the Wiener-Khinchine relations.
4. Define the cross PSD of the random processes $\{X(t)$ and $\{Y(t)\}$.
5. State any 2 properties of the PSD function of a stationary process.
6. What is average power of a WSS process $\{X(t)\}$ and express it in terms of the PSDF of the process.
7. Find the mean-square value (or the average power) of the process $\{X(t)\}$, if its ACF is given by $R(\tau) = e^{-\tau^2/2}$.
8. Prove that the PSDF of a real stationary process is an even function.
9. Prove that the PSDF of a real or complex stationary process is a real function of ω.
10. Prove that the PSDF of a real WSS process is twice the Fourier cosine transform of its ACF.
11. Prove that the ACF of a real WSS process is half the Fourier inverse cosine transform of its PSDF.
12. Find the PSD function of a stationary process whose ACF is $e^{-|\tau|}$.
13. Find the ACF of a stationary process, whose PSDF is given by

$$S(\omega) = \begin{cases} \omega^2, \text{ for } & |\omega| \leq 1 \\ 0, \text{ for } & |\omega| > 1 \end{cases}$$

14. State Wiener–Khinchine theorem.
15. What is the use of Wiener-Khinchine theorem.
16. What do you mean by a system? When is it called (i) a deterministic system and (ii) a stochastic system?
17. Define a system. When is it called a linear system?
18. Define a system. When is it called a time-invariant system?
19. Define a system. When is it called a memoryless system?
20. Define a system. When is it called a causal system?
21. Define system weighting function.
22. If a system is defined as $Y(t) = \frac{1}{T}\int_0^T X(t-u)\,du$, find its weighting function.
23. What is unit impulse response of a system? Why is it called so?

24. If the input $X(t)$ of the system $Y(t) = \int_{-\infty}^{\infty} h(u)\, x(t-u)\, du$ is the unit impulse function, prove that $Y(t) = h(t)$.

25. Express $Y(t) = \frac{1}{2c} \int_{t-c}^{t+c} X(\alpha)\, d\alpha$ as a convolution type of linear system and hence find the unit impulse response of the system.

26. If a system is defined as $Y(t) = \frac{1}{T} \int_{0}^{\infty} X(t-u)\, e^{-u/T}\, du$, find its unit impulse response.

27. Prove that the system $Y(t) = \int_{-\infty}^{\infty} h(u)\, X(t-u)\, du$ is a linear time-invariant system.

28. When is the system $Y(t) = \int_{-\infty}^{\infty} h(u)\, X(t-u)\, du$ said to be stable?

29. If $\{X(t)\}$ and $\{Y(t)\}$ in the system $Y(t) = \int_{-\infty}^{\infty} h(u)\, X(t-u)\, du$ are WSS processes, how are their ACF's related?

30. If the input and output of the system $Y(t) = \int_{-\infty}^{\infty} h(u)\, X(t-u)\, du$ are WSS processes, how are their PSDF's related?

31. Define the power transfer function (or system function) of the system $Y(t) = \int_{-\infty}^{\infty} h(u)\, X(t-u)\, du$.

32. If the system function of a convolution type of linear system is given by

$$h(t) = \begin{cases} \frac{1}{2c}, & \text{for } |t| \le c \\ 0, & \text{for } |t| > c \end{cases}$$

find the relation between PSDF's of the input and output processes.

Part B

33. Calculate the power spectral density of a stationary random process for which the autocorrelation is $R_{xx}(t) = e^{-\alpha|\tau|}$. (MU — Nov. 96)

34. If the autocorrelation function of a WSS process is $R(\tau) = \rho e^{-\rho|\tau|}$, show that its spectral density is given by $S(w) = \dfrac{2}{1+\left(\dfrac{\omega}{\rho}\right)^2}$.

35. Find the power spectral density of the random binary transmission process whose autocorrelation function is

$$R(\tau) = \begin{cases} 1 - \dfrac{|\tau|}{T}, & \text{for } |\tau| \le T \\ 0 \quad , & \text{for elsewhere} \end{cases}$$

36. Find the power spectral density of the random process $\{X(t)\}$, if $E\{X(t)\} = 1$ and $R_{xx}(\tau) = 1 + e^{-\alpha|\tau|}$.

37. If $\{X(t)\}$ is a constant random process with $R(\tau) = m^2$ for all τ, where m is a constant, show that the spectral density of the process is $S(w) = 2\pi m^2 \delta(\omega)$.

38. Find the power spectral density of the random process whose atocorrelation function is $R(\tau) = e^{-\alpha\tau^2} \cos\omega_0\tau$.

39. Find the power spectral density of the random process, if its autocorrelation function is $R(\tau) = e^{-\alpha|\tau|} \cos\beta\tau$.

40. For the process $\{X(t)\}$, where $X(t) = \sum_{i=1}^{n} (A_i \cos p_i t + B_i \sin p_i t)$, where A_i and B_i are uncorrelated RVs with mean zero and variance σ_i^2, show that the autocorrelation function is given by $R(\tau) = \sum_{i=1}^{n} \sigma^2_i \cos p_i\tau$. Prove also, that the power spectrum for this process is given by

$$S(\omega) = \pi \sum_{i=1}^{n} \sigma^2_i \,[\delta(\omega - p_i) + \delta(\omega + p_i)]$$

(**Hint:** Refer to Worked Example 4 in Section 6(c))

41. (i) For the process $\{X(t)\}$, where $X(t) = a\cos(bt + Y)$, where Y is uniformly distributed over $(-\pi, \pi)$ find the autocorrelation function and the spectral density.
(ii) For the process $\{X(t)\}$, where $X(t) = a\sin(bt + Y)$, where Y is uniformly distributed over $(0, 2\pi)$, find the autocorrelation function and the spectral density.

42. Find the autocorrelation function of the process $\{X(t)\}$, for which the power spectral density is given by

$$S(\omega) = \begin{cases} 1 + \omega^2, & \text{for } |\omega| \le 1 \\ 0 \quad , & \text{for } |\omega| > 1 \end{cases}$$

43. Find the average power of the random process $\{X(t)\}$, if its power spectral density is given by

$$S(w) = \frac{10\omega^2 + 35}{(\omega^2 + 4)(\omega^2 + 9)}$$

(i) using $S(\omega)$ directly and
(ii) using the autocorrelation function $R(\tau)$.

[**Hint:** Average power $= \frac{1}{2\pi}\int_{-\infty}^{\infty} S(\omega)\, d\omega$ or $R(0)$.]

44. Find the autocorrelation function of the random process $\{X(t)\}$ for which the power spectral density is given by

(i) $S(\omega) = 2\pi\delta(\omega) + \frac{2\alpha}{\alpha^2 + \omega^2}$ and (ii) $S(\omega) = \frac{1}{\omega^4 + 4}$.

45. $\{X(t)\}$ is a stationary random process with spectral density function $S_{xx}(\omega)$ and $\{Y(t)\}$ is another independent random process where $Y(t) = A\cos(\omega_c t + \theta)$, where θ is a RV uniformly distributed over $(-\pi, \pi)$. Find the spectral density function of $\{Z(t)\}$, where $Z(t) = X(t)\, Y(t)$.
46. If $\{Y(t)\}$ is the moving time average of $\{X(t)\}$ over $\{t - T, t + T\}$, express $S_{yy}(\omega)$ in terms of $S_{xx}(\omega)$. Hence find the autocorrelation function of $\{Y(t)\}$ in terms of that of $\{X(t)\}$.
47. $X(t)$ is the input voltage to a circuit and $Y(t)$ is the output voltage. $\{X(t)\}$ is a stationary random process with $\mu_x = 0$ and $R_{xx}(\tau) = e^{-2|\tau|}$. Find μ_y, $S_{yy}(\omega)$ and $R_{yy}(\tau)$, if the system function is given by $H(\omega) = \frac{1}{\omega + 2i}$.

Answers

Exercise 6(A)

4. (i) When s is fixed, $X(s, t)$ is a time function. (ii) When t is fixed, $X(s, t)$ is a RV.
10. The jpdf's and the joint distribution functions of the member functions of the random process are used to describe it. The first-and second-order distribution functions, namely, $F(x, t) = P\{X(t) \le x\}$ and $F(x_1, x_2; t_1, t_2) = P\{X(t_1) \le x_1; X(t_2) \le x_2\}$ are primarily used to describe the random process $\{X(t)\}$.
11. (i) Markov process, (ii) Process with independent increments and (iii) Stationary process.
12. Poisson process and Wiener process.
13. If $X(t)$ is a representative member function of the random process $\{X(t)\}$, $E\{X(t)\}$ and $\text{Var}\{X(t)\}$ are called the mean and variance of the process.
17. No. Their definitions are entirely different.
19. Two processes $\{X(t)\}$ and $\{Y(t)\}$ are said to be orthogonal, if $E\{X(t_1)\, Y(t_2)\} = 0$.

22. Bernoulli's process.
23. A kth-order stationary process becomes a S S S process when $k \to \infty$.
32. Poisson process.
35. $f_{X(t)}(x) = \dfrac{1}{\sqrt{2\pi\alpha t}} e^{-x^2/2\alpha t}$; $E\{X(t)\} = 0$;

    $\text{Var}\{X(t)\} = at$.
36. No
37. Yes, it is a WSS process.
38. $p + qt$; $\sigma_1^2 + p^2 + (t_1 + t_2) pq + t_1 t_2 (\sigma_2^2 + q^2)$; $\sigma_1^2 + t_1 t_2 \sigma_2^2$; No.
39. $E\{X(t)\} = 0$; $R(t_1, t_2) = \dfrac{1}{2} \cos \omega (t_1 - t_2)$
40. $E\{X(t)\} = 0$; $R(t_1, t_2) = 2 \cos (t_1 - t_2)$
41. $R(t_1, t_2) = (A^2/2) \cos \omega (t_1 - t_2)$
42. $E\{X(t)\} = 0$; $E\{X^2(t)\} = 1/2$; No
44. $E\{X(t)\} = t + (1/2) \sin \pi t$; when $t = 0.25$, $F(x) = 0$, if $x < \dfrac{1}{2}$, $= \dfrac{1}{2}$ if $\dfrac{1}{2}$

    $\le x < \dfrac{1}{\sqrt{2}}$ and $= 1$, if $\dfrac{1}{\sqrt{2}} \le x$; when $t = 0.5$,
    $F(x) = 0$, if $x < 1$ and $= 1$, if $x \ge 1$; when $t = 1$,
    $F(x) = 0$, if $x < 0$, $= 1/2$, if $0 \le x < 2$ and $= 1$, if $x \ge 2$.
45. $E\{X(t)\} = 0$; $R(t_1, t_2) = \sigma^2 \cos (t_1 - t_2)$
46. $E(z) = 0$; $E(z^2) = \sigma^2$
47. $E\{Z\} = E\{W\} = 3$; Var (Z) = Var $(W) = 13$; cov $(Z, W) = 2.195$.
49. $R_{xy}(t_1, t_2) = \sin (t_1 + t_2)$
50. $R_{xy}(t_1, t_2) = 50 \sin 10 (t_1 - t_2)$
51. $R(t_1, t_2) = \dfrac{A^4}{8} \{2 + \cos 2\omega (t_1 - t_2)\}$
52. $\phi(\omega) = e^{-\alpha t \omega^2/2}$
53. $R_{xx}(\tau) = R_{yy}(\tau)$ and $R_{xy}(-\tau) = -R_{xy}(\tau)$; $R_{ww}(\tau) =$
    $R_{xx}(\tau) \cos \omega\tau + R_{yx}(\tau) \sin \omega\tau$ (or) $R_{xx}(\tau) \cos \omega\tau - R_{xy}(\tau) \sin \omega\tau$.

Exercise 6(B)

5. No, it is true only when the stationary process does not contain periodic components.
6. $\mu_x^2 = \lim\limits_{\tau \to \infty} \left(\dfrac{25 + 36/\tau^2}{6.25 + 4/\tau^2} \right) = 4$; $\mu_x = 2$
7. $E[X^2(t)\} = R(0) = 9$;
   $\text{Var}\{X(t)\} = E\{X^2(t)\} - E^2\{X(t)\} = 5$
8. $E\{X(t)\} = \sqrt{2}$; $E\{X^2(t)\} = 6$; $\text{Var}\{X(t)\} = 4$

9. $E\{X(t)\} = 4$; $E\{X^2(t)\} = 25$; $\text{Var}\{X(t)\} = 9$
11. $R_{xy}(\tau) = E[X(t)Y(t-\tau)] = 0$
12. $R_{xy}(\tau) = E[X(t)\,Y(t-\tau)] = E[X(t)]\,E[Y(t-\tau)]$
$= \mu_x\,\mu_y$
13. $E\{X(t)\} = \Sigma\, x_i p_i$ or $\int_{R_X} xf(x)\,dx$; $\overline{X}_T = \dfrac{1}{2T}\int_{-T}^{T} X(t)\,dt$
14. The ensemble average is given by $E\{X(t)\} = \Sigma\; x_i p_i$ or $\int_{R_X} xf(x)\,dx$; the time average is given by $\overline{X}_T = \dfrac{1}{2T}\int_{-T}^{T} X(t)\,dt$. To compute $E\{X(t)\}$, we should know the probability distribution or density function of $X(t)$; To compute $\overline{X}_T$, it is enough we know a single sample function of the process.
16. Stationarity of a random process is the property of the process by which certain probability distributions or averages do not depend on t.
Ergodicity of a random process is the property by which almost every member of the process exhibits the same statistical behaviour as the whole process. Ergodicity is a stronger condition than stationarity and hence all stationary processes are not ergodic.
20. (i) $\lim_{T\to\infty} \{\text{Var}\left(\overline{X}_T\right)\} = 0$ where $\overline{X}_T = \dfrac{1}{2T}\int_{-T}^{T} X(t)\,dt$

(ii) $$\lim_{T\to\infty}\left[\frac{1}{4T^2}\int_{-T}^{T}\int_{-T}^{T} C(t_1, t_2)\,dt_1\,dt_2\right] = 0$$
21. No; $\overline{X}_T = \dfrac{1}{2T}\int_{-T}^{T} X\,dt = X$; $\text{Var}(X_T) = \text{Var}(X) \neq 0$;

$\lim_{T\to\infty} \{\text{Var}\left(\overline{X}_T\right)\} \neq 0$
22. Random binary transmission process
23. $$\text{Var}\left(\overline{X}_T\right) = \frac{1}{T}\int_{0}^{2T} C(\tau)\left\{1 - \frac{|\tau|}{2T}\right\} d\tau;$$

$$\lim_{T\to\infty}\left[\frac{1}{T}\int_{0}^{2T} C(\tau)\left\{1 - \frac{|\tau|}{2T}\right\} d\tau\right] = 0$$
26. $\pm 2, 9, 5$
27. $R_{yy}(t_1, t_2) = R(t_1 + a, t_2 + a) - R(t_1 + a, t_2)$
$- R(t_1, t_2 + a) + R(t_1, t_2)$

29. (i) $a^2 + b^2 R_{xx}(\tau) + c^2 R_{yy}(\tau)$
(ii) $a^2 R_{xx}(\tau) \cdot R_{yy}(\tau)$

31. 80; $5(e^{-20} + 19)$

32. $\sqrt{10}$; $4\int_4^5 e^{-x^2/10}\,dx + 4e^{-2.5} - 14$

33. 2; $20(10e^{-0.1} - 9)$

34. 0; $\frac{1}{2\lambda T} - \frac{1}{8\lambda^2 T^2}(1 - e^{-4\lambda T})$; Yes

35. 1; $\frac{1}{2}(1 + e^{-2})$

36. $\frac{A}{\alpha T}\left\{1 - \frac{(1 - e^{-2\alpha T})}{2\alpha T}\right\}$

40. Random binary transmission process.

Exercise 6(C)

6. Average power of $\{X(t)\} = E\{X^2(t)\} = R(0)$;

$$E\{X^2(t)\} = \int_{-\infty}^{\infty} S(f)\,df \text{ or } \frac{1}{2\pi}\int_{-\infty}^{\infty} S(\omega)\,d\omega$$

7. $E\{X^2(t)\} = R(0) = 1$

12. $S(\omega) = \int_{-\infty}^{\infty} e^{-|\tau|} e^{i\omega\tau}\,d\tau = 2\int_0^{\infty} e^{-\tau}\cos\omega\tau\,d\tau = \frac{2}{1+\omega^2}$

13. $R(\tau) = \frac{1}{2\pi}\int_{-1}^{1} \omega^2 \cos\tau\omega\,d\omega = \frac{1}{\pi}\left(\frac{\sin\tau}{\tau} + \frac{2\cos\tau}{\tau^2} - \frac{2\sin\tau}{\tau^3}\right)$

21. If the output $Y(t)$ of a system is expressed as the convolution of the input $X(t)$ and a function $h(t)$, i.e., $Y(t) = \int_{-\infty}^{\infty} h(u)\,X(t-u)\,du$, then $h(t)$ is called the system weighting function.

22. $h(t) = \begin{cases} \frac{1}{T}, & 0 \le t \le T \\ 0, & \text{elsewhere} \end{cases}$

23. If a system is of the form $Y(t) = \int_{-\infty}^{\infty} h(u)\,X(t-u)\,du$, then the system weighting function $h(t)$ is also called unit impulse response of the system. It is called so because the response (output) $Y(t)$ will be $h(t)$, when the input $X(t)$ = the unit impulse function $\delta(t)$.

25. Put $\alpha = t - u$. Then

$$Y(t) = \int_{-c}^{c} \frac{1}{2c} X(t-u)\, du$$

$$= \int_{-\infty}^{\infty} h(u)\, X(t-u)\, du$$

$$\therefore \quad h(t) = \begin{cases} \dfrac{1}{2c}, & \text{in } |t| \leq c \\ 0, & \text{in } |t| > c \end{cases}$$

26. $h(t) = \begin{cases} \dfrac{1}{T} e^{-t/T}, & \text{for } t \geq 0 \\ 0, & \text{elsewhere} \end{cases}$

29. $R_{yy}(\tau) = R_{xy}(\tau) * h(\tau)$,
where $R_{xy}(\tau) = R_{xx}(\tau) * h(-\tau)$
(or) $R_{yy}(\tau) = R_{xx}(\tau) * h(\tau) * h(-\tau)$

30. $S_{yy}(\omega) = |H(\omega)|^2\, S_{xx}(\omega)$, where $H(\omega)$ is the Fourier transfrom of $h(t)$

32. $H(\omega) = \int_{-c}^{c} \frac{1}{2c} e^{-i\omega t}\, dt = \frac{\sin c\omega}{c\omega}$; $S_{yy}(\omega) = |H(\omega)|^2\, S_{xx}(\omega)$

$$S_{YY}(\omega) = \frac{\sin^2 c\omega}{c^2\omega^2} S_{xx}(\omega)$$

33. $\dfrac{2\alpha}{\alpha^2 + \omega^2}$

35. $4 \sin^2\left(\dfrac{\omega T}{2}\right) \Big/ T\, \omega^2$

36. $2\pi\,\delta(\omega) + \dfrac{2\alpha}{\omega^2 + \alpha^2}$

38. $\sqrt{\dfrac{\pi}{4\alpha}}\, \{e^{-(\omega+\omega_0)^2/4\alpha} + e^{-(\omega-\omega_0)^2/4\alpha}\}$

39. $\dfrac{\alpha}{\alpha^2 + (\omega-\beta)^2} + \dfrac{\alpha}{\alpha^2 + (\omega+\beta)^2}$

41. (i) and (ii) $\dfrac{a^2}{2} \cos b\tau$; $\dfrac{\pi a^2}{2}$; $[\delta(\omega - b) + \delta(\omega + b)]$

42. $\dfrac{1}{\pi\tau^3} [\tau^2 \sin \tau + \tau \cos \tau - \sin \tau]$

43. $R(\tau) = \frac{11}{6} e^{-3|\tau|} - \frac{1}{4} e^{-2|\tau|}$; Average power = $\frac{19}{12}$

44. (i) $1 + e^{-\alpha|\tau|}$

(ii) $\frac{1}{8} e^{-|\tau|} (\cos\tau + \sin|\tau|)$

45. $\frac{A^2}{4} \{S_{xx}(\omega - \omega_c) + S_{xx}(\omega + \omega_c)\}$

46. $S_{yy}(\omega) = \frac{\sin^2 \omega T}{\omega^2 T^2} S_{xx}(\omega)$;

$$R_{yy}(\tau) = \frac{1}{2T} \int_{-2T}^{2T} \left(1 - \frac{|\alpha|}{2T}\right) R_{xx}(\tau - \alpha)\, d\alpha$$

47. 0; $\frac{4}{(\omega^2 + 4)^2}$; $\frac{1}{8}(1 + 2)|\tau| e^{-2|\tau|}$.

7 Special Random Processes

Many random phenomena in physical problems including 'noise' are well approximated by a special class of random process, namely Gaussian random process. A number of processes such as the Wiener process and the shot-noise process can be approximated, as per central limit theorem, by a Gaussian process. Moreover the output of a linear system in which the input is a weighted sum of a large number of independent samples of a random process tends to approach a Gaussian process. Gaussian processes play an important role in the theory and analysis of random phenomena, because they are good approximations to the observations and multivariate Gaussian distributions are analytically simple.

One of the most important uses of the Gaussian process is to model and analyse the effects of thermal noise in electronic circuits used in communication systems. Individual circuits contain resistors, inductors and capacitors as well as semiconductor devices. The resistors and semiconductor elements contain charged particles (free electrons) subjected to random motion due to thermal energy. The random motion of charged particles causes fluctuations in the current waveforms or information bearing signals that flow through these components. These fluctuations are called thermal noise, which are of sufficient strength to disturb a weak signal and to make the recognition of signals a difficult task. Models of thermal noise are used to identify and minimise the effects of noise in signal recognition.

Definition of a Gaussian Process

A real valued random process $\{X(t)\}$ is called a Gaussian process or normal process, if the random variables $X(t_1)$, $X(t_2)$, ..., $X(t_n)$ are jointly normal for every $n = 1, 2, \ldots$ and for any set of t_i's.

The nth order density of a Gaussian process is given by $f(x_1, x_2, \ldots, x_n;$ $t_1, t_2, \ldots, t_n)$

$$= \frac{1}{(2\pi)^{n/2}\,|\Lambda|^{1/2}} \exp\left[-\frac{1}{2|\Lambda|}\sum_{i=1}^{n}\sum_{j=1}^{n} |\Lambda|_{ij}\,(x_i-\mu_i)\,(x_j-\mu_j)\right]$$

where $\mu_i = E\{X(t_i)\}$ and Λ is the nth order square matrix (λ_{ij}), where $\lambda_{ij} = C\{X(t_i), X(t_j)\}$ and $|\Lambda|_{ij}$ = cofactor of λ_{ij} in $|\Lambda|$ (1)

Note *Gaussian process is completely specified by the first and second order moments, viz., means and covariances (variances).*

Note *When we consider the first order density of a Gaussian process,*

$$\Lambda = (\lambda_{11}) = [cov(X(t_1), X(t_1))]$$
$$= [Var\{X(t_1)\}] = (\sigma^2_1)$$
$$\therefore \quad |\Lambda| = \sigma^2_1 \text{ and } |\Lambda|_{11} = 1$$
$$\therefore \quad f(x_1, t_1) = \frac{1}{\sigma_1\sqrt{2\pi}} \exp\{-(x_1-\mu_1)^2/2\sigma_1^2\}$$

Note *When we consider the second order density of a Gaussian process,*

$$\Lambda = \begin{pmatrix} \lambda_{11} & \lambda_{12} \\ \lambda_{21} & \lambda_{22} \end{pmatrix} = \begin{pmatrix} \sigma_1^2 & r_{12}\sigma_1\sigma_2 \\ r_{21}\sigma_1\sigma_2 & \sigma_2^2 \end{pmatrix}$$
$$\therefore \quad |\Lambda| = \sigma^2_1\sigma^2_2(1-r^2), \text{ where } r_{12} = r_{21} = r$$
$$|\Lambda|_{11} = \sigma^2_2,\ |\Lambda|_{12} = -r\sigma_1\sigma_2,\ |\Lambda|_{21} = -r\sigma_1\sigma_2,\ |\Lambda|_{22} = \sigma_1^2$$
$$\therefore f(x_1, x_2; t_1, t_2) = \frac{1}{2\pi\sigma_1\sigma_2\sqrt{1-r^2}} \exp\left[-\frac{1}{2\sigma_1^2\sigma_2^2(1-r^2)}\{\sigma^2_2(x_1-\mu_1)^2 - 2r\sigma_1\sigma_2(x_1-\mu_1)(x_2-\mu_2) + \sigma^2_1(x_2-\mu_2)^2\}\right]$$

$$i.e., \quad f(x_1, x_2; t_1, t_2) = \frac{1}{2\pi\sigma_1\sigma_2\sqrt{1-r^2}} \exp\left[-\frac{1}{2(1-r^2)}\left\{\frac{(x_1-\mu_1)^2}{\sigma_1^2} - \frac{2r(x_1-\mu_1)(x_2-\mu_2)}{\sigma_1\sigma_2} + \frac{(x_2-\mu_2)^2}{\sigma_2^2}\right\}\right]$$

which we have made use of in many problems earlier.

Properties

1. If a Gaussian process is wide-sense stationary, it is also strict-sense stationary. (MU — Apr. 96; BDU — Apr. 96)

Proof The nth order density of a Gaussian process is given by $f(x_1, x_2, \ldots, x_n; t_1, t_2, \ldots, t_n) =$

$$\frac{1}{(2\pi)^{n/2}|\Lambda|^{1/2}} \exp\left[-\frac{1}{2|\Lambda|}\sum_{i=1}^{n}\sum_{j=1}^{n}|\Lambda|_{ij}(x_i-\mu_i)(x_j-\mu_j)\right]$$

where $\mu_i = E(X(t_i))\}$ and Λ is the nth order square matrix (λ_{ij}), $i, j = 1, 2, \ldots, n$, where $\lambda_{ij} = C\{X(t_i), X(t_j)\}$ and $|\Lambda|_{ij}$ is the cofactor of λ_{ij} in $|\Lambda|$.

If the Gaussian process is WSS, then $\lambda_{ij} = C\{X(t_i), X(t_j)\}$ = a function of $(t_i - t_j)$ for all i and j.

Therefore, the nth order densities of $\{X(t_1), X(t_2), \ldots, X(t_n)\}$ and $\{X(t_1 + h), X(t_2 + h), \ldots, X(t_n + h)$ are identical.

Therefore, the Gaussian process is a SSS process.

2. If the member functions of a Gaussian process are uncorrelated, then they are independent.

Proof Consider n member functions $X(t_1), X(t_2), \ldots, X(t_n)$ of the Gaussian process $\{X(t)\}$.

Since the members are uncorrelated,

$$C\{X(t_i), X(t_j)\} = \lambda_{ij} = 0, \text{ for } i \neq j$$
$$= \sigma^2{}_i, \quad \text{for } i = j$$

Therefore, $[\Lambda]$ is a diagonal matrix with terms in the principal diagonal equal to σ_i^2.

$$\therefore \quad |\Lambda| = \sigma_1^2\, \sigma_1^2 \ldots \sigma_n^2; \; |\Lambda|_{ii} = \sigma_1^2\, \sigma_2^2 \ldots \sigma_{i-1}^2\, \sigma_{i+1}^2 \ldots \sigma_n^2$$
$$|\Lambda|_{ij} = 0, \text{ for } i \neq j$$

Hence the nth order density function of the Gaussian process becomes $f(x_1, x_2, \ldots, x_n; t_1, t_2, \ldots, t_n)$.

$$= \frac{1}{(2\pi)^{n/2}\,\sigma_1\sigma_2\ldots\sigma_n} \exp\left[-\frac{1}{2}\sum_{j=1}^{n}(x_j - \mu_j)^2/\sigma_j^2\right]$$

$$= \left\{\frac{1}{\sigma_1\sqrt{2\pi}}\, e^{-(x_1-\mu_1)^2/2\sigma_1^2}\right\}\left\{\frac{1}{\sigma_2\sqrt{2\pi}}\, e^{-(x_2-\mu_2)^2/2\sigma_2^2}\right\}\ldots$$
$$\left\{\frac{1}{\sigma_n\sqrt{2\pi}}\, e^{-(x_n-\mu_n)^2/2\sigma_n^2}\right\}$$

$$= f(x_1, t_1)\, f(x_2, t_2) \ldots f(x_n, t_n)$$

i.e., $X(t_1), X(t_2), \ldots, X(t_n)$ are independent.

3. If the input $\{X(t)\}$ of a linear system is a Gaussian process, the output will also be a Gaussian process. (BU — Apr. 96; MU — Nov. 96)

Proof Let $\{Y(t)\}$ be the output process.

Then $$Y(t) = \int_{-\infty}^{\infty} x(t-u)h(u)\mathrm{d}u \text{ (or) } \int_{-\infty}^{\infty} h(t-u)X(u)\mathrm{d}u$$

$$= \sum_{j=1}^{n}\{h(t-u_j)\,\Delta u\}X(u_j), \text{ as } n \to \infty$$

Consider the n sample functions of the input process $\{X(t)\}$. Let them be $X(t_1), X(t_2), \ldots, X(t_n)$.

Let the corresponding output sample functions be $Y(t_1), Y(t_2), \ldots, Y(t_n)$.

$$\therefore \quad Y(t_i) = \sum_{j=1}^{n} \{h(t_i - u_j)\Delta u\} X(u_j) \text{ in the limit}$$

i.e., $$Y_i = \sum_{j=1}^{n} h_{ij} X_j, \text{ say, where } i = 1, 2, \ldots, n \tag{1}$$

$$\therefore \quad y_i = \sum_{j=1}^{n} h_{ij} x_j, \quad i = 1, 2, \ldots, n \tag{1a}$$

i.e., $$\begin{pmatrix} y_1 \\ y_2 \\ \vdots \\ y_n \end{pmatrix} = \begin{pmatrix} h_{11} & h_{12} & \cdots & h_{1n} \\ h_{21} & h_{22} & \cdots & h_{2n} \\ \vdots & \vdots & \vdots & \vdots \\ h_{n1} & h_{n2} & \cdots & h_{nn} \end{pmatrix} \begin{pmatrix} x_1 \\ x_2 \\ \vdots \\ x_n \end{pmatrix}$$

i.e., $\overline{Y} = H\overline{X}$, say

$$\therefore \quad E(\overline{Y}) = HE(\overline{X}) \text{ (or) } \mu_{\overline{Y}} = H\mu_{\overline{X}}$$

$\therefore \; f_Y(y_1, y_2, \ldots, y_n) = |J(x_1, x_2, \ldots, x_n)| \, f_X(x_1, x_2, \ldots, x_n)$

where $|J(x_1, x_2, \ldots, x_n) = |J(y_1, y_2, \ldots, y_n|^{-1}$.

Now $|J(y_1, y_2 \ldots y_n)|$

$$= \begin{vmatrix} \dfrac{\partial y_1}{\partial x_1} & \dfrac{\partial y_1}{\partial x_2} & \cdots & \dfrac{\partial y_1}{\partial x_n} \\ \vdots & \vdots & \vdots & \vdots \\ \dfrac{\partial y_n}{\partial x_1} & \dfrac{\partial y_n}{\partial x_2} & \cdots & \dfrac{\partial y_n}{\partial x_n} \end{vmatrix} = \begin{vmatrix} h_{11} & h_{12} & \cdots & h_{1n} \\ h_{21} & h_{22} & \cdots & h_{2n} \\ \vdots & \vdots & \vdots & \vdots \\ h_{n1} & h_{n2} & \cdots & h_{nn} \end{vmatrix} = |H|$$

since $\dfrac{\partial y_i}{\partial x_j} = h_{ij} \; (i, j = 1, 2, \ldots, n)$ [from (1)]

$$\therefore \; f_Y(y_1, y_2, \ldots, y_n) = \frac{1}{|H|} f_X(x_1, x_2, \ldots, x_n)$$

$$= \frac{1}{|H|(2\pi)^{n/2} |\Lambda_x|^{1/2}} \exp\left[-\frac{1}{2|\Lambda_x|} \sum_{i=1}^{n} \sum_{j=1}^{n} |\Lambda_x|_{ij} (x_i - \mu_i)(x_j - \mu_j)\right] \tag{2}$$

where Λ_x is the square matrix whose elements are the covariance functions of $\{X(t)\}$, viz.

$$\Lambda_x = \begin{pmatrix} c_{11} & c_{12} & \cdots & c_{1n} \\ c_{21} & c_{22} & \cdots & c_{2n} \\ \vdots & \vdots & \vdots & \vdots \\ c_{n1} & c_{n2} & \cdots & c_{nn} \end{pmatrix}$$

where $c_{ij} = C\{X(t_i), X(t_j)\}$.

If we express the RS of (2) in matrix form, then $f_Y(y_1, y_2, \ldots, y_n)$

$$= \frac{1}{(2\pi)^{n/2} |H| |\Lambda_x|^{1/2}} \exp\left[-\frac{1}{2}(\bar{X} - \mu_{\bar{X}})^T \Lambda_x^{-1} (\bar{X} - \mu_{\bar{X}})\right]$$

$$= \frac{1}{(2\pi)^{n/2} |H| |\Lambda_x|^{1/2}} \exp$$

$$\left[-\frac{1}{2}(H^{-1}\bar{Y} - H^{-1}\mu_{\bar{Y}})^T \Lambda_x^{-1} (H^{-1}\bar{Y} - H^{-1}\mu_{\bar{Y}})\right]$$

$$= \frac{1}{(2\pi)^{n/2} |H| |\Lambda_x|^{1/2}} \exp\left[-\frac{1}{2}\left\{H^{-1}(\bar{Y} - \mu_{\bar{Y}})\right\}^{Y} \Lambda_x^{-1} \left\{H^{-1}(\bar{Y} - \mu_{\bar{Y}})\right\}\right]$$

$$= \frac{1}{(2\pi)^{n/2} |H| |\Lambda_x|^{1/2}} \exp$$

$$\left[-\frac{1}{2}(\bar{Y} - \mu_{\bar{Y}})^T \left\{(H^T)^{-1} \Lambda_x^{-1} H^{-1}\right\} (\bar{Y} - \mu_{\bar{Y}})\right] \quad (3)$$

Now let

$$\begin{aligned} C_{ij}' &= C\{Y(t_i), Y(t_j)\} \\ &= C\{Y_i, Y_j\} \\ &= E[\{Y_i - E(Y_i)\} \{Y_j - E(Y_j)\}] \\ &= E\left[\sum_{r=1}^{n} h_{ir} (X_r - E(X_r)) \sum_{s=1}^{n} h_{js} (X_s - E(X_s))\right] \quad \text{[by (1)]} \\ &= \sum_{r=1}^{n} \sum_{s=1}^{n} h_{ir}\, h_{is}\, c_{rs} \end{aligned}$$

Therefore, $\Lambda_y = H \Lambda_x H^T$, where Λ_y is the covariance matrix of $\{Y(t)\}$ process.

$$\therefore \quad \Lambda_y^{-1} = (H^T)^{-1} \Lambda_x^{-1} H^{-1} \quad (4)$$

and

$$|\Lambda_y| = |H|^2 |\Lambda_x| \quad (5)$$

Using (4) and (5) in (3), we get

$$f_Y(y_1, y_2, \ldots, y_n) = \frac{1}{(2\pi)^{n/2} |\Lambda|^{1/2}} \exp\left[-\frac{1}{2}(\bar{Y} - \mu_{\bar{Y}})^T \Lambda_y^{-1} (\bar{Y} - \mu_{\bar{Y}})\right]$$

which is the nth order density function of a Gaussian process.

Therefore, the output process $(Y(t)\}$ is a Gaussian process.

Note *This property can be interpreted as follows:*

Since $Y(t) = \sum_{j=1}^{n} \{h(t - u_j)\, \Delta u\}\, X(u_j)$, $Y(t)$ is a linear combination of $X(u_j)$'s.

Since each $X(u_j)$ follows a normal distribution, $Y(t)$ also follows a normal distribution.

Processes Depending on Stationary Gaussian Process

(1) Square law detector process

If $\{X(t)\}$ is a zero mean stationary Gaussian process and if $Y(t) = X^2(t)$, then $\{Y(t)\}$ is called a square law detector process.

$$E\{Y(t)\} = E\{X^2(t)\} = \text{Var}\{X(t)\} = R_{xx}(0)$$

$$\begin{aligned} R_{yy}(t_1, t_2) &= E\{Y(t_1)\, Y(t_2)\} \\ &= E\{X^2(t_1)\, X^2(t_2)\} \\ &= E\{X^2(t_1)\}\, E\{X^2(t_2)\} + 2E^2\{X(t_1)\, X(t_2)\} \end{aligned}$$

[since when X and Y are jointly normal,

$$E(X^2Y^2) = E(X^2)\, E(Y^2) + 2E^2(XY) \qquad \text{(refer to Chapter 4)]}$$

$$= R^2_{xx}(0) + 2R^2_{xx}(\tau) \qquad \text{[since } X(t)\} \text{ is stationary]}$$

Since the RS is function of τ, LS is also a function of $\tau = t_1 - t_2$.

i.e., $R_{yy}(\tau) = R^2_{xx}(0) + 2R^2_{xx}\tau$

Therefore, $\{Y(t)\}$ is also a stationary process (at least in the wide-sense).

We note that $E\{Y^2(t)\} = R_{yy}(0) = 3R^2_{xx}(0)$

$\text{Var}\{Y(t)\} = 2R^2_{xx}(0)$ and $C_{yy}(\tau) = 2R^2_{xx}(\tau)$

Power spectral density of $\{Y(t)\}$ is given by

$$S_{yy}(\omega) = \int_{-\infty}^{\infty} \{R^2_{xx}(0) + 2R^2_{xx}(\tau)\}\, e^{-i\tau\omega}\, d\tau$$

$$= 2\pi R^2_{xx}(0)\, \delta(\omega) + 2F\{R_{xx}(\tau)\, R_{xx}(\tau)\} \qquad (1)$$

[since $F^{-1}\{2\pi m^2\, \delta(\omega)\} = m^2$]

Consider $F^{-1}\{s(\omega) * s(\omega)\}$

$$= \frac{1}{2\pi} \int_{-\infty}^{\infty} \int_{-\infty}^{\infty} s(\omega - \alpha)\, s(\alpha)\, e^{i\tau\omega}\, d\alpha\, d\omega \qquad (2)$$

Put $\omega - \alpha = \beta$ and $\alpha = \gamma$

i.e., $\omega = \beta + \gamma \quad \alpha = \gamma$

Then, from calculus,

$$d\omega\, d\alpha = \begin{vmatrix} \dfrac{\partial\omega}{\partial\beta} & \dfrac{\partial\omega}{\partial\gamma} \\ \dfrac{\partial\alpha}{\partial\beta} & \dfrac{\partial\alpha}{\partial\gamma} \end{vmatrix} d\beta\, d\gamma = d\beta\, d\gamma \qquad (3)$$

Using (3) in (2)

$$F^{-1}\{s(\omega) * s(\omega)\} = \frac{1}{2\pi} \int_{-\infty}^{\infty} \int_{-\infty}^{\infty} s(\beta)\, s(\gamma)\, e^{i\tau(\beta+\gamma)} d\beta\, d\gamma$$

$$= 2\pi \frac{1}{2\pi} \int_{-\infty}^{\infty} s(\beta)\, e^{i\tau\beta}\, d\beta\, \frac{1}{2\pi} \int_{-\infty}^{\infty} s(\gamma) e^{i\tau\gamma}\, d\gamma$$

$$= 2\pi R(\tau)\, R(\tau) \tag{4}$$

Using (4) in (1),

$$S_{yy}(\omega) = 2\pi R^2_{xx}(0)\, \delta(\omega) + \frac{1}{\pi}\, s_{xx}(\omega) * s_{xx}(\omega)$$

Two Important Results

We now consider two important results which will be used in the discussion of other processes depending on stationary Gaussian process, that will follow.

(i) If X and Y are two normal RVs with zero means, variances σ_1^2 and σ_2^2 and correlation coefficient r, then the probability that they are of the same sign $= \frac{1}{2} + \frac{1}{\pi}\sin^{-1}(r)$ and the probability that they are of opposite signs $= \frac{1}{2} - \frac{1}{\pi}\sin^{-1}(r)$.

The joint density function of (X, Y) is

$$f_{XY}(x, y) = \frac{1}{2\pi\sigma_1\sigma_2\sqrt{1-r^2}} \exp\left[-\frac{1}{2(1-r^2)}\left\{\frac{x^2}{\sigma_1^2} - \frac{2rxy}{\sigma_1\sigma_2} + \frac{y^2}{\sigma_2^2}\right\}\right]$$

If we put $U = \frac{X}{Y}$ and $V = Y$, then by the usual procedure, we can find that the joint pdf of (U, V) is

$$f_{UV}(u, v) = \frac{1}{2\pi\sigma_1\sigma_2\sqrt{1-r^2}} \exp$$

$$\left[-\frac{1}{2(1-r^2)}\left\{\frac{u^2}{\sigma_1^2} - \frac{2ru}{\sigma_1\sigma_2} + \frac{1}{\sigma_2^2}\right\} v^2\right] |v| \quad (-\infty < u, v < \infty)$$

The pdf of U is simply the marginal pdf, given by

$$f_u(u) = \frac{1}{\pi\sigma_1\sigma_2\sqrt{1-r^2}} \int_0^\infty \exp$$

$$\left[-\frac{1}{2(1-r^2)}\left\{\frac{u^2}{\sigma_1^2} - \frac{2ru}{\sigma_1\sigma_2} + \frac{1}{\sigma_2^2}\right\} v^2\right] v\, dv$$

$$= \frac{\sqrt{1-r^2}}{\pi\sigma_1\sigma_2} \frac{1}{\left(\frac{u}{\sigma_1} - \frac{r}{\sigma_2}\right)^2 + \frac{1-r^2}{\sigma_2^2}}, \; -\infty < u < \infty$$

which is a Cauchy's distribution.

Now, $P\{X$ and Y are of the same sign$\}$

$$= P\left(\frac{X}{Y} > 0\right) = P(U > 0)$$

$$= \frac{\sqrt{1-r^2}}{\pi\sigma_1\sigma_2} \int_0^\infty \frac{du}{\left(\frac{u}{\sigma_1} - \frac{r}{\sigma_2}\right)^2 + \frac{1-r^2}{\sigma_2^2}}$$

$$= \frac{\sqrt{1-r^2}}{\pi\sigma_2} \int_{-\frac{r}{\sigma_2}}^\infty \frac{dz}{z^2 + \left(\sqrt{\frac{1-r^2}{\sigma^2}}\right)^2} \text{ putting } z = \frac{u}{\sigma_1} - \frac{r}{\sigma_2}$$

$$= \frac{1}{\pi}\left(\tan^{-1}\frac{\sigma_2 z}{\sqrt{1-r^2}}\right)_{-\frac{r}{\sigma_2}}^\infty$$

$$= \frac{1}{\pi}\left\{\frac{\pi}{2} + \tan^{-1}\frac{r}{\sqrt{1-r^2}}\right\}$$

$$= \frac{1}{2} + \frac{1}{\pi}\sin^{-1}(r)$$

Similarly, $P\{X \text{ and } Y \text{ are of opposite signs}\}$

$$= P\left(\frac{X}{Y} < 0\right) = P(U < 0)$$

$$= \frac{1}{2} - \frac{1}{\pi}\sin^{-1}(r)$$

(ii) If X and Y are two normal random variables with zero means, variances σ_1^2 and σ_2^2 and correlation coefficient r, then $E\{|XY|\} = \frac{2}{\pi}\sigma_1\sigma_2(\cos\alpha + \alpha\sin\alpha)$, where $\sin\alpha = r$.

We make use of a theorem, called **Price's theorem**, which we state below without proof:

If X and Y are two normal RVs as described above, then

$$\frac{\partial^n}{\partial C^n} E[g(X, Y)] = E\left[\frac{\partial^{2n} g(X, Y)}{\partial X^n \, \partial Y^n}\right], \text{ where } C = C(X, Y)$$

is the covariance of X and Y.

If we take $g(X, Y) = |XY|$ and $n = 1$, then Price's theorem gives

$$\frac{\partial}{\partial C} E\{|XY|\} = E\left\{\frac{\partial^2 |XY|}{\partial X \, \partial Y}\right\} = E\left\{\frac{d|X|}{dX}\frac{d|Y|}{dY}\right\} \quad (1)$$

Now
$$\frac{d|X|}{dX} = \begin{cases} 1 & \text{if } X > 0 \\ -1 & \text{if } X < 0 \end{cases}$$

and
$$\frac{d|Y|}{dY} = \begin{cases} 1 & \text{if } Y > 0 \\ -1 & \text{if } Y < 0 \end{cases}$$

$$\therefore \quad \frac{d}{dX}|X|\frac{d}{dY}|Y| = 1, \text{ if } \quad \frac{X}{Y} > 0$$

$$= -1, \text{ if } \frac{X}{Y} < 0$$

$$\therefore \quad E\left\{\frac{d}{dX}|X|\frac{d}{dY}|Y|\right\} = 1 \times P\left\{\frac{X}{Y} > 0\right\} - 1 \times P\left\{\frac{X}{Y} < 0\right\}$$

$$= \left(\frac{1}{2} + \frac{1}{\pi}\sin^{-1} r\right) - \left(\frac{1}{2} - \frac{1}{\pi}\sin^{-1} r\right)$$

[by the previous result (i)]

$$= \frac{2}{\pi}\sin^{-1}(r)$$

$$= \frac{2}{\pi}\sin^{-1}\left(\frac{C}{\sigma_1\sigma_2}\right) \tag{2}$$

Using (2) in (1), we get,

$$\frac{\partial}{\partial C} E\{|XY|\} = \frac{2}{\pi}\sin^{-1}\left(\frac{C}{\sigma_1\sigma_2}\right)$$

Integrating both sides with respect to C, between 0 and C,

$$E\{|XY|\} = E\{|XY|\}_{C=0} + \frac{2}{\pi}\int_0^C \sin^{-1}\left(\frac{C}{\sigma_1\sigma_2}\right) dC \tag{3}$$

When $C = 0$, X and Y are uncorrelated

$\therefore \quad E\{|XY|\} = E\{|X|\}\ E\{|Y|\}$

$$= \int_{-\infty}^{\infty} |x| \frac{1}{\sigma_1\sqrt{2\pi}} e^{-x^2/2\sigma_1^2} dx \int_{-\infty}^{\infty} |y| \frac{1}{\sigma_2\sqrt{2\pi}} e^{-y^2/2\sigma_2^2} dy$$

$$= \sqrt{\frac{2}{\pi}}\sigma_1 \sqrt{\frac{2}{\pi}}\sigma_2$$

$$\text{i.e.,} E\{|XY|\}_{C=0} = \frac{2}{\pi}\sigma_1\sigma_2 \tag{4}$$

Using (4) in (3), we get,

$$E\{|XY|\} = \frac{2}{\pi}\sigma_1\sigma_2 + \frac{2}{\pi}\left[C\sin^{-1}\frac{C}{\sigma_1\sigma_2} - \int_0^C \frac{C}{\sqrt{1-\frac{C^2}{\sigma_1^2\sigma_2^2}}}\frac{1}{\sigma_1\sigma_2} dC\right]$$

$$= \frac{2}{\pi}\sigma_1\sigma_2 + \frac{2C}{\pi}\sin^{-1}(r) + \frac{2}{\pi}\left(\sqrt{\sigma_1^2\sigma_2^2 - C^2}\right)_0^C$$

$$= \frac{2}{\pi} \sigma_1 \sigma_2 \left(r \sin^{-1} r + \sqrt{1 - r^2} \right)$$

$$= \frac{2}{\pi} \sigma_1 \sigma_2 (\cos \alpha + \alpha \sin \alpha), \text{ putting } r = \sin \alpha$$

(2) Full wave linear detector process

If $\{X(t)\}$ is a zero mean stationary Gaussian process and if $Y(t) = |X(t)|$, then $\{Y(t)\}$ is called a full wave linear detector process.

$$E\{Y(t)\} = E\{|X(t)|\}$$

$$= \int_{-\infty}^{\infty} |X| \frac{1}{\sigma\sqrt{2\pi}} e^{-x^2/2\sigma^2} dx$$

$$= \frac{1}{\sigma} \sqrt{\frac{2}{\pi}} \int_0^{\infty} x e^{-x^2/2\sigma^2} dx$$

$$= \sqrt{\frac{2}{\pi}} \sigma \int_0^{\infty} e^{-t} dt, \text{ putting } \frac{x^2}{2\sigma^2} = t$$

$$= \sqrt{\frac{2}{\pi}} \sigma = \sqrt{\frac{2 R_{xx}(0)}{\pi}}$$

$$R_{yy}(t_1, t_2) = E\{|X(t_1) X(t_2)|\}$$

$$= \frac{2}{\pi} \sigma^2 (\cos \alpha + \alpha \sin \alpha) \qquad \text{(by the previous result)}$$

$$\text{where } \sin \alpha = r = \frac{C\{X(t_1), X(t_2)\}}{\sigma^2}$$

$$= \frac{E\{X(t_1) \cdot X(t_2)\}}{\sigma^2}$$

$$= \frac{R_{xx}(t_1 - t_2)}{\sigma^2} \qquad \text{[since } \{X(t)\} \text{ is stationary]}$$

Therefore, $\{Y(t)\}$ is wide-sense stationary, with $R_{yy}(\tau) = \frac{2}{\pi} R_{xx}(0) (\cos\alpha + \alpha \sin \alpha)$, where

$$\sin \alpha = \frac{R_{xx}(\tau)}{R_{xx}(0)}$$

Now $E\{Y^2(t)\} = R_{yy}(0) = \frac{2}{\pi} R_{xx}(0) \left\{0 + \frac{\pi}{2} 1\right\}$,

$$\text{since} \quad \sin \alpha = \frac{R_{xx}(0)}{R_{xx}(0)} = 1 \text{ and } \alpha = \frac{\pi}{2}$$

$$\therefore \quad E\{Y^2(t)\} = R_{xx}(0) \text{ and } \operatorname{Var}\{Y(t)\} = \left(1 - \frac{2}{\pi}\right) R_{xx}(0)$$

(3) Half-wave linear detector process

If $\{X(t)\}$ is a zero mean stationary Gaussian process and if

$$Z(t) = \begin{cases} X(t), & \text{for } X(t) \geq 0 \\ 0, & \text{for } X(t) < 0 \end{cases}$$

then $\{Z(t)\}$ is called a half-wave linear detector process.

$Z(t)$ can be rewritten as $Z(t) = \frac{1}{2}\{X(t) + |X(t)|\}$

$$\therefore \quad E\{Z(t)\} = \frac{1}{2}[E(X(t)\} + E\{|X(t)|\}]$$

$$= \frac{1}{2}\left[0 + \sqrt{\frac{2}{\pi}}\, R_{xx}(0)\right] \text{ (refer to the previous process)}$$

$$= \sqrt{\frac{R_{xx}(0)}{2\pi}}$$

$$E\{Z(t)\, Z(t-\tau)\} = E[E\{Z(t)\, Z(t-\tau)/X(t)\, X(t-\tau)\}] \quad (1)$$

Now $Z(t)\, Z(t-\tau)/X(t)\, X(t-\tau)$

$$= \frac{1}{2}\{X(t)\, X(t-\tau) + |X(t)\, X(t-\tau)|\} \text{ (or) } = 0$$

The first value is assumed, when $X(t)\, X(t-\tau) > 0$, i.e., when $X(t)$ and $X(t-\tau)$ are both positive or both negative.

$\therefore$ P\{The first value of assumed\} $= \frac{1}{4} + \frac{1}{4} = \frac{1}{2}$

Similarly, P\{the second value is assumed\} $= \frac{1}{2}$

$\therefore$ $E\{Z(t)\, Z(t-\tau)/X(t)\, X(t-\tau)\}$

$$= \frac{1}{4}\{X(t)\, X(t-\tau) + |X(t)\, X(t-\tau)|\} \quad (2)$$

Using (2) in (1), we get,

$$E\{Z(t)\, Z(t-\tau) = \frac{1}{4}[E\{X(t)\, X(t-\tau)\} + E\{|X(t)\, X(t-\tau)|\}]$$

$$= \frac{1}{4}[R_{xx}(\tau) + R_{yy}(\tau)]$$

[where $\{Y(t)\}$is the full-wave linear detector process]

i.e., $$R_{zz}(\tau) = \frac{1}{4}[R_{xx}(\tau) + \frac{2}{\pi}R_{xx}(0)(\cos\alpha + \alpha\sin\alpha)],$$

where $\sin\alpha = \frac{R_{xx}(\tau)}{R_{xx}(0)}$

Therefore, the process $\{Z(t)\}$ is wide-sense stationary.

Now $E\{Z^2(t)\} = R_{zz}(0) = \frac{1}{2}R_{xx}(0)$

$$\therefore \qquad \text{Var}\{Z(t)\} = \frac{1}{2}R_{xx}(0) - \frac{1}{2\pi}R_{xx}(0)$$

$$= \frac{1}{2}\left(1 - \frac{1}{\pi}\right)R_{xx}(0)$$

(4) Hard limiter process

If $\{X(t)\}$ is a zero mean stationary Gaussian process and if

$$Y(t) = \begin{cases} +1 & \text{for } X(t) \geq 0 \\ -1 & \text{for } X(t) < 0 \end{cases}$$

then $\{Y(t)\}$ is called a hard limiter process or ideal limiter process.

$$E\{Y(t)\} = 1 \times P\{X(t) \geq 0\} - 1 \times P\{X(t) < 0\}$$

$$= 0$$

Now $\quad Y(t)\, Y(t - \tau) = 1, \quad \text{if } X(t)\, X(t - \tau) \geq 0$

$$= -1, \text{ if } X(t)\, X(t - \tau) < 0$$

i.e., $P\{Y(t)\, Y(t - \tau) = 1\} = P\{X(t)\, X(t - \tau) \geq 0$

$$= \frac{1}{2} + \frac{1}{\pi}\sin^{-1}(r_{xx}) \text{ [by result (i) above]}$$

$$= \frac{1}{2} + \frac{1}{\pi}\sin^{-1}\left\{\frac{R_{xx}(\tau)}{R_{xx}(0)}\right\}$$

and $P\{Y(t)\, Y(t - \tau) = -1\} = P\{X(t)\, X(t - \tau) < 0\}$

$$= \frac{1}{2} - \frac{1}{\pi}\sin^{-1}\left\{\frac{R_{xx}(\tau)}{R_{xx}(0)}\right\} \text{ [by result (1)\}}$$

$$\therefore \; E\{Y(t)\, Y(t - \tau)\} = \frac{2}{\pi}\sin^{-1}\left\{\frac{R_{xx}(\tau)}{R_{xx}(0)}\right\}$$

i.e., $$R_{yy}(\tau) = \frac{2}{\pi}\sin^{-1}\left\{\frac{R_{xx}(\tau)}{R_{xx}(0)}\right\} \qquad (1)$$

[(1) is called the **arcsine law**]

Therefore $\{Y(t)\}$ is wide-sense stationary.

Also $E\{Y^2(t)\} = 1$ and $\text{Var}\{Y(t)\} = 1$

Band Pass Process (Signal)

If the power spectrum of a random process $\{X(t)\}$ is zero outside a certain band (an interval in the ω-axis),

i.e., $$S_{xx}(\omega) \neq 0, \text{ in } |\omega - \omega_0| \leq \frac{\omega_B}{2} \text{ and in } |\omega + \omega_0| \leq \frac{\omega_B}{2}$$

$$= 0, \text{ in } |\omega - \omega_0| > \frac{\omega_B}{2} \text{ and in } |\omega + \omega_0| > \frac{\omega_B}{2}$$

then $\{X(t)\}$ is called a ***band pass process***.

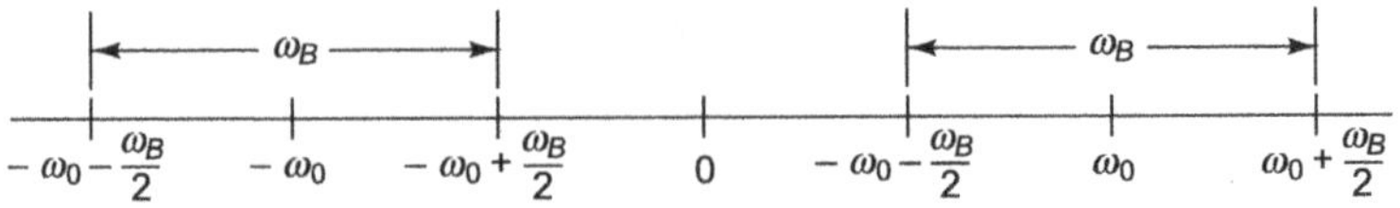

If $\quad S_{xx}(\omega) \neq 0, \text{in } |\omega| \le \omega_B$

$\qquad = 0, \text{ in } |\omega| > \omega_B$

then $\{X(t)\}$ is called a ***low pass process*** or ***ideal low pass process***.

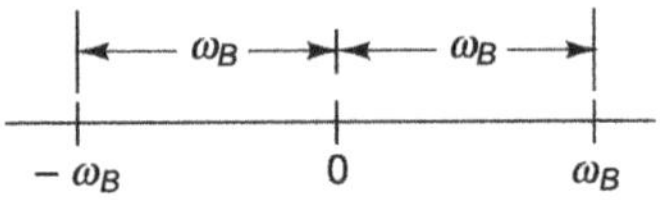

If the bandwith ω_B of a band pass process is small compared with the centre frequency ω_0, the process is called ***narrow band process*** or ***quasimonochromatic***.

If the power spectrum $S_{xx}(\omega)$ of a bandpass process $\{X(t)\}$ is an impulse function, then the process is called ***monochromatic***.

Narrow-Band Gaussian Process

In communication system, information bearing signals are often narrow-band Gaussian processes. When such signals are viewed on an oscilloscope, they appear like a sine wave with slowly varying amplitude and phase. Hence a narrow-band Gaussian process $\{X(t)\}$ is often represented as

$$X(t) = R_X(t)\cos\left[\omega_0 \pm \theta_X(t)\right] \tag{1}$$

$R_X(t)$ and $\theta_X(t)$, which are low pass processes, are called the ***envelope*** and ***phase*** of the process $\{X(t)\}$ respectively. (1) can be rewritten as

$$X(t) = [R_X(t)\cos\theta_X(t)]\cos\omega_0 t \mp [R_X(t)\sin\theta_X(t)]\sin\omega_0 t \tag{2}$$

$R_X(t)\cos\theta_Y(t)$ is called the ***inphase component*** of the process $\{X(t)\}$ and denoted as $X_c(t)$ or $I(t)$. $R_X(t)\sin\theta_X(t)$ is called the ***quadrature component*** of $\{X(t)\}$ and denoted as $X_s(t)$ or $Q(t)$.

Both $X_c(t)$ and $X_s(t)$ are low pass processes.

Property

The envelope of a narrow-band Gaussian process follows a Rayleigh distribution and the phase follows a uniform distribution in $(0, 2\pi)$.

We note that

$$\sqrt{X_c^2(t) + X_s^2(t)} = R_X(t) \text{ and } \tan^{-1}\left\{\frac{X_s(t)}{X_c(t)}\right\} = \theta_X(t)$$

Refer to Worked Example 25 in Chapter 3, in which we have proved the following result:

If X and Y are two independent $N(0, \sigma)$ then $R = \sqrt{X^2 + Y^2}$ follows a Rayleigh distribution and $\phi = \tan^{-1}\dfrac{Y}{X}$ follows a uniform distribution in $(0, 2\pi)$.

According to this problem, the required property follows.

Quadrature Representation of a WSS Process

In order to represent a process in the quadrature form, it need not be a narrow-band Gaussian process. Any arbitrary zero mean WSS process $\{X(t)\}$ can be represented in the quadrature form as proved below.

Let $\{A(t)\}$ and $\{B(t)\}$ be any two zero mean, jointly WSS processes, α be a constant and $\{Y(t)\}$ be the 'dual' process of the given process $\{X(t)\}$, defined below:

$Y(t)$ is so chosen that

$$X(t) + iY(t) = [A(t) + iB(t)\}e^{+i\alpha t} \tag{1}$$

or,

$$A(t) + iB(t) = \{X(t) + iY(t)\}\, e^{-i\alpha t} \tag{2}$$

Then

$$A(t) = X(t)\cos\alpha t + Y(t)\sin\alpha t \tag{3}$$

and

$$B(t) = Y(t)\cos\alpha t - X(t)\sin\alpha t \tag{4}$$

It is easily verified, using (3) and (4), that $E\{A(t)\,A(t-\tau)\}$, $E\{B(t)\,B(t-\tau)\}$ and $E\{A(t)\,B(t-\tau)\}$ are independent of t, i.e., functions of τ,

only if

$$R_{xx}(\tau) = R_{yy}(\tau) \tag{5}$$

and

$$R_{xy}(\tau) = -R_{yx}(\tau) \tag{6}$$

Since $\{A(t)\}$ and $\{B(t)\}$ are jointly WSS, conditions (5) and (6) must be satisfied by $\{X(t)\}$ and its dual $\{Y(t)\}$.

When (5) and (6) are true, from (1) we get,

$$X(t) = A(t)\cos\alpha t - B(t)\sin\alpha t \tag{7}$$

$$Y(t) = A(t)\sin\alpha t + B(t)\cos\alpha t \tag{8}$$

Thus, if we can get a suitable $Y(t)$ and α, satisfying (5) and (6), $X(t)$ can be expressed in form (7), which is the quadrature form.

Consider the linear time-invariant system

$$Z(t) = \int_{-\infty}^{\infty} X(\alpha)\,h(t-\alpha)\,d\alpha \tag{9}$$

where $h(t) = \dfrac{1}{\pi t}$ (10)

We note that $F\{h(t)\} = \dfrac{1}{\pi}\displaystyle\int_{-\infty}^{\infty} \frac{e^{-i\omega t}}{t}\,dt$

$$= \frac{1}{\pi}\left[\int_{-\infty}^{\infty} \frac{\cos\omega t}{t}\,dt - i\int_{-\infty}^{\infty} \frac{\sin\omega t}{t}\,dt\right]$$

$$= \frac{1}{\pi}(0 - i\pi)$$

(using well-known results in contour integration if $\omega > 0$)

$= -i$, if $\omega > 0$

and $F[h(t)] = i$, if $\omega < 0$

i.e., $$H(\omega) = \begin{cases} i, \text{ if } & \omega < 0 \\ -i, \text{ if } & \omega > 0 \end{cases}$$

Using (10) in (9) we get

$$Z(t) = \frac{1}{\pi} \int_{-\infty}^{\infty} \frac{X(\alpha)}{t - \alpha} \, d\alpha$$

which is called the ***Hilbert transform*** of $X(t)$ and denoted as $\breve{X}(t)$.

$$\therefore \qquad \breve{X}(t) = X(t) * \frac{1}{\pi t}$$

Therefore, Property 3 of the linear time-invariant system,

$$S_{X\breve{X}}(\omega) = H^*(\omega)\, S_{xx}(\omega)$$

$$= \begin{cases} -i S_{xx}(\omega), & \text{if } \omega < 0 \\ +i S_{xx}(\omega), & \text{if } \omega > 0 \end{cases} \qquad (11)$$

and $$S_{\breve{X}X}(\omega) = H(\omega)\, S_{xx}(\omega)$$

$$= \begin{cases} i S_{xx}(\omega), & \text{if } \omega < 0 \\ -i S_{xx}(\omega), & \text{if } \omega > 0 \end{cases} \qquad (12)$$

From (11) and (12), $S_{X\breve{X}}(\omega) = -\, S_{\breve{X}X}(\omega)$ (13)

Also $S_{\breve{X}\breve{X}}(\omega) = |H(\omega)|^2\, S_{xx}(\omega)$

i.e., $S_{\breve{X}\breve{X}}(\omega) = S_{xx}(\omega)$ (14)

From (13) and (14), it follows that

$$R_{X\breve{X}}(\tau) = -\, R_{\breve{X}X}(\tau) \qquad (15)$$

and $$R_{\breve{X}\breve{X}}(\tau) = R_{xx}(\tau) \qquad (16)$$

Thus conditions (5) and (6) are satisfied by $Y(t) = \breve{X}(t)$.

Therefore, we can use $\breve{X}(t)$ in the place of $Y(t)$ for the quadrature representation of $X(t)$.

Now let $X(t) + i\breve{X}(t) = \{I(t) + iQ(t)\}\, e^{i\omega_0 t}$ (17)

Then, replacing $A(t)$, $B(t)$, α, by $I(t)$, $Q(t)$ and ω_0 respectively in (7), we get $X(t) = I(t) \cos \omega_0 t - Q(t) \sin \omega_0 t$, which is the required quadrature representation of $\{X(t)\}$.

Note 1. *I (t) and Q(t) are called the inphase and quadrature components of {X(t)}.*

2. *The quadrature representation of {X(t)} is not unique.*
3. *The quadrature representation is* ***useful****, only when {X(t)} is a zero mean WSS bandpass process.*

Noise in Communication Systems

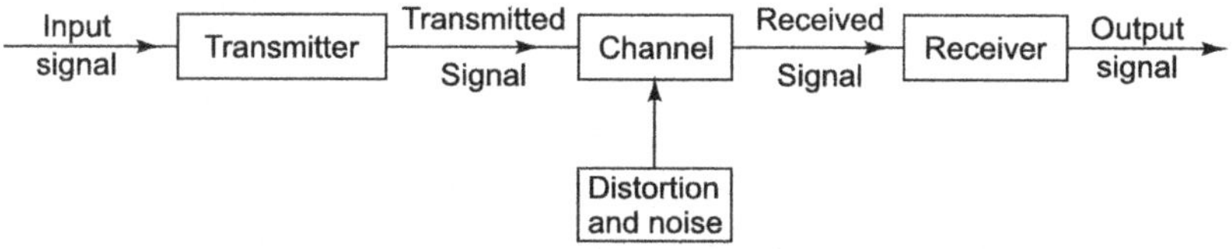

In communication systems, the message to be transmitted to a far-off location is first converted into an electrical waveform called input signal, before being sent into the transmitter. The transmitter processes and modifies the input signal for efficient transmission. The transmitter output is then sent through the channel which is just a medium such as wire, coaxial cable or optical fibre. The channel output or the received signal is then reprocessed by the receiver which sends out the output signal. The output signal is converted to its original form, namely the message.

When the message is communicated in this manner, the signal is not only distorted by the channel but also contaminated along the path by undesirable signals that are generally referred to by the term *noise*. The noise can come from many external and internal sources and take many forms.

External noise includes interfering signals from nearby sources, man-made noise generated by faulty contact switches for the electrical equipment, by ignition radiation, fluorescent lights, natural noise from lighting and extraterrestrial radiation etc. Internal noise results from thermal motion of electrons in conductors, random emission and diffusion or recombination of charged carriers in electronic devices. By careful engineering techniques, the effects of many unwanted signals can be eliminated or minimised. But there always remain certain inescapable random signals that set a limit to system performance, i.e., on the efficiency of communication.

One of the main reasons for introducing probability theory in the study of 'Signal Analysis' is the random nature of noise. Because of this randomness, it is usual to describe noise as a random process and hence in terms of a probabilitic model. Such a model describes the noise amplitude or any other parameter by means of a probability density function $f(x)$ [x represents voltage]. For many important types of noise, the density function can be determined theoretically and for others it has been estimated empirically.

Certain properties of noise, such as mean value, mean square value and the root-mean square value can be found by using the probability density function.

However the probability density function does not describe a noise waveform sufficiently so as to determine its effect on the performance of a communication system. To achieve this, it is necessary to know how the noise changes with time. This information is provided by a mean-square voltage spectrum, called the power spectrum or spectral density, that represents the distribution of signal power as a function of frequency.

Thermal Noise

Thermal noise is the noise because of the random motion of free electrons in conducting media such as a resistor. Thermal noise generated in resistors and semiconductors is assumed to be a zero mean, stationary Gaussian random process $\{N(t)\}$ with a power spectral density that is flat over a very wide range of frequencies, i.e., the graph of $S_{NN}(\omega)$ is a straight line parallel to the ω-axis. Since $S_{NN}(\omega)$ contains all frequencies in equal amount, the noise is also called ***white Gaussian noise*** or simply ***white noise*** in analogy to white light which consists of all colours.

It is customary to denote the constant spectral density of white noise by $\frac{N_0}{2}\left(\text{or } \frac{\eta}{2}\right)$.

i.e., $$S_{NN}(\omega) = \frac{N_0}{2}$$

The autocorrelation function of the white noise is given by

$$R_{NN}(\tau) = \frac{N_0}{2}\,\delta(\tau) \text{ since}$$

$$\int_{-\infty}^{\infty} \frac{N_0}{2}\,\delta(\tau)\, e^{-i\omega\tau}\, d\tau = \frac{N_0}{2}$$

The average power of the white noise $\{N(t)\}$ is given by

$$R_{NN}(0) = \int_{-\infty}^{\infty} S_{NN}(\omega)\, d\omega = \int_{-\infty}^{\infty} \frac{N_0}{2}\, d\omega \to \infty$$

Therefore, the spectral density of $\{N(t)\}$ is not physically realisable. However, since the bandwidths of real processes are always finite and since

$$\int_{-\omega_B}^{\omega_B} S_{NN}(\omega)\, d\omega = N_0\omega_B < \infty$$

for any finite bandwidth, the spectral density $S_{NN}(\omega)$ can be used over finite bandwidths.

Definition: Noise having a nonzero and constant spectral density over a finite frequency band and zero elsewhere is called ***band-limited white noise***.

i.e., if $\{N(t)\}$ is a band-limited white noise then

$$S_{NN}(\omega) = \begin{cases} \frac{N_0}{2}, & |\omega| \le \omega_B \\ 0, & \text{elsewhere} \end{cases}$$

We give below a few properties of the band-limited white noise which can be easily verified by the reader.

1. $E\{N^2(t)\} = \dfrac{N_0\,\omega_B}{2\pi}$

2. $R_{NN}(\tau) = \dfrac{N_0\,\omega_B}{2\pi}\left(\dfrac{\sin \omega_B\tau}{\omega_B\tau}\right)$

3. $N(t)$ and $N\left(t + \dfrac{k\pi}{\omega_B}\right)$ are independent, where k is a nonzero integer.

Filters

Filtering is commonly used in electrical systems to reject undesirable signals and noise and to select the desired signal. A simple example of filtering occurs when we 'tune' in a particular radio to 'select' one of many signals.

Filtering actually means selecting carefully the transfer function $H(\omega)$ in a stable, linear, time-invariant system, so as to modify the spectral components of the input signal. The system function $H(\omega)$ or the linear system itself is referred to as filter, when it does the filtering.

The commonly used filters are narrow-band filters, i.e., band pass and low pass filters.

If the system function $H(\omega)$ is defined as

$$H(\omega) \neq 0, \text{ for } \omega_0 - \varepsilon/2 < \omega < \omega_0 + \varepsilon/2$$
$$= 0, \text{ otherwise}$$

then the filter is called a ***band pass filter***.

If
$$H(\omega) \neq 0, \text{ for } -\varepsilon/2 < \omega < \varepsilon/2$$
$$= 0, \quad \text{otherwise}$$

then the filter is called a ***low pass filter***.

The equation $S_{yy}(\omega) = |H(\omega)|^2 S_{xx}(\omega)$ shows that the spectral properties of a signal can be modified by passing it through a linear time-invariant system with the appropriate transfer function. By carefully choosing $H(\omega)$, we can remove or ***filter out*** certain spectral components in the input. For example, let the input $X(t) = S(t) + N(t)$, where $S(t)$ is the signal of interest and $N(t)$ is an unwanted noise process. If the spectral densities of $\{S(t)\}$ and $\{N(t)\}$ are non-overlapping in the frequency domain, the noise $N(t)$ can be removed by passing $X(t)$ through a filter $H(\omega)$ that has a response of 1 for the range of frequencies occupied by the signal and a response of 0 for the range of frequencies occupied by the noise. But in most practical situations there is spectral overlap and the design of optimum filters to separate signal and noise is somewhat difficult. The discussion of this problem and the various optimum filters in common use such as matched filter and Wiener filter may be found in textbooks on Random Signal Analysis. It is beyond the scope of this book.

Worked Examples 7(A)

Example 1 *If $\{X(t)\}$ is a Gaussian process with $\mu(t) = 10$ and $C(t_1, t_2) = 16\, e^{-|t_1 - t_2|}$ find the probability that (i) $X(10) \le 8$ and (ii) $|X(10) - X(6)| \le 4$.*

Solution Since $\{X(t)\}$ is a Gaussian process, any member of the process is a normal RV.

Therefore, $X(10)$ is a normal RV with mean $\mu(10) = 10$ and variance $C(10, 10) = 16$.

$$P\{X(10) \le 8\} = P\left\{\frac{X(10) - 10}{4} \le -0.5\right\}$$

$$= P\{Z \le -0.5\} \qquad \text{(where } Z \text{ is the standard normal RV)}$$

$$= 0.5 - P\{0 \le Z \le 0.5\}$$

$$= 0.5 - 0.1915 \text{ (from normal tables)}$$

$$= 0.3085$$

$X(10) - X(6)$ is also a normal RV with mean $\mu(10) - \mu(6) = 10 - 10 = 0$.

$$\text{Var}\{X(10) - X(6)\} = \text{Var}\{X(10)\} + \text{Var}\{X(6)\} - 2\text{ Covar }\{X(10), X(6)\}$$

$$= C(10, 10) + C(6, 6) - 2C(10, 6)$$

$$= 16 + 16 - 2 \times 16e^{-4}$$

$$= 31.4139$$

Now $$P\{|X(10) - X(6)| \le 4\} = P\left\{\frac{|X(10) - X(6)|}{5.6048} \le \frac{4}{5.6048}\right\}$$

$$= P\{|Z| \le 0.7137\}$$

$$= 2 \times 0.2611$$

$$= 0.5222$$

Example 2 *The process $\{X(t)\}$ is normal with $\mu_t = 0$ and $R_x(\tau) = 4e^{-3|\tau|}$. Find a memoryless system $g(x)$ such that the first order density $f_y(y)$ of the resulting output $Y(t) = g[X(t)]$ is uniform in the interval (6, 9).* (MKU — Apr. 96)

Solution Since $\{X(t)\}$ is a normal process, a sample function $X(t)$ follows a normal distribution with mean 0 and variance given by $R_x(0) = 4$.

$$\therefore \qquad f_X(x) = \frac{1}{2\sqrt{2\pi}}\, e^{-x^2/8}, \quad -\infty < x < \infty$$

Now $Y(t)$ is to be uniform in (6, 9)

$$\therefore \qquad f_Y(y) = \frac{1}{3}, \; 6 < y < 9$$

Therefore, the distribution function of Y is given by

$$F_Y(y) = \int_6^y f_Y(y)\, dy$$

$$= \frac{1}{3}(y - 6) \tag{1}$$

Now $$F_Y(y) = P\{Y(t) \le y\}$$

$$= P\{g[X(t)] \le y\}$$

$$= P\{X(t) \le g^{-1}(y)\}$$

$$= P\{X(t) \le x\} \qquad \text{[since } y = g(x)\text{]}$$

$$= F_X(x)$$

But, from (1), $F_Y\{g(x)\} = \frac{1}{3}\{g(x) - 6\}$

$$\therefore \quad \frac{1}{3}\{g(x) - 6\} = F_X(x)$$

$$\therefore \quad g(x) = 6 + 3F_X(x)$$

$$= 6 + 3\int_{-\infty}^{x} \frac{1}{2\sqrt{2\pi}} e^{-x^2/8}\, dx$$

Example 3 Let Z and θ be independent RVs such that Z has a density function

$$f(z) = \begin{cases} 0 & , \text{ in } \quad z < 0 \\ z\, e^{-z^2/2} & , \text{ in } \quad z > 0 \end{cases}$$

and θ is uniformly distributed in $(0, 2\pi)$. Show that $\{X_t; -\infty < t < \infty\}$ is a Gaussian process, if $X_t = Z\cos(2\pi t + \theta)$. (MU — Apr. 96)

Solution Let us first find the density function of

$$W = \cos(2\pi t + \theta), \text{ where } f_\theta(\theta) = \frac{1}{2\pi}$$

Since $w = \cos(2\pi t + \theta)$, $\theta = \cos^{-1}(w) - 2\pi t$.

There are only two values of θ in $(0, 2\pi)$ for a given value of w. Let them be θ_1 and θ_2.

By the transformation rule (refer Chapter 3)

$$f_W(w) = f_\theta(\theta_1)\left|\frac{d\theta_1}{dw}\right| + f_\theta(\theta_2)\left|\frac{d\theta_2}{dw}\right|$$

$$= 2 \times \frac{1}{2\pi} \times \left| -\frac{1}{\sqrt{1-w^2}} \right|$$

$$= \frac{1}{\pi\sqrt{1-w^2}}, \quad |w| < 1$$

Let us now find the first order density of $X = ZW$, where X_t has been taken as X.

Introduce the auxiliary variable $Y = W$, so that we may find the joint pdf of (X, Y)

$$x = zw \text{ and } y = w$$

i.e.,

$$z = \frac{x}{y} \text{ and } w = y$$

$$\therefore \quad f_{XY}(x, y) = |J|\, f_{ZW}(z, w)$$

where $$J = \begin{vmatrix} \frac{\partial z}{\partial x} & \frac{\partial z}{\partial y} \\ \frac{\partial w}{\partial x} & \frac{\partial w}{\partial y} \end{vmatrix} = \begin{vmatrix} \frac{1}{y} & -\frac{x}{y^2} \\ 0 & 1 \end{vmatrix} = \frac{1}{|y|}$$

$\therefore$ $f_X(x)$ is the marginal density function of X.

i.e., $$f_X(x) = \int_{-1}^{1} \frac{1}{|y|} f_Z(z) f_W(w)\, dy, \text{ where } z = \frac{x}{y} \text{ and } w = y$$

(by independence of Z and W)

$$= \int_{-1}^{1} \frac{1}{|y|} \frac{x}{y} e^{-x^2/2y^2} \frac{1}{\pi\sqrt{1-y^2}}\, dy, \quad \frac{x}{y} > 0$$

$$= \begin{cases} \frac{1}{\pi}\int_{-1}^{0} -\frac{x}{y^2} e^{-x^2/2y^2} \frac{1}{\sqrt{1-y^2}}\, dy, \text{ where } x < 0 \text{ and } y < 0 & (1) \\ \frac{1}{\pi}\int_{0}^{1} \frac{x}{y^2} e^{-x^2/2y^2} \frac{1}{\sqrt{1-y^2}}\, dy, \text{ where } x > 0 \text{ and } y > 0 & (2) \end{cases}$$

Changing y to $-y'$, we note that the integral in (1) becomes

$$\frac{1}{\pi}\int_{0}^{1} \frac{x}{y'^2} e^{-x^2/2y'^2} \frac{1}{\sqrt{1-y'^2}}\, dy', \text{ which is the same as integral (2).}$$

$$\therefore f_X(x) = \frac{1}{\pi}\int_{0}^{1} \frac{x}{y^2} e^{-x^2/2y^2} \frac{1}{\sqrt{1-y^2}}\, dy, \; -\infty < x < \infty \qquad (3)$$

Put $\frac{x^2}{2y^2} = t$ in (3), treating x as a parameter,

Then $$f_X(x) = \frac{1}{\pi}\int_{x^2/2}^{\infty} \frac{1}{\sqrt{2t - x^2}} e^{-t}\, dt \qquad (4)$$

Put $t - \frac{x^2}{2} = u$ in (4), treating x as a parameter,

Then $$f_X(x) = \frac{1}{\pi\sqrt{2}}\int_{0}^{\infty} e^{-x^2/2} u^{-1/2} e^{-u}\, du$$

$$= \frac{1}{\pi\sqrt{2}} e^{-x^2/2} \Gamma\left(\frac{1}{2}\right)$$

$$= \frac{1}{\sqrt{2\pi}} e^{-x^2/2}, \quad \infty < x < \infty$$

Thus each member of the process $\{X_t\}$ follows a normal distribution with mean zero and variance 1.

Therefore, $a_1 X_{t_1} + a_2 X_{t_2} + \ldots + a_n X_{t_n}$ also follows a normal distribution, for any set of $a_1, a_2, \ldots a_n$.

Therefore $\{X_{t_1}, X_{t_2}, \ldots, X_{t_n}\}$ are jointly normal for any n

[see the Problem 2 given in Exercises 8 (a)].

Therefore, the process $\{X_t\}$ is Gaussian.

Example 4 *It is given that $R_x(\tau) = e^{-|\tau|}$ for a certain stationary Gaussian random process $\{X(t)\}$. Find the joint pdf of the RVs $X(t)$, $X(t+1)$, $X(t+2)$.*

Solution Let us denote the given RVs by $X(t_1)$, $X(t_2)$, $X(t_3)$.

The joint pdf of $\{X(t_1), X(t_2), X(t_3)\}$ is given by

$F(x_1, x_2, x_3, t_1, t_2, t_3) =$

$$\frac{1}{(2\pi)^{3/2}|\Lambda|^{1/2}} \exp\left[-\frac{1}{2|\Lambda|}\sum_{i=1}^{3}\sum_{j=1}^{3}|\Lambda|_{ij}(x_i-\mu_i)(x_j-\mu_j)\right]$$

where $\mu_i = E\{X(t_i)\}$ and Λ is the third order square matrix (λ_{ij}), where $\lambda_{ij} = C\{X(t_i), X(t_j)\}$ and $|\Lambda|_{ij}$ = cofactor of λ_{ij} in $|\Lambda|$.

$$E\{X(t)\} = \sqrt{\lim_{\tau\to\infty} R_x(\tau)} = \sqrt{\lim_{\tau\to\infty} e^{-|\tau|}} = 0$$

$$\therefore \quad \lambda_{ij} = C\{X(t_i)\, X(t_j)\} = R(t_i - t_j)$$

$$\therefore \quad \lambda_{11} = R(t_1, t_1) = R(t, t) = R(0) = 1$$

$$\lambda_{12} = R(t, t+1) = R(1) = e^{-1} \text{ etc.}$$

$$\therefore \quad \Lambda = \begin{pmatrix} 1 & \frac{1}{e} & \frac{1}{e^2} \\ \frac{1}{e} & 1 & \frac{1}{e} \\ \frac{1}{e^2} & \frac{1}{e} & 1 \end{pmatrix} \text{ and } |\Lambda| = \left(1-\frac{1}{e^2}\right)^2$$

$$|\Lambda|_{11} = 1 - \frac{1}{e^2} \quad |\Lambda|_{12} = -\frac{1}{e} + \frac{1}{e^3} \quad |\Lambda|_{13} = 0 \text{ etc.}$$

Therefore, the required joint pdf is given by

$$f(x_1, x_2, x_3) = \frac{1}{(2\pi)^{3/2}\left(1-\frac{1}{e^2}\right)} \exp$$

$$\left[-\frac{1}{2\left(1-\frac{1}{e^2}\right)^2}\left\{\left(1-\frac{1}{e^2}\right)x_1^2 - \frac{2}{e}\left(1-\frac{1}{e^2}\right)x_1x_2 + \left(1-\frac{1}{e^4}\right)x_2^2\right.\right.$$

$$\left.\left. - \frac{2}{e}\left(1-\frac{1}{e^2}\right)x_2x_3 + \left(1-\frac{1}{e^2}\right)x_3^2\right\}\right]$$

i.e., $$f(x_1, x_2, x_3) = \frac{1}{(2\pi)^{3/2}\left(1 - \frac{1}{e^2}\right)} \exp$$

$$\left[-\frac{1}{2\left(1 - \frac{1}{e^2}\right)} \left\{ x_1^2 - \frac{2}{e} x_1 x_2 + \left(1 + \frac{1}{e^2}\right) x_2^2 - \frac{2}{e} x_2 x_3 + x_3^2 \right\} \right]$$

Example 5 *If $\{Y(t)\}$ is the square law detector process and if $Z(t) = Y(t) - E\{Y(t)\}$, show that the spectral density of $\{Z(t)\}$ is given by $S_{zz}(\omega) = \frac{1}{\pi}\int_{-\infty}^{\infty} S_{xx}(\alpha)\, S_{xx}(\omega - \alpha)\, d\alpha$, where $S_{xx}(\omega)$ is the input spectral density.*

Note *$Z(t)$ is called the fluctuation of the square law detector.*

$$E\{Z(t)\, Z(t-\tau)\} = E[\{(Y(t) - E[Y(t)]\}\, \{Y(t-\tau) - E[Y(t-\tau)\}]$$
$$= E\{Y(t)\, Y(t-\tau)\} - E\{Y(t)\}\, E\{Y(t-\tau)\}$$

i.e., $$R_{zz}(\tau) = R_{yy}(\tau) - E\{Y(t)\}\, E\{Y(t-\tau)\}$$
$$= R_{xx}^2(0) + 2R_{xx}^2(\tau) - R_{xx}^2(0)$$

[see 'square law detector process']

$$= 2R_{xx}^2(\tau)$$

Taking Fourier transforms,

$$S_{zz}(\omega) = \frac{1}{\pi}\int_{-\infty}^{\infty} S_{xx}(\alpha)\, S_{xx}(\omega - \alpha)\, d\alpha$$

Example 6 *If $\{X(t)\}$ is a band pass process, prove that its inphase and quadrature components are low pass processes (refer to the book-work on quadrature representation of a WSS process).*

Solution If we take the input of a system as $X(t)$ and the impulse response function

$$h(t) = \delta(t) + i\frac{1}{\pi t}$$

then the output of the system is given by

$$Z(t) = \int_{-\infty}^{\infty} x(\alpha)\left[\delta(t-\alpha) + i\frac{1}{\pi(t-\alpha)}\right] d\alpha$$

$$= \int_{-\infty}^{\infty} X(t-\alpha)\, \delta(\alpha)\, d\alpha + \frac{i}{\pi}\int_{-\infty}^{\infty} \frac{X(\alpha)}{t-\alpha}\, d\alpha$$

$$= X(t) + i\breve{X}(t),$$

[by the property of $\delta(t)$ and by the definition of Hilbert transform]

$$\therefore \quad S_{zz}(\omega) = |H(\omega)|^2 S_{xx}(\omega)$$

where $$H(\omega) = F\{h(t)\}$$

$$= F[\delta(t)] + iF\left[\frac{1}{\pi t}\right]$$

$$= \begin{cases} 1 + i(i) & \text{, if } \omega < 0 \\ 1 + i(-i) & \text{, if } \omega > 0 \end{cases}$$

$$= \begin{cases} 0 & \text{, if } \omega < 0 \\ 2 & \text{, if } \omega > 0 \end{cases}$$

$$= 2U(\omega),$$

where $U(\omega)$ is the unit step function.

$$\therefore \quad S_{zz}(\omega) = 4S_{xx}(\omega)\, U(\omega) \tag{1}$$

Also $Z(t)\, Z^*(t - \tau) = \{X(t) + i\breve{X}(t)\}\, \{X(t - \tau) - i\breve{X}(t - \tau)\}$

$$= X(t)X(t - \tau) + \breve{X}(t)\breve{X}(t - \tau) + i\{\breve{X}(t)X(t - \tau) - X(t)\breve{X}(t - \tau)\}$$

$$\therefore \quad R_{zz}(\tau) = R_{xx}(\tau) + R_{\breve{X}\breve{X}}(\tau) + i\{R_{\breve{X}X}(\tau) - R_{X\breve{X}}(\tau)\}$$

$$= 2\{R_{xx}(\tau) + iR_{\breve{X}X}(\tau)\}, \tag{2}$$

by steps (15) and (16) of the book-work

$$\therefore \quad S_{zz}(\omega) = 2\{S_{xx}(\omega) + \mathrm{i}S_{\breve{x}x}(\omega)\} \tag{3}$$

From step (17) of the book-work

$$I(t) + iQ(t) = \{X(t) + i\breve{X}(t)\}e^{-i\omega_0 t}$$

$$\therefore \quad I(t) = X(t)\cos\omega_0 t + \breve{X}(t)\sin\omega_0 t \tag{4}$$

$$Q(t) = \breve{X}(t)\cos\omega_0 t - X(t)\sin\omega_0 t \tag{5}$$

From (4) and (5),

$$R_{II}(\tau) = R_{xx}(\tau)\cos\omega_0\tau + R_{\breve{X}X}(\tau)\sin\omega_0\tau$$

$$R_{QI}(\tau) = -R_{xx}(\tau)\sin\omega_0\tau + R_{\breve{X}X}(\tau)\cos\omega_0\tau$$

$$\therefore \quad R_{II}(\tau) + iR_{QI}(\tau) = [R_{xx}(\tau) + iR_{\breve{X}X}(\tau)]\, e^{-i\omega_0\tau}$$

$$= \frac{1}{2} R_{zz}(\tau)\, e^{-i\omega_0\tau}$$

$$\therefore \quad R_{II}(\tau) = \frac{1}{2} Rl\, \{R_{zz}(\tau)\, e^{-i\omega_0\tau}\}$$

$$\therefore \quad S_{II}(\omega) = \frac{1}{2} Rl\, \{S_{zz}(\omega + \omega_0)\}$$

$$= S_{xx}(\omega + \omega_0) \tag{6}$$

Changing τ into $-\tau$ in (2),

$$R_{zz}(-\tau) = 2\{R_{xx}(-\tau) + i\,R_{\breve{X}X}(-\tau)\}$$

$$= 2\{R_{xx}(\tau) - i\,R_{\breve{X}X}(\tau)\} \tag{7}$$

From (2) and (7)

$$R_{xx}(\tau) = \frac{1}{4}\{R_{zz}(\tau) + R_{zz}(-\tau)\}$$

$$\therefore \qquad S_{xx}(\omega) = \frac{1}{4}\{S_{zz}(\omega) + S_{zz}(-\omega)\} \tag{8}$$

Using (8) in (6),

$$S_{II}(\omega) = \frac{1}{4}\{S_{zz}(\omega + \omega_0) + S_{zz}(-\omega + \omega_0)$$

$$= S_{xx}(\omega + \omega_0)U(\omega + \omega_0) + S_{xx}(-\omega + \omega_0)$$

$$U(-\omega + \omega_0) \quad [\text{by (1)}] \tag{9}$$

Since $\qquad R_{QQ}(\tau) = R_{II}(\tau),\ S_{QQ}(\omega) = S_{II}(\omega).$

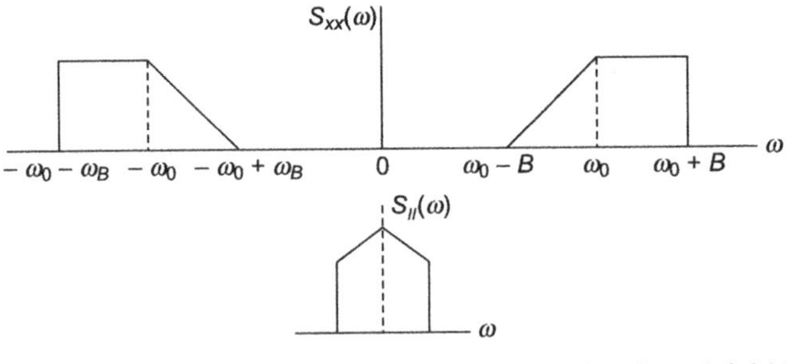

Therefore, when $\{X(t)\}$ is a band pass process, both $\{I(t)\}$ and $\{Q(t)\}$ are low pass processes as given in the above figures.

Example 7 Obtain the autocorrelation for an ideal low pass stochastic process. (MSU — Apr. 96; Nov. 96)

Solution Let the spectral density function of the low pass process $\{X(t)\}$ be $S_{xx}(\omega)$, in $|\omega| < \omega_B$.

Let the complex form of Fourier series of $S_{xx}(\omega)$ in $(-\omega_B, \omega_B)$ be

$$S_{xx}(\omega) = \sum_{n=-\infty}^{\infty} c_n\, e^{in\pi\omega/\omega_B} \tag{1}$$

where c_n is given by

$$c_n = \frac{1}{2\omega_B}\int_{-\omega_B}^{\omega_B} S_{xx}(\omega)\, e^{-in\pi\omega/\omega_B}\, d\omega \tag{2}$$

Taking the inverse Fourier transform of (1)

$$R_{xx}(\tau) = \frac{1}{2\pi}\int_{-\infty}^{\infty} \sum c_n \, e^{in\pi\omega'/\omega_B} \, e^{i\tau\omega'} \, d\omega'$$

$$= \sum \frac{1}{2\pi}\int_{-\omega_B}^{\omega_B} \frac{1}{2\omega_B}\int_{-\omega_B}^{\omega_B} S_{xx}(\omega)\, e^{-in\pi\omega/\omega_B}\, d\omega$$

$$e^{i\left(\frac{n\pi}{\omega_B}+\tau\right)\omega'} \, d\omega' \qquad \text{[since } \{X(t)\} \text{ is low pass]}$$

$$= \sum \frac{1}{2\omega_B}\int_{-\omega_B}^{\omega_B} R_{xx}\left(-\frac{n\pi}{\omega_B}\right) e^{i\left(\frac{n\pi}{\omega_B}\right)\omega'} \, \omega' \, d\omega'$$

$$= \sum R_{xx}\left(-\frac{n\pi}{\omega_B}\right)\frac{1}{\omega_B}\left\{\frac{\sin\left(\frac{n\pi}{\omega_B}+\tau\right)\omega'}{\left(\frac{n\pi}{\omega_B}+\tau\right)}\right\}_0^{\omega_B}$$

$$= \sum_{n=-\infty}^{\infty} R_{xx}\left(-\frac{n\pi}{\omega_B}\right)\frac{\sin\left(\frac{n\pi}{\omega_B}+\tau\right)\omega_B}{\left(\frac{n\pi}{\omega_B}+\tau\right)\omega_B}$$

$$= \sum_{n=-\infty}^{\infty} R_{xx}\left(\frac{n\pi}{\omega_B}\right)\frac{\sin\left(\tau-\frac{n\pi}{\omega_B}\right)\omega_B}{\left(\tau-\frac{n\pi}{\omega_B}\right)\omega_B} \qquad \text{(Changing } n \text{ to } -n\text{)}$$

Let us assume that the values of $X(t)$ at $t = nT$, $n = \ldots -3, -2, -1, 0, 1, 2, 3, \ldots$ are given, using which we can construct $X(t)$, where $T = \dfrac{\pi}{\omega_B}$.

$$\therefore \qquad R_{xx}(\tau) = \sum_{n=-\infty}^{\infty} R_{xx}(nT)\,\frac{\sin \omega_B(\tau - nT)}{\omega_B(\tau - nT)}$$

Thus, when $\{X(t)\}$ is a low pass process, its autocorelation is found out by summation.

Example 8 *If $\{X(t)\}$ is a band limited process such that $S_{xx}(\omega) = 0$, when $|\omega| > \sigma$, prove that*

$2[R_{xx}(0) - R_{xx}(\tau)] \le \sigma^2 \tau^2 R_{xx}(0)$.

Solution $\quad R_{xx}(\tau) = \dfrac{1}{2\pi}\displaystyle\int_{-\infty}^{\infty} S_{xx}(\omega)\, e^{i\tau\omega}\, d\omega$

$$= \frac{1}{2\pi} \int_{-\infty}^{\infty} S_{xx}(\omega) \cos \tau\omega \, d\omega \qquad \text{[since } S_{xx}(\omega) \text{ is even]}$$

$$R_{xx}(0) - R_{xx}(\tau) = \frac{1}{2\pi} \int_{-\sigma}^{\sigma} S_{xx}(\omega)(1 - \cos \tau\omega)\, d\omega \, \text{[since } \{X(t)\} \text{ is band limited]}$$

$$= \frac{1}{2\pi} \int_{-\sigma}^{\sigma} S_{xx}(\omega) \times 2 \sin^2\left(\frac{\tau\omega}{2}\right) d\omega \qquad (1)$$

From trigonometry, $|\sin \theta| \leq \theta$

$\therefore \qquad \sin^2 \theta \leq \theta^2$

$$\therefore \qquad 2 \sin^2\left(\frac{\tau\omega}{2}\right) \leq \frac{\tau^2\omega^2}{2} \qquad (2)$$

Inserting (2) in (1)

$$R_{xx}(0) - R_{xx}(\tau) \leq \frac{1}{2\pi} \int_{-\sigma}^{\sigma} S_{xx}(\omega) \frac{\tau^2\omega^2}{2} d\omega$$

$$\leq \frac{\sigma^2\tau^2}{4\pi} \int_{-\sigma}^{\sigma} S_{xx}(\omega)\, d\omega$$

$$\leq \frac{\sigma^2\tau^2}{4\pi} \int_{-\infty}^{\infty} S_{xx}(\omega)\, d\omega$$

i.e., $$\leq \frac{\sigma^2\tau^2}{2} R_{xx}(0)$$

Example 9 *Consider a white Gaussian noise of zero mean and power spectral density $N_0/2$ applied to a low pass RC filter whose transfer function is $H(f) = \frac{1}{1 + i2\pi f RC}$. Find the autocorrelation function of the output random process.*

(BU — Nov. 96)

Solution The simple RC-circuit for which the transfer function is given is a linear time-invariant system.

The power spectral densities of the input $\{X(t)\}$ and the output $\{Y(t)\}$ of a linear system are connected by

$$S_{YY}(\omega) = |H(\omega)|^2 S_{xx}(\omega)$$

In the problem the transfer function is expressed in terms of the frequency f. Therefore, the above relation is

$$S_{yy}(f) = |H(f)|^2 S_{xx}(f)$$

$$= \frac{1}{1 + 4\pi^2 f^2 R^2 C^2} \frac{N_0}{2} \text{(since the input is a white noise)}$$

$$\therefore \quad R_{yy}(\tau) = \frac{N_0}{2} \int_{-\infty}^{\infty} \frac{e^{i2\pi\tau f}}{1 + 4\pi^2 f^2 R^2 C^2} \, df$$

$$= \frac{N_0}{8\pi^2 R^2 C^2} \int_{-\infty}^{\infty} \frac{e^{i(2\pi\tau)f} \, df}{\left(\frac{1}{2\pi RC}\right)^2 + f^2} \tag{1}$$

Compare the integral in (1) with $\int_{-\infty}^{\infty} \frac{e^{imx} dx}{a^2 + x^2}$, which can be evaluated by contour integration technique [refer to the worked example in the section 'power spectral density'].

$$\int_{-\infty}^{\infty} \frac{e^{imx} dx}{a^2 + x^2} = \frac{\pi}{a} e^{-|m|a} \tag{2}$$

Using (2) in (1),

$$R_{yy}(\tau) = \frac{N_0}{8\pi^2 R^2 C^2} \pi \times 2\pi RC \, e^{-|2\pi\tau|/2\pi RC}$$

$$= \frac{N_0}{4RC} e^{-|\tau|/RC}$$

The mean square value of $\{Y(t)\}$ is given by

$$E\{Y^2(t)\} = R_{yy}(0) = \frac{N_0}{4RC}$$

Example 10 *If $Y(t) = A \cos(\omega_0 t + \theta) + N(t)$, where A is a constant, θ is a random variable with a uniform distribution in $(-\pi, \pi)$ and $\{N(t)\}$ is a band-limited Gaussian white noise with a power spectral density*

$$S_{NN}(\omega) = \begin{cases} \frac{N_0}{2}, & \text{for } |\omega - \omega_0| < \omega_B \\ 0, & \text{elsewhere} \end{cases}$$

find the power spectral density of $\{Y(t)\}$. Assume that $N(t)$ and θ are independent.

Solution $Y(t_1)\,Y(t_2) = \{A\cos(\omega_0 t_1 + \theta) + N(t_1)\}\,\{A\cos(\omega_0 t_2 + \theta) + N(t_2)\}$

$$= A^2 \cos(\omega_0 t + \theta)\cos(\omega_0 t_2 + \theta) + N(t_1)\,N(t_2) + A\cos(\omega_0 t_1 + \theta)\,N(t_2) + A\cos(\omega_0 t_2 + \theta)\,N(t_1)$$

$$\therefore \quad R_{YY}(t_1, t_2) = A^2 E\{\cos(\omega_0 t_1 + \theta)\cos(\omega_0 t_2 + \theta) + R_{NN}(t_1, t_2) + AE\{\cos(\omega_0 t_1 + \theta)\}\, E\{N(t_2)\} + AE\{\cos(\omega_0 t_2 + \theta)\}\, E\{N(t_1)\} \text{ (by independence)}$$

i.e., $$R_{YY}(\tau) = \frac{A^2}{2}\cos\omega_0\tau + R_{NN}(\tau) \qquad [\text{since } \{N(t)\} \text{ is stationary}]$$

$$\therefore \quad S_{YY}(\omega) = \frac{A^2}{2}\int_{-\infty}^{\infty} \cos \omega_0 \tau\, e^{-i\omega\tau}\, d\tau + S_{NN}(\omega)$$

$$= \frac{\pi A^2}{2}\{\delta(\omega - \omega_0) + \delta(\omega + \omega_0)\} + S_{NN}(\omega)$$

where $S_{NN}(\omega)$ is given in the problem.

Exercise 7(A)

Part A (Short answer questions)

1. Define a Gaussian process.
2. When is a random process said to be normal?
3. What is the output of (i) a dc meter and (ii) a true rms meter when it measures a voltage $V(t)$ which is a Gaussian ergodic random process with a mean of zero and variance of 4 V?
4. Give the nth order density of Gaussian process (or) the nth order normal density function.
5. What is the importance of Gaussian process in signal analysis?
6. State the properties of a Gaussian process.
7. If a Gaussian process is WSS, prove that it is also SSS.
8. If the member functions of a Gaussian process are uncorrelated, prove that they are independent.
9. Name a few random processes that are defined in terms of stationary Gaussian process.
10. Define square law detector process.
11. Define Full-wave linear detector process.
12. Define Half-wave linear detector process.
13. Define Hard limiter process.
14. Define a band pass process or band pass signal.
15. Define a low pass process or an ideal low pass process.
16. Define a narrow band process (or) when is a random process said to be quasimonochromatic?
17. When is a random process called monochromatic?
18. Give the representation of a narrow band Gaussian process $\{X(t)\}$ that is frequently used in signal analysis.
19. What is the justification for the representation of a narrow band Gaussian process $\{X(t)\}$ in the form
   $X(t) = R_x(t) \cos \{\omega_0 \pm \theta_x(t)\}$
20. Define the envelope and phase of a narrow-band Gaussian process.
21. What are the distributions followed by the envelope and phase of a narrow-band Gaussian process?

22. Define the inphase and quadrature components of a narrow-band Gaussian process.
23. What is the nature of the inphase and quadrature components of a band pass process?
24. What do you mean by the quadrature representation of a WSS process?
25. What kind of random processes can be represented in the quadrature form?
26. For what kind of random processes is the quadrature representation useful in communication theory?
27. What do you mean by 'noise' in signal transmission?
28. How are external and internal noises caused?
29. What is thermal noise? By what type of random processes is it represented?
30. If $\{N(t)\}$ is a thermal noise, what is the nature of the graph of $S_{NN}(\omega)$?
31. Why is thermal noise called white noise?
32. If the input to a linear time-invariant system is white noise $\{N(t)\}$, what is PSDF of the output?
33. If the PSD of white noise is $\frac{N_0}{2}$, find its ACF.
34. Find the average power (or) the mean square value of the white noise $\{N(t)\}$.
35. Why is the spectral density of the white noise $\{N(t)\}$ not physically realisable? How is this difficulty overcome?
36. Define band-limited white noise.
37. State a few properties of band-limited white noise.
38. Find the ACF of the band-limited white noise.
39. Find the average power of the band-limited white noise.
40. What is meant by a filter in electrical systems.
41. Explain band pass and low pass filters.
42. How is filtering done in electrical systems?

Part B

43. Suppose that $\{X(t)\}$ is a normal process with $\eta\,(t) = 3$ and $C(t_1, t_2) = 4e^{-0.2|t_1 - t_2|}$.
    (i) Find the probability that $X(5) \le 2$ and
    (ii) Find the probability that $|X(8) - X(5)| \le 1$ (BDU — Apr. 96)
44. Prove that the RVs $X_1, X_2, \ldots, X_n$ are jointly normal, if the sum $a_1X_1 + a_2X_2 + \ldots + a_nX_n$ is a normal RV for any set of constants $a_1, a_2, \ldots, a_n$.
    [**Hint:** From the given condition

$$Z = \omega_1 X_1 + \omega_2 X_2 + \ldots + \omega_n X_n \text{ is a normal RV}$$

$$E(Z) = 0 \text{ and } E(Z^2) = \sigma_z^2 \text{ if } E(X_i) = 0$$

$$\phi(Z) = e^{-\sigma_z^2/2}]$$

45. Let $Y_n = \sum_{k=1}^{n} X_k$, where the X_k's are a set of independent RVs each normally distributed with mean μ and variance σ^2. Show that $\{Y_n; n = 1, 2, \ldots\}$ is a normal (Gaussian) process. (BDU — Nov. 96)
46. Given a normal process $\{X(t)\}$ with $\mu_x = 0$ and $R_x(t) = 4e^{-2|\tau|}$, we form the RVs $Z = X(t + 1)$, $W = X(t - 1)$.
    (i) Find $E(ZW)$ and $E\{(Z + W)^2\}$ and
    (ii) Find $f_z(z)$, $P\{Z < 1\}$ and $f_{zw}(z, w)$.
47. Show that, if the RVs X, Y, Z are jointly normal and independent in pairs, then they are independent.
    [**Hint:** Prove that $f_{xyz}(x, y, z) = f_x(x)\, f_y(y)\, f_z(z)$.]
48. A voltage $V(t)$, which is Gaussian ergodic random process with a mean of zero and a variance of 4 V, is measured (i) by a dc meter, (ii) a true rms meter and (iii) a meter which first squares $V(t)$ and then read its dc component. Find the output of each meter and justify your answer. (BU — Apr. 96)
    [**Hint:** If $\{X(t)\}$ is an ergodic process such as a voltage waveform, μ_x can be measured by using a dc voltmeter and σ_x can be measured by using a true rms (ac coupled) voltmeter.]
49. $\{X(t)\}$ is a Gaussian random process with mean $\mu_x(t)$ and autocorrelation function $R_{xx}(t_1, t_2)$. Find $E\{X(t_2)/X(t_1)\}$, $t_1 < t_2$.
50. If $Z(t) = X \cos \omega_0 t + Y \sin \omega_0 t$, where X and Y are independent Gaussian RVs with zero mean and unit variance and ω_0 is a constant,
    (i) Show that $\{Z(t)\}$ is a Gaussian random process.
    (ii) Find the joint pdf of $Z(t_1)$ and $Z(t_2)$.
    (iii) Is the process WSS?
    (iv) Is the process SSS?
    (v) Find $E\{Z(t_2)/Z(t_1)\}$, $t_2 > t_1$.
51. If $\{X(t)\}$ is a Gaussian process with $\mu_x = 0$ and $R_{xx}(\tau) = 0$ for $|\tau| > a$, prove that it is correlation-ergodic.
52. If $\{X(t)\}$ is a zero mean low pass process with a bandwidth of ω_B, prove that
$$E\{X(t + \tau) - X(t)\}^2 \le \omega^2{}_B\, \tau^2\, E\{X^2(t)\}$$
    [**Hint:** Refer to Worked Example 8.]
53. If $\{X(t)\}$ is a band limited low pass process with bandwidth ω_B, prove that
$$R(\tau) \ge R(0) \cos \omega_B \tau, \text{ for } |\tau| < \frac{\pi}{2\omega_B}.$$
54. For a narrow band process $X(t) = X_c(t) \cos \omega_0 t + X_s(t) \sin \omega_0 t$, where $\{X_c(t)\}$ and $\{X_s(t)\}$ are stationary, uncorrelated, lowpass processes with
$$S_{x_c x_c}(\omega) = S_{x_s x_s}(\omega) = \begin{cases} g(\omega), & |\omega| < \omega_B \\ 0, & |\omega| > \omega_B \end{cases}$$

show that $S_{xx}(\omega) = \frac{1}{2}\{g(\omega - \omega_0\} + g(\omega + \omega_0)\}$.

55. Determine the autocorrelation of white noise. (BDU — Apr. 97)
(See the property of white noise)

56. Show that the narrow band noise $\{n(t)\}$ can be represented as $n(t) = n_c(t) \cos\omega_c t - n_s(t) \sin\omega_c t$, where $n_c(t)$ and $n_s(t)$ are inphase and quadrature phase components of $n(t)$ and ω_c is the centre frequency of the band.
(BU — Apr. 96; MU — Nov. 96; BU — Nov. 96)
[**Hint:** This is the same as the book-work on 'quadrature representation' of a WSS process. $X(t)$, $I(t)$ and $\theta(t)$ in the book-work must be replaced by $n(t)$, $n_c(t)$ and $n_s(t)$ respectively].

57. (i) If $\{N(t)\}$ is a band limited white noise such that

$$S_{NN}(\omega) = \frac{N_0}{2}, \text{ for } |\omega| < \omega_B$$
$$= 0, \quad \text{elsewhere}$$

find the autocorrelation of $\{N(t)\}$.

(ii) If $\{N(t)\}$ is a band limited white noise centered at a carrier frequency ω_0 such that

$$S_{NN}(\omega) = \frac{N_0}{2}, \text{ for } |\omega - \omega_0| < \omega_B$$
$$= 0, \quad \text{elsewhere}$$

find the autocorrelation of $\{N(t)\}$.

58. A source of noise is a Gaussian with a mean of 0.4 V and a standard deviation of 0.15 V. For what percentage of time would you expect the measured noise voltage to exceed 0.7 V?
[**Hint:** Noise is a normal RV with the given parameters.]

59. If $\{N(t)\}$ is a band-pass white noise and $\{N_c(t)\}$ and $\{N_s(t)\}$ are its quadrature components, prove that rms values of $\{N(t)\}$, $\{N_c(t)\}$ and $\{N_s(t)\}$ are equal.

60. If $X(t) = A \cos \omega_0 t + B \sin \omega_0 t$, find the Hilbert transform of $X(t)$.

61. If the input to a linear time-invariant system is a zero mean, white Gaussian process $\{N(t)\}$ and $\{Y(t)\}$ is the output, prove that (*i*) $E\{Y(t)\} = 0$,

(ii) $R_{YY}(\tau) = \frac{N_0}{2} \delta(t) * h(\tau) * h(-\tau)$, and

(iii) $S_{YY}(\omega) = \frac{N_0}{2} |H(\omega)|^2$.

62. The impulse response of a low pass filter is $\alpha e^{-\alpha t} U(t)$; where $\alpha = \frac{1}{RC}$.
If a zero mean, white Gaussian process $\{N(t)\}$ is input into this filter, find the mean square value and autocorrelation function of the output.

Poisson Process

There are many practical situations where the random times of occurrences of some specific events are of primary interest. For example, we may want to study the times at which components fail in a large system or the times at which jobs enter the queue in a computer system or the times of arrival of phone calls at an exchange or the times of emission of electrons from the cathode of a vacuum tube. In these examples, our main interest may not be the event itself but the sequence of random time instants at which the events occur. An ensemble of discrete sets of points from the time domain called a ***point process*** is used to model and analyse phenomena such as the ones mentioned above. An independent increments point process, i.e., a point process with the property that the number of occurrences in any finite collection of nonoverlapping time intervals are independent RVs, leads to a Poisson process.

Definition: If $X(t)$ represents the number of occurrences of a certain event in $(0, t)$, then the discrete random process $\{X(t)\}$ is called the Poisson process, provided the following postulates are satisfied:

(i) $P[1 \text{ occurrence in } (t, t + \Delta t)] = \lambda \Delta t + 0(\Delta t)$

(ii) $P[0 \text{ occurrence in } (t, t + \Delta t)] = 1 - \lambda \Delta t + 0(\Delta t)$

(iii) $P[2 \text{ or more occurrences in } (t, t + \Delta t) = 0(\Delta t)$

(iv) $X(t)$ is independent of the number of occurrences of the event in any interval prior and after the interval $(0, t)$.

(v) The probability that the event occurs a specified number of times in $(t_0, t_0 + t)$ depends only on t, but not on t_0.

Probability Law for the Poisson Process {X(t)}

Let λ be the number of occurrences of the event in unit time.

Let $\quad P_n(t) = P\{X(t) = n\}$

$$\therefore \quad P_n(t + \Delta t) = P\{X(t + \Delta t) = n\}$$

$$= P\{(n-1) \text{ calls in } (0, t) \text{ and } 1 \text{ call in } (t, t + \Delta t)\} + P\{n \text{ calls in } (0, t) \text{ and no call in } (t, t + \Delta t)\}$$

$$= P_{n-1}(t)\, \lambda \Delta t + P_n(t)\,(1 - \lambda \Delta t) \text{ (by the postulates)}$$

$$\therefore \frac{P_n(t + \Delta t) - P_n(t)}{\Delta t} = \lambda \{P_{n-1}(t) - P_n(t)\}$$

Taking the limits as $\Delta t \to 0$

$$\frac{d}{dt} P_n(t) = \lambda \{P_{n-1}(t) - P_n(t)\} \tag{1}$$

Let the solution of the equation (1) be

$$P_n(t) = \frac{(\lambda t)^n}{\lfloor n} f(t) \tag{2}$$

Differentiating (2) with respect to t,

$$P_n'(t) = \frac{\lambda^n}{\lfloor n} \{nt^{n-1} f(t) + t^n f'(t)\} \tag{3}$$

Using (2) and (3) in (1),

$$\frac{\lambda^n}{\lfloor n} t^n f'(t) = -\lambda \frac{(\lambda t)^n}{\lfloor n} f(t)$$

i.e., $$f'(t) = -\lambda f(t)$$

$\therefore$ $$f(t) = ke^{-\lambda t} \tag{4}$$

From (2), $$f(0) = P_0(0) = P\{X(0) = 0\}$$
$$= P\{\text{no event occurs in } (0, 0)\}$$
$$= 1 \tag{5}$$

Using (5) in (4), we get $k = 1$ and hence

$$f(t) = e^{-\lambda t} \tag{6}$$

Using (6) in (2),

$$P_n(t) = P\{X(t) = n\} = \frac{e^{-\lambda t} (\lambda t)^n}{\lfloor n}, \; n = 0, 1, 2, \ldots, \infty$$

Thus the probability distribution of $X(t)$ is the Poisson distribution with parameter λt.

Note *We have assumed that the rate of occurrence of the event λ is a constant, but it can be function of t also. When λ is a constant, the process is called a* ***homogeneous Poisson process****. Unless specified otherwise, the Poisson process will be assumed homogeneous.*

Second-Order Probability Function of a Homogeneous Poisson Process

$$P[X(t_1) = n_1, X(t_2) = n_2]$$
$$= P[X(t_1) = n_1]\, P[X(t_2) = n_2/X(t_1) = n_1], \quad t_2 > t_1$$
$$= P[X(t_1) = n_1]\, P[\text{the event occurs } (n_2 - n_1) \text{ times in the interval of } (t_2 - t_1)]$$
$$= \frac{e^{-\lambda t_1} (\lambda t_1)^{n_1}}{\lfloor n_1} \frac{e^{-\lambda(t_2 - t_1)} \{\lambda(t_2 - t_1\}^{n_2 - n_1}}{\lfloor n_2 - n_1}, \text{ if } n_2 \geq n_1$$

$$= \begin{cases} \dfrac{e^{-\lambda t_2} \lambda_{t_1}^{n_1} t_1^{n_1} (t_2 - t_1)^{n_2 - n_1}}{\lfloor n_1 \cdot \lfloor n_2 - n_1}, & n_2 \geq n_1 \\ 0, & \text{otherwise} \end{cases}$$

Proceeding similarly, we can get the third-order probability function as

$$P[X(t_1) = n_1, X(t_2) = n_2, X(t_3) = n_3]$$

$$= \begin{cases} \dfrac{e^{-\lambda t_3} \lambda^{n_3} t_1^{n_1} (t_2 - t_1)^{n_2 - n_1} (t_3 - t_2)^{n_3 - n_2}}{\lfloor n_1 \lfloor n_2 - n_1 \lfloor n_3 - n_2}, & n_3 \geq n_2 \geq n_1 \\ 0 & , \text{otherwise} \end{cases}$$

Mean and Autocorrelation of the Poisson Process

The probability law of the Poisson process $\{X(t)\}$ is the same as that of a Poisson distribution with parameter λt.

$$\therefore \quad E\{X(t)\} = \text{Var}\{X(t)\} = \lambda t$$

$$\therefore \quad E\{X_2(t)\} = \lambda t + \lambda^2 t^2 \qquad (1)$$

$$R_{xx}(t_1, t_2) = E\{X(t_1)\, X(t_2)\}$$

$$= E[X(t_1)\, \{X(t_2) - X(t_1) + X(t_1)\}]$$

$$= E[X(t_1)\, \{X(t_2) - X(t_1)\}] + E\{X^2(t_1)\}$$

$$= E[X(t_1)]\, E[X(t_2) - X(t_1)] + E\{X^2(t_1)\},$$

since $\{X(t)\}$ is a process of independent increments.

$$= \lambda t_1, \lambda\,(t_2 - t_1) + \lambda t_1 + \lambda t^2{}_1, \text{ if } t_2 \geq t_1 \qquad \text{[by (1)]}$$

$$= \lambda^2 t_1 t_2 + \lambda t_1, \text{ if } t_2 \geq t_1$$

or $\quad R_{xx}(t_1, t_2) = \lambda^2 t_1 t_2 + \lambda \min\,(t_1, t_2)$

$$C_{xx}(t_1, t_2) = R_{xx}(t_1, t_2) - E\{X(t_1)\}\, E\{X(t_2)\}$$

$$= \lambda^2 t_1 t_2 + \lambda t_1 - \lambda^2 t_1 t_2$$

$$= \lambda t_1, \text{ if } t_2 \geq t_1$$

or $\quad = \min\,(t_1, t_2)$

$$r_{xx}(t_1, t_2) = \frac{C_{xx}(t_1, t_2)}{\sqrt{\text{var}\,\{X(t_1)\}\ \text{var}\,\{X(t_2)\}}}$$

$$= \frac{\lambda t_1}{\sqrt{\lambda t_1\, \lambda t_2}} = \sqrt{\frac{t_1}{t_2}}, \text{ if } t_2 \geq t_1$$

Note *Poisson process is not a stationary process.*

Properties of Poisson Process

1. The Poisson process is a Markov process.

Proof Consider $P[X(t_3) = n_3 / X(t_2) = n_2, X(t_1) = n_1]$

$$= \frac{P[X(t_1) = n_1, X(t_2) = n_2, X(t_3) = n_3]}{P[X(t_1) = n_1, X(t_2) = n_2]}$$

$$= \frac{e^{-\lambda(t_3 - t_2)} \lambda^{n_3 - n_2} (t_3 - t_2)^{n_3 - n_2}}{\lfloor n_3 - n_2}$$

[refer to the second-and third-order probability functions of the Poisson process]

$$= P[X(t_3) = n_3/X(t_2) = n_2]$$

This means that the conditional probability distribution of $X(t_3)$ given all the past values $X(t_1) = n_1$, $X(t_2) = n_2$ depends only on the most recent value $X(t_2) = n_2$.

That is, the Poisson process possesses the Markov property. Hence the result.

2. **Additive property:** Sum of two independent Poisson processes is a Poisson process.

Proof We have already derived in Chapter IV the characteristic function of a Poisson distribution with parameter λ as $e^{-\lambda(1 - e^{i\omega})}$

Therefore, the characteristic functions of $X_1(t)$ and $X_2(t)$ are given by

$$\phi_{X_1(t)}(\omega) = e^{-\lambda_1 t(1 - e^{i\omega})} \text{ and } \phi_{X_2(t)}(\omega) = e^{-\lambda_2 t(1 - e^{i\omega})}$$

Since $X_1(t)$ and $X_2(t)$ are independent,

$$\phi_{X_1(t) + X_2(t)}(\omega) = \phi_{X_1(t)}(\omega)\, \phi_{X_2(t)}(\omega)$$

$$= e^{-(\lambda_1 + \lambda_2)t(1 - e^{i\omega})}$$

which is the characteristic function of Poisson distribution with parameter $(\lambda_1 + \lambda_2)t$.

Therefore, $\{X_1(t) + X_2(t)\}$ is a Poisson process.

Alternative proof

Let $\quad X(t) = X_1(t) + X_2(t)$.

$$P\{X(t) = n\} = \sum_{r=0}^{n} P\{X_1(t) = r\}\, P\{X_2(t) = n - r\}$$

$$= \sum_{r=0}^{n} \frac{e^{-\lambda_1 t} (\lambda_1 t)^r}{\lfloor r} \frac{e^{-\lambda_2 t} (\lambda_2 t)^{n-r}}{\lfloor n - r}$$

$$= e^{-(\lambda_1 + \lambda_2)t} \frac{1}{\lfloor n} \sum_{r=0}^{n} nC_r (\lambda_1 t)^r (\lambda_2 t)^{n-r}$$

$$= e^{-(\lambda_1 + \lambda_2)t} \left(\lambda_1 + \lambda_2\right) n / \lfloor n$$

Therefore, $X_1(t) + X_2(t)\}$ is a Poisson process with parameter $(\lambda_1 + \lambda_2)t$.

Note *The additive property holds good for any number of independent Poisson processes.*

3. Difference of two independent Poisson processes is not a Poisson process.

Proof Let $X(t) = X_1(t) - X_2(t)$

$$E\{X(t)\} = E\{X_1(t)\} - E\{X_2(t)\}$$
$$= (\lambda_1 - \lambda_2)t$$
$$E\{X^2(t)\} = E\{X_1^2(t)\} + E\{X_2^2(t)\} - 2E\{X_1(t)\}\ E(X_2(t)\}$$

(by independence)

$$= (\lambda_1^2\, t^2 + \lambda_1\, t) + (\lambda_2^2\, t^2 + \lambda_2\, t) - 2(\lambda_1\, t)\, (\lambda_2\, t)$$
$$= (\lambda_1 + \lambda_2)t + (\lambda_1 - \lambda_2)^2\, t^2$$
$$\neq (\lambda_1 - \lambda_2)t + (\lambda_1 - \lambda_2)^2\, t^2$$

Recall that $E\{X^2(t)\}$ for a Poisson process $\{X(t)\}$ with parameter λ is given by $E\{+ X^2_{(t)}\} = \lambda t + \lambda^2 t^2$.

Therefore, $\{X_1(t) - X_2(t)\}$ is not a Poisson process.

4. The interarrival time of a Poisson process, i.e., the interval between two successive occurrences of a Poisson process with parameter λ has an exponential distribution with mean $1/\lambda$.

Proof Let two consecutive occurrences of the event be E_i and E_{i+1}.

Let E_i take place at time instant t_i and T be the interval between the occurrences of E_i and E_{i+1}. T is a continuous RV.

$$P(T > t) = P\{E_{i+1} \text{ did not occur in } (t_i, t_i + t)\}$$
$$= P\{\text{No event occurs in an interval of length } t\}$$
$$= P\{X(t) = 0\}$$
$$= e^{-\lambda t}$$

Therefore, the cdf of T is given by

$$F(t) = P\{T \leq t\} = 1 - e^{-\lambda t}$$

Therefore, the pdf of T is given by

$$f(t) = \lambda \quad e^{-\lambda t} \quad (t \geq 0)$$

which is an exponential distribution with mean $1/\lambda$.

5. If the number of occurrences of an event E is an interval of length t is a Poisson process $\{X(t)\}$ with parameter λ and if each occurrence of E has a constant probability p of being recorded and the recordings are independent of each other, then the number $N(t)$ of the recorded occurrences in t is also a Poisson process with parameter λp.

Proof $P\{N(t) = n\} = \sum_{r=0}^{\infty} P\{E$ occurs $(n + r)$ times in t and n of them are recorded$\}$

$$= \sum_{r=0}^{\infty} \frac{e^{-\lambda t}(\lambda t)^{n+r}}{\lfloor n+r} (n+r)C_n p^n q^r, \quad q = 1 - p$$

$$= \sum_{r=0}^{\infty} \frac{e^{-\lambda t}(\lambda t)^{n+r}}{\lfloor n+r} \frac{\lfloor n+r}{\lfloor n \lfloor r} p^n q^r$$

$$= \frac{e^{-\lambda t}(\lambda pt)^n}{\lfloor n} \sum_{r=0}^{\infty} \frac{(\lambda qt)^r}{\lfloor r}$$

$$= \frac{e^{-\lambda t}(\lambda pt)^n}{\lfloor n} e^{\lambda qt}$$

$$= \frac{e^{-\lambda pt}(\lambda pt)^n}{\lfloor n}$$

Worked Examples 7(B)

Example 1 *Suppose that customers arrive at a bank according to a Poisson process with a mean rate of 3 per minute; find the probability that during a time interval of 2 min (i) exactly 4 customers arrive and (ii) more than 4 customers arrive.* (MU — Apr. 96)

Solution Mean of the Poisson process = λt

Mean arrival rate = mean number of arrivals per minute (unit time) = λ

Given $\lambda = 3$

$$P\{X(t) = k\} = \frac{e^{-\lambda t}(\lambda t)^k}{\lfloor k}$$

$$\therefore \quad P\{X(2) = 4\} = \frac{e^{-6} 6^4}{\lfloor 4} = 0.133$$

$$P\{X(2) > 4\} = 1 - [P\{(X(2) = 0\} + P\{(X(2) = 1\} + P\{X(2) = 2\} + P\{X(2) = 3\} + P\{X(2) = 4\}]$$

$$= 1 - \sum_{k=0}^{4} e^{-6} 6^k / \lfloor k$$

$$= 0.715$$

Example 2 *A machine goes out of order, whenever a component fails. The failure of this part follows a Poisson process with a mean rate of 1 per week. Find the probability that 2 weeks have elapsed since last failure. If there are 5 spare parts of this component in an inventory and that the next supply is not due in 10 weeks, find the probability that the machine will not be out of order in the next 10 weeks.*

Solution (i) Here the unit time is 1 week.

Mean failure rate = mean number of failures in a week = $\lambda = 1$.

$P\{\text{no failures in the 2 weeks since last failure}\}$

$= P\{X(2) = 0\}$

$$= \frac{e^{-2\lambda}(2\lambda)^0}{\lfloor 0} = e^{-2} = 0.135$$

(ii) There are only 5 spare parts and the machine should not go out of order in the next 10 weeks.

$P\{\text{for this event}\} = P\{X(10) \le 5\}$

$$= \sum_{k=0}^{5} \frac{e^{-10}\, 10^k}{\lfloor k}$$

$= 0.068$

Example 3 *If $\{N_1(t)\}$ and $\{N_2(t)\}$ are 2 independent Poisson processes with parameters λ_1 and λ_2 respectively, show that*

$P[N_1(t) = k/\{N_1(t) + N_2(t) = n\}] = nC_k\ p^k q^{n-k}$, *where*

$P = \dfrac{\lambda_1}{\lambda_1 + \lambda_2}$ *and* $q = \dfrac{\lambda_2}{\lambda_1 + \lambda_2}$.

Solution Required conditional probability

$$= \frac{P[\{N_1(t) = k\} \cap \{N_1(t) + N_2(t) = n\}]}{P\{N_1(t) + N_2(t) = n\}}$$

$$= \frac{P[\{N_1(t) = k\} \cap \{N_2(t) = n - k\}]}{P\{N_1(t) + N_2(t) = n\}}$$

$$= \frac{\dfrac{e^{-\lambda_1 t}(\lambda_1 t)^k}{\lfloor k} \times \dfrac{e^{-\lambda_2 t}(\lambda_2 t)^{n-k}}{\lfloor n-k}}{\dfrac{e^{-(\lambda_1+\lambda_2)t}\{(\lambda_1 + \lambda_2)t\}^n}{\lfloor n}}$$

(by indepence and additive property)

$$= \frac{\lfloor n}{\lfloor k \lfloor n-k} \frac{(\lambda_1 t)^k (\lambda_2 t)^{n-k}}{\{(\lambda_1 + \lambda_2)t\}^n}$$

$$= nC_k \left(\frac{\lambda_1}{\lambda_1 + \lambda_2}\right)^k \left(\frac{\lambda_2}{\lambda_1 + \lambda_2}\right)^{n-k}$$

$= nC_k\ p^k q^{n-k}$

Example 4 *If customers arrive at a counter in accordance with a Poisson process with a mean rate of 2 per minute, find the probability that the interval between 2 consecutive arrivals is (i) more than 1 min, (ii) between 1 min and 2 min and (iii) 4 min or less.*

Solution Refer to Property 4 of Poisson processes.

The interval T between 2 consecutive arrivals follows an exponential distribution with parameter $\lambda = 2$.

$$\text{(i)}\ P(T > 1) = \int_1^\infty 2e^{-2t}\, dt = e^{-2} = 0.135$$

$$\text{(ii)}\ P(1 < T < 2) = \int_1^2 2e^{-2t}\, dt = e^{-1} - e^{-2} = 233$$

$$\text{(iii)}\ P(T \le 4) = \int_0^4 2e^{-2t}\, dt = 1 - e^{-8} = 0.999$$

Example 5 *A radioactive source emits particles at a rate of 5 per minute in accordance with Poisson process. Each particle emitted has a probability 0.6 of being recorded. Find the probability that 10 particles are recorded in 4-min period.*

Solution Refer to Property 5 of Poisson processes. The number of recorded particles $N(t)$ follows a Poisson process with parameter λp.
Here $\lambda = 5$ and $p = 0.6$.

$$\therefore \quad P\{N(t) = k\} = \frac{e^{-3t}(3t)^k}{\lfloor k}$$

$$\therefore \quad P\{N(4) = 10\} = \frac{e^{-12}(12)^{10}}{\lfloor 10}$$

$$= 0.104$$

Example 6 *The number of accidents in a city follows a Poisson process with a mean of 2 per day and the number X_i of people involved in the ith accident has the distribution (independent) $P\{X_i = k\} = \frac{1}{2^k}$ $(k \ge 1)$. Find the mean and variance of the number of people involved in accidents per week.*

(MU — Apr. 96)

Solution The mean and variance of the distribution $P\{X_i = k\} = \frac{1}{2^k}$, $k = 1, 2, 3, \ldots, \infty$ can be obtained as 2 and 2.

Let the number of accidents on any day be assumed as n.

The numbers of people involved in these accidents be $X_1, X_2, \ldots, X_n$.

$X_1, X_2, \ldots, X_n$ are independent and identically distributed RVs with mean 2 and variance 2.

Therefore, by central limit theorem, $(X_1 + X_2 + \ldots + X_n)$ follows a normal distribution with mean $2n$ and variance $2n$, i.e., the total number of people involved in all the accidents on a day with n accidents = $2n$.

If N denotes the number of people involved in accidents on any day, then

$$P\{N = 2n\} = P\{X(t) = n\} \text{ [where } X(t) \text{ is the number of accidents]}$$

$$= \frac{e^{-2t}(2t)^n}{\lfloor n} \text{ (by data)}$$

$$\therefore \quad E\{N\} = \sum_{n=0}^{\infty} \frac{2n\, e^{-2t}(2t)^n}{\lfloor n}$$

$$= 2E\{X(t)\} = 4t$$

$$\text{Var}\{N\} = E\{N^2\} - E^2(N)$$

$$= \sum_{n=0}^{\infty} \frac{4n^2\, e^{-2t}(2t)^n}{\lfloor n} - 16t^2$$

$$= 4E\{X^2(t)\} - 16t^2$$

$$= 4[\text{Var}(X(t)\} + E^2\{X(t)\}] - 16t^2$$

$$= 4[2t + 4t^2] - 16t^2 = 8t$$

Therefore, mean and variance of the number of people involved in accidents per week are 28 and 56 respectively.

Example 7 *If T_n is the RV denoting the time of occurrence of the nth event in a Poisson process with parameter λ, show that the distribution function $F_n(t)$ of T_n is given by*

$$F_n(t) = \begin{cases} 1 - \displaystyle\sum_{k=0}^{n-1} \frac{(\lambda t)^k}{\lfloor k} e^{-\lambda t}, & \text{if } \quad t \geq 0 \\ 0 & , \text{if } \quad t < 0 \end{cases}$$

Deduce the density function $f_n(t)$ of T_n

Solution $\quad F_n(t) = P\{T_n \leq t\}$

$$= 1 - P\{T_n > t\}$$

When $T_n > t$, i.e., the time of occurrence of the nth event $> t$, $(n - 1)$ or less events must have occurred in $(0, t)$.

$$\therefore \quad F_n(t) = 1 - P\{X(t) \leq n - 1\}$$

$$= 1 - \sum_{k=0}^{n-1} \frac{(\lambda t)^k}{\lfloor k} e^{-\lambda t}, \text{ when } t \geq 0$$

Differentiating both sides with respect to t and noting that $F_n'(t) = f_n(t)$

$$f_n(t) = -\sum_{k=0}^{n-1} \left\{ -\lambda \frac{(\lambda t)^k}{\lfloor k} e^{-\lambda t} + \lambda \frac{(\lambda t)^{k-1}}{\lfloor k-1} e^{-\lambda t} \right\}$$

$$= \lambda e^{-\lambda t} \sum_{k=0}^{n-1} \left\{ \frac{(\lambda t)^k}{\lfloor k} - \frac{(\lambda t)^{k-1}}{\lfloor k-1} \right\}$$

$$= \lambda e^{-\lambda t} \times \left[1 + \left\{ \frac{\lambda t}{\lfloor 1} - 1 \right\} + \left\{ \frac{(\lambda t)^2}{\lfloor 2} - \frac{\lambda t}{\lfloor 1} \right\} + \ldots + \left\{ \frac{(\lambda t)^{n-1}}{\lfloor n-1} - \frac{(\lambda t)^{n-2}}{\lfloor n-2} \right\} \right]$$

$$= \frac{\lambda^n t^{n-1} e^{-\lambda t}}{\lfloor n-1}, \quad t \geq 0$$

Example 8 *If $\{X(t)\}$ is a Poisson process, prove that*

$$P\{X(s) = r/X(t) = n\} = nC_r \left(\frac{s}{t}\right)^r \left(1 - \frac{s}{t}\right)^{n-r} \text{ where } s < t$$

Solution $P\{X(s) = r/X(t) = n\} = \dfrac{P[\{X(s) = r\} \cap \{X(t) = n\}]}{P\{X(t) = n\}}$

$$= \frac{P\{X(s) = r \cap X(t - s) = n - r\}}{P\{X(t) = n\}}$$

$$= \frac{P\{X(s) = r\} \, P\{X(t - s) = n - r\}}{P\{X(t) = n\}} \quad \text{(by independence)}$$

$$= \frac{e^{-\lambda s}(\lambda s)^r / \lfloor r \; e^{-\lambda(t - s)} [\lambda(t - s)]^{n-r} / \lfloor n - r}{e^{-\lambda t}(\lambda t)^n / \lfloor n}$$

$$= \frac{\lfloor n}{\lfloor r \lfloor n - r} \; \frac{s^r (t - s)^{n-r}}{t^n}$$

$$= nC_r \left(\frac{s}{t}\right)^r \left(1 - \frac{s}{t}\right)^{n-r}$$

Exercise 7(B)

Part A (Short answer questions)

1. What is a point process? Give an example.
2. Define a Poisson process.
3. What are the postulates of a Poisson process?
4. State the probability law of a Poisson process.
5. When is a Poisson process said to be homogeneous?
6. If $\{X(t)\}$ is a homogeneous Poisson process, find
$$P\{X(t_1) = n_1, X(t_2) = n_2\}, \; t_2 > t_1$$
7. Find the autocorrelation $R_{xx}(t_1, t_2)$ of the Poisson process $\{X(t)\}$.
8. State and prove the additive property of a Poisson process.
9. Prove that the difference of 2 independent Poisson processes is not a Poisson process.
10. Prove that the interarrival time of a Poisson process follows an exponential distribution.

11. If the customers arrive at a bank according to a Poisson process with mean rate of 2 per minute, find the probability that, during an 1-min interval, no customer arrives.

Part B

12. Find the first-order characteristic function of a Poisson process.
(MKU — Apr. 96)
13. If particles are emitted from a radioactive source at the rate of 20 per hour, find the probability that exactly 5 particles are emitted during a 15-min period.
14. On the average, a submarine on patrol sights 6 enemy ships per hour. Assuming that the number of ships sighted in a given length of time is a Poisson variate, find the probability of sighting
    (i) 6 ships in the next half-an-hour,
    (ii) 4 ships in the next 2 h,
    (iii) at least 1 ship in the next 15 min and
    (iv) at least 2 ships in the next 20 min.
15. Patients arrive randomly and independently at a doctor's consulting room from 8 A.M. at an average rate of one in 5 min. The waiting room can hold 12 persons. What is the probability that the room will be full when the doctor arrives at 9 A.M.?
16. Messages arrive at a telegraph office in accordance with the laws of a Poisson process with a mean rate of 3 messages per hour.
    (i) What is the probability that no message will have arrived during the morning hours, i.e., between 8 A.M. and 12 noon?
    (ii) What is the distribution of the time at which the first afternoon message arrives?
17. Assume that a circuit has an IC whose time to failure is an exponentially distributed RV with expected lifetime of 3 months. If there are 10 spare IC's and time from failure to replacement is zero, what is the probability that the circuit can be kept operational for at least 1 year?
18. Assume that an office switchboard has 5 telephone lines and that starting at 8 A.M. on Monday, the time that a call arrives on each line is an exponential RV with parameter λ. Also assume that the calls arrive independently on the lines. Show that the time of arrival of the first call (irrespective of which line it arrives on) is exponential with parameter 5λ.
19. A radioactive source emits particles at a rate of 6 per minute in accordance with Poisson process. Each particle emitted has a probability of 1/3 of being recorded. Find the probability that at least 5 particles are recorded in a 5-min period.
20. Suppose that customers arrive at a counter independently from 2 different sources. Arrivals occur in accordance with a Poisson process with mean rate of 6 per hour from the first source and 4 per hour from the second source. Find the mean interval between any 2 successives arrivals.

21. Assume that a device fails when a cumulative effect of k shocks occur. If the shocks occur according to a Poisson process with parameter λ, find the density function for the life T of the device.
[**Hint:** Refer to Worked Example 7]
22. In the case of a Poisson process, show that the conditional probability that events have occurred at $\tau_1, \tau_2, \ldots, \tau_n$, given that n events have occurred in $(0, t)$, is given by $\dfrac{\lfloor n}{t^n}$. (BDU — Nov. 96)
23. Passengers arrive at a terminal for boarding the next bus. The times of their arrival are Poisson with an average arrival rate of 1 per minute. The times of departure of each bus are Poisson with an average departure rate of 2 per hour. Assume that the capacity of the bus is large. Find the average number of passengers in (i) each bus and (ii) the first bus that leaves after 9 A.M.
24. Passengers arrive at a terminal after 9 A.M. The times of their arrival are Poisson with mean density $\lambda = 1$ per minute. The time interval from 9 A.M. to the departure of the next bus is RV T. Find the mean number of passengers in this bus (i) if T has an exponential density with mean 30 min and (ii) if T is uniform between 0 and 60 min.

Markov Process

Another interesting model of a random process is the one in which the value of the random process depends only upon the most recent previous value and is independent of all values in the more distant past. Such a model is called a Markov model and is often described by saying that a Markov process is one in which the future value is independent of the past values, given the present value. Models in which the future depends only upon the present are common among electrical engineering models.

Consider the experiment of tossing a fair coin a number of times. The possible outcomes at each trial are two—'head' with probability 1/2 and 'tail' with probability 1/2. If we denote the outcome of the nth toss, which is a RV, by X_n and the outcomes 'head' and 'tail' by 1 and 0 respectively, then

$$P\{X_n = 1\} = \frac{1}{2} \text{ and } P\{X_n = 0\} = \frac{1}{2}; \; n = 1, 2, \ldots$$

Thus we have a sequence of independent RVs $X_1, X_2, \ldots$, since the trials are independent and hence the outcome of the nth trial does not depend in any way on the previous trials.

Consider now the RV that represents the total number of heads in the first n trials and is given by $S_n = X_1 + \ldots + X_n$. The possible values of S_n are $0, 1, 2, \ldots, n$. If $S_n = k (k = 0, 1, \ldots, n)$, then the RV S_{n+1} $(= S_n + X_{n+1})$ can assume only 2 possible values, namely $k + 1$ [if the $(n + 1)$th trial results in a head] and k [if the $(n + 1)$th trial results in a tail].

Thus $P\{S_{n+1} = k + 1/S_n = k\} = \frac{1}{2}$

$$P\{S_{n+1} = k/S_n = k\} = \frac{1}{2}$$

These probabilities are not at all affected by the values of the RVs $S_1, S_2, \ldots, S_{n-1}$. Also the conditional probability of S_{n+1} given S_n depends on the value of S_n and not on the manner in which the value of S_n was reached. This is a simple example of a Markov chain. Random processes $\{X(t)\}$ (with Markov property) which take discrete values, whether t is discrete or continuous, are called ***Markov chains***. Poisson process, discussed earlier, is a continuous time Markov chain. In this section, we will discuss discrete time Markov chains.

Definition of a Markov Chain

If, for all n, $P\{X_n = a_n/X_{n-1} = a_{n-1}, X_{n-2} = a_{n-2}, \ldots, X_0 = a_0\} = P\{X_n = a_n/X_{n-1} = a_{n-1}\}$, then the process $\{X_n\}$, $n = 0, 1, \ldots$, is called a Markov chain.

$(a_1, a_2, \ldots, a_n, \ldots)$ are called the states of the Markov chain. The conditional probability $P\{X_n = a_j/X_{n-1} = a_i\}$ is called the ***one-step transition probability*** from state a_i to state a_j at the nth step (trial) and is denoted by $p_{ij}(n-1, n)$.

If the one-step transition probability does not depend on the step, i.e., $p_{ij}(n-1, n) = p_{ij}(m-1, m)$ the Markov chain is called a ***homogeneous Markov chain*** or the chain is said to have stationary transition probabilities. The use of the word 'stationary' does not imply a stationary random sequence.

When the Markov chain is homogeneous, the one-step transition probability is denoted by p_{ij}. The matrix $P = \{p_{ij}\}$ is called (one-step) ***transition probability matrix***, shortly, tpm.

Note *The tpm of a Markov chain is a stochastic matrix, since $p_{ij} \geq 0$ and $\sum_j p_{ij} = 1$ for all i, i.e., the sum of all the elements of any row of the tpm is 1. This is obvious because the transition from state a_i to any one of the states (including a_i itself) is a certain event.*

The conditional probability that the process is in state a_j at step n, given that it was in state a_i at step 0, i.e., $P\{X_n = a_j/X_0 = a_i\}$ is called the ***n-step transition probability*** and denoted by $p_{ij}(n)$.

Note $p_{ij}^{(1)} = p_{ij}$.

Let us consider an example in which we explain how the tpm is formed for a Markov chain. Assume that a man is at an integral point of the x-axis between the origin and the point $x = 3$. He takes a unit step either to the right with probability 0.7 or to the left with probability 0.3, unless he is at the origin when he takes a step to the right to reach $x = 1$ or he is at the point $x = 3$, when he takes a step to the left to reach $x = 2$. The chain is called 'Random walk with reflecting barriers'.

The tpm is given below:

$$\begin{array}{c c} & \text{States of } X_n \\ & \begin{array}{cccc} 0 & 1 & 2 & 3 \end{array} \\ \text{States of } X_{n-1} \begin{array}{c} 0 \\ 1 \\ 2 \\ 3 \end{array} & \begin{pmatrix} 0 & 1 & 0 & 0 \\ 0.3 & 0 & 0.7 & 0 \\ 0 & 0.3 & 0 & 0.7 \\ 0 & 0 & 1 & 0 \end{pmatrix} \end{array}$$

Note *p_{23} = the element in the 2nd row, 3rd column of this tpm = 0.7. This means that, if the process is at state 2 at step $(n - 1)$, the probability that it moves to state 3 at step $n = 0.7$, where n is any positive integer.*

Definition: If the probability that the process is in state a_i is p_i $(i = 1, 2, \ldots, k)$ at any arbitrary step, then the row vector $p = (p_1, p_2, \ldots, p_k)$ is called ***the probability distribution of the process*** at that time. In particular, $p^{(0)} = \{p_1^{(0)}, p_2^{(0)}, \ldots, p_k^{(0)}\}$ is the initial probability distribution.

[**Remark:** The transition probability matrix together with the initial probability distribution completely specifies a Markov chain $\{X_n\}$. In the example given above, let us assume that the initial probability distribution of the chain is

$$p^{(0)} = \left(\frac{1}{4}, \frac{1}{4}, \frac{1}{4}, \frac{1}{4}\right).$$

i.e., $\quad P\{X_0 = i\} = 1/4, \ i = 0, 1, 2, 3$

Then we have, for the example given above,

$$P\{X_1 = 2/X_0 = 1\} = 0.7; \quad P\{X_2 = 1/X_1 = 2\} = 0.3,$$

$$\begin{aligned} P\{X_2 = 1, X_1 = 2/X_0 = 1\} &= P\{X_2 = 1/X_1 = 2\} \times P\{X_1 = 2/X_0 = 1\} \\ &= 0.3 \times 0.7 = 0.21 \qquad (1) \end{aligned}$$

$$\begin{aligned} P\{X_2 = 1, X_1 = 2, X_0 = 1\} &= P\{X_0 = 1\} \times P\{X_2 = 1, X_1 = 2/X_0 = 1\} \\ &= 1/4 \times 0.21 = 0.0525 \quad [\text{by } (1)] \qquad (2) \end{aligned}$$

$$\begin{aligned} P\{X_3 = 3, X_2 = 1, X_1 = 2, X_0 = 1\} &= P\{X_2 = 1, X_1 = 2, X_0 = 1\} \\ &\quad \times P\{X_3 = 3/X_2 = 1, X_1 = 2, X_0 = 1\} \\ &= 0.0525\, P\{X_3 = 3/X_2 = 1\} \\ &\qquad (\text{Markov property}) \ [\text{by } (2)] \\ &= 0.0525 \times 0 = 0 \end{aligned}$$

Chapman–Kolmogorov Theorem

If P is the tpm of a homogeneous Markov chain, then the n-step tpm $P^{(n)}$ is equal to P^n.

i.e., $= [p_{ij}^{(n)}] = [p_{ij}]^n$

Proof $P_{ij}^{(2)} = P\{X_2 = j/X_0 = i\}$, since the chain is homogeneous.

The state j can be reached from the state i in 2 steps through some intermediate state k.

$$\begin{aligned}\text{Now} \quad p_{ij}^{(2)} &= P\{X_2 = j/X_0 = i\} = P\{X_2 = j, X_1 = k/X_0 = i\}\\ &= P\{X_2 = j/X_1 = k, X_0 = i\}\ P\{X_1 = k/X_0 = i\}\\ &= p_{kj}^{(1)}\ p_{ik}^{(1)}\\ &= p_{ik}\, p_{kj}\end{aligned}$$

Since the transition from state i to state j in 2 steps can take place through any one of the intermediate states, k can assume the values 1, 2, 3, …. The transitions through various intermediate states are mutually exclusive.

$$\text{Hence} \quad p_{ij}^{(2)} = \sum_k p_{ik}\, p_{kj}$$

i.e., the ij-th element of 2 step tpm =
the ij-th element of the product of the 2 one-step tpm's

i.e., $P^{(2)} = P^2$

$$\begin{aligned}\text{Now} \quad p_{ij}^{(3)} &= P\{X_3 = j/X_0 = i\}\\ &= \sum_k P\{X_3 = j/X_2 = k\}\ P\{X_2 = k/X_0 = i\}\\ &= \sum_k p_{kj}\, p_{ik}^{(2)}\\ &= \sum_k p_{ik}^{(2)}\, p_{kj}\end{aligned}$$

$$\text{Similarly} \quad p_{ij}^{(3)} = \sum_k p_{ik}\, p_{kj}^{(2)}$$

i.e., $P^{(3)} = P^2\, P = P\, P^2 = P^3$

Proceeding further in a similar way, we get

$$P^{(n)} = P^n$$

For example, consider the problem of Random walk with reflecting barriers, discussed above, for which the tpm is

$$P = \begin{pmatrix} 0 & 1 & 0 & 0\\ 0.3 & 0 & 0.7 & 0\\ 0 & 0.3 & 0 & 0.7\\ 0 & 0 & 1 & 0 \end{pmatrix}$$

$$P^2 = \begin{pmatrix} 0.3 & 0 & 0.7 & 0 \\ 0 & 0.51 & 0 & 0.49 \\ 0.09 & 0 & 0.91 & 0 \\ 0 & 0.3 & 0 & 0.7 \end{pmatrix}$$

From this matrix, we see that $p_{11}^{(2)} = 0.51$. This is so, because

$$\begin{aligned} p_{11}^{(2)} &= p_{10}\,p_{01} + p_{11}P_{11} + p_{12}\,p_{21} + p_{13}\,p_{31} \\ &= (0.3)\,(1) + (0)\,(0) + (0.7)\,(0.3) + (0)\,(0) = 0.51 \end{aligned}$$

Definition: A stochastic matrix P is said to be a ***regular matrix***, if all the entries of P^m (for some positive integer m) are positive. A homogeneous Markov chain is said to be ***regular*** if its tpm is regular.

We state below two theorems without proof:

1. If $p = \{p_i\}$ is the state probability distribution of the process at an arbitrary time, then that after one step is pP, where P is the tpm of the chain and that after n steps in pP^n.
2. If a homogeneous Markov chain is regular, then every sequence of state probability distributions approaches a unique fixed probability distribution called the ***stationary (state) distribution*** or ***steady-state distribution*** of the Markov chain.

   That is, $\lim\limits_{n \to \infty} \{p^{(n)}\} = \pi$, where the state probability distribution at step n, $p^{(n)} = (p_1^{(n)}, p_2^{(n)}, \ldots, p_k^{(n)})$ and the stationary distribution $\pi = (\pi_1, \pi_2, \ldots, \pi_k)$ are row vectors.
3. Moreover, if P is the tpm of the regular chain, then $\pi P = \pi$ (π is a row vector). Using this property of π, it can be found out, as in the worked examples given below:

Classification of States of a Markov Chain

If $p_{ij}^{(n)} > 0$ for some n and for all i and j, then every state can be reached from every other state. When this condition is satisfied, the Markov chain is said to be ***irreducible***. The tpm of an irreducible chain is an irreducible matrix. Otherwise, the chain is said to be ***nonirreducible*** or ***reducible***.

State i of a Markov chain is called a ***return state***, if $p_{ii}^{(n)} > 0$ for some $n > 1$.

*The **period** d_i* of a return state i is defined as the greatest common divisor of all m such that $p_{ii}^{(m)} > 0$, i.e., $d_i = \text{GCD}\,\{m: p_{ii}^{(m)} > 0\}$. State i is said to be ***periodic*** with period d_i if $d_i > 1$ and ***aperiodic*** if $d_i = 1$.

Obviously state i is aperiodic if $p_{ii} \neq 0$. The probability that the chain returns to state i, having started from state i, for the first time at the nth step (or after n transitions) is denoted by $f_{ii}^{(n)}$ and *called the **first return time probability*** or *the **recurrence time probability***. $\{n, f_{ii}^{(n)}\}$, $n = 1, 2, 3, \ldots$, is the distribution of recurrence times of the state i.

If $F_{ii} = \sum\limits_{n=1}^{\infty} f_{ii}^{(n)} = 1$, the return to state i is certain.

$\mu_{ii} = \sum_{n=1}^{\infty} n f_{ii}^{(n)}$ is called *the **mean recurrence time*** of the state i.

A state i is said to be ***persistent*** or ***recurrent*** if the return to state i is certain, i.e., if $F_{ii} = 1$. The state i is said to be ***transient*** if the return to state i is uncertain, i.e., if $F_{ii} < 1$. The state i is said to be ***nonnull persistent*** if its mean recurrence time μ_{ii} is finite and ***null persistent***, if $\mu_{ii} = \infty$.

A nonnull persistent and aperiodic state is called ***ergodic***.

We give below two theorems, without proof, which will be helpful to classify the states of a Markov chain.

1. If a Markov chain is irreducible, all its states are of the same type. They are all transient, all null persistent or all nonnull persistent. All its states are either aperiodic or periodic with the same period.
2. If a Markov chain is finite irreducible, all its states are nonnull persistent.

Worked Examples 7(C)

Example 1 *The transition probability matrix of a Markov chain $\{X_n\}$, $n = 1, 2, 3, \ldots$, having 3 states 1, 2 and 3 is*

$$P = \begin{pmatrix} 0.1 & 0.5 & 0.4 \\ 0.6 & 0.2 & 0.2 \\ 0.3 & 0.4 & 0.3 \end{pmatrix}$$

and the initial distribution is $p^{(0)} = (0.7, 0.2, 0.1)$.
Find (i) $P\{X_2 = 3\}$ and (ii) $P\{X_3 = 2, X_2 = 3, X_1 = 3, X_0 = 2\}$.

Solution

$$P^{(2)} = P^2 = \begin{pmatrix} 0.1 & 0.5 & 0.4 \\ 0.6 & 0.2 & 0.2 \\ 0.3 & 0.4 & 0.3 \end{pmatrix}\begin{pmatrix} 0.1 & 0.5 & 0.4 \\ 0.6 & 0.2 & 0.2 \\ 0.3 & 0.4 & 0.3 \end{pmatrix} = \begin{pmatrix} 0.43 & 0.31 & 0.26 \\ 0.24 & 0.42 & 0.34 \\ 0.36 & 0.35 & 0.29 \end{pmatrix}$$

(i) $$P\{X_2 = 3\} = \sum_{i=1}^{3} P\{X_2 = 3/X_0 = i\} \times P\{X_0 = i\}$$
$$= p_{13}^{(2)} P\{(X_0 = 1) + p_{23}^{(2)} P(X_0 = 2) + p_{33}^{(2)} P(X_0 = 3)$$
$$= 0.26 \times 0.7 + 0.34 \times 0.2 + 0.29 \times 0.1$$
$$= 0.182 + 0.068 + 0.029$$
$$= 0.279$$

(ii) $$P\{X_1 = 3/X_0 = 2\} = p_{23} = 0.2 \quad (1)$$
$$P\{X_1 = 3, X_0 = 2\} = P\{X_1 = 3/X_0 = 2\} \times P\{X_0 = 2\}$$
$$= 0.2 \times 0.2 = 0.04 \text{ [by (1)]} \quad (2)$$
$$P\{X_2 = 3, X_1 = 3, X_0 = 2\} = P\{X_2 = 3/X_1 = 3, X_0 = 2\} \times P\{X_1 = 3, X_0 = 2\}$$

$$= P\{X_2 = 3/X_1 = 3\} \times P\{X_1 = 3, X_0 = 2\}$$

(by Markov property)

$$= 0.3 \times 0.04 \text{ [by (2)]}$$
$$= 0.012 \qquad (3)$$

$$P\{X_3 = 2, X_2 = 3, X_1 = 3, X_0 = 2\}$$
$$= P\{X_3 = 2/X_2 = 3, X_1 = 3, X_0 = 2\} \times P\{X_2 = 3, X_1 = 3, X_0 = 2\}$$
$$= P\{X_3 = 2/X_2 = 3\} \times P\{X_2 = 3, X_1 = 3, X_0 = 2\}$$

(by Markov property)

$$= 0.4 \times 0.012 \text{ [by (3)]}$$
$$= 0.0048$$

Example 2 *A fair dice is tossed repeatedly. If X_n denotes the maximum of the numbers occuring in the first n tosses, find the transition probability matrix P of the Markov chain $\{X_n\}$.*

Find also P^2 and $P(X_2 = 6)$

Solution State space: {1, 2, 3, 4, 5, 6}

The tpm is formed using the following analysis.

Let X_n = the maximum of the numbers occurring in the first n trials = 3, say

Then X_{n+1} = 3, if the (n + 1)th trial results in 1, 2 or 3

= 4, if the (n + 1)th trial results in 4

= 5, if the (n + 1)th trial results in 5

= 6, if the (n + 1)th trial results in 6

$$\therefore \quad P\{X_{n+1} = 3/X_n = 3\} = \frac{1}{6} + \frac{1}{6} + \frac{1}{6} = \frac{3}{6}$$

$$P\{X_{n+1} = i/X_n = 3\} = \frac{1}{6}, \text{ when } i = 4, 5, 6$$

Therefore, the transition probability matrix of the chain is

$$P = \begin{pmatrix} 1/6 & 1/6 & 1/6 & 1/6 & 1/6 & 1/6 \\ 0 & 2/6 & 1/6 & 1/6 & 1/6 & 1/6 \\ 0 & 0 & 3/6 & 1/6 & 1/6 & 1/6 \\ 0 & 0 & 0 & 4/6 & 1/6 & 1/6 \\ 0 & 0 & 0 & 0 & 5/6 & 1/6 \\ 0 & 0 & 0 & 0 & 0 & 1 \end{pmatrix}$$

$$P^2 = \frac{1}{36}\begin{pmatrix} 1 & 3 & 5 & 7 & 9 & 11 \\ 0 & 4 & 5 & 7 & 9 & 11 \\ 0 & 0 & 9 & 7 & 9 & 11 \\ 0 & 0 & 0 & 16 & 9 & 11 \\ 0 & 0 & 0 & 0 & 25 & 11 \\ 0 & 0 & 0 & 0 & 0 & 36 \end{pmatrix}$$

Initial state probability distribution is $p^{(0)} = \left(\frac{1}{6}, \frac{1}{6}, \frac{1}{6}, \frac{1}{6}, \frac{1}{6}, \frac{1}{6}\right)$ since all the values 1, 2, ..., 6 are equally likely.

$$P\{X_2 = 6\} = \sum_{i=1}^{6} P\{X_2 = 6/X_0 = i\} \times P\{X_0 = i\}$$

$$= \frac{1}{6} \sum_{i=1}^{6} p_{i6}^{(2)}$$

$$= \frac{1}{6} \times \frac{1}{36} \times (11 + 11 + 11 + 11 + 11 + 36)$$

$$= \frac{91}{216}$$

Example 3 *A man either drives a car or catches a train to go to office each day. He never goes 2 days in a row by train but if he drives one day, then the next day he is just as likely to drive again as he is to travel by train. Now suppose that on the first day of the week, the man tossed a fair dice and drove to work if and only if a 6 appeared. Find (i) the probability that he takes a train on the third day and (ii) the probability that he drives to work in the long run.*

Solution The travel pattern is a Markov chain, with state space = (train, car)

The tpm of the chain is

$$P = \begin{matrix} & T & C \\ T & 0 & 1 \\ C & 1/2 & 1/2 \end{matrix}$$

The initial state probability distribution is $p^{(1)} = \left(\frac{5}{6}, \frac{1}{6}\right)$,

since P(travelling by car) = P(getting 6 in the toss of the dice)

$$= \frac{1}{6}$$

and P(travelling by train) $= \frac{5}{6}$

$$p^{(2)} = p^{(1)}P = \left(\frac{5}{6}, \frac{1}{6}\right)\begin{pmatrix} 0 & 1 \\ 1/2 & 1/2 \end{pmatrix} = \left(\frac{1}{12}, \frac{11}{12}\right)$$

$$p^{(3)} = p^{(2)}P = \left(\frac{1}{12}, \frac{11}{12}\right)\begin{pmatrix} 0 & 1 \\ 1/2 & 1/2 \end{pmatrix} = \left(\frac{11}{24}, \frac{13}{24}\right)$$

$\therefore$ P(the man travels by train on the third day) $= \frac{11}{24}$

Let $\pi = (\pi_1, \pi_2)$ be the limiting form of the state probability distribution or stationary state distribution of the Markov chain.

By the property of π, $\pi P = \pi$

i.e., $$(\pi_1, \pi_2)\begin{pmatrix} 0 & 1 \\ 1/2 & 1/2 \end{pmatrix} = (\pi_1, \pi_2)$$

i.e., $$\frac{1}{2}\pi_2 = \pi_1 \qquad (1)$$

and $$\pi_1 + \frac{1}{2}\pi_2 = \pi_2 \qquad (2)$$

Equations (1) and (2) are one and the same.

Therefore, consider (1) or (2) with $\pi_1 + \pi_2 = 1$, since π is a probability distribution.

Solving, $\pi_1 = \frac{1}{3}$ and $\pi_2 = \frac{2}{3}$

$\therefore$ $P\{\text{the man travels by car in the long run}\} = \frac{2}{3}$.

Example 4 *Consider a communication system which transmits the 2 digits 0 and 1 through several stages. Let X_n $(n \ge 1)$ be the digit leaving the nth stage of the system and X_0 be the digit entering the first stage (or leaving the 0th stage). At each stage there is a constant probability q that the digit which enters will be transmitted unchanged (i.e., the digit will remain unchanged when it leaves) and the probability p otherwise (i.e., the digit changes when it leaves), where $p + q = 1$. Write down the tpm P of the homogeneous two-state Markov chain $\{X_n\}$. Find P^m, P^∞ and the conditional probability that the digit entering the first stage is 0, given that the digit leaving the mth stage is 0. Assume that the initial state probability distribution is $p^{(0)} = (a, 1 - a)$.*

Solution State space = (0, 1); $P \equiv$ State of X_n $\begin{array}{c} \\ 0 \\ 1 \end{array}\!\!\begin{array}{c} \text{State of } X_{n+1} \\ \begin{matrix} 0 & 1 \end{matrix} \\ \begin{pmatrix} q & p \\ p & q \end{pmatrix} \end{array}$

Now $$P^2 = \begin{pmatrix} q & p \\ p & q \end{pmatrix}\begin{pmatrix} q & p \\ p & q \end{pmatrix}$$

$$= \begin{pmatrix} p^2 + q^2 & 2pq \\ 2pq & p^2 + q^2 \end{pmatrix}$$

$$= \begin{pmatrix} \frac{1}{2}\left[(q+p)^2 + (q-p)^2\right] & \frac{1}{2}\left[(q+p)^2 - (q-p)^2\right] \\ \frac{1}{2}\left[(q+p)^2 - (q-p)^2\right] & \frac{1}{2}\left[(q+p)^2 + (q-p)^2\right] \end{pmatrix}$$

$$= \begin{pmatrix} \frac{1}{2} + \frac{1}{2}r^2 & \frac{1}{2} - \frac{1}{2}r^2 \\ \frac{1}{2} - \frac{1}{2}r^2 & \frac{1}{2} + \frac{1}{2}r^2 \end{pmatrix}, \text{ where } q - p = r$$

$$P^3 = \begin{pmatrix} \frac{1}{2}+\frac{1}{2}r^3 & \frac{1}{2}-\frac{1}{2}r^3 \\ \frac{1}{2}-\frac{1}{2}r^3 & \frac{1}{2}+\frac{1}{2}r^3 \end{pmatrix}$$

The values of P^2 and P^3 make us guess that

$$P^m = \begin{pmatrix} \frac{1}{2}+\frac{1}{2}r^m & \frac{1}{2}-\frac{1}{2}r^m \\ \frac{1}{2}-\frac{1}{2}r^m & \frac{1}{2}+\frac{1}{2}r^m \end{pmatrix}$$

It is correct as can be proved by induction as follows:

$$P^{m+1} = \begin{pmatrix} q & p \\ p & q \end{pmatrix} \begin{pmatrix} \frac{1}{2}+\frac{1}{2}r^m & \frac{1}{2}-\frac{1}{2}r^m \\ \frac{1}{2}-\frac{1}{2}r^m & \frac{1}{2}+\frac{1}{2}r^m \end{pmatrix}$$

$$= \begin{pmatrix} \frac{q}{2}+\frac{q}{2}r^m+\frac{p}{2}-\frac{p}{2}r^m & \frac{q}{2}-\frac{q}{2}r^m+\frac{p}{2}+\frac{p}{2}r^m \\ \frac{p}{2}+\frac{p}{2}r^m+\frac{q}{2}-\frac{q}{2}r^m & \frac{p}{2}-\frac{p}{2}r^m+\frac{q}{2}+\frac{q}{2}r^m \end{pmatrix}$$

$$= \begin{pmatrix} \frac{1}{2}+\frac{1}{2}r^{m+1} & \frac{1}{2}-\frac{1}{2}r^{m+1} \\ \frac{1}{2}-\frac{1}{2}r^{m+1} & \frac{1}{2}+\frac{1}{2}r^{m+1} \end{pmatrix}$$

$$\therefore \quad P^m = \begin{pmatrix} \frac{1}{2}+\frac{1}{2}r^m & \frac{1}{2}-\frac{1}{2}r^m \\ \frac{1}{2}-\frac{1}{2}r^m & \frac{1}{2}+\frac{1}{2}r^m \end{pmatrix}, \text{ where } m \text{ is a positive integer} \geq 1$$

$$p^{\infty} = \lim_{m \to \infty} \left(P^m\right) = \begin{pmatrix} \frac{1}{2} & \frac{1}{2} \\ \frac{1}{2} & \frac{1}{2} \end{pmatrix} \quad \text{since } |r| < 1$$

Now $\quad P\{X_m = 0, X_0 = 0\} = P\{X_m = 0/X_0 = 0\} \times P\{X_0 = 0)$

$\qquad = ap_{00}^{(m)}$

and $\quad P\{X_m = 0, X_0 = 1\} = bp_{10}^{(m)}\ b = 1 - a$

Now $\quad P\{X_0 = 0/X_m = 0\} = \dfrac{p\{X_0 = 0\} \times P\{X_m = 0/X_0 = 0\}}{p\{X_0 = 0\} \times p_{00}^{(m)} + p\{X_0 = 1\} \times P_{10}^{(m)}}$

(by Baye's theorem)

$$= \frac{a\left\{\frac{1}{2}+\frac{1}{2}r^m\right\}}{a\left\{\frac{1}{2}+\frac{1}{2}r^m\right\}+b\left\{\frac{1}{2}-\frac{1}{2}r^m\right\}}$$

$$= \frac{a\left(1+r^m\right)}{1+(a-b)r^m}, \text{ where } b = 1 - a$$

Example 5 *A gambler has Rs 2/-. He bets Re 1 at a time and wins Re 1 with probability 1/2. He stops playing if he loses Rs 2 or wins Rs 4 (a) What is the tpm of the related Markov chain? (b) What is the probability that he has lost his money at the end of 5 plays? (c) What is the probability that the game lasts more than 7 plays?*

Solution Let X_n represent the amount with the player at the end of the nth round of the play.

State space of $\{X_n\}$ = (0, 1, 2, 3, 4, 5, 6), as the game ends, if the player loses all the money ($X_n = 0$) or wins Rs. 4, i.e., has Rs 6 ($X_n = 6$). The tpm of the Markov chain is

$$P = \begin{array}{c} \\ 0 \\ 1 \\ 2 \\ 3 \\ 4 \\ 5 \\ 6 \end{array}\begin{array}{c} \begin{array}{ccccccc} 0 & 1 & 2 & 3 & 4 & 5 & 6 \end{array} \\ \begin{pmatrix} 1 & 0 & 0 & 0 & 0 & 0 & 0 \\ 1/2 & 0 & 1/2 & 0 & 0 & 0 & 0 \\ 0 & 1/2 & 0 & 1/2 & 0 & 0 & 0 \\ 0 & 0 & 1/2 & 0 & 1/2 & 0 & 0 \\ 0 & 0 & 0 & 1/2 & 0 & 1/2 & 0 \\ 0 & 0 & 0 & 0 & 1/2 & 0 & 1/2 \\ 0 & 0 & 0 & 0 & 0 & 0 & 1 \end{pmatrix} \end{array}$$

Note *This is called a random walk with* ***absorbing barriers*** *at 0 and 6, since the chain cannot come out of the states 0 and 6, once it has entered them.*

The initial probability distribution of $\{X_n\}$ is $p(0)$ = (0, 0, 1, 0, 0, 0, 0), as the player has got Rs 2/- to start with.

$$p^{(1)} = p^{(0)} P = (0, 1/2, 0, 1/2. 0, 0, 0)$$
$$p^{(2)} = p^{(1)} P = (1/4, 0, 1/2, 0, 1/4, 0, 0)$$
$$p^{(3)} = p^{(2)} P = (1/4, 1/4, 0, 3/8, 0, 1/8, 0)$$
$$p^{(4)} = p^{(3)} P = (3/8, 0, 5/16, 0, 1/4, 0, 1/16)$$
$$p^{(5)} = p^{(4)} P = (3/8, 5/32, 0, 9/32, 0, 1/8, 1/16)$$

P\{the man has lost his money at the end of 5 plays\}

$= P\{X_5 = 0\}$ = the entry corresponding to state 0 in $p^{(5)}$

$= 3/8$

Again $$p^{(6)} = p^{(5)}P = \left(\frac{29}{64}, 0, \frac{7}{32}, 0, \frac{13}{64}, 0, \frac{1}{8}\right)$$

$$p^{(7)} = p^{(6)}P = \left(\frac{29}{64}, \frac{7}{64}, 0, \frac{27}{128}, 0, \frac{13}{128}, \frac{1}{8}\right)$$

$P\{$the game lasts more than 7 rounds$\} = P\{$the system is neither in state 0 nor in 6 at the end of the seventh round$\}$

$$= P\{X_7 = 1, 2, 3, 4 \text{ or } 5\}$$

$$= \frac{7}{64} + 0 + \frac{27}{128} + 0 + \frac{13}{128} = \frac{27}{64}$$

Example 6 *There are 2 white marbles in urn A and 3 red marbles in urn B. At each step of the process, a marble is selected from each urn and the 2 marbles selected are interchanged. Let the state a_i of the system be the number of red marbles in A after i changes. What is the probability that there are 2 red marbles in A after 3 steps? In the long run, what is the probability that there are 2 red marbles in urn A?*

Solution State space of the chain $\{X_n\} = (0, 1, 2)$, since the number of balls in the urn A is always 2.

Let the tpm of the chain $\{X_n\}$ be

$$P = \begin{matrix} & 0 & 1 & 2 \\ 0 & p_{00} & p_{01} & p_{02} \\ 1 & p_{10} & p_{11} & p_{12} \\ 2 & p_{20} & p_{21} & p_{22} \end{matrix}$$

$p_{00} = 0$ (since the state cannot remain at 0 after interchange of marbles)

$p_{02} = p_{20} = 0$ (since the number of red marbles in urn cannot increase or decrease by 2 in one interchange)

To start with, A contains 0 red marble. After an interchange, A will contain 1 red marble (and 1 white marble) certainly.

$\therefore \quad p_{01} = 1.$

Let $X_n = 1$, i.e., A contains 1 red marble (and 1 white marble) and B contains 1 white and 2 red marbles.

Then $X_{n+1} = 0$, if A contains 0 red marble (and 2 white marbles) and B contains 3 red marbles, i.e., if 1 red marble is chosen from A and 1 white marble is chosen from B and interchanged.

$$\therefore P\{X_{n+1} = 0/X_n = 1\} = p_{10} = \frac{1}{2} \times \frac{1}{3} = \frac{1}{6}$$

Similarly, we can find $p_{12} = \frac{1}{2} \times \frac{2}{3} = \frac{1}{3}$

Since P is a stochastic matrix,

$p_{10} + p_{11} + p_{12} = 1$

$\therefore \qquad p_{11} = \frac{1}{2}$

Similarly, $p_{21} = \frac{2}{3}$ and $p_{22} = 1 - (p_{20} + p_{21}) = \frac{1}{3}$

$$\therefore \qquad P = \begin{pmatrix} 0 & 1 & 0 \\ 1/6 & 1/2 & 1/3 \\ 0 & 2/3 & 1/3 \end{pmatrix}$$

Now $\quad p^{(0)} = (1, 0, 0)$, as there is no red marble in A in the beginning.

$p^{(1)} = p^{(0)} P = (0, 1, 0)$

$$p^{(2)} = p^{(1)} P = \left(\frac{1}{6}, \frac{1}{2}, \frac{1}{3}\right)$$

$$p^{(3)} = p^{(1)} P = \left(\frac{1}{12}, \frac{23}{36}, \frac{5}{18}\right)$$

$\therefore \quad P\{$there are 2 red marbles in A after 3 steps$\}$

$$= P\{X_3 = 2\} = p_2^{(3)} = \frac{5}{18}$$

Let the stationary probability distribution of the chain be $\pi = (\pi_0, \pi_1, \pi_2)$. By the property of π, $\pi P = \pi$ and $\pi_0 + \pi_1 + \pi_2 = 1$

$$\text{i.e., } (\pi_0, \pi_1, \pi_2) \begin{pmatrix} 0 & 1 & 0 \\ 1/6 & 1/2 & 1/3 \\ 0 & 2/3 & 1/3 \end{pmatrix} = (\pi_0, \pi_1, \pi_2)$$

i.e., $\quad \frac{1}{6}\pi_1 = \pi_0$

$$\pi_0 + \frac{1}{2}\pi_1 + \frac{2}{3}\pi_2 = \pi_1$$

$$\frac{1}{3}\pi_1 + \frac{1}{3}\pi_2 = \pi_2$$

and $\quad \pi_1 + \pi_2 + \pi_3 = 1$

Solving, $\quad \pi_0 = \frac{1}{10}, \pi_1 = \frac{6}{10}, \pi_2 = \frac{3}{10}$

$\therefore$ $P\{$there are 2 red marbles in A in the long run$\} = 0.3$.

Example 7 Find the nature of the states of the Markov chain with the tpm

$$P = \begin{array}{c} \\ 0 \\ 1 \\ 2 \end{array} \begin{array}{c} \begin{array}{ccc} 0 & 1 & 2 \end{array} \\ \begin{pmatrix} 0 & 1 & 0 \\ 1/2 & 0 & 1/2 \\ 0 & 1 & 0 \end{pmatrix} \end{array}$$

Solution $P^2 = \begin{pmatrix} 1/2 & 0 & 1/2 \\ 0 & 1 & 0 \\ 1/2 & 0 & 1/2 \end{pmatrix}; P^3 = P$

$\therefore \quad P^4 = P^2$

and so on. In general, $P^{2n} = P^2$, $P^{2n+1} = P$

We note that $p_{00}^{(2)} > 0, p_{01}^{(1)} > 0, p_{02}^{(2)} > 0$

$p_{10}^{(1)} > 0, p_{11}^{(2)} > 0, p_{12}^{(1)} > 0$

$p_{20}^{(2)} > 0, p_{21}^{(1)} > 0, p_{22}^{(2)} > 0$

Therefore, the Markov chain is irreducible.

Also $p_{ii}^{(2)} = p_{ii}^{(4)} = p_{ii}^{(6)} \ldots > 0$, for all i, all the states of the chain are periodic, with period 2.

Since the chain is finite and irreducible, all its states are nonnull persistent. All states are not ergodic.

Example 8 *Three boys A, B and C are throwing a ball to each other. A always throws the ball to B and B always throws the ball to C, but C is just as likely to throw the ball to B as to A. Show that the process is Markovian. Find the transition matrix and classify the states.* (MKU — Nov. 96)

Solution The transition probability matrix of the process $\{X_n\}$ is given below:

State of X_{n-1} \ State of X_n	A	B	C
A	0	1	0
B	0	0	1
C	1/2	1/2	0

$$p \equiv \begin{matrix} A \\ B \\ C \end{matrix}\begin{pmatrix} 0 & 1 & 0 \\ 0 & 0 & 1 \\ 1/2 & 1/2 & 0 \end{pmatrix}$$

States of X_n depend only on states of X_{n-1}, but not on states of $X_{n-2}, X_{n-3}, \ldots$, or earlier states. Therefore, $\{X_n\}$ is a Markov chain.

Now $P^2 = \begin{pmatrix} 0 & 0 & 1 \\ 1/2 & 1/2 & 0 \\ 0 & 1/2 & 1/2 \end{pmatrix}; P^3 = \begin{pmatrix} 1/2 & 1/2 & 0 \\ 0 & 1/2 & 1/2 \\ 1/4 & 1/4 & 1/2 \end{pmatrix}$

$p_{11}^{(3)} > 0, p_{13}^{(2)} > 0, p_{21}^{(2)} > 0, P_{22}^{(2)} > 0, p_{33}^{(2)} > 0$ and all other $p_{ij}^{(1)} > 0$. Therefore, the chain is irreducible.

$$P^4 = \begin{pmatrix} 0 & 1/2 & 1/2 \\ 1/4 & 1/4 & 1/2 \\ 1/4 & 1/2 & 1/4 \end{pmatrix}; \quad P^5 = \begin{pmatrix} 1/4 & 1/4 & 1/2 \\ 1/4 & 1/2 & 1/4 \\ 1/8 & 3/8 & 1/2 \end{pmatrix}; \quad P^6 = \begin{pmatrix} 1/4 & 1/2 & 1/4 \\ 1/4 & 3/8 & 1/2 \\ 1/8 & 3/8 & 3/8 \end{pmatrix}$$

and so on.

We note that $p_{ii}^{(2)}, p_{ii}^{(3)}, p_{ii}^{(5)}; p_{ii}^{(6)}$ etc are > 0 for $i = 2, 3$, and GCD of 2, 3, 5, 6, … = 1.

Therefore, the states 2 and 3 (i.e., B and C) are periodic with period 1. i.e., aperiodic.

We note that $p_{11}^{(3)}, p_{11}^{(5)}, p_{11}^{(6)}$ etc. are > 0 and GCD of 3, 5, 6, … = 1

Therefore, the state 1 (i.e., state A) is periodic with period 1, i.e., aperiodic.

Since the chain is finite and irreducible, all its states are nonnull persistent. Moreover all the states are ergodic.

Exercise 7(C)

Part A (Short answer questions)

1. Define a Markov process.
2. Define a Markov chain and give an example of a Markov chain.
3. Prove that the Poisson process is a Markov process.
4. When is a Markov chain called homogeneous?
5. When is a homogeneous Markov chain said to be regular?
6. Define transition probability matrix of a Markov chain.
7. What is a stochastic matrix? When is it said to be regular?
8. Prove that the tpm of a Markov chain is a stochastic matrix.
9. Define n-step transition probability in a Markov chain.
10. State Chapman-Kolmogorov theorem.
11. What do you mean by probability distribution of a Markov chain?
12. When is a Markov chain completely specified?
13. What is meant by steady-state distribution of a Markov chain?
14. Write down the relation satisfied by the steady-state distribution and the tpm of a regular Markov chain.
15. If the tpm of a Markov chain is $\begin{pmatrix} 0 & 1 \\ 1/2 & 1/2 \end{pmatrix}$ find the steady-state distribution of the chain.
16. When is a Markov chain said to be irreducible or ergodic?
17. Prove that the matrix $\begin{pmatrix} 0 & 1 & 0 \\ 0 & 0 & 1 \\ 1/2 & 1/2 & 0 \end{pmatrix}$ is the tpm of an irreducible Markov chain.
18. What do you mean by an absorbing Markov chain. Give an example.
19. If the initial state probability distribution of a Markov chain is $p^{(0)} = \left(\frac{5}{6}, \frac{1}{6}\right)$ and the tpm of the chain is $\begin{pmatrix} 0 & 1 \\ 1/2 & 1/2 \end{pmatrix}$, find the probability distribution of the chain after 2 steps.

Part B

20. The tpm of a Markov chain with three states 0, 1, 2 is

$$P = \begin{pmatrix} 3/4 & 1/4 & 0 \\ 1/4 & 1/2 & 1/4 \\ 0 & 3/4 & 1/4 \end{pmatrix}$$

and the initial state distribution of the chain is $P\{X_0 = i\} = \frac{1}{3}$, $i = 0, 1, 2$.

Find (i) $P\{X_2 = 2\}$ and (ii) $P\{X_3 = 1, X_2 = 2, X_1 = 1, X_0 = 2\}$.

21. A man is at an integral point on the x-axis between the origin and the point 3. He takes a unit step to the right with probability $\frac{1}{3}$ or to the left with probability 2/3, unless he is at the origin, where he takes a step to the right to reach the point 1 or is at the point 3, where he takes a step to the left to reach the point 2. What is the probability that (i) he is at the point 1 after 3 walks? and (ii) he is at the point 1 in the long run?

22. Suppose that the probability of a dry day (state 0) following a rainy day (state 1) is $\frac{1}{3}$ and that the probability of a rainy day following a dry day is $\frac{1}{2}$. Given that May 1 is a dry day, find the probability that (i) May 3 is also a dry day and (ii) May 5 is also a dry day.

23. A gambler has Rs 3/-. At each play of the game, he loses Re 1 with probability $\frac{3}{4}$, but wins Rs 2/- with probability $\frac{1}{4}$. He stops playing if he has lost his initial amount of Rs 3/- or he has won at least Rs 3/-. Write down the tpm of the associated Markov chain. Find the probability that there are at least 4 rounds to the game.

24. A communication source can generate 1 of 3 possible messages 1, 2 and 3. Assume that the generation can be described by a homogeneous Markov chain with the following transition probability matrix

Current message	*Next message*		
	1	2	3
1	0.5	0.3	0.2
2	0.4	0.2	0.4
3	0.3	0.3	0.4

and the initial state probability distribution $p^{(0)} = (0.3, 0.3, 0.4)$. Find $p^{(3)}$.

25. Assume that the weather in a certain locality can be modeled as the homogeneous markov chain whose transition probability matrix is given below.

Today's weather	Tomorrow's weather		
	Fair	Cloudy	Rainy
Fair	0.8	0.15	0.05
Cloudy	0.5	0.3	0.2
Rainy	0.6	0.3	0.1

If the initial state distribution is given by $p^{(0)} = (0.7, 0.2, 0.1)$, find $p^{(2)}$ and $\lim_{n\to\infty} p^{(n)}$.

26. A fair coin is tossed until 3 heads occur in a row. Let X_n be the sequence of heads ending at the nth trial. What is the probability that there are at least 8 tosses of the coin?
27. There are 2 white marbles in urn A and 4 red marbles in urn B. At each step of the process, a marble is selected from each urn and the 2 marbles selected are interchanged. The state of the related Markov chain is the number of red balls in A after the interchange. What is the probability that there are 2 red balls in urn A (i) after 3 steps and (ii) in the long run?
28. A student's study habits are as follows: If he studies one night, he is 70% sure not to study the next night. On the other hand, if he does not study one night, he is 60% sure not to study the next night as well. In the long run, how often does he study?
29. A salesman's territory consists of 3 cities A, B and C. He never sells in the same city on successive days. If he sells in city A, then the next day he sells in B. However, if he sells either in B or C, then the next day he is twice as likely to sell in city A as in the other city. How often does he sell in each of the cities in the steady state?
30. A housewife buys 3 kinds of cereals, A, B and C. She never buys the same cereal in successive weeks. If she buys cereal A, the next week she buys cereal B. However if she buys B or C, the next week she is 3 times as likely to buy A as the other cereal. In the long run, how often she buys each of the three cereals?
31. Two boys B_1, B_2 and two girls G_1, G_2 are throwing a ball from one to another. Each boy throws the ball to the other boy with probability 1/2 and to each girl with probability 1/4. On the other hand, each girl throws the ball to each boy with probability 1/2 and never to the other girl. In the long run, how often does each receive the ball?
32. A gambler's luck follows a pattern. If he wins a game, the probability of his winning the next game is 0.6. However if he loses a game, the probability of his losing the next game is 0.7. There is an even chance that the gambler wins the first game. What is the probability that he wins (i) the second game, (ii) the third game and (iii) in the long run?
33. The three-state Markov chain is given by the tpm

$$P = \begin{pmatrix} 0 & 2/3 & 1/3 \\ 1/2 & 0 & 1/2 \\ 1/2 & 1/2 & 0 \end{pmatrix}$$

Prove that the chain is irreducible and all the states are aperiodic and non-null persistent. Find also the steady–state distribution of the chain.

34. A man tosses a fair coin until 3 heads occur in a row. Let $X_n = k$, if at the nth trial, the last tail occurred at the $(n - k)$th trial; i.e., X_n denotes the longest string of heads ending at the nth trial. Show that the process is Markovian. Find the transition matrix and classify the states.

(MKU — Apr. 97)

Answers

Exercise 7(A)

3. DC metre measures the mean of the input process;
   $\therefore$ output of the dc meter = 0 V
   True rms metre measures the SD of the input process.
   $\therefore$ output of the true rms meter = 2 V
5. Gaussian process is used to model and analyse the effects of thermal noise in electronic circuits used in communication system.
9. Square law detector process, Full-wave linear detector process, Half-wave linear detector process and Hard limiter process.
18. $X(t) = R_x(t) \cos \{\omega_0 \pm \theta_x(t)\}$
19. In communication systems, information bearing signals are often narrow-band Gaussian processes. When such signals are viewed on an oscilloscope, they appear like a sine wave with slowly varying amplitude and phase. Hence the representation.
21. $R_x(t)$ (envelope) follows a Rayleigh distribution and $\theta_x(t)$ (phase) follows a uniform distribution in $(0, 2\pi)$.
23. They are low pass processes.
24. A zero mean WSS process $\{X(t)\}$ can be represented in the form $X(t) = I(t) \cos \omega_0 t - Q(t) \sin \omega_0 t$. This kind of representation is called the quadrature representation.
25. WSS processes with zero mean can be represented in the quadrature form.
26. The quadrature representation is useful in communication theory, only when $\{X(t)\}$ is a zero mean, WSS, bandpass process.
30. The graph of $S_{NN}(\omega)$ is a straight line parallel to the ω-axis.
31. $S_{NN}(\omega)$ is a constant for all values of ω, i.e., $S_{NN}(\omega)$ contains all frequencies in equal amount. White noise is called so in analogy to white light which consists of all colours.

32. $S_{YY}(\omega) = \dfrac{N_0}{2}\,|H(\omega)|^2$, where $\{Y(t)\}$ is the output process and $H(\omega)$ is the power transfer function.

33. $$F\left\{\frac{N_0}{2}\,\delta(\tau)\right\} = \int_{-\infty}^{\infty} \frac{N_0}{2}\,\delta(\tau)\,e^{-i\omega\tau}\,d\tau = \frac{N_0}{2}$$

$$\therefore\ R_{NN}(\tau) = F^{-1}\,\frac{N_0}{2} = \frac{N_0}{2}\,\delta(\tau)$$

34. $$E\{N^2(t)\} = R_{NN}(0) = \int_{-\infty}^{\infty} S_{NN}(\omega)\,d\omega = \int_{-\infty}^{\infty} \frac{N_0}{2}\,d\omega \to \infty$$

35. Since $\int_{-\infty}^{\infty} S_{NN}(\omega)d\omega \to \infty$, it is not physically realisable. If the frequency band is taken to be finite, say $(-\omega_B,\ \omega_B)$, then $\int_{-\omega_B}^{\omega_B} S_{NN}(\omega)\,d\omega = N_0\,\omega_B < \infty$. Thus the spectral density becomes realisable.

38. $$S_{NN}(\omega) = \begin{cases} \dfrac{N_o}{2}, & \text{in } |\omega| \le \omega_B \\ 0, & \text{elsewhere} \end{cases}$$

$$R_{NN}(\tau) = \frac{1}{2\pi}\int_{-\omega_B}^{\omega_B} \frac{N_0}{2}\,e^{i\tau\omega}d\omega = \frac{N_0}{2\pi}\int_{0}^{\omega_B} \cos\tau\omega\,d\omega = \frac{N_0}{2\pi}\,\frac{\sin\tau\,\omega_B}{\tau}$$

39. $$E\{N^2(t)\} = R_{NN}(0) = \frac{N_0\,\omega_B}{2\pi}\,\lim_{\tau\to 0}\left(\frac{\sin\tau\,\omega_B}{\tau\,\omega B}\right) = \frac{N_0\,\omega_B}{2\pi}$$

42. Filtering in a stable, linear, time-invariant system is done by selecting carefully the power transfer function $H(\omega)$ so that certain undesired spectral components of the input signal are removed or filtered out.

43. (i) 0.309; (ii) 0.4

46. $4\,e^{-4}$; $8\,(1 + e^{-4})$; $\dfrac{1}{2\sqrt{2\pi}}\,e^{-z^2/8}\quad -\infty < z < \infty$; 0.6915;

$$\frac{1}{8\pi\sqrt{1-e^{-8}}}\,\exp\left[-\frac{1}{8\,(1-e^{-8})}\,(z^2 - 2e^{-4}zw + \omega^2)\right] \quad -\infty < z,\ \omega < \infty$$

48. 0, 2, 4 volts

49. $\dfrac{\{R_{xx}(\tau)-\mu_x^2\}}{R_x(0)-\mu_x^2}\,X(t_1)$, where $\tau = |t_1 - t_2|$

50. (i) $Z(t)$ is a Gaussian R.V. for all t and hence $\{Z(t)\}$ is a Gaussian process.

(ii) $f(z_1, z_2) = \dfrac{1}{2\pi\sqrt{1-r^2}} \exp\left\{-\dfrac{1}{2(1-r^2)}\, z_1^2 - 2rz_1z_2 + z_2^2\right\}$

where $r = \cos\omega(t_1 - t_2)$.

(iii) Yes (iv) Yes

(v) $R_{zz}(\tau)\cdot z(t_1)$ where $\tau = |t_1 - t_2|$

55. $\dfrac{N_0}{2}\,\delta(\tau)$

57. (i) $\dfrac{N_0\omega_B}{2\pi}\left(\dfrac{\sin\omega_B\tau}{\omega_B\tau}\right)$ (ii) $\dfrac{N_0\omega_B}{2\pi}\left(\dfrac{\sin\omega_B\tau}{\omega_B\tau}\right)\cos\omega_0\tau$

58. 2.3% of the time

60. $\breve{X}(t) = A\sin\omega_0 t - B\cos\omega_0 t$

62. $\dfrac{\alpha N_0}{4}$; $\dfrac{\alpha N_0}{4}\,e^{-\alpha\tau}$, $\tau \ge 0$

Exercise 7(B)

1. An ensemble of discrete sets of points from the time domain is called a point process, e.g. the times at which components fail in a large system; the times at which phone calls arrive at an exchange.

11. $P\{X(t) = k\} = \dfrac{e^{-\lambda t}(\lambda t)^k}{k!}$, where $\lambda = 2$; $P\{X(1) = 0\} = e^{-2}$

12. $e^{-\lambda t(1 - e^{i\omega})}$

13. 0.1755

14. (i) 0.0504 (ii) 0.0054 (iii) 0.7769 (iv) 0.5941

15. 0.1144

16. (i) e^{-12} (ii) $1 - e^{-3(t-12)}$, $t \ge 12$

17. 0.9972

19. 0.9707

20. 6 minutes

21. $f_T(t) = \dfrac{\lambda^k \cdot t^{k-1} e^{-\lambda t}}{\lfloor k-1}$, $t > 0$

23. (i) 30 (ii) 60

24. (i) 30 (ii) 30

Exercise 7(C)

7. A square matrix, in which the sum of all the elements of each row is 1, is called a stochastic matrix. A stochastic matrix P is said to be regular if all the entries of P^m (for some positive integer m) are positive.
12. A Markov chain is completely specified when the initial probability distribution and the tpm are given.
14. If $\pi = (\pi_1, \pi_2, \ldots, \pi_n)$ is the steady-state distribution of the chain whose tpm is the nth order square matrix P, then $\pi P = \pi$.
15. If (π_1, π_2) is the steady-state distribution of the chain, $(\pi_1, \pi_2).\begin{pmatrix} 0 & 1 \\ \frac{1}{2} & \frac{1}{2} \end{pmatrix}$

$= (\pi_1, \pi_2); \frac{1}{2}\pi_2 = \pi_1$ and $\pi_1 + \frac{1}{2}\pi_2 = \pi_2$, i.e., $2\pi_1 = \pi_2$. Also $\pi_1 + \pi_2 = 1$

$\therefore \pi_1 = \frac{1}{3}, \pi_2 = \frac{2}{3}$.

18. A state i of a markov chain is said to be an absorbing state if $p_{ii} = 1$, i.e., if it is impossible to leave it.
A Markov chain is said to be absorbing if it has at least one absorbing state.

$$\textbf{Example :}\quad \begin{array}{c} \\ 1 \\ 2 \\ 3 \\ 4 \end{array}\begin{array}{c} \begin{array}{cccc} 1 & 2 & 3 & 4 \end{array} \\ \begin{pmatrix} 1 & 0 & 0 & 0 \\ 1/2 & 0 & 1/2 & 0 \\ 0 & 1/2 & 0 & 1/2 \\ 0 & 0 & 0 & 1 \end{pmatrix} \end{array}$$

19. $p^{(1)} = P^{(0)}P;\ P^{(2)} = p^{(1)}P = \left(\frac{11}{14}, \frac{13}{24}\right)$.
20. 1/6; 3/64
21. $P = \begin{pmatrix} 0 & 1 & 0 & 0 \\ 2/3 & 0 & 1/3 & 0 \\ 0 & 2/3 & 0 & 1/3 \\ 0 & 0 & 1 & 0 \end{pmatrix}$

(i) $\frac{22}{27}$ (ii) $\frac{3}{7}$

22. (i) 5/12; (ii) 173/432
23. 27/64
24. (• 4083, • 2727, • 3190)

25. (• 7245, • 1920, • 0835); $\left(\frac{114}{157}, \frac{30}{157}, \frac{13}{157}\right)$

26. 81/128

27. 3/8; 2/5

28. 4/11 of the nights

29. $P = \begin{pmatrix} 0 & 1 & 0 \\ 2/3 & 0 & 1/3 \\ 2/3 & 1/3 & 0 \end{pmatrix}$; (40%, 45%, 15%)

30. $P = \begin{pmatrix} 0 & 1 & 0 \\ 3/4 & 0 & 1/4 \\ 3/4 & 1/4 & 0 \end{pmatrix}$; $\left(\frac{15}{35}, \frac{16}{35}, \frac{4}{35}\right)$

31. $\left(\frac{1}{3}, \frac{1}{3}, \frac{1}{6}, \frac{1}{6}\right)$

32. $\frac{9}{20}, \frac{87}{200}, \frac{3}{7}$

33. $\left(\frac{9}{27}, \frac{10}{27}, \frac{8}{27}\right)$

34. $$X_n \begin{matrix} \\ 0 \\ 1 \\ 2 \\ 3 \end{matrix} \overset{X_{n+1}}{\begin{pmatrix} 0 & 1 & 2 & 3 \\ 1/2 & 1/2 & 0 & 0 \\ 1/2 & 0 & 1/2 & 0 \\ 1/2 & 0 & 0 & 1/2 \\ 0 & 0 & 0 & 1 \end{pmatrix}}$$

The chain is not irreducible. State 3 is absorbring and other states are aperiodic.

8 Tests of Hypotheses

Every statistical investigation aims at collecting information about some aggregate or collection of individuals or of their attributes, rather than the individuals themselves. In statistical language, such a collection is called a *population* or *universe*. For example, we have the population of products turned out by a machine, of lives of electric bulbs manufactured by a company etc. A population is finite or infinite, according as the number of elements is finite or infinite. In most situations, the population may be considered infinitely large. A finite subset of a population is called *a sample* and the process of selection of such samples is called *sampling*. The basic objective of the theory of sampling is to draw inference about the population, using the information of the sample.

Parameters and Statistics

Generally in statistical investigations, our ultimate interest will lie in one or more characteristics possessed by the members of the population. If there is only one characteristic of importance, it can be assumed to be a variable x. If x_i be the value of x for the ith member of the sample, then $(x_1, x_2, \ldots, x_n)$ are referred to as sample observations. Our primary interest will be to know the values of different statistical measures such as mean and variance of the population distribution of x. Statistical measures calculated on the basis of population values of x are called *parameters*. Corresponding measures computed on the basis of sample observations are called *statistics*.

Sampling Distribution

If a number of samples, each of size n (i.e. each containing n elements) are drawn from the same population and if for each sample the value of some statistic, say, mean is calculated, a set of values of the statistic will be obtained.

Note: *The values of the statistic will normally vary from one sample to another, as the values of the population members included in different samples, though drawn from the same population, may be different. These differences in the values of a statistic are said to be due to sampling fluctuations.*

If the number of samples is large, the values of the statistic may be classified in the form of a frequency table. The probability distribution of the statistic that would be obtained if the number of samples, each of same size, were infinitely large is called the *sampling distribution* of the statistic. If we adopt random sampling technique that is, the most popular and frequently used method of sampling (the discussion of which is beyond the scope of this book), the nature of the sampling distribution of a statistic can be obtained theoretically, using the theory of probability, provided the nature of the population distribution is known.

Like any other distribution, a sampling distribution will have its mean, standard deviation and moments of higher order. The standard deviation of the sampling distribution of a statistic is of particular importance in tests of hypotheses and is called *the standard error* of the statistic.

Estimation and Testing of Hypotheses

In sampling theory, we are primarily concerned with two types of problems:

(i) Some characteristic or feature of the population in which we are interested may be completely unknown to us and we may like to make a guess about this characteristic entirely on the basis of a random sample drawn from the population. This type of problem is known as the problem of *estimation*.

(ii) Some information regarding the characteristic or feature of the population may be available to us and we may like to know whether the information is tenable (or can be accepted) in the light of the random sample drawn from the population and if it can be accepted, with what degree of confidence it can be accepted. This type of problem is known as the problem of *testing of hypotheses*.

Tests of Hypotheses and Tests of Significance

When we attempt to make decisions about the population on the basis of sample information, we have to make assumptions or guesses about the nature of the population involved or about the value of some parameter of the population. Such assumptions, which may or may not be true, are called *statistical hypotheses*. Very often, we set up a hypothesis which assumes that there is no significant difference between the sample statistic and the corresponding population parameter or between two sample statistics. Such a hypothesis of no difference is called a *null hypothesis* and is denoted by H_0. A hypothesis that is different from (or complementary to) the null hypothesis is called an *alternative hypothesis* and is denoted by H_1. A procedure for deciding whether to accept or to reject a null hypothesis (and hence to reject or to accept the alternative hypothesis respectively) is called the *test of hypothesis*.

If θ_0 is a parameter of the population and θ is the corresponding sample statistic, usually there will be some difference between θ_0 and θ since θ is based on sample observations and is different for different samples. Such a difference which is caused due to sampling fluctuations is called *insignificant difference*. The difference that arises either because the sampling procedure is not purely random or that the sample has not been drawn from the given population is known as *significant difference*. This procedure of testing whether the difference between θ_0 and θ is significant or not is called the *test of significance*.

Critical Region and Level of Significance

If we are prepared to accept that the difference between a sample statistic and the corresponding parameter is significant when the sample statistic lies in a certain region or interval, then that region is called the *critical region* or *region of rejection*. The region complementary to the critical region is called the *region of acceptance*.

In the case of large samples, the sampling distributions of many statistics tend to become normal distributions. If t is a statistic in large samples, then t follows a normal distribution with mean $E(t)$, which is the corresponding population parameter, and SD equal to SE (t). Hence $Z = \dfrac{t - E(t)}{\text{SE}(t)}$ is a standard normal variate i.e., Z (called the *test statistic*) follows a normal distribution with mean zero and SD unity.

It is known from the study of normal distribution, that the area under the standard normal curve between $t = -1.96$ and $t = +1.96$ is 0.95. Equivalently the area under the general normal curve of t between $[E(t) - 1.96\ \text{SE}(t)]$ and $[E(t) + 1.96\ \text{SE}(t)]$ is 0.95. In other words, 95% of the values of t will lie between $[E(t) \mp 1.96\ \text{SE}(t)]$ or only 5% of values of t will lie outside this interval.

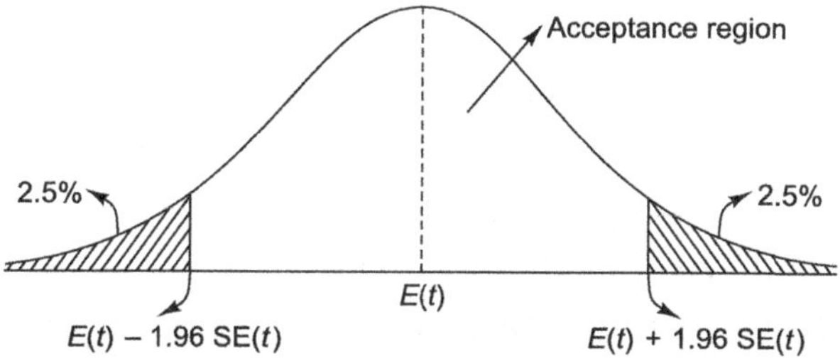

Fig. 8.1

If we are prepared to accept that the difference between t and $E(t)$ is significant when t lies in either of the regions $[-\infty, E(t) - 1.96\text{SE}(t)]$ and $[E(t) + 1.96\text{SE}(t), \infty]$ then these two regions constitute the critical region of t.

The probability that a random value (α) of the statistic lies in the critical region is called the *level of significance* (LOS) and is usually expressed as a percentage. In the other words, the total area of the critical region expressed as $\alpha\%$ is the LOS.

From the study of normal distributions, it is known that

$$P\{E(t) - 1.96\ \text{SE}(t) < t < E(t) + 1.96\ \text{SE}(t)\} = 0.95$$

i.e. $$P\left\{\left|\frac{t - E(t)}{\mathrm{SE}(t)}\right| < 1.96\right\} = 0.95$$

i.e. $$P\{|Z| > 1.96\} = 0.05 \quad \text{or} \quad 5\%$$

Thus when t lies in either of the two regions constituting the critical region given above, the LOS is 5%.

Note *The level of significance can also be defined as the maximum probability with which we are prepared to reject H_0 when it is true. In other words, the total area of the region of rejection expressed as a percentage is called the level of significance.*
(The specification of critical region and the choice of LOS will depend upon the nature of the problem and is a matter of judgement for those who carry out the investigation. Usually the LOS are taken as 5, 2 or 1%.)

Errors in Testing of Hypotheses

The LOS is fixed by the investigator and as such it may be fixed at a higher level by his wrong judgement. Because of this, the region of rejection becomes larger and the probability of rejecting a null hypothesis, when it is true, becomes greater. The error committed in rejecting H_0, when it is really true, is called *Type I error*. This is similar to a good product being rejected by the consumer and hence Type I error is also known as *producer's risk*. The error committed in accepting H_0, when it is false, is called *Type II error*. As this error is similar to that of accepting a product of inferior quality, it is also known as *consumer's risk.*

The probabilities of committing Type I and II errors are denoted by α and β respectively. It is to be noted that the probability α of committing Type I error is the LOS.

One-Tailed and Two-Tailed Tests

If θ_0 is a population parameter and θ is the corresponding sample statistic and if we set up the null hypothesis $H_0 : \theta = \theta_0$, then the alternative hypothesis which is complementary to H_0 can be any one of the following:

(i) $H_1 : \theta \neq \theta_0$, i.e. $\theta > \theta_0$ or $\theta < \theta_0$

(ii) $H_1 : \theta > \theta_0$

(iii) $H_1 : \theta < \theta_0$

H_1 given in (i) is called a two-tailed alternative hypothesis, whereas H_1 given in (ii) is called a right-tailed alternative hypothesis and H_1 given in (iii) is called a left-tailed alternative hypothesis.

When H_0 is tested while H_1 is a one-tailed alternative (right or left), the test of hypothesis is called *a one-tailed test.*

When H_0 is tested while H_1 is two-tailed alternative, the test of hypothesis is called a *two-tailed test.*

The application of one-tailed or two-tailed test depends upon the nature of the alternative hypothesis. The choice of the appropriate alternative hypothesis depends on the situation and the nature of the problem concerned.

Critical Values or Significant Values

The value of the test statistic z for which the critical region and acceptance region are separated is called the *critical value* or the *significant value* of z and denoted by z_α, when α is the LOS. It is clear that the value of z_α depends not only on α but also on the nature of alternative hypothesis.

When $z = \dfrac{t - E(t)}{\text{SE}(t)}$, we have seen that

$P(|z| < 1.96) = 95\%$ and $P(|z| > 1.96) = 5\%$.

Thus $z = \pm 1.96$ separates the critical region and the acceptance region at 5% LOS for a two-tailed test. That is the critical values of z in this case are ± 1.96.

In general, the critical value z_α for the LOS α is a given by the equation $P(|z| > z_\alpha) = \alpha$ for a two-tailed test, by the equation $P(z > z_\alpha) = \alpha$ for the right-tailed test and by the equation.

$P(z < -z_\alpha) = \alpha$ for the left-tailed test. These equations are solved by using the normal tables.

Note *If z_α is the critical value of z corresponding to the LOS α in the right-tailed test, then $P(z > z_\alpha) = \alpha$.*

By symmetry of the standard normal distribution followed by z, $P(z < -z_\alpha) = \alpha$.

$$\therefore \quad P(|z| > z_\alpha) = P\{(z > z_\alpha) + (z < -z_\alpha)\}$$
$$= P\{z > z_\alpha) + P(z < -z_\alpha)$$
$$= 2\alpha$$

That is z_α is the critical value of z corresponding to the LOS 2α.

Thus the critical value of z for a single-tailed test (right or left) at LOS α is the same as that for a two-tailed test of LOS 2α.

The critical values for some standard LOSs are given in the following table both for two-tailed and one-tailed tests.

Nature of test	*LOS* 1% (0.01)	2% (0.02)	5% (0.05)	10% (0.1)
Two-tailed	$\|z_\alpha\| = 2.58$	$\|z_\alpha\| = 2.33$	$\|z_\alpha\| = 1.96$	$\|z_\alpha\| = 1.645$
Right-tailed	$z_\alpha = 2.33$	$z_\alpha = 2.055$	$z_\alpha = 1.645$	$z_\alpha = 1.28$
Left-tailed	$z_\alpha = -2.33$	$z_\alpha = -2.055$	$z_\alpha = -1.645$	$z_\alpha = -1.28$

Procedure for Testing of Hypothesis

1. Null hypothesis H_0 is defined.
2. Alternative hypothesis H_1 is also defined after a careful study of the problem and also the nature of the test (whether one-tailed or two-tailed) is decided.

3. LOS α is fixed or taken from the problem if specified and z_α is noted.
4. The test-statistic $z = \dfrac{t - E(t)}{\text{SE}(t)}$ is computed.
5. Comparison is made between $|z|$ and z_α. If $|z| < z_\alpha$, H_0 is accepted or H_1 is rejected, i.e. it is concluded that the difference between t and $E(t)$ is not significant at α% LOS.

On the other hand, if $|z| > z_\alpha$, H_0 is rejected or H_1 is accepted, i.e. it is concluded that the difference between t and $E(t)$ is significant at α % LOS.

Interval Estimation of Population Parameters

It was pointed out that the objective of the theory of sampling is to estimate population parameters with the help of the corresponding sample statistics. Estimation of a parameter by single value is referred to as *point estimation*, the study of which is beyond the scope of this book. However, an alternative procedure is to give an interval within which the parameter may be supposed to lie. This is called *interval estimation*. The interval within which the parameter is expected to lie is called the *confidence interval* for that parameter. The end points of the confidence interval are called *confidence limits* or *fiducial limits*.

We have already seen that

$$P\{|z| \le 1.96\} = 0.95$$

i.e. $$P\left\{\left|\frac{t - E(t)}{\text{SE}(t)}\right| \le 1.96\right\} = 0.95$$

i.e. $P\{t - 1.96\text{SE}(t) \le E(t) \le t + 1.96\text{SE}(t)\} = 0.95$

This means that we can assert, with 95% confidence, that the parameter $E(t)$ will lie between $t - 1.96$ SE(t) and $t + 1.96$SE(t). Thus $\{t - 1.96\text{SE}(t), t + 1.96\ \text{SE}(t)$ are the 95% confidence limits for $E(t)$.

Similarly $\{t - 2.58\ \text{SE}(t), t + 2.58\ \text{SE}(t)\}$ is the 99% confidence interval for $E(t)$.

Tests of Significance for Large Samples

It is generally agreed that, if the size of the sample exceeds 30, it should be regarded as a large sample. The tests of significance used for large samples are different from the ones used for small samples for the reason that the following assumptions made for large samples do not hold for small samples:

1. The sampling distribution of a statistic is approximately normal, irrespective of whether the distribution of the population is normal or not.
2. Sample statistics are sufficiently close to the corresponding population parameters and hence may be used to calculate the standard error of the sampling distribution.

Test 1

Test of significance of the difference between sample proportion and population proportion.

Let X be the number of successes in n independent Bernoulli trials in which the probability of success for each trial is a constant, (say) = P. Then it is known that X follows a binomial distribution with mean $E(X) = nP$ and variance $V(X) = nPQ$.

When n is large, X follows $N(nP, \sqrt{nPQ})$, i.e. a normal distribution with mean nP and SD $\sqrt{nPQ}$, where $Q = 1 - P$.

$$\therefore \quad \frac{X}{n} \text{ follows } N\left\{\frac{nP}{n}, \sqrt{\frac{nPQ}{n^2}}\right\}$$

Now $\frac{X}{n}$ is the proportion of successes in the sample consisting of n trials, which is denoted by p. Thus the sample proportion p follows $N\left(P, \sqrt{\frac{PQ}{n}}\right)$.

Therefore, the test statistic z is given by $z = \dfrac{p - P}{\sqrt{\dfrac{PQ}{n}}}$.

If $|z| \le z_\alpha$, the difference between the sample proportion p and the population proportion P is not significant at α % LOS.

Note 1. *If P is not known, we assume that p is nearly equal to P, and hence, SE(p) is taken as* $\sqrt{\frac{pq}{n}}$. *Thus*

$$z = \frac{p - P}{\sqrt{\dfrac{pq}{n}}}$$

2. *95% confidence limits for P are then given by*

$$\frac{|P - p|}{\sqrt{\dfrac{pq}{n}}} \le 1.96,$$

i.e. they are $\left(p - 1.96\sqrt{\frac{pq}{n}}, p + 1.96\sqrt{\frac{pq}{n}}\right)$.

Test 2

Test of significance of the difference between two sample proportions.

Let p_1 and p_2 be the proportions of successes in two large samples of size n_1 and n_2, respectively, drawn from the same population or from two populations with the same proportion P.

Then p_1 follows $N\left(P, \sqrt{\frac{PQ}{n_1}}\right)$ and p_2 follows $N\left(P, \sqrt{\frac{PQ}{n_2}}\right)$.

Therefore $p_1 - p_2$, which is a linear combination of two normal variables, also follows a normal distribution.

Now $\quad E(p_1 - p_2) = E(p_1) - E(p_2) = P - P = 0$

$$V(p_1 - p_2) = V(p_1) + V(p_2) \quad (\because \text{ the two samples are independent})$$

$$= PQ\left(\frac{1}{n_1} + \frac{1}{n_2}\right)$$

Therefore, $(p_1 - p_2)$ follows $N\left\{0, \sqrt{PQ\left(\frac{1}{n_1} + \frac{1}{n_2}\right)}\right\}$.

Therefore, The test statistic z is given by

$$z = \frac{p_1 - p_2}{\sqrt{PQ\left(\frac{1}{n_1} + \frac{1}{n_2}\right)}}.$$

If P is not known, an unbiased estimate of P based on both samples, given by

$$\hat{P} = \frac{n_1 p_1 + n_2 p_2}{n_1 + n_2},$$

is used in the place of P.

As before, if $|z| \le z_\alpha$, the difference between the two sample proportions p_1 and p_2 is not significant at α % LOS.

Note *A sample statistic θ is said to be an unbiased estimate of the parameter θ_0, if $E(\theta) = \theta_0$. In the present case,*

$$E\left\{\frac{n_1 p_1 + n_2 p_2}{n_1 + n_2}\right\} = \frac{1}{n_1 + n_2}\{n_1 E(p_1) + n_2 E(p_2)\}$$

$$= \frac{1}{n_1 + n_2}(n_1 P + n_2 P) = P$$

Therefore, An unbiased estimate of P is $\left(\frac{n_1 p_1 + n_2 p_2}{n_1 + n_2}\right)$.

Test 3

Test of significance of the difference between sample mean and population mean.

Let $X_1, X_2, \ldots, X_n$ be the sample observations in a sample of size n, drawn from a population that is $N(\mu, \sigma)$.

Then each X_i follows $N(\mu, \sigma)$.

It is known that if X_i $(i = 1, 2, \ldots, n)$ are independent normal variates with mean μ_i and variance σ_i^2, then $\Sigma c_i X_i$ is a normal variate with mean $\mu = \Sigma c_i \mu_i$ and variance $\sigma^2 = \Sigma c_i^2 \sigma_i^2$.

Now putting $c_i = \frac{1}{n}$, $\mu_i = \mu$ and $\sigma_i = \sigma$, we get

$$\Sigma c_i X_i = \frac{1}{n} \Sigma X_i = \overline{X} \quad \Sigma c_i \mu_i = \frac{1}{n}\mu + \frac{1}{n}\mu + \ldots + \frac{1}{n}\mu \ (n \text{ terms}) = \mu$$

and

$$\Sigma c_i^2 \sigma_i^2 = \frac{1}{n^2}\sigma^2 + \frac{1}{n^2}\sigma^2 + \ldots + \frac{1}{n^2}\sigma^2 \ (n \text{ terms})$$

$$= \frac{\sigma^2}{n}.$$

Thus, if X_i are n independent normal variates with the same mean μ and same variance σ^2, then their mean $\overline{X}$ follows a $N\left(\mu, \frac{\sigma}{\sqrt{n}}\right)$. Even if the population, from which the sample is drawn, is non-normal, it is known (from central limit theorem) that the above result holds good, provided n is large.

Therefore, the test statistic z is given by

$$z = \frac{\overline{X} - \mu}{\sigma / \sqrt{n}}.$$

As usual, if $|z| \le z_\alpha$, the difference between the sample mean $\overline{X}$ and the population mean μ is not significant at α% LOS.

Note *1. If σ is not known, the sample SD s can be used in its place, as s is nearly equal to σ when n is large.*

2. 95% confidence limits for μ are given by $\frac{|\mu - \overline{X}|}{\sigma/\sqrt{n}} \le 1.96$, i.e.

$\left(\overline{X} - 1.96 \frac{\sigma}{\sqrt{n}}, \overline{X} + 1.96 \frac{\sigma}{\sqrt{n}}\right)$, if σ is known. If σ is not known, then

the 95% confidence interval is $\left(\overline{X} - \frac{1.96 s}{\sqrt{n}}, \overline{X} + \frac{1.96 s}{\sqrt{n}}\right)$.

Test 4

Test of significance of the difference between the means of two samples.

Let $\overline{X}_1$ and $\overline{X}_2$ be the means of two large samples of sizes n_1 and n_2 drawn from two populations (normal or non-normal) with the same mean μ and variances σ_1^2 and σ_2^2 respectively.

Then $\overline{X}_1$ follows $N\left(\mu, \frac{\sigma_1}{\sqrt{n_1}}\right)$ and $\overline{X}_2$ follows $N\left(\mu, \frac{\sigma_2}{\sqrt{n_1}}\right)$ either exactly or approximately.

Therefore, $\overline{X}_1 - \overline{X}_2$ also follows a normal distribution.

$$E(\overline{X}_1 - \overline{X}_2) = E(\overline{X}_1) - E(\overline{X}_2) = \mu - \mu = 0.$$

$$V(\overline{X}_1 - \overline{X}_2) = V(\overline{X}_1) + V(\overline{X}_2)$$

($\because$ $\overline{X}_1$ and $\overline{X}_2$ are independent, as the samples are independent)

$$= \frac{\sigma_1^2}{n_1} + \frac{\sigma_2^2}{n_2}$$

Thus $(\overline{X}_1 - \overline{X}_2)$ follows $N\left\{0, \sqrt{\frac{\sigma_1^2}{n_1} + \frac{\sigma_2^2}{n_2}}\right\}$.

Therefore, The test statistic z is given by

$$z = \frac{\overline{X}_1 - \overline{X}_2}{\sqrt{\frac{\sigma_1^2}{n_1} + \frac{\sigma_2^2}{n_2}}} \quad (1)$$

If $|z| \le z_\alpha$, the difference between $(\overline{X}_1 - \overline{X}_2)$ and 0 or the difference between $\overline{X}_1$ and $\overline{X}_2$ is not significant at α % LOS.

Note 1. *If the samples are drawn from the same population, i.e. if $\sigma_1 = \sigma_2 = \sigma$, then*

$$z = \frac{\overline{X}_1 - \overline{X}_2}{\sigma\sqrt{\frac{1}{n_1} + \frac{1}{n_2}}} \quad (2)$$

2. *If σ_1 and σ_2 are not known and $\sigma_1 \neq \sigma_2$, σ_1 and σ_2 can be approximated by the sample SDs s_1 and s_2. Hence, in such a situation [from (1)],*

$$z = \frac{\overline{X}_1 - \overline{X}_2}{\sqrt{\frac{s_1^2}{n_1} + \frac{s_2^2}{n_2}}} \quad (3)$$

3. *If σ_1 and σ_2 are equal and not known, then $\sigma_1 = \sigma_2 = \sigma$ is approximated by* $\sigma^2 = \dfrac{n_1 s_1^2 + n_2 s_2^2}{n_1 + n_2}$ *. Hence, in such a situation, [from (2)],*

$$z = \frac{\bar{X}_1 - \bar{X}_2}{\sqrt{\left(\dfrac{n_1 s_1^2 + n_2 s_2^2}{n_1 + n_2}\right)\left(\dfrac{1}{n_1} + \dfrac{1}{n_2}\right)}},$$

i.e.
$$z = \frac{\bar{X}_1 - \bar{X}_2}{\sqrt{\dfrac{s_1^2}{n_2} + \dfrac{s_2^2}{n_1}}} \qquad (4)$$

4. *The difference in the denominators of the values of z given in (3) and (4) may be noted.*

Test 5

Test of significance of the difference between sample SD and population SD.

Let s be the SD of a large sample of size n drawn from a normal population with SD σ. Then it is known that s follows a $N\left(\sigma, \dfrac{\sigma}{\sqrt{2n}}\right)$ approximately.

Therefore, The test statistic z is given by

$$z = \frac{s - \sigma}{\sigma/\sqrt{2n}}$$

As before, the significance of the difference between s and σ is tested.

Test 6

Test of significance of the difference between SDs of two large samples.

Let s_1 and s_2 be the SDs of two large samples of sizes n_1 and n_2 drawn from a normal population with SD σ.

s_1 follows $N\left(\sigma, \dfrac{\sigma}{\sqrt{2n_1}}\right)$ and s_2 follows $N\left(\sigma, \dfrac{\sigma}{\sqrt{2n_2}}\right)$.

Therefore, $(s_1 - s_2)$ follows $N\left\{0, \sigma\sqrt{\dfrac{1}{2n_1} + \dfrac{1}{2n_2}}\right\}$.

Therefore, the test statistic z is given by

$$z = \frac{s_1 - s_2}{\sigma\sqrt{\dfrac{1}{2n_1} + \dfrac{1}{2n_2}}}.$$

As usual, the significance of the difference between s_1 and s_2 is tested.

Note: If σ is not known, it is approximated by $\sqrt{\left(\dfrac{n_1 s_1^2 + n_2 s_2^2}{n_1 + n_2}\right)}$, when n_1 and n_2 are large. In this situation,

$$z = \frac{s_1 - s_2}{\sqrt{\left(\dfrac{n_1 s_1^2 + n_2 s_2^2}{n_1 + n_2}\right)\left(\dfrac{1}{2n_1} + \dfrac{1}{2n_2}\right)}}$$

i.e.

$$z = \frac{s_1 - s_2}{\sqrt{\dfrac{s_1^2}{2n_2} + \dfrac{s_2^2}{2n_1}}}$$

Worked Examples 8(A)

Example 1 *Experience has shown that 20% of a manufactured product is of top quality. In one day's production of 400 articles, only 50 are of top quality. Show that either the production of the day chosen was not a representative sample or the hypothesis of 20% was wrong. Based on the particular day's production, find also the 95% confidence limits for the percentage of top quality product.*

Solution $H_0 : P = \dfrac{1}{5}$, i.e. 20% of the products manufactured is of top quality.

$H_1 : P \neq \dfrac{1}{5}$.

p = Proportion of top quality products in the sample

$$= \frac{50}{400} = \frac{1}{8}$$

From the alternative hypothesis H_1, we note that two-tailed test is to be used. Let LOS be 5%. Therefore, $z_\alpha = 1.96$.

$$z = \frac{p - P}{\sqrt{\dfrac{PQ}{n}}} = \frac{\dfrac{1}{8} - \dfrac{1}{5}}{\sqrt{\dfrac{1}{5} \times \dfrac{4}{5} \times \dfrac{1}{400}}}$$

since the size of the sample is equal to 400.

i.e $$z = -\frac{3}{40} \times 50 = -3.75$$

Now $|z| = 3.75 > 1.96$.

The difference between p and P is significant at 5% level.

Also H_0 is rejected. Hence H_0 is wrong or the production of the particular day chosen is not a representative sample.

95% confidence limits for P are given by

$$\frac{|p-P|}{\sqrt{\frac{pq}{n}}} \le 1.96$$

Note *We have taken* $\sqrt{\frac{pq}{n}}$ *in the denominator, because P is assumed to be unknown, for which we are trying to find the confidence limits and P is nearly equal to p.*

i.e. $$p - \sqrt{\frac{pq}{n}} \times 1.96 \le P \le p + \sqrt{\frac{pq}{n}} \times 1.96$$

i.e. $$0.125 - \sqrt{\frac{1}{8} \times \frac{7}{8} \times \frac{1}{400}} \times 1.96 \le P \le 0.125 + \sqrt{\frac{1}{8} \times \frac{7}{8} \times \frac{1}{400}} \times 1.96$$

i.e. $0.093 \le P \le 0.157$

Therefore, 95% confidence limits for the percentage of top quality product are 9.3 and 15.7.

Example 2 *The fatality rate of typhoid patients is believed to be 17.26%. In a certain year 640 patients suffering from typhoid were treated in a metropolitan hospital and only 63 patients died. Can you consider the hospital efficient?*

Solution $H_0 : p = P$, i.e. the hospital is not efficient.

$H_1 : p < P$.

One-tailed (left-tailed) test is to be used.

Let LOS be 1%. Therefore, $z_\alpha = -2.33$.

$$z = \frac{p-P}{\sqrt{\frac{PQ}{n}}}, \text{ where } p = \frac{63}{640} = 0.0984$$

$$P = 0.1726, \qquad Q = 0.8274$$

$$z = \frac{0.0984 - 0.1726}{\sqrt{\frac{0.1726 \times 0.8274}{640}}} = -4.96$$

∴ $|z| > |z_\alpha|$

Therefore, difference between p and P is significant. i.e., H_0 is rejected and H_1 is accepted.

That is, the hospital is efficient in bringing down the fatality rate of typhoid patients.

Example 3 *A salesman in a departmental store claims that at most 60 percent of the shoppers entering the store leave without making a purchase. A random sample of 50 shoppers showed that 35 of them left without making a purchase. Are these sample results consistent with the claim of the salesman? Use an LOS of 0.05.*

Solution Let P and p denote the population and sample proportions of shoppers not making a purchase.

$$H_0 : p = P$$

$$H_1 : p > P, \text{ since } p = 0.7 \quad \text{and} \quad P = 0.6$$

One-tailed (right-tailed) test is to be used.

Let LOS be 5%. Therefore, $z_\alpha = 1.645$.

$$z = \frac{p - P}{\sqrt{\dfrac{PQ}{n}}} = \frac{0.7 - 0.6}{\sqrt{\dfrac{0.6 \times 0.4}{50}}} = 1.443$$

$$\therefore \qquad |z| < z_\alpha$$

Therefore, the difference between p and P is not significant at 5% level, i.e., H_0 is accepted and H_1 is rejected.

That is, the sample results are consistent with the claim of the salesman.

Example 4 *Show that for a random sample of size 100, drawn with replacement, the standard error of sample proportion cannot exceed 0.05.*

Solution The items of the sample are drawn one after another with replacement.

Therefore, the proportion (probability) of success in the population, i.e. P-remains a constant.

We know that the sample proportion p follows $N\left(P, \sqrt{\dfrac{PQ}{n}}\right)$,

$$\text{i.e. SE of } p = \sqrt{\frac{PQ}{n}} = \frac{1}{10}\sqrt{PQ} \quad (\because n = 100) \tag{1}$$

$$\text{Now} \qquad \left(\sqrt{P} - \sqrt{Q}\right)^2 \geq 0$$

$$\text{i.e.} \qquad P + Q - 2\sqrt{PQ} \geq 0$$

$$\text{i.e.} \qquad 1 - 2\sqrt{PQ} \geq 0 \quad \text{or} \quad \sqrt{PQ} \leq \frac{1}{2} \tag{2}$$

Using (2) in (1), we get,

SE of $p \leq \dfrac{1}{20}$,

i.e. SE of p cannot exceed 0.05.

Example 5 *A cubical die is thrown 9000 times and a throw of 3 or 4 is observed 3240 times. Show that the die cannot be regarded as an unbiased one, and find the extreme limits between which the probability of a throw of 3 or 4 lies.*

Solution H_0 : the die is unbiased, i.e. $P = \dfrac{1}{3}$ (= the probability of getting 3 or 4).

$H_1 : P \neq \dfrac{1}{3}$.

Two-tailed test is to be used.

Let LOS be 5%. Therefore, $z_\alpha = 1.96$.

Although we may test the significance of the difference between the sample and population proportions, we shall test the significance of the difference between the number of successes X in the sample and that in the population.

When n is large, X follows $np, \sqrt{nPQ}$ [refer to Test 1].

$$\therefore \qquad z = \frac{X - nP}{\sqrt{nPQ}} = \frac{3240 - \left(9000 \times \frac{1}{3}\right)}{\sqrt{9000 \times \frac{1}{3} \times \frac{2}{3}}} = 5.37$$

$\therefore \qquad |z| > z_\alpha$

Therefoe, the difference between X and nP is significant, i.e., H_0 is rejected. That is the dice cannot be regarded as unbiased.

If X follows a $N(\mu, \sigma)$, then the reader can easily verify that $P(\mu - 3\sigma \leq X \leq \mu + 3\sigma) = 0.9974$.

The limits $\mu \pm 3\sigma$ are considered as the extreme (confidence) limits within which X lies.

Accordingly, the extreme limits for P are given by

$$\frac{|P - p|}{\sqrt{\frac{pq}{n}}} \leq 3 \qquad \text{[refer to Example 1]}$$

i.e.
$$p - 3\sqrt{\frac{pq}{n}} \leq P \leq p + 3\sqrt{\frac{pq}{n}}$$

i.e.
$$0.36 - 3\sqrt{\frac{0.36 \times 0.64}{9000}} \leq P \leq 0.36 + 3\sqrt{\frac{0.36 \times 0.64}{9000}}$$

i.e.
$$0.345 \leq P \leq 0.375$$

Example 6 *In a large city A, 20% of a random sample of 900 school boys had a slight physical defect. In another large city B, 18.5 percent of a random sample of 1600 school boys had the same defect. Is the difference between the proportions significant?*

Solution $p_1 = 0.2$, $p_2 = 0.185$, $n_1 = 900$ and $n_2 = 1600$.

H_0: $p_1 = p_2$.

H_1: $p_1 \neq p_2$.

Two-tailed test is to be used.

Let LOS be 5%. Therefore, $z_\alpha = 1.96$.

$$z = \frac{p_1 - p_2}{\sqrt{PQ\left(\frac{1}{n_1} + \frac{1}{n_2}\right)}} \tag{1}$$

Since P, the population proportion, is not given, we estimate it as

$$\hat{P} = \frac{n_1 p_1 + n_2 p_2}{n_1 + n_2} = \frac{180 + 296}{900 + 1600} = 0.1904.$$

Using in (1), we have

$$z = \frac{0.2 - 0.185}{\sqrt{0.1904 \times 0.8096 \times \left(\frac{1}{900} + \frac{1}{1600}\right)}} = 0.92$$

$\therefore$ $|z| \leq z_\alpha$.

Therefore The difference between p_1 and p_2 is not significant at 5% level.

Example 7 *Before an increase in excise duty on tea, 800 people out of a sample of 1000 were consumers of tea. After the increase in duty, 800 people were consumers of tea in a sample of 1200 persons. Find whether there is significant decrease in the consumption of tea after the increase in duty.*

Solution Let p_1 and p_2 be the proportions of the consumers before and after the increase in duty respectively.

Then $p_1 = \frac{800}{1000} = \frac{4}{5}$ and $p_2 = \frac{800}{1200} = \frac{2}{3}$.

H_0: $p_1 = p_2$.

H_1: $p_1 > p_2$.

One-tailed (right-tailed) test is to be used.

Let LOS be 1%. Therefore, $z_\alpha = 2.33$.

$$z = \frac{p_1 - p_2}{\sqrt{PQ\left(\frac{1}{n_1} + \frac{1}{n_2}\right)}},$$

where $$P \simeq \frac{n_1 p_1 + n_2 p_2}{n_1 + n_2} = \frac{800 + 800}{2200} = 0.7273,$$

$$= \frac{0.8 - 0.67}{\sqrt{0.7273 \times 0.2727 \times \left(\frac{1}{1000} + \frac{1}{1200}\right)}}$$

$$= \frac{0.13 \times \sqrt{1000 \times 1200}}{\sqrt{0.7273 \times 0.2727 \times 2200}} = 6.82$$

$\therefore$ $|z| > z_\alpha$

Therefore the difference between p_1 and p_2 is significant at 1% level, i.e., H_0 is rejected and H_1 is accepted.

That is, there is significant decrease in the consumption of tea after the increase in duty.

Example 8 *15.5% of a random sample of 1600 undergraduates were smokers, whereas 20% of a random sample of 900 postgraduates were smokers in a state. Can we conclude that less number of undergraduates are smokers than the postgraduates?*

Solution Here $p_1 = 0.155$, $p_2 = 0.2$, $n_1 = 1600$ and $n_2 = 900$.

H_0: $p_1 = p_2$.

H_1: $p_1 < p_2$.

One-tailed (left-tailed) test is to be used.

Let LOS be 5%. Therefore, $z_\alpha = -1.645$.

$$z = \frac{p_1 - p_2}{\sqrt{PQ\left(\frac{1}{n_1} + \frac{1}{n_2}\right)}},$$

where $$P \simeq \frac{n_1 p_1 + n_2 p_2}{n_1 + n_2} = 0.1712,$$

$$= \frac{0.155 - 0.2}{\sqrt{0.1712 \times 0.8288 \times \left(\frac{1}{1600} + \frac{1}{900}\right)}}$$

$$= \frac{-0.045 \times 1200}{\sqrt{0.1712 \times 0.8288 \times 2500}},$$

$$= -2.87$$

$\therefore$ $|z| > |z_\alpha|$

Therefore the difference between p_1 and p_2 is significant, i.e. H_0 is rejected and H_1 is accepted.

That is, the habit of smoking is less among the undergraduates than among the postgraduates.

Example 9 *A sample of 100 students is taken from a large population. The mean height of the students in this sample is 160 cm. Can it be reasonably regarded that, in the population, the mean height is 165 cm, and the SD is 10 cm?*

Solution Here $\bar{x} = 160$, $n = 100$, $\mu = 165$ and $\sigma = 10$.

H_0: $\bar{x} = \mu$ (i.e. the difference between $\bar{x}$ and μ is not significant).

H_1: $\bar{x} \neq \mu$.

Two-tailed test is to be used.

Let LOS be 1%. Therefore, $z_\alpha = 2.58$.

$$z = \frac{\bar{x} - \mu}{\sigma / \sqrt{n}} = \frac{160 - 165}{10 / \sqrt{100}} = -5$$

$\therefore$ $|z| > z_\alpha$.

Therefore, the difference between $\bar{x}$ and μ is significant at 1% level, i.e., H_0 is rejected.

That is, it is not statistically correct to assume that $\mu = 165$.

Example 10 *The mean breaking strength of the cables supplied by a manufacturer is 1800 with a SD of 100. By a new technique in the manufacturing process, it is claimed that the breaking strength of the cable has increased. To test this claim, a sample of 50 cables is tested and it is found that the mean breaking strength is 1850. Can we support the claim at 1% LOS?*

Solution $\bar{x} = 1850$, $n = 50$, $\mu = 1800$ and $\sigma = 100$.

H_0: $\bar{x} = \mu$.

H_1: $\bar{x} > \mu$.

One-tailed (right-tailed) test is to be used.

Let LOS = 1%. Therefore, $z_\alpha = 2.33$.

$$z = \frac{\bar{x} - \mu}{\sigma / \sqrt{n}} = \frac{1850 - 1800}{100 / \sqrt{50}} = 3.54$$

$\therefore$ $|z| > z_\alpha$

Therefore the difference between $\bar{x}$ and μ is significant at 1% level, i.e., H_0 is rejected and H_1 is accepted.

That is, based on the sample data, we may support the claim of increase in breaking strength.

Example 11 *The mean value of a random sample of 60 items was found to be 145, with a SD of 40. Find the 95% confidence limits for the population mean. What size of the sample is required to estimate the population mean within 5 of its actual value with 95% or more confidence, using the sample mean?*

Solution Since the population SD σ too is not given, we can approximate it by the sample SDs. Therefore 95% confidence limits for μ are given by

$$\frac{|\mu - \bar{x}|}{s / \sqrt{n}} \leq 1.96,$$

i.e. $$\bar{x} - 1.96\frac{s}{\sqrt{n}} \le \mu \le \bar{x} + 1.96\frac{s}{\sqrt{n}}.$$

i.e. $$145 - \frac{1.96 \times 40}{\sqrt{60}} \le \mu \le 145 + \frac{1.96 \times 40}{\sqrt{60}}$$

i.e. $$134.9 \le \mu \le 155.1$$

We have to find the value of n such that

$$P\{\bar{x} - 5 \le \mu \le \bar{x} + 5\} \ge 0.95$$

i.e. $$P\{-5 \le \mu - \bar{x} \le 5\} \ge 0.95$$

i.e. $$P\{|\mu - \bar{x}| \le 5\} \ge 0.95 \quad \text{or}$$

$$P\{|\bar{x} - \mu| \le 5\} \ge 0.95$$

$$\therefore\ P\left\{\frac{|\bar{x} - \mu|}{\sigma/\sqrt{n}} \le \frac{5}{\sigma/\sqrt{n}}\right\} \ge 0{\cdot}95$$

i.e. $$P\left\{|z| \le \frac{5\sqrt{n}}{\sigma}\right\} \ge 0{\cdot}95,$$

where z is the standard normal variate. (1)

We know that $P\{|z| \le 1.96\} = 0{\cdot}95$. Therefore, the least value of $n = n_L$ that will satisfy (1) is given by

$$\frac{5\sqrt{n_L}}{\sigma} = 1.96$$

i.e. $$\sqrt{n_L} = \frac{1.96\, s}{5} \quad (\because\ \sigma \simeq s)$$

i.e. $$n_L = \left(\frac{1.96 \times 40}{5}\right)^2$$

i.e. $$n_L = 245.86$$

Therefore, least size of the sample is 246.

Example 12 *A normal population has a mean of 0.1 and SD of 2.1. Find the probability that the mean of a sample of size 900 drawn from this population will be negative.*

Solution Since $\bar{x}$ follows $N\left(\mu, \frac{\sigma}{\sqrt{n}}\right)$, $z = \frac{\bar{x} - \mu}{\sigma/\sqrt{n}}$ is the standard normal variate.

Now $$P(\bar{x} < 0) = P\{\bar{x} - 0.1 < -0.1\}$$

$$= P\left\{\frac{\bar{x} - 0.1}{(2.1)/\sqrt{900}} < \frac{-0.1}{(2.1)/\sqrt{900}}\right\}$$

$$= P\{z < -1.43\}$$
$$= P\{z > 1.43\},$$
(by symmetry of the standard normal distribution)
$$= 0.5 - P\{0 < z < 1.43\}$$
$$= 0.5 - 0.4236 \text{ (from the normal tables)}$$
$$= 0.0764$$

Example 13 *In a random sample of size 500, the mean is found to be 20. In another independent sample of size 400, the mean is 15. Could the samples have been drawn from the same population with SD 4?*

Solution Here $\bar{x}_1 = 20$, $n_1 = 500$, $\bar{x}_2 = 15$, $n_2 = 400$; $\sigma = 4$

H_0: $\bar{x}_1 = \bar{x}_2$, i.e. the samples have been drawn from the same population.

H_1: $\bar{x}_1 \neq \bar{x}_2$.

Two-tailed test is to be used.

Let LOS be 1%. Therefore, $z_\alpha = 2.58$.

$$z = \frac{\bar{x}_1 - \bar{x}_2}{\sigma\sqrt{\frac{1}{n_1} + \frac{1}{n_2}}} \quad \text{(refer to Note 1 under Test 4)}$$

$$= \frac{20 - 15}{4\sqrt{\frac{1}{500} + \frac{1}{400}}} = 18.6$$

$\therefore \quad |z| > z_\alpha$

Therefore, the difference between $\bar{x}_1$ and $\bar{x}_2$ is significant at 1% level, i.e., H_0 is rejected

That is, the samples could not have been drawn from the same population.

Example 14 *A simple sample of heights of 6400 English men has a mean of 170 cm and a SD of 6.4 cm, while a simple sample of heights of 1600 Americans has a mean of 172 cm and a SD of 6.3 cm. Do the data indicate that Americans are, on the average, taller than the Englishmen?*

Solution Here $n_1 = 6400$, $\bar{x}_1 = 170$ and

$s_1 = 6.4$; $n_2 = 1600$, $\bar{x}_2 = 172$ and $s_2 = 6.3$.

H_0: $\mu_1 = \mu_2$ or $\bar{x}_1 = \bar{x}_2$,

i.e. the samples have been drawn from two different populations with the same mean.

H_1: $\bar{x}_1 < \bar{x}_2$ or $\mu_1 < \mu_2$.

Left-tailed test is to be used.

Let LOS be 1%. Therefore, $z_\alpha = -2.33$.

$$z = \frac{\bar{x}_1 - \bar{x}_2}{\sqrt{\frac{\sigma_1^2}{n_1} + \frac{\sigma_2^2}{n_2}}} = \frac{\bar{x}_1 - \bar{x}_2}{\sqrt{\frac{s_1^2}{n_1} + \frac{s_2^2}{n_2}}}$$

[$\because \sigma_1 \simeq s_1$ and $\sigma_2 \simeq s_2$. Refer to Note 2 under Test 4]

$$= \frac{170 - 172}{\sqrt{\frac{(6.4)^2}{6400} + \frac{(6.3)^2}{1600}}} = -11.32$$

$\therefore \qquad |z| > |z_\alpha|$

Therefore, the difference between $\bar{x}_1$ and $\bar{x}_2$ (or μ_1 and μ_2) is significant at 1% level, i.e., H_0 is rejected and H_1 is accepted.

That is, the Americans are, on the average, taller than the Englishmen.

Example 15 *Test the significance of the difference between the means of the samples, drawn from two normal populations with the same SD using the following data:*

	Stize	*Mean*	*SD*
Sample 1	100	61	4
Sample 2	200	63	6

Solution H_0: $\bar{x}_1 = \bar{x}_2$ or $\mu_1 = \mu_2$.

H_1: $\bar{x}_1 \neq \bar{x}_2$ or $\mu_1 \neq \mu_2$.

Two-tailed test is to be used.

Let LOS be 5%. Therefore, $z_\alpha = 1.96$

$$z = \frac{\bar{x}_1 - \bar{x}_2}{\sqrt{\frac{s_1^2}{n_2} + \frac{s_2^2}{n_1}}}$$

(refer to Note 3 under Test 4; the populations have the same SD)

$$= \frac{61 - 63}{\sqrt{\frac{4^2}{200} + \frac{6^2}{100}}} = -3.02$$

$\therefore \qquad |z| > z_\alpha$

Therefore, the difference between $\bar{x}_1$ and $\bar{x}_2$ (or μ_1 and μ_2) is significant at 5% level, i.e., H_0 is rejected and H_1 is accepted.

That is, the two normal populations, from which the samples are drawn, may not have the same mean, though they may have the same SD.

Example 16 *The average marks scored by 32 boys is 72 with a SD of 8, while that for 36 girls is 70 with a SD of 6. Test at 1% LOS whether the boys perform better than girls.*

Solution H_0: $\bar{x}_1 = \bar{x}_2$ (or $\mu_1 = \mu_2$).

H_1: $\bar{x}_1 > \bar{x}_2$.

Right-tailed test is to be used.

Let LOS be 1%. Therefore, $z_\alpha = 2.33$.

$$z = \frac{\bar{x}_1 - \bar{x}_2}{\sqrt{\frac{s_1^2}{n_1} + \frac{s_2^2}{n_2}}}$$

(The two populations are assumed to have SD s $\sigma_1 \simeq s_1$ and $\sigma_2 \simeq s_2$)

$$= \frac{72 - 70}{\sqrt{\frac{8^2}{32} + \frac{6^2}{36}}} = 1.15$$

$\therefore$ $|z| < z_\alpha$

Therefore, the difference between $\bar{x}_1$ and $\bar{x}_2$ (μ_1 and μ_2) is not significant at 1% level, i.e. H_0 is accepted and H_1 is rejected.

That is, statistically we cannot concluede that boys perform better than girls.

Example 17 *The heights of men in a city are normally distributed with mean 171 cm and SD 7 cm., while the corresponding values for women in the same city are 165 cm and 6 cm respectively. If a man and a woman are chosen at random from this city, find the probability that the woman is taller than the man.*

Solution Let $\bar{x}_1$ and $\bar{x}_2$ denote the mean heights of men and women respectively.

Then $\bar{x}_1$ follows a N (171, 7) and $\bar{x}_2$ follows a N (165, 6). Therefore, $\bar{x}_1 - \bar{x}_2$ also follows a normal distribution.

$$E(\bar{x}_1 - \bar{x}_2) = E(x_1) - E(\bar{x}_2) = 171 - 165 = 6$$

$$V(\bar{x}_1 - \bar{x}_2) = V(\bar{x}_1) + V(\bar{x}_2) = 49 + 36 = 85 \qquad \text{(refer to Test 4)}$$

Therefore, SD of $(\bar{x}_1 - \bar{x}_2) = \sqrt{85} = 9.22$.

Thus, $\bar{x}_1 - \bar{x}_2$ follows a N (6, 9.22).

$$\therefore \quad P(\bar{x}_2 > \bar{x}_1) = P(\bar{x}_1 - \bar{x}_2 < 0)$$

$$= P\left\{\frac{(\bar{x}_1 - \bar{x}_2) - 6}{9.22} < \frac{-6}{9.22}\right\}$$

$= P\{z < -0.65\}$ (where z is the standard normal variate).

$= P\{z > 0.65\}$ (by symmetry).

$$= 0.5 - P\,(0 < z < 0.65)$$
$$= 0.5 - 0.2422 = 0.2578.$$

Example 18 *Two populations have the same mean, but the SD of one is twice that of the other. Show that in samples, each of size 500, drawn under simple random conditions, the difference of the means will, in all probability, not exceed 0.3 σ, where σ is the smaller SD.*

Solution Let $\bar{x}_1$ and $\bar{x}_2$ be the means of the samples of size 500 each. Let their SD's be σ and 2σ respectively.

Then $\bar{x}_1$ follows a $N\left(\mu, \dfrac{\sigma}{\sqrt{500}}\right)$ and $\bar{x}_2$ follows $N\left(\mu, \dfrac{2\sigma}{\sqrt{500}}\right)$.

Therefore, $\bar{x}_1 - \bar{x}_2$ also follows a normal distribution.

$$E\,(\bar{x}_1 - \bar{x}_2) = E\,(\bar{x}_1) - E\,(\bar{x}_2) = \mu - \mu = 0$$

$$V\,(\bar{x}_1 - \bar{x}_2) = V\,(\bar{x}_1) + V\,(\bar{x}_2)$$

$$= \frac{\sigma^2}{500} + \frac{4\sigma^2}{500} = \frac{\sigma^2}{100}$$

Therefore, SD of $(\bar{x}_1 - \bar{x}_2) = \dfrac{\sigma}{10}$

Thus $(\bar{x}_1 - \bar{x}_2)$ follows $N\left(0, \dfrac{\sigma}{10}\right)$.

$\therefore$ $$P\,\{|\bar{x}_1 - \bar{x}_2| \le 0.3\,\sigma\}$$

$$= P\left\{\frac{|(\bar{x}_1 - \bar{x}_2) - 0|}{\sigma/10} \le \frac{0.3\,\sigma}{\sigma/10}\right\}$$

$= P\,\{|z| \le 3\}$. (where z is the standard normal variate)

$= 0.9974 \simeq 1$

Therefore, $|\bar{x}_1 - \bar{x}_2|$ will not exceed $0.3\,\sigma$ almost certainly.

Example 19 *A manufacturer of electric bulbs, according to a certain process, finds the SD of the life of lamps to be 100 h. He wants to change the process, if the new process results in a smaller variation in the life of lamps. In adopting a new process, a sample of 150 bulbs gave an SD of 95 h. Is the manufacturer justified in changing the process?*

Solution Here $\sigma = 100$, $n = 150$ and $s = 95$

H_0: $s = \sigma$

H_1: $s < \sigma$

Left-tailed test is to be used.

Let LOS be 5%. Therefore, $z_\alpha = -1.645$

$$z = \frac{s-\sigma}{\sigma/\sqrt{2n}} = \frac{95-100}{100/\sqrt{300}} = -0.866$$

$\therefore \qquad |z| < |z_\alpha|$

Therefore, the difference between s and σ is not significant at 5% level, i.e., H_0 is accepted and H_1 is rejected.

That is, the manufacturer is not justified in changing the process.

Example 20 *The SD of a random sample of 1000 is found to be 2.6 and the SD of another random sample of 500 is 2.7. Assuming the samples to be independent, find whether the two samples could have come from populations with the same SD.*

Solution Here $n_1 = 1000$, $s_1 = 2.6$; $n_2 = 500$ and $s_2 = 2.7$.

H_0: $s_1 = s_2$ (or $\sigma_1 = \sigma_2$).

H_1: $s_1 \neq s_2$.

Two-tailed test is to be used.

Let LOS be 5%. Therefore, $z_\alpha = 1.96$.

$$z = \frac{s_1 - s_2}{\sqrt{\dfrac{s_1^2}{2n_2} + \dfrac{s_2^2}{2n_1}}} \quad \text{(since } \sigma \text{ is not known)}$$

$$= \frac{2.6 - 2.7}{\sqrt{\dfrac{(2.6)^2}{1000} + \dfrac{(2.7)^2}{2000}}} = -0.98$$

$\therefore \qquad |z| < z_\alpha$

Therefore, the difference between s_1 and s_2 (and hence between σ_1 and σ_2) is not significant at 5% level.

i.e., H_0 is accepted.

That is, the two samples could have come from populations with the same SD.

Exercise 8(A)

Part A (Short answer questions)

1. What is the difference between population and sample?
2. Distinguish between parameter and statistic.
3. What do you mean by sampling distribution?
4. What is meant by standard error?

5. What do you mean by estimation?
6. What is meant by testing of hypothesis?
7. Define null hypothesis and alternative hypothesis.
8. What is meant by test of significance?
9. What do you mean by critical region and acceptance region?
10. Define level of significance.
11. Give the general form of a test statistic.
12. Define type I and type II errors.
13. Define producer's risk and consumer's risk.
14. What is the relation between type I error and LOS?
15. Define one-tailed and two-tailed tests.
16. Define critical value of a test statistic.
17. What is the relation between the critical value and LOS?
18. What is the relation between the critical values for a single-tailed test and a two-tailed test?
19. Write down the 1% and 5% critical values for right-tailed and two-tailed tests for large samples.
20. What do you mean by interval estimation and confidence limits?
21. Write down the general form of 95% confidence limits of a population parameter in terms of the corresponding sample statistic.
22. What is the standard error of the sample proportion when the population proportion is (i) known and (ii) not known?
23. What is the standard error of the difference between two sample proportions when the population proportion is (i) known and (ii) not known?
24. What do you mean by unbiased estimate? Give an example.
25. Write down the form of the 98% confidence interval for the population mean in terms of (i) population SD; and (ii) Sample SD.
26. What is the standard error of the difference between the means of two large samples drawn from different populations with (i) known SDs and; (ii) unknown SDs?
27. What is the standard error of the difference between the means of two large samples drawn from the same population with (i) known SD and (ii) unknown SD?
28. What is the standard error of the difference between the SD's of two large samples drawn from the same population with (i) known SD and (ii) unknown SD?

Part B

29. Out of 200 individuals, 40% show a certain trait and the number expected on a certain theory is 50%. Find whether the number observed differs significantly from expectation.
30. A coin is thrown 400 times and is found to result in 'Head' 245 times. Test whether the coin is a fair one.

31. A manufacturer of light bulbs claims that on the average 2 % of the bulbs manufactured by his firm are defective. A random sample of 400 bulbs contained 13 defective bulbs. On the basis of the this sample, can you support the manufacturer's claim at 5% LOS?
32. 100 people were affected by cholera and out of them only 90 survived. Would you reject the hypothesis that the survival rate, if affected by cholera, is 85% in favour of the hypothesis that it is more at 5% LOS?
33. A random sample of 400 mangoes was taken from a big consignment and 40 were found to be bad. Prove that the percentage of bad mangoes in the consignment will, in all probability, lie between 5.5 and 14.5.
34. A random sample of 64 articles produced by a machine contained 14 defectives. Is it reasonable to assume that only 10% of the articles produced by the machine are defective? If not, find the 99% confidence limits for the percentage of defective articles produced by the machine.
35. Certain crosses of the pea gave 5321 yellow and 1804 green seeds. The expectation is 25% of green seeds based on a certain theory. Is this divergence significant or due to sampling fluctuations?
36. During a countrywide investigation, the incidence of TB was found to be 1% . In a college with 400 students, 5 are reported to be affected whereas in another college of 1200 students, 10 are found to be affected. Does this indicate any significant difference?
37. A random sample of 600 men chosen from a certain city contained 400 smokers. In another sample of 900 men chosen from another city, there were 450 smokers. Do the data indicate that (i) the cities are significantly different with respect to smoking habit among men? and (ii) the first city contains more smokers than the second?
38. A sample of 300 spare parts produced by a machine contained 48 defectives. Another sample of 100 spare parts produced by another machine contained 24 defectives. Can you conclude that the first machine is better than the second?
39. In two large populations, there are 30 and 25%, respectively, of fair-haired people. Is this difference likely to be hidden in samples of sizes 1200 and 900, respectively, drawn from the two populations?

$$\left[\textbf{Hint: } H_0 : P_1 - P_2 = 0, \text{ or } P_1 = P_2; \quad H_1 : P_1 \neq P_2 \quad z = \frac{p_1 - p_2}{\sqrt{\dfrac{P_1 Q_1}{n_1} + \dfrac{P_2 Q_2}{n_2}}}\right]$$

40. A machine produces 16 defective bolts in a batch of 500 bolts. After the machine is overhauled, it produces 3 defective bolts in a batch of 100 bolts. Has the machine improved?
41. There were 956 births in a year in town A, of which 52.5% were males, while in towns A and B combined together this proportion in a total of

1406 births was 0.496. Is there any significant difference in the proportion of male births in the two towns?

42. A cigarette manufacturing company claims that its brand A cigarettes outsells its brand B by 8%. It is found that 42 out of a sample of 200 smokers prefer brand A and 18 out of another sample of 100 smokers prefer brand B. Test at 5% LOS. whether the 8% difference is a valid claim.

[**Hint**: $H_0 : P_1 - P_2 = .08$; $H_1 : P_1 - P_2 \neq .08$ and

$$z = \frac{(p_1 - p_2) - (P_1 - P_2)}{\sqrt{PQ\left(\frac{1}{n_1} + \frac{1}{n_2}\right)}}, \quad \text{where } P = \frac{n_1 p_1 + n_2 p_2}{n_1 + n_2}$$
]

43. A sample of 900 items is found to have a mean of 3.47 cm. Can it be reasonably regarded as a simple sample from a population with mean 3.23 cm and SD 2.31 cm?
44. A sample of 400 observations has mean 95 and SD 12. Could it be a random sample from a population with mean 98? What should be the maximum value of the population mean so that the sample can be regarded as one drawn from it almost certainly?
45. A manufacturer claims that, the mean breaking strength of safety belts for air passengers produced in his factory is 1275 kgs. A sample of 100 belts was tested and the mean breaking strength and SD were found to be 1258 and 90 kg respectively. Test the manufacturer's claim at 5% LOS.
46. An IQ test was given to a large group of boys in the age group of 18–20 years, who scored an average of 62.5 marks. The same test was given to a fresh group of 100 boys of the same age group. They scored an average of 64.5 marks with a SD 12.5 marks. Can we conclude that the fresh group of boys have better IQ?
47. The guaranteed average life of a certain brand of electric bulb is 1000 h, with a SD of 125 h. It is decided to sample the output so as to ensure that 90% of the bulbs do not fall short of the guaranteed average by more than 2.5%. What should be the minimum sample size?
48. A random sample of 100 students gave a mean weight of 58 kg with a SD of 4 kg. Find the 95% and 99% confidence limits of the mean of the population.
49. The means of two simple samples of 1000 and 2000 items are 170 cm and 169 cm respectively. Can the samples be regarded as drawn from the same population with SD 10 at 5% LOS?

50. The mean and SD of a sample of size 400 are 250 and 40 respectively. Those of another sample of size 400 are 220 and 55. Test at 1% LOS whether the means of the two populations from which the samples have been drawn are equal.
51. Intelligence tests were given to two groups of boys and girls of the same age group chosen from the same college and the following results were got:

	Size	*Mean*	*SD*
Boys	100	73	10
Girls	60	75	8

Examine whether the difference between the means is significant or not.
52. A sample of 100 bulbs of brand *A* gave a mean lifetime of 1200 h, with an SD of 70 h, while another sample of 120 bulbs of brand *B* gave a mean lifetime of 1150 h, with a SD of 85 h. Can we conclude that brand *A* bulbs are superior to brand *B* bulbs?
53. In a college, 60 junior students are found to have a mean height of 171.5 cm and 50 senior students are found to have a mean height of 173.8 cm. Can we conclude, based on these data, that the juniors are shorter than seniors at (i) 5% LOS and (ii) 1% LOS, assuming that the SD of students of that college is 6.2 cm?
54. Two samples drawn from two different populations gave the following results:

	Size	*Mean*	*SD*
Sample I	400	124	14
Sample II	250	120	12

Find the 95% confidence limits for the difference of the population means.

$$\left[\textbf{Hint: } (\bar{x}_1 - \bar{x}_2) - 1.96\sqrt{\frac{s_1^2}{n_1} + \frac{s_2^2}{n_2}} \le \mu_1 - \mu_2 \le (\bar{x}_1 - \bar{x}_2) + 1.96\sqrt{\frac{s_1^2}{n_1} + \frac{s_2^2}{n_2}}\right]$$

55. Two samples drawn from two different populations gave the following results:

	Size	*Mean*	*SD*
Sample I	100	582	24
Sample II	100	540	28

Test the hypothesis, at 5% LOS, that the difference of the means of the populations is 35.

$$\left[\textbf{Hint}: z = \frac{(\bar{x}_1 - \bar{x}_2) - 35}{\sqrt{\dfrac{s_1^2}{n_1} + \dfrac{s_2^2}{n_2}}} \right]$$

56. Two populations have their means equal, but the SD of one is twice the other. Show that in samples, each of size 2000, drawn one from each, the difference of the means will, in all probability, not exceed 0.15σ, where σ is the smaller SD.
57. In a certain random sample of 72 items, the SD is found to be 8. Is it reasonable to suppose that it has been drawn from a population with SD 7?
58. In a random sample of 200 items, drawn from a population with SD 0.8, the sample SD is 0.7. Can we conclude that the sample SD is less than the population SD at 1% LOS?
59. The SD of a random sample of 900 members is 4.6 and that of another independent sample of 1600 members is 4.8. Examine if the two samples could have been drawn from a population with SD 4.0?
60. Examine whether the two samples for which the data are given in the following table could have been drawn from populations with the same SD:

	Size	*SD*
Sample I	100	5
Sample II	200	7

Tests of Significance for Small Samples

The tests of significance discussed in the previous section hold good only for large samples, i.e. only when the size $n \geq 30$. When the sample is small, i.e. $n < 30$, the sampling distributions of many statistics are not normal, even though the parent populations may be normal. Moreover the assumption of near equality of population parameters and the corresponding sample statistics will not be justified for small samples. Consequently we have to develop entirely different tests of significance that are applicable to small samples.

Student's t-Distribution

A random variable T is said to follow student's t-distribution or simply t-distribution, if its probability density function is given by

$$f(t) = \frac{1}{\sqrt{v}\,\beta\left(\frac{v}{2}, \frac{1}{2}\right)} \cdot \left(1 + \frac{t^2}{v}\right)^{-(v+1)/2}, \quad -\infty < t < \infty,$$

where v denoteds the number of degrees of freedom of the t-distribution.

Note *t-distribution was defined by the mathematician W.S.D Gosset, whose pen name is Student.)*

Properties of t-Distribution

1. The probability curve of the t-distribution is similar to the standard normal curve, and is symmetric about $t = 0$, bell-shaped and asymptotic to the t-axis, as shown in the fig. 8.2.
2. For sufficiently large value of v, the t-distribution tends to the standard normal distribution.
3. The mean of the t-distribution is zero.
4. The variance of the t-distribution is $\frac{v}{v-2}$, if $v > 2$ and is greater than 1, but it tends to 1 as $v \to \infty$.

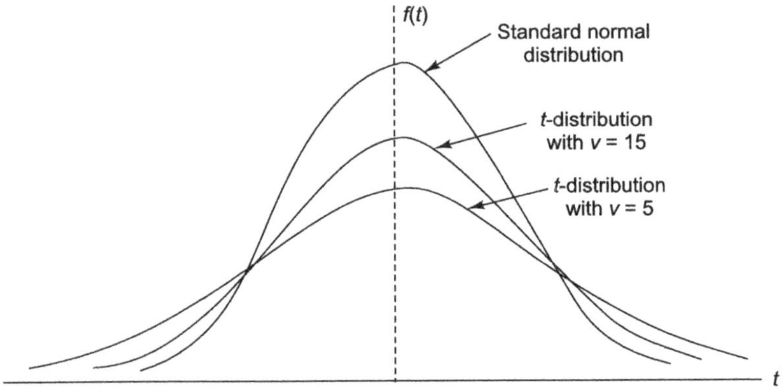

Fig. 8.2

Uses of t-Distribution

The t-distribution is used to test the significance of the difference between

1. The mean of a small sample and the mean of the population,
2. The means of two small samples and
3. The coefficient of correlation in the small sample and that in the population, assumed zero.

Note on Degree of Freedom

The number of degrees of freedom, usually denoted by the Greek alphabet v, can be interpreted as the number of useful bits of information generated by a sample of given size for estimating a population parameter. Suppose we wish to find the mean of a sample with observations $x_1, x_2, \ldots, x_n$. We have to use all the n values taken by the variable with full freedom (i.e. without putting any constraint or restriction on them) for computing $\bar{x}$. Hence $\bar{x}$ is said to have n degrees of freedom.

Suppose we wish to further compute the SDs of this sample, using the formula $s^2 = \frac{1}{n}\sum\left(x_i - \bar{x}\right)^2$. Though we use the n values $x_1 - \bar{x}$, $x_2 - \bar{x}$, ..., $x_n - \bar{x}$ for this computation, they do not have n degrees of freedom, as they depend on $\bar{x}$, which has already been calculated and fixed. Since there is one restriction regarding the value of $\bar{x}$, s is said to have $(n - 1)$ degrees of freedom.

If we compute another statistic of the sample based on $\bar{x}$ and s, then that statistic will be assumed to have $(n - 2)$ degrees of freedom and so on.

Thus the number of independent variates used to compute the test statistic is known as the number of degrees of freedom of that statistic. In general, the number of degrees of freedom is given by $v = n - k$, where n is the number of observations in the sample and k is the number of constraints imposed on them or k is the number of values that have been found out and specified by prior calculations.

Critical Values of t and the t-Table

The critical value of t at level of significance (LOS) α and degrees of freedom v is given by $P\{|t| > t_v(\alpha)\} = \alpha$ for two-tailed test, as in the case of normal distribution and large samples, and by $P\{t > t_v(\alpha)\} = \alpha$ for the right-tailed test, also as in the case of normal distribution. The critical value of t for a single (right-or-left) tailed test at LOS α corresponding to v degrees of freedom is the same as that for a two-tailed test at LOS 2α corresponding to the same degrees of freedom.

Critical values $t_v(\alpha)$ of the t-distribution for two-tailed tests, corresponding to a few important LOS and a range of values of v, have been published by Prof. R.A. Fisher in the form of a table, called the t-table, which is given in the end of this chapter.

Test 1

Test of significance of the difference between sample mean and population mean.

If $\bar{x}$ is the mean of a sample of size n, drawn from a population $N(\mu, \sigma)$, we have seen that $z = \frac{\bar{x} - \mu}{\sigma / \sqrt{n}}$ follows $N(0, 1)$.

If σ, the SD of the population is not known, we have to estimate it using the sample SDs. From the theory of estimation, it is known that $s\sqrt{\frac{n}{n-1}}$ is an unbiased estimate of σ with $(n - 1)$ degrees of freedom. When n is large, $\frac{n}{n-1} \simeq 1$ and hence s was taken as a satisfactory estimate of σ and hence $z = \frac{\bar{x} - \mu}{s / \sqrt{n}}$ was assumed to follow a $N(0, 1)$. But when n is small, we cannot use s as an estimate of σ, since

$$\frac{\bar{x}-\mu}{\sigma/\sqrt{n}} = \frac{\bar{x}-\mu}{s\sqrt{\frac{n}{n-1}\cdot\frac{1}{\sqrt{n}}}} = \frac{\bar{x}-\mu}{s/\sqrt{n-1}}$$

Now $\frac{\bar{x}-\mu}{s/\sqrt{n-1}}$ does not follow a normal distribution, but follows a t-distribution with number of degrees of freedom $\nu = n - 1$. Hence $\frac{\bar{x}-\mu}{s/\sqrt{n-1}}$ is denoted by t and is taken as the test statistic.

Sometimes $t = \frac{\bar{x}-\mu}{s/\sqrt{n-1}}$ is also taken as $t = \frac{\bar{x}-\mu}{S/\sqrt{n}}$,

where $S^2 = \frac{1}{n-1}\sum_{r=1}^{n}(x_r - \bar{x})^2$ and is called student's t. We shall use only $t = \frac{\bar{x}-\mu}{s/\sqrt{n-1}}$, where s is the sample SD.

We get the value of $t_\nu(\alpha)$ for the LOS α and $\nu = n - 1$ from the t-table.

If the calculated value of t satisfies $|t| < t_\nu(\alpha)$, the null hypothesis H_0 is accepted at LOS α otherwise H_0 is rejected at LOS α.

Note *95% confidence interval of μ is given by*

$$\left|\frac{\bar{x}-\mu}{s/\sqrt{n-1}}\right| \le t_{0.05}, \text{ since } P\left\{\left|\frac{\bar{x}-\mu}{s/\sqrt{n-1}}\right| \le t_{0.05}\right\} = 0.95$$

i.e. by $\bar{x} - t_{0.05}\frac{s}{\sqrt{n-1}} \le \mu \le \bar{x} + t_{0.05} \times \frac{s}{\sqrt{n-1}}$, *where* $t_{0.05}$ *is the 5% critical value of t for* $\nu\,(= n - 1)$ *degrees of freedom for a two-tailed test.*

Test 2

Test of significance of the difference between means of two small samples drawn from the same normal population.

In Test 4 for large samples, the test statistic used to test the significance of the difference between the means of two samples from the same normal population was taken as

$$z = \frac{\bar{x}_1 - \bar{x}_2}{\sigma\sqrt{\frac{1}{n_1}+\frac{1}{n_2}}},$$

which follows a $N(0, 1)$ (1)

If σ is not known, we may assume that $\sigma \simeq \sqrt{\dfrac{n_1 s_1^2 + n_2 s_2^2}{n_1 + n_2}}$, when n_1 and n_2 are large, where s_1 and s_2 are the sample SDs. This assumption no longer holds good when n_1 and n_2 are small.

In fact, it is known from the theory of estimation, that an estimate of σ is $\sqrt{\dfrac{n_1 s_1^2 + n_2 s_2^2}{n_1 + n_2 - 2}}$ with $(n_1 + n_2 - 2)$ degrees of freedom, when n_1 and n_2 are small.

Using this value of σ in (1), the test statistic becomes

$$\frac{\bar{x}_1 - \bar{x}_2}{\sqrt{\left(\dfrac{n_1 s_1^2 + n_2 s_2^2}{n_1 + n_2 - 2}\right)\left(\dfrac{1}{n_1} + \dfrac{1}{n_2}\right)}},$$

which does not follow a $N(0, 1)$, but follows a t-distribution, with $\nu = (n_1 + n_2 - 2)$ degrees of freedom. Hence t-test is applied in this case.

Note 1. *If f $n_1 = n_2 = n$ and if the samples are independent i.e., the observations in the two samples are not at all related, then the test statistic is given by*

$$t = \frac{\bar{x}_1 - \bar{x}_2}{\sqrt{\dfrac{s_1^2 + s_2^2}{n-1}}} \quad \text{with} \quad \nu = 2n - 2 \qquad (2)$$

2. *If $n_1 = n_2 = n$ and if the pairs of values of x_1 and x_2 are associated in some way (or correlated), the formula (2) for t in Note (1) should not be used. In this case, we shall assume that $H_0 : \bar{d} (= \bar{x} - \bar{y}) = 0$ and test the significance of the difference between $\bar{d}$ and 0, using the test statistic $t = \dfrac{\bar{d}}{s/\sqrt{n-1}}$ with $\nu = n - 1$, where $d_i = x_i - y_i$ $(i = 1, 2, \ldots\ldots, n)$, $\bar{d} = \bar{x} - \bar{y}$; and $s = SD\ of\ d = \dfrac{1}{n}\sum_{i=1}^{n}\left(d_i - \bar{d}\right)^2$.*

Snedecor's F-Distribution

A random variable F is said to follow snedecor's F-distribution or simply F-distribution if its probability density function is given by

$$f(F) = \frac{(v_1/v_2)^{v_1/2}}{\beta\left(\frac{v_1}{2}, \frac{v_2}{2}\right)} \cdot \frac{F^{v_1/2-1}}{\left(1 + \frac{v_1 F}{v_2}\right)^{(v_1+v_2)/2}}, \quad F > 0.$$

Note *The mathematical variable corresponding to the random variable F is also taken as F. v_1 and v_2 used in $f(F)$ are the degrees of freedom associated with the F-distribution.*

Properties of the F-Distribution

1. The probability curve of the F-distribution is roughly sketched in Fig. 8.3.

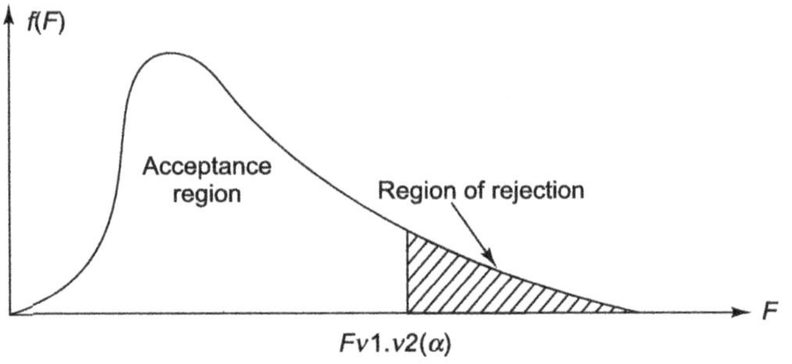

Fig. 8.3

2. The square of the t-variate with n degrees of freedom follows a F-distribution with 1 and n degrees of freedom.

3. The mean of the F-distribution is $\dfrac{v_2}{v_2 - 2}$ $(v_2 > 2)$.

4. The variance of the F-distribution is

$$\frac{2v_2^2(v_1 + v_2 - 2)}{v_1(v_2 - 2)^2(v_2 - 4)} \quad (v_2 > 4)$$

Use of F-Distribution

F-distribution is used to test the equality of the variance of the populations from which two small samples have been drawn.

F-test

F-test of significance of the difference between population variances and F-table.

To test the significance of the difference between population variances, we shall first find their estimates, $\hat{\sigma}_1^2$ and $\hat{\sigma}_2^2$ based on the sample variances s_1^2 and s_2^2 and then test their equality. It is known that $\hat{\sigma}_1^2 = \dfrac{n_1 s_1^2}{n_1 - 1}$ with the number of degrees of freedom $\nu_1 = n_1 - 1$ and $\hat{\sigma}_2^2 = \dfrac{n_2 s_2^2}{n_2 - 1}$ with the number of degrees of freedom $\nu_2 = n_2 - 1$.

It is also known that $F = \dfrac{\hat{\sigma}_1^2}{\hat{\sigma}_2^2}$ follows a F-distribution with ν_1 and ν_2 degrees of freedom. If $\hat{\sigma}_1^2 = \hat{\sigma}_2^2$, then $F = 1$. Hence our aim is to find how far any observed value of F can differ from unity as a result of fluctuations of sampling.

Snedecor has prepared tables that give, for different values of ν_1 and ν_2, the 5% and 1% critical values of F. An extract from these tables is given at the end of this chapter. If F denotes the observed (calculated) value and $F_{\nu_1,\nu_2}(\alpha)$ denotes the critical (tabulated) value of F at LOS α, then $P\{F > F_{\nu_1,\nu_2}(\alpha)\} = \alpha$.

(**Note:** F-test is not a two-tailed test and is always a right-tailed test, since F cannot be negative. Thus if $F > F_{\nu_1,\nu_2}(\alpha)$, then the difference between F and 1, i.e. the difference between $\hat{\sigma}_1^2$ and $\hat{\sigma}_2^2$ is significant at LOS α. In other words, the samples may not be regarded as drawn from the same population or from populations with the same variance. If $F < F_{\nu_1,\nu_2}(\alpha)$, the difference is not significant at *LOS* α).

Note

1. *We should always make $F > 1$. This is done by taking the larger of the two estimates of σ^2 as σ_1^2 and by assuming that the corresponding degree of freedom as ν_1.*
2. *To test if two small samples have been drawn from the same normal population, it is not enough to test if their means differ significantly or not, because in this test we assumed that the two samples came from the same population or from populations with equal variance. So, before applying the t-test for the significance of the difference of two sample means, we should satisfy ourselves about the equality of the population variances by F-test.*

Worked Examples 8(B)

Example 1 *Tests made on the breaking strength of 10 pieces of a metal gave the following results: 578, 572, 570, 568, 572, 570, 570, 572, 596 and 584 kg. Test if the mean breaking strength of the wire can be assumed as 577 kg.*

Solution Let us first compute sample mean $\bar{x}$ and sample SDs and then test if $\bar{x}$ differs significantly from the population mean $\mu = 577$.

We take the assumed mean $A = \dfrac{568 + 596}{2} = 582$

$$d_i = x_i - A$$

$\therefore$ $$x_i = d_i + A$$

$\therefore$ $$\bar{x} = \frac{1}{n}\sum x_i = \frac{1}{n}\sum d_i + A$$

$$= \frac{1}{10} \times (-68) + 582 = 575.2 \text{ (see the table below)}$$

$$s^2 = \frac{1}{n}\sum d_i^2 - \left(\frac{1}{n}\sum d_i\right)^2 \text{ (see the table below)}$$

$$= \frac{1}{10} \times 1144 - \left(\frac{1}{10} \times (-68)\right)^2 = 68.16$$

$\therefore$ $$s = 8.26$$

Now $$t = \frac{\bar{x} - \mu}{s/\sqrt{n-1}} = \frac{575 \cdot 2 - 577}{8 \cdot 26/\sqrt{9}}$$

$$= -0.65$$

and $$\nu = n - 1 = 9.$$

x_i	$d_i = x_i - A$	d_i^2
578	– 4	16
572	– 10	100
570	– 12	144
568	– 14	196
572	– 10	100
570	– 12	144
570	– 12	144
572	– 10	100
596	14	196
584	2	4
Total	– 68	1144

H_0: $\bar{x} = \mu$ and H_1: $\bar{x} \neq \mu$.

Two tailed test is to be used. Let LOS be 5%.

From the t-table, for $\nu = 9$, $t_{0.05} = 2.26$. Since $|t| < t_{0.05}$, the difference between $\bar{x}$ and μ is not significant or H_0 is accepted. Therefore, the mean breaking strength of the wire can be assumed as 577 kg at 5% LOS.

Example 2 *A machinist is expected to make engine parts with axle diameter of 1.75 cm. A random sample of 10 parts shows a mean diameter 1.85 cm with an SD of 0.1 cm. On the basis of this sample, would you say that the work of the machinist is inferior?*

Solution Here $\bar{x} = 1.85$, $s = 0.1$, $n = 10$ and $\mu = 1.75$.

H_0: $\bar{x} = \mu$; H_1: $\bar{x} \neq \mu$.

Two-tailed test is to be used. Let LOS be 5%.

$$t = \frac{\bar{x} - \mu}{s/\sqrt{n-1}} = \frac{0{\cdot}10}{0{\cdot}1/\sqrt{9}} = 3 \quad \text{and} \quad \nu = n - 1 = 9$$

From the t-table, for $\nu = 9$, $t_{0.05} = 2.26$ and $t_{0.01} = 3.25$.

$\therefore$ $|t| > t_{0.05}$ and $|t| < t_{0.01}$

Therefore, H_0 is rejected and H_1 is accepted at 5% level, but H_0 is accepted and H_1 is rejected at 1% level. That is at 5% LOS the work of the machinist can be assumed to be inferior, but at 1% LOS the work cannot be assumed to be inferior.

Example 3 *A certain injection administered to each of 12 patients resulted in the following increases of blood pressure:*

5, 2, 8, – 1, 3, 0, 6, – 2, 1, 5, 0, 4.

Can it be concluded that the injection will be, in general, accompanied by an increase in BP?

Solution The mean of the sample is given by

$$\bar{x} = \frac{1}{n}\sum x = \frac{31}{12} = 2.58$$

The SDs of the sample is given by

$$s^2 = \frac{1}{n}\sum x^2 - \left(\frac{1}{n}\sum x\right)^2 = \frac{1}{12} \times 185 - (2.58)^2 = 8.76$$

$\therefore$ $s = 2.96$

H_0: $\bar{x} = \mu$,

where $\mu = 0$, i.e. the injection will not result in increase in BP.

H_1: $\bar{x} > \mu$.

Right-tailed test is to be used. Let LOS be 5%. Now $t_{0.05\%}$ for one-tailed test for $(\nu = 11) = t_{10\%}$ for two-tailed test for $(\nu = 11) = 1.80$ (from t-table)

Now $$t = \frac{\bar{x} - \mu}{s/\sqrt{n-1}} = \frac{2{\cdot}58 - 0}{2{\cdot}96/\sqrt{11}} = 2.89$$

We see that $|t| > t_{10\%}$ $(\nu = 11)$.

Therefore, H_0 is rejected and H_1 is accepted. That is, we may conclude that the injection is accompanied by an increase in BP.

Example 4 *The mean lifetime of a sample of 25 bulbs is found as 1550 hours with a SD of 120 h. The company manufacturing the bulbs claims that the average life of their bulbs is 1600 h. Is the claim acceptable at 5% LOS?*

Solution Here $\bar{x} = 1550$, $s = 120$, $n = 25$ and $\mu = 1600$.

H_0: $\bar{x} = \mu$ and H_1: $\bar{x} < \mu$.

Left-tailed test is to be used. LOS be 5%.

Now $$t = \frac{\bar{x} - \mu}{s/\sqrt{n-1}} = \frac{-50\sqrt{24}}{120} = -2.04 \quad \text{and} \quad \nu = 24$$

$t_{0.05}$ for one-tailed test = $t_{0.1}$ for two-tailed test (for $v = 24$) = 1.71.

We see that $|t| > |t_{0.1}|$.

Therefore, H_0 is rejected and H_1 is accepted at 5% LOS. That is, the claim of the company cannot be accepted at 5% LOS.

Example 5 *The heights of 10 males of a given locality are found to be 175, 168, 155, 170, 152, 170, 175, 160, 160 and 165 cms. Based on this sample, find the 95% confidence limits for the height of males in that locality.*

Solution We shall first find the mean $\bar{x}$ and SDs of the sample, by taking the assumed mean $A = 165$ (see the following table).

$$d_i = x_i - A$$

$$\therefore \quad \bar{x} = A + \bar{d}$$

$$= 165 + \frac{1}{10} \times 0 = 165$$

$$s^2 = \frac{1}{n}\sum d_i^2 - \left(\frac{1}{n}\sum d_i\right)^2$$

$$= \frac{1}{10} \times 578 = 57.8$$

$$\therefore \quad s = 7.6$$

From the t-table,

$$t_{0.05}\ (v = 9) = 2.26.$$

The 95% confidence limits for μ are

$$\left(\bar{x} - 2{\cdot}26\frac{s}{\sqrt{n-1}},\ \bar{x} + 2{\cdot}26\frac{s}{\sqrt{n-1}}\right)$$

i.e.
$$\left(165 - \frac{2{\cdot}26 \times 7{\cdot}6}{\sqrt{9}},\ 165 + \frac{2{\cdot}26 \times 7{\cdot}6}{\sqrt{9}}\right)$$

i.e. (159.3, 170.7)

i.e., the heights of males in the locality are likely to lie within 159.3 cm and 170.7cm.

x_i	$d_i = x_i - A$	d_i^2
175	10	100
168	3	9
155	– 10	100
170	5	25
152	– 13	169
170	5	25
175	10	100
160	– 5	25
160	– 5	25
165	0	0
Total	0	578

Example 6 *Two independent samples of sizes 8 and 7 contained the following values:*

Sample 1 :	*19*	*17*	*15*	*21*	*16*	*18*	*16*	*14*
Sample 2 :	*15*	*14*	*15*	*19*	*15*	*18*	*16*	

Is the difference between the sample means significant?

Solution

Sample 1			Sample 2		
x_1	$d_1 = x_1 - 18$	d_1^2	x_2	$d_2 = x_2 - 16$	d_2^2
19	1	1	15	−1	1
17	−1	1	14	−2	4
15	−3	9	15	−1	1
21	3	9	19	3	9
16	−2	4	15	−1	1
18	0	0	18	2	4
16	−2	4	16	0	0
14	−4	16			
Total	−8	44	Total	0	20

For sample 1,

$$\bar{x}_1 = 18 + \bar{d}_1 = 18 + \frac{1}{8}\sum d_1$$

$$= 18 + \frac{1}{8} \times (-8) = 17.$$

$$s_1^2 = \frac{1}{n_1}\sum d_1^2 - \left(\frac{1}{n_1}\sum d_1\right)^2$$

$$= \frac{1}{8} \times 44 - \left[\frac{1}{8} \times (-8)\right]^2 = 4.5$$

$\therefore \quad s_1 = 2.12.$

For sample 2,

$$\bar{x}_2 = 16 + \bar{d}_2 = 16 + \frac{1}{7}\sum d_2 = 16$$

$$s_2^2 = \frac{1}{n_2}\sum d_2^2 - \left(\frac{1}{n_2}\sum d_2\right)^2$$

$$= \frac{1}{7} \times 20 - \left(\frac{1}{7} \times 0\right)^2 = 2.857$$

$\therefore \quad s_2 = 1.69$

$$H_0: \bar{x}_1 = \bar{x}_2 \quad \text{and} \quad H_1: \bar{x}_1 \neq \bar{x}_2$$

Two-tailed test is to be used. Let LOS be 5%.

$$t = \frac{\bar{x}_1 - \bar{x}_2}{\sqrt{\left(\frac{n_1 s_1^2 + n_2 s_2^2}{n_1 + n_2 - 2}\right)\left(\frac{1}{n_1} + \frac{1}{n_2}\right)}} = \frac{17 - 16}{\sqrt{\left(\frac{8 \times 4{\cdot}5 + 7 \times 2{\cdot}857}{13}\right)\left(\frac{1}{8} + \frac{1}{7}\right)}}$$

$$= 0.93$$

Also $\quad \nu = n_1 + n_2 - 2 = 13.$

From the t-table, $t_{0.05}\ (\nu = 13) = 2.16$

Since $|t| < t_{0.05}$, H_0 is accepted and H_1 is rejected.

That is, the two sample means do not differ significantly at 5% LOS.

Example 7 *The following data represent the biological values of protein from cow's milk and buffalo's milk at a certain level.*

Cow's milk	:	*1.82*	*2.02*	*1.88*	*1.61*	*1.81*	*1.54*
Buffalo's milk	:	*2.00*	*1.83*	*1.86*	*2.03*	*2.19*	*1.88*

Examine if the average values of protein in the two samples significantly differ.

Solution $\quad n = 6$

$$\bar{x}_1 = \frac{1}{6} \times 10.68 = 1.78$$

$$s_1^2 = \frac{1}{6} \times \Sigma x_1^2 - (\bar{x}_1)^2 = \frac{1}{6} \times 19.167 - (1.78)^2 = 0.0261$$

$$\bar{x}_2 = \frac{1}{6} \times 11.79 = 1.965$$

$$s_2^2 = \frac{1}{6} \times \Sigma x_2^2 - (\bar{x}_2)^2 = \frac{1}{6} \times 23.2599 - (1.965)^2 = 0.0154$$

As the two samples are independent, the test statistic is given by

$$t = \frac{\bar{x}_1 - \bar{x}_2}{\sqrt{\frac{s_1^2 + s_2^2}{n-1}}}$$

with $\nu = 2n - 2$ (refer to Note 2 under Test 2)

$$t = \frac{1{\cdot}78 - 1{\cdot}965}{\sqrt{\frac{0{\cdot}0261 + 0{\cdot}0154}{5}}} = \frac{-0{\cdot}185}{\sqrt{{\cdot}0083}} = -2.03 \text{ and } \quad \nu = 10$$

$$H_0: \bar{x}_1 = \bar{x}_2 \quad \text{and} \quad H_1: \bar{x}_1 \neq \bar{x}_2.$$

Two-tailed test is to be used. Let LOS be 5%.

From t-table, $t_{0.05}\ (\nu = 10) = 2.23$.

Since $|t| < t_{0.05}$ ($\nu = 10$). H_0 is accepted. That is, the difference between the mean protein values of the two varieties of milk is not significant at 5% level.

Example 8 *Samples of two types of electric bulbs were tested for length of life and the following data were obtained.*

	Size	*Mean*	*SD*
Sample 1	*8*	*1234 h*	*36 h*
Sample 2	*7*	*1036 h*	*40 h*

Is the difference in the means sufficient to warrant that type 1 bulbs are superior to type 2 bulbs?

Solution *Here* $\bar{x}_1 = 1234$, $s_1 = 36$, $n_1 = 8$; $\bar{x}_2 = 1036$, $s_2 = 40$, $n_2 = 7$

H_0: $\bar{x}_1 = \bar{x}_2$; H_1: $\bar{x}_1 > \bar{x}_2$.

Right-tailed test is to be used. Let LOS be 5%.

$$t = \frac{\bar{x}_1 - \bar{x}_2}{\sqrt{\left(\frac{n_1 s_1^2 + n_2 s_2^2}{n_1 + n_2 - 2}\right)\left(\frac{1}{n_1} + \frac{1}{n_2}\right)}} = \frac{198}{\sqrt{\left(\frac{21568}{13}\right)\left(\frac{1}{8} + \frac{1}{7}\right)}} = \frac{198}{21 \cdot 0807}$$

$$= 9.39$$

$$\nu = n_1 + n_2 - 2 = 13$$

$t_{0.05}$ ($\nu = 13$) for one-tailed test = $t_{10\%}$ ($\nu = 13$) for two-tailed test = 1.77 (from t-table).

Now $t > t_{0.1}$ ($\nu = 13$). Therefore, H_0 is rejected and H_1 is accepted. That is, type 1 bulbs may be regarded superior to type 2 bulbs at 5% LOS.

Example 9 *The mean height and the SD height of 8 randomly chosen soliders are 166.9 and 8.29 cm respectively. The corresponding values of 6 randomly chosen sailors are 170.3 and 8.50 cm respectively. Based on this data, can we conclude that soldiers are, in general, shorter than sailors?*

Solution Here $\bar{x}_1 = 166.9$, $s_1 = 8.29$, $n_1 = 8$; $\bar{x}_2 = 170.3$, $s_2 = 8.50$, $n_2 = 6$.

H_0: $\bar{x}_1 = \bar{x}_2$; H_1: $\bar{x}_1 < \bar{x}_2$.

Left-tailed test is to be used. Let LOS be 5%.

$$t = \frac{\bar{x}_1 - \bar{x}_2}{\sqrt{\left(\frac{n_1 s_1^2 + n_2 s_2^2}{n_1 + n_2 - 2}\right)\left(\frac{1}{n_1} + \frac{1}{n_2}\right)}} = \frac{-3 \cdot 4}{\sqrt{\left(\frac{983 \cdot 29}{12}\right)\left(\frac{1}{8} + \frac{1}{6}\right)}}$$

$$= -0.695$$

$$\nu = n_1 + n_2 - 2 = 12$$

$t_{0.05}$ (v = 12) for one-tailed test = $t_{0.1}$ (v = 12) for two-tailed test = 1.78 (from t-table)

Now $|t| < t_{0.1}$ (v = 12). Therefore, H_0 is accepted and H_1 is rejected. That is, based on the given data, we cannot conclude that soldiers are, in general, shorter than sailors.

Example 10 *The following data relate to the marks obtained by 11 students in 2 tests, one held at the beginning of a year and the other at the end of the year after intensive coaching.*

Test 1:	19	23	16	24	17	18	20	18	21	19	20
Test 2:	17	24	20	24	20	22	20	20	18	22	19

Do the data indicate that the students have benefited by coaching?

Solution The given data relate to the marks obtained in 2 tests by the same set of students. Hence the marks in the 2 tests can be regarded as correlated and so the t-test for paired values should be used.

Let $d = x_1 - x_2$, where x_1, x_2 denote the marks in the 2 tests.

Thus the values of d are 2, – 1 – 4, 0, – 3, – 4, 0, – 2, 3, – 3, 1.

$$\Sigma d = -11 \quad \text{and} \quad \Sigma d^2 = 69$$

$$\therefore \quad \bar{d} = \frac{1}{n}\Sigma d = \frac{1}{11} \times (-11) = -1$$

$$s^2 = s^2{}_d = \frac{1}{n}\Sigma d^2 - (\bar{d})^2 = \frac{1}{11} \times 69 - (-1)^2 = 5.27$$

$$\therefore \quad s = 2.296$$

H_0: $\bar{d} = 0$, i.e. the students have not benefitted by coaching; H_1: $\bar{d} < 0$ (i.e. $\bar{x}_1 < \bar{x}_2$).

One-tailed test is to be used. Let LOS be 5%.

$$t = \frac{\bar{d}}{s/\sqrt{n-1}} = \frac{-1}{2 \cdot 296/\sqrt{10}} = -1.38 \quad \text{and} \quad v = 10$$

$t_{0.05}$ (v = 10) for one-tailed test = $t_{0.1}$ (v = 10) for two-tailed test = 1.81 (from t-table).

Now $|t| < t_{10}$ (v = 10). Therefore, H_0 is accepted and H_1 is rejected, i.e., there is no significant difference between the two sets of marks. That is, the students have not benefitted by coaching.

Example 11 *A sample of size 13 gave an estimated population variance of 3.0, while another sample of size 15 gave an estimate of 2.5. Could both samples be from populations with the same variance?*

Solution Here, $n_1 = 13$, $\hat{\sigma}_1^2 = 3.0$ and $v_1 = 12$, $n_2 = 15$, $\hat{\sigma}_2^2 = 2.5$ and $v_2 = 14$.

H_0: $\hat{\sigma}_1^2 = \hat{\sigma}_2^2$, i.e. the two samples have been drawn from populations with the same variance. $H_1 : \hat{\sigma}_1^2 \neq \hat{\sigma}_2^2$.

Let LOS. be 5%.

$$F = \frac{\hat{\sigma}_1^2}{\hat{\sigma}_2^2} = \frac{3 \cdot 0}{2 \cdot 5} = 1.2$$

$$\nu_1 = 12 \quad \text{and} \quad \nu_2 = 14.$$

$F_{0.05\%}$ (ν_1 = 12, ν_2 = 14) = 2.53, from the F-table. Since $F < F_{0.05}$, H_0 is accepted. That is the two samples could have come from two normal populations with the same variance.

Example 12 *Two samples of sizes 9 and 8 gave the sums of squares of deviations from their respective means equal to 160 and 91 respectively. Can they be regarded as drawn from the same normal population?*

Solution $n_1 = 9$, $\Sigma(x_i - \bar{x})^2 = 160$, i.e. $n_1 s_1^2 = 160$

$n_2 = 8$, $\Sigma(y_i - \bar{y})^2 = 91$, i.e. $n_2 s_2^2 = 91$

$$\hat{\sigma}_1^2 = \frac{n_1 s_1^2}{n_1 - 1} = \frac{1}{8} \times 160 = 20 \; ; \quad \hat{\sigma}_2^2 = \frac{n_2 s_2^2}{n_2 - 1} = \frac{1}{7} \times 91 = 13$$

Since $\hat{\sigma}_1^2 > \hat{\sigma}_2^2$, $\nu_1 = n_1 - 1 = 8$ and $\nu_2 = n_2 - 1 = 7$

H_0: $\hat{\sigma}_1^2 = \hat{\sigma}_2^2$ and H_1: $\hat{\sigma}_1^2 \neq \hat{\sigma}_2^2$.

Let the LOS be 5%.

$$F = \frac{\hat{\sigma}_1^2}{\hat{\sigma}_2^2} = \frac{20}{13} = 1.54$$

$F_{0.05}$ ($\nu_1 = 8$, $\nu_2 = 7$) = 3.73, from the F-table. Since $F < F_{0.05\%}$, H_0 is accepted.

That is, the two samples could have come from two normal populations with the same variance.

We cannot say that the samples have come from the same population, as we are unable to test whether the means of the samples differ significantly or not.

Example 13 *Two independent samples of 8 and 7 items respectively had the following values of the variable.*

Sample 1 :	*9*	*11*	*13*	*11*	*15*	*9*	*12*	*14*
Sample 2 :	*10*	*12*	*10*	*14*	*9*	*8*	*10*	

Do the two estimates of population variance differ significantly at 5% LOS?

Solution For the first sample, $\Sigma x_1 = 94$ and $\Sigma x_1^2 = 1138$.

$$\therefore \quad s_1^2 = \frac{1}{n_1} \sum x_1^2 - \left(\frac{1}{n_1} \sum x_1 \right)^2$$

$$= \frac{1}{8} \times 1138 - \left(\frac{1}{8} \times 94\right)^2 = 4.19$$

For the second sample, $\Sigma x_2 = 73$ and $\Sigma x_2^2 = 785$.

$$\therefore \quad s_2^2 = \frac{1}{n_2} \sum x_2^2 - \left(\frac{1}{n_2} \sum x_2\right)^2$$

$$= \frac{1}{7} \times 785 - \left(\frac{1}{7} \times 73\right)^2 = 3.39$$

$$\hat{\sigma}_1^2 = \frac{n_1}{n_1 - 1} s_1^2 = 4.79 \quad \text{and} \quad \hat{\sigma}_2^2 = \frac{n_2}{n_2 - 1} s_2^2 = 3.96$$

since $\hat{\sigma}_1^2 > \hat{\sigma}_2^2$, $\nu_1 = 7$ and $\nu_2 = 6$

$$H_0: \hat{\sigma}_1^2 = \hat{\sigma}_2^2 \quad \text{and} \quad H_1: \hat{\sigma}_1^2 \neq \hat{\sigma}_2^2$$

$$F = \frac{\hat{\sigma}_1^2}{\hat{\sigma}_2^2} = \frac{4\cdot 79}{3\cdot 96} = 1.21$$

$F_{0.05}$ $(\nu_1 = 7, \nu_2 = 6) = 4.21$, from the F-table. Since $F < F_{0.05}$, H_0 is accepted. That is, $\hat{\sigma}_1^2$ and $\hat{\sigma}_2^2$ do not differ significantly at 5% LOS.

Example 14 *Two random samples gave the following data:*

Sample no.	*Size*	*Mean*	*Variance*
1	*8*	*9.6*	*1.2*
2	*11*	*16.5*	*2.5*

Can we conclude that the two samples have been drawn from the same normal population?

Solution Refer to Note 2 under F-test. To conclude that the two samples have been drawn from the same population, we have to check first that the variances of the populations do not differ significantly and then check that the sample means (and hence the population means) do not differ significantly.

$$\hat{\sigma}_1^2 = \frac{8 \times 1\cdot 2}{7} = 1.37\,; \quad \hat{\sigma}_2^2 = \frac{11 \times 2\cdot 5}{10} = 2.75$$

$$F = \frac{\hat{\sigma}_2^2}{\hat{\sigma}_1^2} = \frac{2\cdot 75}{1\cdot 37} = 2.007 \text{ with } \nu_1 = 10 \text{ and } \nu_2 = 7.$$

From the F-table, $F_{0.05}$ $(10, 7) = 3.64$.

If H_0: $\hat{\sigma}_1^2 = \hat{\sigma}_2^2$ and H_1: $\hat{\sigma}_1^2 \neq \hat{\sigma}_2^2$, then, H_0 is accepted, since $F < F_{0.05}$.

That is, the variances of the populations from which samples are drawn may be regarded as equal.

$$t = \frac{\bar{x}_1 - \bar{x}_2}{\sqrt{\left(\frac{n_1 s_1^2 + n_2 s_2^2}{n_1 + n_2 - 2}\right)\left(\frac{1}{n_1} + \frac{1}{n_2}\right)}} = \frac{-6{\cdot}9}{\sqrt{\left(\frac{9{\cdot}6 + 27{\cdot}5}{17}\right)\left(\frac{1}{8} + \frac{1}{11}\right)}}$$

$$= -\frac{6{\cdot}9}{0{\cdot}6864} = -10.05$$

and $v = n_1 + n_2 - 2 = 17$

From the t-table $t_{0.05}$ $(v = 17) = 2.11$.

If H_0: $\bar{x}_1 = \bar{x}_2$ and H_1: $\bar{x}_1 \neq \bar{x}_2$, then H_0 is rejected, since $|t| > t_{0.05}$.

That is, the means of two samples (and so the populations) differ significantly.

Therefore, the two samples could not have been drawn from the same normal population.

Example 15 *The nicotine contents in two random samples of tobacco are given below.*

Sample 1 :	*21*	*24*	*25*	*26*	*27*	
Sample 2 :	*22*	*27*	*28*	*30*	*31*	*36*

Can you say that the two samples came from the same population?

Solution $\bar{x}_1$ = Mean of sample 1 = $\frac{123}{5} = 24.6$

$\bar{x}_2$ = Mean of sample 2 = $\frac{174}{6} = 29.0$

s_1^2 = Variance of sample 1 = $\frac{1}{5}\,\Sigma(x_i - 24.6)^2 = 4.24$

s_2^2 = Variance of sample 2 = $\frac{1}{6}\,\Sigma(x_i - 29.0)^2 = 18.0$

$$\hat{\sigma}_1^2 = \frac{5}{4} \times 4.24 = 5.30 \text{ and } v = 4; \ \hat{\sigma}_2^2 = \frac{6}{5} \times 18.0 = 21.60 \text{ and } v = 5$$

H_0: $\hat{\sigma}_1^2 = \hat{\sigma}_2^2$; H_1: $\hat{\sigma}_1^2 \neq \hat{\sigma}_2^2$

$$F = \frac{\hat{\sigma}_2^2}{\hat{\sigma}_1^2} = \frac{21{\cdot}60}{5{\cdot}30} = 4.07$$

$F_{0.05}$ (5, 4) = 6.26.

Since $F < F_{0.05}$, H_0 is accepted. Therefore, the variances of the two populations can be regarded as equal.

$$t = \frac{\bar{x}_1 - \bar{x}_2}{\sqrt{\left(\frac{n_1 s_1^2 + n_2 s_2^2}{n_1 + n_2 - 2}\right)\left(\frac{1}{n_1} + \frac{1}{n_2}\right)}} = \frac{-4{\cdot}4}{\sqrt{\left(\frac{21{\cdot}2 + 108{\cdot}0}{9}\right)\left(\frac{1}{5} + \frac{1}{6}\right)}}$$

$$= \frac{-4 \cdot 4}{2 \cdot 2943} = -1.92$$

and $\nu = 9$.

From t-table, $F_{0.05}$ $(\nu = 9) = 2.26$.

If H_0: $\bar{x}_1 = \bar{x}_2$ and H_1: $\bar{x}_1 \neq \bar{x}_2$, H_0 is accepted, since $|t| < F_{0.05}$.

That is, the means of two samples (and hence the populations) do not differ significantly. Therefore, the two samples could have been drawn from the same normal population.

Exercise 8(B)

Part A (Short answer questions)

1. Write down the probability density of student's t-distribution.
2. State the important properties of the t-distribution.
3. Give any two uses of t-distribution.
4. What do you mean by degrees of freedom?
5. How will you get the critical value of t for a single-tailed test at LOS α?
6. What is the test statistic used to test the significance of the difference between small sample mean and population mean?
7. Give the 95% confidence interval of the population mean in terms of the mean and SD of a small sample.
8. What is the test statistic used to test the significance of the difference between the means of two small samples?
9. Give an estimate of the population variance in terms of variances of two small samples. What is the associated number of degrees of freedom?
10. What is the test statistic used to test the significance of the difference between the means of two small samples of the same size? What is the associated number of degree of freedom?
11. What is the test statistic used to test the significance of the difference between the means of two small samples of the same size, when the sample items are correlated?
12. Write down the probability density function of the F-distribution.
13. State the important properties of the F-distribution.
14. What is the use of F-distribution?
15. Why is the F-distribution associated with two numbers of degrees of freedom?

Part B

16. A random sample of 10 boys had the following IQs: 70, 120, 110, 101, 88, 83, 95, 98, 107, 100. Does the data support the assumption of a population

mean IQ of 100? Find a reasonable range in which most of the mean IQ values of samples of ten boys lie.

17. A random sample of 16 values from a normal population showed a mean of 103.75 cm and the sum of the squares of deviations from this mean is equal to 843.75 cm^2. Show that the assumption of a mean of 108.75 cm for the population is not reasonable. Obtain 95% and 99% fiducial limits for the same.
18. The mean weekly sales of soap bars in departmental stores is 145 bars store. After an advertising campaign, the mean weekly sales in 17 stores for a typical week increased to 155 and showed an SD of 16. Was the advertising campaign successful?
19. The annual rainfall at a certain place is normally distributed with mean 30. If the rainfalls during the past 8 years are 31.1, 30.7, 24.3, 28.1, 27.9, 32.2, 25.4 and 29.1, can we conclude that average rainfall during the past 8 years is less than the normal rainfall?
20. A machine is expected to produce nails of length 7 cm. A random sample of 10 nails was found to measure the following lengths: 7.2, 7.3, 7.1, 6.9, 6.8, 6.5, 6.9, 6.8, 7.1 and 7.2 cm. On the basis of this sample, what can be said about the reliability of the machine?
21. A random sample of 8 envelopes is taken from the letter box of a post office and their weights in grams are found to be 12.2, 11.9, 12.5, 12.3, 11.6, 11.7, 12.2 and 12.4. Find the 95% and 99% confidence limits for the mean weight of the envelopes in the letter box.
22. The average production of 16 workers in a factory was 107 with an SD of 9, while 12 workers in another comparable factory had an average production of 111 with an SD of 10. Can we say that the production rate of workers in the latter factory is more than that in the former factory?
23. Two different types of drugs A and B were tried on certain patients for increasing weight. Five persons were given drug A and 7 persons were given drug B. The increase in weight (in kg.) is given below:

Drug A :	3.6	5.5	5.9	4.1	1.4		
Drug B :	4.5	3.6	5.5	6.8	2.7	3.6	5.0

Do the two drugs differ significantly with regard to their effect in increasing weight?

24. Samples of 12 foremen in one division and 10 foremen in another division of a factory were selected at random and the following data were obtained:

	Division 1	*Division 2*
Sample size	12	10
Average monthly salary of foremen (Rs)	5250	4900
SD of salary (Rs)	152	165

Can you conclude that foremen in Division 1 get more salary than foremen in division 2?

25. Two independent groups of 10 children were tested to find how many digits they could repeat from memory after hearing them. The results are as follows:

Group A:	8	6	5	7	6	8	7	4	5	6
Group B:	10	6	7	8	6	9	7	6	7	7

Is the difference between mean scores of the two groups significant?

26. The following data represent the marks obtained by 12 students in 2 tests, one held before coaching and the other after coaching:

Test 1:	55	60	65	75	49	25	18	30	35	54	61	72
Test 2:	63	70	70	81	54	29	21	38	32	50	70	80

Does the data indicate that the coaching was effective in improving the performance of students?

27. In one sample of 8 items, the sum of the squares of deviations of the sample values from the sample mean was 84.4, and in another sample of 10 observations it was 102.6. Test whether this difference is significant at 5% level.

28. Two random samples drawn from two normal populations gave the following observations.

Sample 1:	20	16	26	27	23	22	18	24	25	19		
Sample 2:	17	23	32	25	22	24	28	18	31	33	20	27

Test whether the two populations have the same variance.

29. Two random samples gave the following results:

Sample no.	*Size*	*Mean*	*Variance*
1	16	440	40
2	25	460	42

Test whether the samples have been drawn from the same normal population.

30. Two random samples gave the following results:

Sample no.	*Size*	*Sum of the values*	*Sum of the squares of values*
1	10	150	2340
2	12	168	2460

Test whether the samples have been drawn from the same normal population.

Chi-Square Distribution

If $X_1, X_2, \ldots, X_n$ are normally distributed independent random variables, then it is known that $(X_1^2 + X_2^2 + \ldots + X_n^2)$ follows a probability distribution, called *chi-*

square (χ^2-distribution) distribution with n degrees of freedom. The probability density function of the χ^2-distribution is given by

$$f(\chi^2) = \frac{1}{2^{v/2}\sqrt{\left(\frac{v}{2}\right)}} \cdot (\chi^2)^{v/2-1} \; e^{-\chi^2/2} 0 < \chi^2 < \infty,$$

where v is the number of degrees of freedom.

Properties of x²-Distribution

1. A rough sketch of the probability curve of the χ^2-distribution for $v = 3$ and $v = 6$ is given in Fig. 8.4.
2. As v becomes smaller and smaller, the curve is skewed more and more to the right. As v increases, the curve becomes more and more symmetrical.
3. The mean and variance of the χ^2-distribution are v and $2v$ respectively.

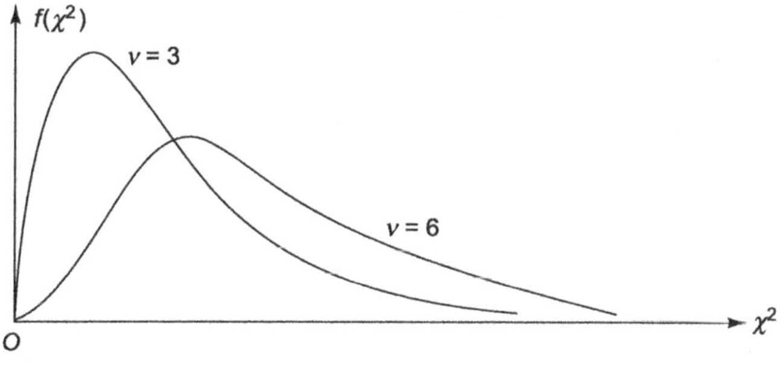

Fig. 8.4

4. As v tends to ∞, the χ^2-distribution becomes a normal distribution.

Uses of χ²-Distribution

1. χ^2-distribution is used to test the goodness of fit. i.e., it is used to judge whether a given sample may be reasonably regarded as a simple sample from a certain hypothetical population.
2. It is used to test the independence of attributes. That is, if a population is known to have two attributes (or traits), then χ^2-distribution is used to test whether the two attributes are associated or independent, based on a sample drawn from the population.

χ²-Test of Goodness of Fit

On the basis of the hypothesis assumed about the population, we find the expected frequencies $E_i(i = 1, 2, \ldots, n)$, corresponding to the observed frequencies O_i $(i = 1, 2, \ldots, n)$ such that $\sum E_i = \sum O_i$. It is known that

$$\chi^2 = \sum_{i=1}^{n} \frac{(O_i - E_i)^2}{E_i}$$

follows approximately a χ^2-distribution with degrees of freedom equal to the number of independent frequencies. To test the goodness of fit, we have to determine how far the differences between O_i and E_i can be attributed to fluctuations of sampling and when can we assert that the differences are large enough to conclude that the sample is not a simple sample from the hypothetical population. In other words, we have to determine how large a value of χ^2 we can get so as to assume that the sample is a simple sample from the hypothetical population.

The critical value of χ^2 for ν degrees of freedom at α LOS, denoted by $\chi^2_\nu(\alpha)$, is given by

$$P[\chi^2 > \chi^2_\nu(\alpha)] = \alpha$$

Critical values of the χ^2-distribution corresponding to a few important LOS and a range of values of ν are available in the form of a table called χ^2-table, which is given at the end of the chapter.

If the calculated $\chi^2 < \chi^2_\nu(\alpha)$, we will accept the null hypothesis H_0 which assumes that the given sample is one drawn from the hypothetical population, i.e. we will conclude that the difference between the observed and expected frequencies is not significant at α % LOS. If $\chi^2 > \chi^2_\nu(\alpha)$, we will reject H_0 and conclude that the difference is significant.

Conditions for the Validity of χ^2-Test

1. The number of observations N in the sample must be reasonably large, say ≥ 50.
2. Individual frequencies must not be too small, i.e. $O_i \geq 10$. In case $O_i < 10$, it is combined with the neighbouring frequencies, so that the combined frequency is ≥ 10.
3. The number of classes n must be neither too small nor too large, i.e. $4 \leq n \leq 16$.

χ^2-Test of Independence of Attributes

If the population is known to have two major attributes A and B, then A can be divided into m categories $A_1, A_2, \ldots, A_m$ and B can be divided into n categories $B_1, B_2, \ldots, B_n$. Accordingly the members of the population, and hence those of the sample, can be divided into mn classes. In this case, the sample data may be presented in the form of a matrix containing m rows and n columns and hence mn cells, and showing the observed frequencies O_{ij}, in the various cells, where $i = 1, 2, \ldots, m$ and $j = 1, 2, \ldots, n$. O_{ij} means the number of observed frequencies possessing the attributes A_i and B_j. The matrix or tabular form of the sample data, called an $(m \times n)$ *contingency table* is given below:

A	B						Row Total
	B_1	B_2	-	B_j	-	B_n	
A_1	O_{11}	O_{12}	-	O_{1j}	-	O_{1n}	O_{1*}
A_2	O_{21}	O_{22}	-	O_{2j}	-	O_{2n}	O_{2*}
:	-	-	-	-	-	-	-
A_i	O_{i1}	O_{i2}	-	O_{ij}	-	O_{in}	O_{i*}
:	-	-	-	-	-	-	-
A_m	O_{m1}	O_{m2}		O_{mj}	-	O_{mn}	O_{m*}
Column Total	O_{*1}	O_{*2}	-	O_{*j}	-	O_{*n}	N

Now, based on the null hypothesis H_0, i.e. the assumption that the two attributes A and B are independent, we compute the expected frequencies E_{ij} for various cells, using the following formula:

$$E_{ij} = \frac{O_{i*} \times O_{*j}}{N}, \quad i = 1, 2, \ldots, m; \quad j = 1, 2, \ldots, n$$

i.e.

$$E_{ij} = \left(\frac{\text{Total of observed frequencies in the } ith \text{ row} \times \text{total of a observed frequencies in the } j\text{th column}}{\text{Total of all cell frequencies}} \right)$$

Then we compute

$$\chi^2 = \sum_{i=1}^{m} \sum_{j=1}^{n} \left\{ \frac{\left(O_{ij} - E_{ij}\right)^2}{E_{ij}} \right\}$$

The number of degrees of freedom for this χ^2 computed from the $(m \times n)$ contingency table is $\nu = (m - 1)(n - 1)$.

If $\chi^2 < \chi^2_\nu(\alpha)$, H_0 is accepted at α % LOS, i.e. the attributes A and B are independent.

If $\chi^2 > \chi^2_\nu(\alpha)$, H_0 is rejected at α % LOS, i.e. A and B are not independent.

Worked Examples 8(C)

Example 1 *The following table shows the distribution of digits in the numbers chosen at random from a telephone directory:*

Digit:	*0*	*1*	*2*	*3*	*4*	*5*	*6*	*7*	*8*	*9*	*Total*
Frequency:	*1026*	*1107*	*997*	*966*	*1075*	*933*	*1107*	*972*	*964*	*853*	*10,000*

Test whether the digits may be taken to occur equally frequently in the directory.

Solution H_0: The digits occur equally frequently, i.e. they follow a uniform distribution.

Based on H_0, we compute the expected frequencies.

The total number of digits = 10,000.

If the digits occur uniformly, then each digit will occur $\frac{10,000}{10} = 1000$ times.

O_i: 1026, 1107, ..., 853

E_i: 1000, 1000, ..., 1000

$$\chi^2 = \sum \frac{(O_i - E_i)^2}{E_i}$$

$$= \frac{1}{1000} [(26)^2 + (107)^2 + (-3)^2 + (-34)^2 + (75)^2 + (-67)^2 + (107)^2 + (-28)^2 + (-36)^2 + (-147)^2]$$

$$= 58.542$$

Since ΣE_i was taken equal to ΣO_i (i.e. an information from the sample), $\nu = n - 1 = 10 - 1 = 9$. From the χ^2–table,

$\chi^2_{0.05}\ (n = 9) = 16.919.$

Since $\chi^2 > \chi^2_{0.05}$, H_0 is rejected. That is, the digits do not occur uniformly in the directory.

Example 2 *The following data give the number of air-craft accidents that occurred during the various days of a week.*

Day:	*Mon*	*Tues*	*Wed*	*Thu*	*Fri*	*Sat*
No. of accidents:	*15*	*19*	*13*	*12*	*16*	*15*

Test whether the accidents are uniformly distributed over the week.

Solution H_0: Accidents occur uniformly over the week.

Total number of accidents = 90

Based on H_0, the expected number of accidents on any day $= \frac{90}{6} = 15$.

O_i :	15	19	13	12	16	15
E_i :	15	15	15	15	15	15

$$\chi^2 = \sum \frac{(O_i - E_i)^2}{E_i} = \frac{1}{15}(0 + 16 + 4 + 9 + 1 + 0) = 2$$

Since $\sum E_i = \sum O_i$, $\nu = 6 - 1 = 5$.

From the χ^2-table, $\chi^2_{0.05}\ (\nu = 5) = 11.07$.

Since $\chi^2 < \chi^2_{0.05}$, H_0 is accepted. That is, accidents may be regarded to occur uniformly over the week.

Example 3 *The following data show defective articles produced by 4 machines:*

Machine:	*A*	*B*	*C*	*D*
Production time:	*1*	*1*	*2*	*3*
No. of detectives:	*12*	*30*	*63*	*98*

Do the figures indicate a significant difference in the performance of the machines?

Solution H_0: Production rates of the machines are the same.

Total number of detectives = 203.

Based on H_0, the expected numbers of defectives produced by the machines are

	E_i:	$\frac{1}{7} \times 203$	$\frac{1}{7} \times 203$	$\frac{2}{7} \times 203$	$\frac{3}{7} \times 203$
i.e.	E_i:	29	29	58	87
	O_i:	12	30	63	98

$$\chi^2 = \sum \frac{(O_i - E_i)^2}{E_i} = \frac{17^2}{29} + \frac{1^2}{29} + \frac{5^2}{58} + \frac{11^2}{87} = 11.82$$

Since $\sum E_i = \sum O_i$, $v = 4 - 1 = 3$.

From the χ^2-table, $\chi^2_{0.05}$ $(v = 3) = 7.815$.

Since $\chi^2 > \chi^2_{0.05}$, H_0 is rejected. That is, there is significant difference in the performance of machines.

Example 4 *The following data represent the monthly sales (in Rs) of a certain retail store in a leap year. Examine if there is any seasonality in the sales.*
6100, 5600, 6350, 6050, 6250, 6200, 6300, 6250, 5800, 6000, 6150 and 6150.

Solution H_0: There is no seasonability in the sales, i.e. the daily sales are uniform throughout the year or the daily sales follow a uniform distribution.

Based on H_0, we compute the expected frequencies.

The total sales in the year = Rs 73,200.

If the daily sales are uniform, then the sales on each day $= \frac{73{,}200}{366}$ = Rs 200.

O_i: 6100, 5600, 6350, 6050, 6250, 6200, 6300, 6250, 5800, 6000, 6150, 6150.

Assuming that the months are taken in the usual calendar order, namely, January, February, etc., the expected monthly sales are:

E_i: 6200, 5800, 6200, 6000, 6200, 6000, 6200, 6200, 6000, 6200, 6000, 6200

Then $$\chi^2 = \sum \frac{(O_i - E_i)^2}{E_i}$$

$$= \frac{(-100)^2}{6200} + \frac{(-200)^2}{5800} + \dots + \frac{(-50)^2}{6200} = 38.913$$

Since $\sum E_i$ was found as $\sum O_i$ from the sample, $v = n - 1 = 12 - 1 = 11$.

From the χ^2-table $\chi^2_{0.05}$ $(v = 11) = 19.675$.

Since $\chi^2 > \chi^2_{0.05}$, H_0 is rejected. That is, the daily sales are not uniform throughout the year.

Example 5 *Theory predicts that the proportion of beans in 4 groups A, B, C, D should be 9 : 3 : 3 : 1. In an experiment among 1600 beans, the numbers in the 4 groups were 882, 313, 287 and 118. Does the experiment support the theory?*

Solution H_0: The experiment supports the theory, i.e. the numbers of beans in the 4 groups are in the ratio 9 : 3 : 3 : 1.

Based on H_0, the expected numbers of beans in the 4 groups are as follows:

	E_i:	$\frac{9}{16} \times 1600$	$\frac{3}{16} \times 1600$	$\frac{3}{16} \times 1600$	$\frac{1}{16} \times 1600$
i.e.	E_i:	900	300	300	100
	O_i:	882	313	287	118

$$\chi^2 = \sum \frac{(O_i - E_i)^2}{E_i} = \frac{18^2}{900} + \frac{13^2}{300} + \frac{13^2}{300} + \frac{18^2}{100} = 4.73$$

Since $\sum E_i = \sum O_i$, $v = 4 - 1 = 3$.

From the χ^2-table, $\chi^2_{0.05}(v = 3) = 7.82$.

Since $\chi^2 < \chi^2_{0.05}$, H_0 is accepted. That is, the experimental data support the theory.

Example 6 *A survey of 320 families with 5 children revealed the following distribution:*

No. of boys :	*0*	*1*	*2*	*3*	*4*	*5*
No. of girls :	*5*	*4*	*3*	*2*	*1*	*0*
No. of families:	*12*	*40*	*88*	*110*	*56*	*14*

Is this result consistent with the hypothesis that male and female births are equally probable?

Solution H_0: Male and female births are equally probable, i.e. P (male birth) = $p = 1/2$ and P (female birth) = $q = 1/2$.

Based on H_0, the probability that a family of 5 children has r male children

$$= 5C_r \left(\frac{1}{2}\right)^5 \text{ (by binomial law).}$$

$\therefore$ Expected number of families having r male children $= 320 \times 5C_r \times \frac{1}{2^5}$

$$= 10 \times 5C_r.$$

Thus	E_i:	10	50	100	100	50	10
	O_i:	12	40	88	110	56	14

$$\chi^2 = \sum \frac{(O_i - E_i)^2}{E_i} = \frac{2^2}{10} + \frac{10^2}{50} + \frac{12^2}{100} + \frac{10^2}{100} + \frac{6^2}{50} + \frac{4^2}{10}$$

$$= 7.16$$

We have used the sample data to get $\sum E_i$ only. The values of p and q have not been found by using the sample data.

$\therefore$ $\quad v = n - 1 = 6 - 1 = 5$ and $\chi^2_{0.05}(v = 5) = 11.07$

Since $\chi^2 < \chi^2_{0.05}$, H_0 is accepted. That is, male and female births are equally probable.

Example 7 *Twelve dice were thrown 4096 times and a throw of 6 was considered a success. The observed frequencies were as given below:*

No. of successes:	*0*	*1*	*2*	*3*	*4*	*5*	*6*	*7 and over*
Frequency:	*447*	*1145*	*1180*	*796*	*380*	*115*	*25*	*8*

Test whether the dice were unbiased.

Solution H_0: All the dice were unbiased,

i.e. P (getting 6) $= p = \frac{1}{6}$. Therefore, $q = \frac{5}{6}$.

Based on H_0, the probability of getting exactly r successes

$$= 12C_r\, p^r\, q^{12-r} \quad (r = 0, 1, 2, \ldots, 12)$$

Therefore, expected number of times in which r successes are obtained:

$$E_r = 4096 \times 12C_r \left(\frac{1}{6}\right)^r \cdot \left(\frac{5}{6}\right)^{12-r}$$

$$= 4096 \times 12C_r \times \frac{5^{12-r}}{6^{12}} \quad (r = 0, 1, 2, \ldots, 12)$$

i.e. $\quad E_0 = N$ (0 success) $= N(r = 0) = 459.39$

$E_1 = N(r = 1) = 1102.54$

$E_2 = N(r = 2) = 1212.80$

$E_3 = N(r = 3) = 808.53$

$E_4 = N(r = 4) = 363.84$

$E_5 = N(r = 5) = 116.43$

$E_6 = N(r = 6) = 27.17$

$E_7 = N(r \geq 7) = 5.30$

Converting E_i's into whole numbers subject to the condition that $\sum E_i = 4096$, we get

E_i:	459	1103	1213	809	364	116	27	5
O_i:	497	1145	1180	796	380	115	25	8

Since E and O corresponding to the last class i.e. 5 and 8 are less than 10, we combine the last 2 classes and consider as a single class.

$$\chi^2 = \sum \frac{(O_i - E_i)^2}{E_i} = \frac{12^2}{459} + \frac{42^2}{1103} + \frac{33^2}{1213} + \frac{13^2}{809} + \frac{16^2}{364} + \frac{1^2}{116} + \frac{1^2}{32}$$

$= 3.76$

$v = n - 1$ (since only $\sum E_i$ has been found using the sample data).

$= 7 - 1$ (n must be taken as the number of classes after combination of end classes, if any)

$= 6$

and $\chi^2_{0.05}$ $(v = 6) = 12.59$, from the χ^2-table. Since $\chi^2 < \chi^2_{0.05}$, H_0 is accepted, i.e. the dice were unbiased.

Example 8 *Fit a binomial distribution for the following data and also test the goodness of fit.*

x:	0	1	2	3	4	5	6	Total
f:	5	18	28	12	7	6	4	80

To find the binomial distribution $N(q + p)^n$, which fits the given data, we require p.

Solution We know that the mean of the binomial distribution is np, from which we can find p. Now the mean of the given distribution is found out and is equated to np.

x :	0	1	2	3	4	5	6	Total
f :	5	18	28	12	7	6	4	80
fx :	0	18	56	36	28	30	24	192

$$\bar{x} = \frac{\sum fx}{\sum f} = \frac{192}{80} = 2.4$$

i.e. $np = 2.4$ or $6p = 2.4$, since the maximum value taken by x is n.

$\therefore$ $p = 0.4$ and hence $q = 0.6$

Therefore, the expected frequencies are given by the successive terms in the expansion of $80\,(0.6 + 0.4)^6$.

Thus E_i: 3.73 14.93 24.88 22.12 11.06 2.95 0.33

Converting the E_i's into whole number such that $\sum E_i = \sum O_i = 80$, we get

E_i: 4 15 25 22 11 3 0

Let us now proceed to test the goodness of binomial fit.

O_i: 5 18 28 12 7 6 4

The first class is combined with the second and the last 2 classes are combined with the last but second class in order to make the expected frequency in each class greater than or equal to 10. Thus, after regrouping, we have

E_i :	19	25	22	14
O_i :	23	28	12	17

$$\chi^2 = \sum \frac{(O_i - E_i)^2}{E_i} = \frac{4^2}{19} + \frac{3^2}{25} + \frac{10^2}{22} + \frac{3^2}{14} = 6.39$$

We have used the given sample to find $\sum E_i (= \sum O_i)$ and p through its mean.

Hence $n = n - k = 4 - 2 = 2$ and $\chi^2_{0.05}$ $(\nu = 2) = 5.99$, from the χ^2-table.

Since $\chi^2 > \chi^2_{0.05}$, H_0, which assumes that the given distribution is approximately a binomial distribution, is rejected, i.e. the binomial fit for the given distribution is not satisfactory.

Example 9 *Fit a Poisson distribution for the following distribution and also test the goodness of fit.*

x :	0	1	2	3	4	5	Total
f :	142	156	69	27	5	1	400

Solution *To find the Poisson distribution whose probability law is*

$$P(X = r) = \frac{e^{-\lambda} \lambda^r}{r!}, \; r = 0, 1, 2, \ldots$$

we require λ, which is the mean of the Poisson distribution.
We find the mean of the given distribution and assume it as λ.

x :	0	1	2	3	4	5	Total
f :	142	156	69	27	5	1	400
fx :	0	156	138	81	20	5	400

$$\bar{x} = \frac{\sum fx}{\sum f} = \frac{400}{400} = 1 = \lambda$$

The expected frequencies are given by

$$\frac{N \cdot e^{-\lambda} \lambda^r}{r!} \text{ or } \frac{400 \times e^{-1}}{r!}, \; r = 0, 1, 2, \ldots, \infty$$

Thus

E_i:	147.15	147.15	73.58	24.53	6.13	1.23

The values of E_i are very small for $i = 6, 7, \ldots$ and hence neglected.

Converting the values of E_i's into whole numbers such that $\sum E_i = 400$, we get

E_i:	147	147	74	25	6	1

Let us now proceed to test the goodness of Poisson fit.

O_i:	142	156	69	27	5	1

The last 3 classes are combined into one, so that the expected frequency in that class may be ≥ 10. Thus, after regrouping, we have

O_i:	142	156	69	33
E_i:	147	147	74	32

$$\chi^2 = \sum \frac{(O_i - E_i)^2}{E_i} = \frac{5^2}{147} + \frac{9^2}{147} + \frac{5^2}{74} + \frac{1^2}{32}$$

$$= 1.09$$

We have used the sample data to find $\sum E_i$ and λ. Hence, $n = n - k = 4 - 2 = 2$

From the χ^2-table, $\chi^2_{0.05}$ $(v = 2) = 5.99$.

Since $\chi^2 < \chi^2_{00.5}$, H_0, which assumes that the given distribution is nearly Poisson, is accepted. That is, the Poisson fit for the given distribution is satisfactory.

Example 10 *Test the normality of the following distribution by using χ^2-test of goodness of fit*

x:	*125*	*135*	*145*	*155*	*165*	*175*	*185*	*195*	*205*	*Total*
f:	*1*	*1*	*14*	*22*	*25*	*19*	*13*	*3*	*2*	*100*

Solution Let us first fit a normal distribution to the given data and then test the goodness of fit.

To fit a normal distribution and hence find the expected frequencies, we require the density function of the normal distribution which involves the mean and SD. Let us now compute the mean $\bar{x}$ and SD s of the sample distribution and assume them as μ and σ.

x	f	$d = \frac{x - 165}{10}$	fd	fd^2
125	1	– 4	– 4	16
135	1	– 3	– 3	9
145	14	– 2	– 28	56
155	22	– 1	– 22	22
165	25	0	0	0
175	19	1	19	19
185	13	2	26	52
195	3	3	9	27
205	2	4	8	32
Total	100	–	5	233

$$\bar{x} = A + \frac{c}{N}\sum fd = 165 + \frac{10}{100} \times 5 = 165.5$$

$$s^2 = c^2\left\{\frac{1}{N}\sum fd^2 - \left(\frac{1}{N}\sum fd\right)^2\right\} = 10^2\,(2.33 - 0.0025)$$

$$= 232.75$$

$\therefore \qquad s = 15.26$

Therefore, the density function of the normal distribution which fits the given distribution is $f(x) = \dfrac{1}{15 \cdot 26\sqrt{2\pi}} e^{-(x-165 \cdot 5)^2/465 \cdot 5}$.

To find the expected frequency corresponding to a given x, we find $y = f(x)$ and multiply y by the class-width and then by the total frequency N.

Note $y = \dfrac{1}{\sigma\sqrt{2\pi}} e^{-(x-\mu)^2/2\sigma^2}$. *If we put* $\dfrac{x-\mu}{\sigma} = z$,

then $y = \dfrac{1}{\sigma}\left\{\dfrac{1}{\sqrt{2\pi}} e^{-z^2/2}\right\} = \dfrac{\phi(z)}{\sigma}$, *where* $\phi(z)$ *is density function of the standard normal distribution. Values of* $\phi(z)$ *are obtained from the normal table given at the end of this chapter.*

x	$y = \dfrac{x - 165 \cdot 5}{15 \cdot 26}$	$\phi(z)$	$\dfrac{c \cdot \phi(z)}{\sigma} = \dfrac{10\phi(z)}{15 \cdot 26}$	*Expected frequency* $= N\phi(z)/\sigma$
125	– 2.65	0.0119	0.0078	0.78
135	– 2.00	0.0540	0.0354	3.54
145	– 1.34	0.1626	0.1066	10.66
155	– 0.69	0.3144	0.2060	20.60
165	– 0.03	0.3988	0.2613	26.13
175	0.62	0.3292	0.2157	21.57
185	1.28	0.1758	0.1152	11.52
195	1.93	0.0620	0.0406	4.06
205	2.59	0.0139	0.0091	0.91

Converting the expected frequencies as whole numbers such that $\sum E_i = 100$, we get

E_i: 1 3 11 21 26 22 11 4 1

Let us now proceed to test the goodness of normal fit.

Combining the end classes so as to make the individual frequencies greater than 10,

E_i:	15	21	26	22	16
O_i:	16	22	25	19	18

$$\chi^2 = \sum \frac{(O_i - E_i)^2}{E_i} = \frac{1^2}{15} + \frac{1^2}{21} + \frac{1^2}{26} + \frac{3^2}{22} + \frac{2^2}{16}$$

$$= 0.82$$

We have used the sample data to find $\sum E_i$, μ and σ.

Hence $v = n - k = 5 - 3 = 2$.

From the χ^2-table, $\chi^2_{0.05}$ ($v = 2$) = 5.99.

Since $\chi^2 < \chi^2_{0.05}$, H_0, which assumes that the given distribution is nearly normal, is accepted. That is, the normal fit for the given distribution is satisfactory.

Example 11 *The following data is collected on two characters. Based on this, can you say that there is no relation between smoking and literacy?*

	Smokers	*Non-smokers*
Literates:	*83*	*57*
Illiterates:	*45*	*68*

Solution H_0: Literacy and smoking habit are independent

	Smokers	*Non-smokers*	*Total*
Literates	83	57	140
Illiterates	45	68	113
Total	128	125	253

O	*E*	*E (rounded)*	$(O-E)^2/E$
83	$\frac{128 \times 140}{253} = 70.83$	71	$12^2/71 = 2.03$
57	$\frac{125 \times 140}{253} = 69.17$	69	$12^2/69 = 2.09$
45	$\frac{128 \times 113}{253} = 57.17$	57	$12^2/57 = 2.53$
68	$\frac{125 \times 113}{253} = 55.83$	56	$12^2/56 = 2.57$
			$\chi^2 = 9.22$

$$v = (m - 1)(n - 1)$$
$$= (2 - 1 (2 - 1) = 1.$$

From the χ^2-table, $\chi^2_{0.05}$ ($v = 1$) = 3.84

Since $\chi^2 > \chi^2_{0.05}$, H_0 is rejected. That is, there is some association between literacy and smoking.

Example 12 *Prove that the value of χ^2 for the 2 × 2 contingency table*

a	*b*
c	*d*

is given by

$$\chi^2 = \frac{N(ad - bc)^2}{(a+b)(c+d)(a+c)(b+d)}, \text{ where } N = a + b + c + d.$$

Hence compute χ^2 for the 2×2 contingency table given in Example 11.

Solution The value of E corresponding to the cell in which $O = a$ is given by

$$E = \frac{(a+b)(a+c)}{(a+b+c+d)}.$$

Therefore, the value of χ^2 corresponding to this cell is given by

$$\chi^2 = \left\{a - \frac{(a+b)(a+c)}{a+b+c+d}\right\}^2 \div \frac{(a+b)(a+c)}{(a+b+c+d)}$$

$$= \frac{\{a(a+b+c+d) - (a+b)(a+c)\}^2}{N(a+b)(a+c)}$$

$$= \frac{(ad-bc)^2}{N(a+b)(a+c)}$$

Similarly the values of χ^2 are found out for the other three cells.
$\therefore$ χ^2 for the table

$$= \frac{(ad-bc)^2}{N} \left[\frac{1}{(a+b)(a+c)} + \frac{1}{(a+b)(b+d)} + \frac{1}{(a+c)(c+d)} + \frac{1}{(b+d)(c+d)}\right]$$

$$= \frac{(ad-bc)^2}{N(a+b)(c+d)(a+c)(b+d)} [(b+d)(c+d) + (a+c)(c+d) + (a+b)(b+d) + (a+b)(a+c)]$$

$$= \frac{(ad-bc)^2\left[\sum a^2 + 2\sum ab\right]}{N(a+b)(c+d)(a+c)(b+d)} = \frac{(ad-bc)^2(a+b+c+d)^2}{N(a+b)(c+d)(a+c)(b+d)}$$

$$= \frac{N(ad-bc)^2}{(a+b)(c+d)(a+c)(b+d)} \qquad (1)$$

Using (1) for the contingency table in Example 11, we get

$$\chi^2 = \frac{253(83 \times 68 - 45 \times 57)^2}{140 \times 113 \times 128 \times 125} = 9.48$$

Example 13 *Two batches each of 12 animals are taken for test of inoculation. One batch was inoculated and the other batch was not inoculated. The numbers of dead and surviving animals are given in the following table for both cases. Can the inoculation be regarded as effective against the disease? Make Yate's correction for continuity of χ^2.*

	Dead	Survived	Total
Inoculated	2	10	12
Not inoculated	8	4	12
Total	10	14	24

Note on Yate's correction for continuity of χ^2.

Solution The χ^2-table was prepared using the theoretical χ^2-distribution which is continuous, whereas the approximate values of χ^2 that we are using are discrete. To rectify this defect, Yates has shown that, when

$$\chi^2 = \sum \left[\frac{\left\{ \left| 0_i - E_i \right| - \frac{1}{2} \right\}^2}{E_i} \right]$$

is used, the χ^2-approximation is improved. Yate's correction is used only when $v = 1$ and hence for a 2×2 contingency table. It is used only when some cell frequency is small, i.e. less than 5.

In the present problem, two cell frequencies are less than 5 each. Hence we apply Yate's correction:

O	E	$\lvert O - E \rvert - 0.5$	$\{\lvert O - E \rvert - 0.5\}^2/E$
2	$\frac{12 \times 10}{24} = 5$	2.5	6.25/5 = 1.25
10	$\frac{12 \times 14}{24} = 7$	2.5	6.25/7 = 0.89
8	$\frac{12 \times 10}{24} = 5$	2.5	6.25/5 = 1.25
4	$\frac{12 \times 14}{24} = 7$	2.5	6.25/7 = 0.89
	$v = (2 - 1)(2 - 1) = 1$		$\chi^2 = 4.28$

From the χ^2-table, $\chi^2_{0.05}$ ($v = 1$) = 3.84.

If H_0: Inoculation and effect on the diseases are independent, then H_0 is rejected as $\chi^2 > \chi^2_{0.05}$ i.e. Inoculation can be regarded as effective against the disease.

Note *Even if Yate's correction is not made, we would have arrived at the same conclusion.*

Example 14 *A total number of 3759 individuals were interviewed in a public opinion survey on a political proposal. Of them, 1872 were men and the rest women. A total of 2257 individuals were in favour of the proposal and 917 were opposed to it. A total of 243 men were undecided and 442 women were opposed* *to the proposal. Do you justify or contradict the hypothesis that there is no association between sex and attitude?*

Solution A careful analysis of the problem results in the following contingency table:

	Favoured	*Opposed*	*Undecided*	*Total*
Men	1154	475	243	1872
Women	1103	442	342	1887
Total	2257	917	585	3759

H_0: Sex and attitude are independent, i.e. there is no association between sex and attitude.

O	*E (rounded E)*	$(O-E)^2/E$
1154	$\frac{1872 \times 2257}{3759} \simeq 1124$	$30^2/1124 = 0.80$
475	$\frac{1872 \times 917}{3759} \simeq 457$	$18^2/457 = 0.71$
243	$\frac{1872 \times 585}{3759} \simeq 291$	$48^2/291 = 7.92$
1103	$\frac{1887 \times 2257}{3759} \simeq 1133$	$30^2/1133 = 0.79$
442	$\frac{1887 \times 917}{3759} \simeq 460$	$18^2/460 = 0.70$
342	$\frac{1887 \times 585}{3759} \simeq 294$	$48^2/294 = 7.84$
	$v = (3-1)(2-1) = 2$	$\chi^2 = 18.76$

From the χ^2-table, $\chi^2_{0.05}$ $(v = 2) = 5.99$.

Since $\chi^2 > \chi^2_{0.05}$, H_0 is rejected. Therefroe, sex and attitude are not independent, i.e. there is some association between sex and attitude.

Example 15 *The following table gives for a sample of married women, the level of education and the marriage adjustment score:*

Level of	*Marriage adjustment*				
Education	*Very low*	*Low*	*High*	*Very high*	*Total*
College	*24*	*97*	*62*	*58*	*241*
High school	*22*	*28*	*30*	*41*	*121*
Middle school	*32*	*10*	*11*	*20*	*73*
Total	*78*	*135*	*103*	*119*	*435*

Can you conclude from the above data that the higher the level of education, the greater is the degree of adjustment in marriage?

Solution H_0: There is no relation between the level of education and adjustment in marriage.

$$\nu = (4 - 1)(3 - 1) = 6$$

$$\chi^2_{0.05}(\nu = 6) = 12.59$$

O	E (rounded)	$(O-E)^2/E$
24	43	$19^2/43 = 8.40$
97	75	$22^2/75 = 6.45$
62	57	$5^2/57 = 0.44$
58	66	$8^2/66 = 0.97$
22	22	$0^2/22 = 0.00$
28	37	$9^2/37 = 2.19$
30	29	$1^2/29 = 0.03$
41	33	$8^2/33 = 1.94$
32	13	$19^2/13 = 27.77$
10	23	$13^2/23 = 7.35$
11	17	$6^2/17 = 2.12$
20	20	$0^2/20 = 0.00$
$\chi^2_{0.05}(\nu = 6) = 12.59$		$\chi^2 = 57.66$

Since $\chi^2 > \chi^2_{0.05}$, H_0 is rejected. That is, the level of education and adjustment in marriage are associated.

Thus we may conclude that the higher the level of education, the greater is the degree of adjustment in marriage.

Exercise 8(C)

Part A (Short answer questions)

1. Define chi-square distribution.
2. Write down the probability density function of the χ^2-distribution.
3. State the important properties of χ^2-distribution.
4. Give 2 uses of χ^2-distribution.
5. What is χ^2-test of goodness of fit?
6. State the conditions under which χ^2-test of goodness of fit is valid.
7. What is χ^2-test of independence of attributes?
8. What is contingency table?
9. Write down the value of χ^2 for a 2×2 contingency table with cell frequencies a, b, c and d.
10. What is Yate's correction for continuity of χ^2?

Part B

11. In 250 digits from the lottery numbers, the frequencies of the digits were as follows:

Digit:	0	1	2	3	4	5	6	7	8	9
Frequency:	23	25	20	23	23	22	29	25	33	27

Test the hypothesis that the digits were randomly drawn.

12. The following table gives the number of fatal road accidents that occurred during the 7 days of the week. Find whether the accidents are uniformly distributed over the week.

Day:	Sun	Mon	Tues	Wed	Thu	Fri	Sat
Number:	8	14	16	12	11	14	9

13. In 120 throws of a single die, the following distribution of faces are obtained:

Face:	1	2	3	4	5	6
Frequency:	30	25	18	10	22	15

Do these results support the equal probability hypothesis?

14. The number of demands for a particular spare part in a shop was found to vary from day to day. In a sample study, the following information was obtained:

Day:	Mon	Tues	Wed	Thu	Fri	Sat
No. of demands:	124	125	110	120	126	115

Test the hypothesis that the number of parts demanded does not depend on the day of the week.

15. According to genetic theory, children having one parent of blood type *M* and the other of blood type *N* will always be one of the three types-*M*, *MN* and *N* and the average proportions of these types will be 1 : 2 : 1. Out of 300 children, having one *M* parent and one N parent, 30 per cent were found to be of type *M*, 45 per cent of type *MN* and the remaining of type *N*. Test the genetic theory by χ^2-test.

16. Five coins are tossed 256 times. The number of heads observed is given below. Examine if the coins are true.

No. of heads :	0	1	2	3	4	5
Frequency :	5	35	75	84	45	12

17. Five dice were thrown 243 times and the numbers of times 1 or 2 was thrown (x) are given below:

x:	0	1	2	3	4	5
Frequency:	30	75	76	47	13	2

Examine if the dice were unbiased.

18. Fit a binomial distribution for the following data and also test the goodness of fit.

x:	0	1	2	3	4
f:	5	29	36	25	5

19. Fit a binomial distribution for the following data and also test the goodness of fit

x:	0	1	2	3	4	5	6	7	8	9
f:	3	8	11	15	16	14	12	11	9	1

20. Fit a Poisson distribution for the following distribution and also test the goodness of fit.

x:	0	1	2	3	4	5	6	7
f:	314	335	204	86	29	9	3	0

21. Fit a Poisson distribution for the following distribution and also test the goodness of fit.

x:	0	1	2	3	4
f:	123	59	14	3	1

22. The figures given below are (i) the observed frequencies of a distribution and (ii) the expected frequencies of the normal distribution having the same mean, SD and total frequency as in (i).
(i) 1, 12, 66, 220, 495, 792, 924, 792, 495, 220, 66, 12, 1
(ii) 2, 15, 66, 210, 484, 799, 943, 799, 484, 210, 66, 15, 2
Do you think that the normal distribution provides a good fit to the data?
23. Fit a normal distribution to the following data and find also the goodness of fit.

x :	4	6	8	10	12	14	16	18	20	22	24
f:	1	7	15	22	35	43	38	20	13	5	1

24. In an epidemic of certain disease, 92 children contacted the disease. Of these 41 received no treatment and of these 10 showed after-effects. Of the remainder who did receive the treatment, 17 showed after-effects. Test the hypothesis that the treatment was not effective.
25. Out of 1660 candidates who appeared for a competitive examination, 422 were successful. Out of these, 256 had attended a coaching class and 150 of them came out successful. Examine whether coaching was effective as regards the success in the examination.
26. In a pre-poll survey, out of 1000 rural voters, 620 favoured A and the rest B. Out of 1000 urban voters, 450 favoured B and the rest A. Examine whether the nature of the area is related to voting preference.
27. The following information was obtained in a sample of 40 small general shops:

	Shops in urban areas	*Shops in rural areas*
Owned by men	17	18
Owned by women	3	12

Can it be said that there are more women owners in rural areas than in urban areas? Use Yate's correction for continuity.
28. A certain drug is claimed to be effective in curing cold. In an experiment on 500 persons with cold, half of them were given the drug and half of them were given the sugar pills. The patients' reaction to the treatment are recorded in the following table:

	Helped	*Harmed*	*No effect*
Drug	150	30	70
Sugar pills	130	40	80

On the basis of this data, can it be concluded that the drug and sugar pills differ significantly in curing cold?

29. A survey of radio listeners' preference for two types of music under various age groups gave the following information.

Type of music	*Age group*		
	19-25	26-35	Above 36
Carnatic music	80	60	90
Film music	210	325	44
Indifferent	16	45	132

Is preference for type of music influenced by age?

30. The table given below shows the results of a survey in which 250 respondents were classified according to the levels of education and attitude towards students' agitation in a certain town. Test whether the two criteria of classification are independent.

	Attitude		
Education	*Against*	*Neutral*	*For*
Middle school:	40	25	5
High school:	40	20	5
College:	30	15	30
Postgraduate:	15	15	10

Test whether the two criteria of classification are independent.

Table of normal ordinates and areas:

z	*Ordinate*	*Area* = $\int_0^z$	*z*	*Ordinate*	*Area* = $\int_0^z$
0.00	0.3989	0.0000	1.55	0.1200	0.4394
0.05	0.3984	0.0199	1.60	0.1109	0.4452
0.10	0.3970	0.0398	1.65	0.1023	0.4505
0.15	0.3945	0.0596	1.70	0.941	0.4554
0.20	0.3910	0.0793	1.75	0.0863	0.4099
0.25	0.3867	0.0987	1.80	0.0790	0.4641
0.30	0.3814	0.1179	1.85	0.0721	0.4678
0.35	0.3752	0.1368	1.90	0.0656	0.4719
0.40	0.3683	0.1554	1.95	0.0596	0.4744
0.45	0.3605	0.1736	2.00	0.0540	0.4773
0.50	0.3521	0.1910	2.05	0.0488	0.4798
0.55	0.3429	0.2088	2.10	0.0440	0.4821
0.60	0.3332	0.2258	2.15	0.0396	0.4842
0.65	0.3230	0.2422	2.20	0.0355	0.4861

(*Contd*)

z	*Ordinate*	*Area* = $\int_0^z$	z	*Ordinate*	*Area* = $\int_0^z$
0.70	0.3123	0.2080	2.25	0.0317	0.4878
0.75	0.3011	0.2734	2.30	0.0283	0.4893
0.80	0.2897	0.2882	2.35	0.0252	0.4906
0.85	0.2780	0.3023	2.40	0.0224	0.4918
0.90	0.2661	0.3159	2.45	0.0224	0.4929
0.95	0.2541	0.3289	2.50	0.0198	0.4938
1.00	0.2420	0.3413	2.55	0.0175	0.4946
1.05	0.2299	0.3531	2.60	0.0155	0.4953
1.10	0.2179	0.3643	2.65	0.0136	0.4960
1.15	0.2059	0.3749	2.70	0.0119	0.4965
1.20	0.1942	0.3849	2.75	0.0104	0.4970
1.25	0.1827	0.3944	2.80	0.0091	0.4974
1.30	0.1714	0.4032	2.85	0.0079	0.4978
1.35	0.1604	0.4115	2.90	0.0060	0.4981
1.40	0.1497	0.4192	2.95	0.0051	0.4984
1.45	0.1394	0.4265	3.00	0.0044	0.4987
1.50	0.1295	0.4332	3.05	0.0038	0.4989

t-Table

	Probability				
n	0.9	0.1	0.05	0.02	0.01
1	0.158	6.314	12.706	31.821	63.657
2	0.142	2.920	4.303	6.965	9.925
3	0.137	2.353	3.182	4.541	5.841
4	0.134	2.132	2.776	3.747	4.604
5	0.132	2.015	2.571	3.365	4.032
6	0.131	1.943	2.447	3.143	3.707
7	0.130	1.895	2.365	2.998	3.496
8	0.129	1.860	2.306	2.896	3.355
9	0.129	1.833	2.262	2.821	3.250
10	0.129	1.812	2.228	2.764	3.169
11	0.128	1.796	2.201	2.718	3.106
12	0.128	1.782	2.179	2.681	3.055
13	0.128	1.771	2.160	2.650	3.012
14	0.128	1.761	2.145	2.624	2.977
15	0.128	1.753	2.131	2.602	2.947
16	0.128	1.746	2.120	2.583	2.921
17	0.128	1.740	2.110	2.567	2.898
18	0.127	1.734	2.101	2.552	2.878
19	0.127	1.729	2.093	2.539	2.861
20	0.127	1.725	2.086	2.528	2.845
21	0.127	1.721	2.080	2.518	2.831
22	0.127	1.717	2.074	2.508	2.819
23	0.127	1.714	2.069	2.500	2.807
24	0.127	1.711	2.064	2.492	2.797

(Contd)

	Probability				
n	0.9	0.1	0.05	0.02	0.01
25	0.127	1.708	2.060	2.485	2.787
30	0.127	1.697	2.042	2.457	2.750
40	0.126	1.684	2.021	2.423	2.704
60	0.126	1.671	2.000	2.390	2.660
120	0.126	1.658	1.980	2.358	2.617
∞	0.126	1.645	1.960	2.326	2.576

χ^2-Table

	Probability					
n	0.99	0.95	0.10	0.05	0.02	0.01
1	0.000157	0.00393	2.706	3.841	5.412	6.635
2	0.0201	0.103	4.605	5.991	7.824	9.210
3	0.115	2.352	6.251	4.815	9.837	11.345
4	0.297	0.711	7.779	9.488	11.668	13.277
5	0.554	1.145	9.236	11.070	13.388	15.086
6	0.872	1.635	10.645	12.592	15.033	16.812
7	1.238	2.167	12.017	14.067	16.622	18.475
8	1.646	2.733	13.362	15.507	18.168	20.090
9	2.088	3.325	14.684	16.919	19.670	21.666
10	2.558	3.940	15.987	18.307	21.161	23.209
11	3.053	4.575	17.275	19.675	22.618	24.725
12	3.571	5.226	18.549	21.026	24.054	26.217
13	4.107	5.892	19.812	22.362	25.472	27.688
14	4.660	6.571	21.064	23.685	26.873	29.141
15	5.229	7.261	23.307	24.996	28.259	30.578
16	5.812	7.962	23.542	26.296	29.633	32.000
17	6.408	8.672	24.768	27.587	30.995	33.409
18	7.015	9.390	25.989	28.869	32.346	34.805
19	7.633	10.117	27.204	30.114	33.687	36.191
20	8.260	10.851	28.412	31.410	35.020	37.566
21	8.897	11.581	29.615	32.671	36.343	38.932
22	9.542	12.338	30.813	33.924	37.659	40.289
23	10.196	13.091	32.007	35.172	38.968	41.638
24	10.856	13.848	33.196	36.415	40.270	42.980
25	11.524	14.611	34.382	37.652	41.566	44.314
26	12.198	15.379	35.563	38.885	42.856	45.642
27	12.879	16.151	36.741	40.113	44.140	46.963
28	13.565	16.928	37.916	41.337	45.419	48.278
29	14.256	17.708	39.087	42.557	46.693	49.588
30	14.953	18.493	40.256	43.773	47.962	50.892

For larger values of n, the expression $\sqrt{2\chi^2} - \sqrt{2n-1}$ may be used as a normal variate with unit variance.

Table of F (Variance ratio)—1 Per Cent Points

ν_2 \ ν_1	1	2	3	4	5	6	8	12	24	∞
1	4052	4999	5403	5625	5764	5859	5981	6106	6234	6366
2	98.49	99.01	99.17	99.25	99.30	99.33	99.36	99.42	99.46	99.50
3	34.12	30.81	29.46	28.71	28.24	27.91	27.49	27.05	26.60	26.12
4	21.20	18.00	16.69	15.98	15.52	15.21	14.80	14.37	13.93	13.46
5	16.26	13.27	12.06	11.39	10.97	10.67	10.27	9.89	9.47	9.02
6	13.74	10.92	9.78	9.15	8.75	8.47	8.10	7.72	7.31	6.88
7	12.25	9.55	8.45	7.85	7.46	7.19	6.84	6.47	6.07	5.65
8	11.26	8.65	7.59	7.01	6.63	6.37	6.03	5.67	5.28	4.86
9	10.56	8.02	6.99	6.42	6.06	5.80	5.47	5.11	4.73	4.31
10	10.04	7.56	6.55	5.99	5.64	5.39	5.06	4.71	4.33	3.91
11	9.65	7.20	6.22	5.67	5.32	5.07	4.74	4.40	4.02	3.60
12	9.33	6.93	5.95	5.41	5.06	4.82	4.50	4.16	3.78	3.36
13	9.07	6.70	5.74	5.20	4.86	4.62	4.30	3.96	3.59	3.16
14	8.86	6.51	5.56	5.03	4.69	4.46	4.14	3.80	3.43	3.00
15	8.68	6.36	5.42	4.89	4.56	4.32	4.00	3.67	3.29	2.87
16	8.53	6.23	5.29	4.77	4.44	4.20	3.89	3.55	3.18	2.75
17	8.40	6.11	5.18	4.67	4.34	4.10	3.79	3.45	3.08	2.65
18	8.28	6.01	5.09	4.58	4.25	4.01	3.71	3.37	3.00	2.57
19	8.18	5.93	5.01	4.50	4.17	3.94	3.63	3.30	2.92	2.49
20	8.10	5.85	4.94	4.43	4.10	3.87	3.56	3.23	2.86	2.42
21	8.02	5.78	4.87	4.37	4.04	3.81	3.51	3.17	2.80	2.36
22	7.94	5.72	4.82	4.31	3.99	3.76	3.45	3.12	2.75	2.31
23	7.88	5.66	4.76	4.26	3.94	3.71	3.41	3.07	2.70	2.26
24	7.82	5.61	4.72	4.22	3.90	3.67	3.36	3.03	2.66	2.21
25	7.77	5.57	4.68	4.18	3.86	3.63	3.32	2.99	2.62	2.17
26	7.72	5.53	4.64	4.14	3.82	3.59	3.29	2.96	2.58	2.13
27	7.68	5.49	4.60	4.11	3.79	3.56	3.26	2.93	2.55	2.10
28	7.64	5.45	4.57	4.07	3.76	3.53	3.23	2.90	2.52	2.06
29	7.60	5.42	4.54	4.04	3.73	3.50	3.20	2.87	2.49	2.03
30	7.56	5.39	4.51	4.02	3.70	3.47	3.17	2.84	2.47	2.01
40	7.31	5.18	4.31	3.83	3.51	3.29	2.99	2.66	2.29	1.81
60	7.08	4.98	4.13	3.65	3.34	3.12	2.82	2.50	2.12	1.60
120	6.85	4.79	3.95	3.48	3.17	2.96	2.66	2.34	1.95	1.38
∞	6.64	4.60	3.78	3.32	3.02	2.80	2.51	2.18	1.79	1.00

Table of F—5 Per Cent Points

ν_2 \ ν_1	1	2	3	4	5	6	8	12	24	∞
1	161.4	199.5	215.7	224.6	230.2	234.0	238.9	243.9	249.0	253.4
2	18.51	19.00	19.16	19.25	19.30	19.33	19.37	19.41	19.45	19.50
3	10.13	9.55	9.28	9.12	9.01	8.94	8.84	8.74	8.64	8.53
4	7.71	6.94	6.59	6.39	6.26	6.16	6.04	5.91	5.77	5.63
5	6.61	5.79	5.41	5.19	5.05	4.95	4.82	4.68	4.53	4.36

(Contd)

	v_1									
v_2	1	2	3	4	5	6	8	12	24	∞
6	5.99	5.14	4.76	4.53	4.39	4.28	4.15	4.00	3.84	3.67
7	5.59	4.74	4.35	4.12	3.97	3.87	3.73	3.57	3.41	3.23
8	5.32	4.46	4.07	3.87	3.69	3.58	3.44	3.28	3.12	2.93
9	5.12	4.26	3.86	3.63	3.48	3.37	3.23	3.07	2.90	2.71
10	4.96	4.10	3.71	3.48	3.33	3.22	3.07	2.91	2.74	2.54
11	4.84	3.98	3.59	3.36	3.20	3.09	2.95	2.79	2.61	2.40
12	4.75	3.88	3.49	3.26	3.11	3.00	2.85	2.69	2.50	2.30
13	4.67	3.80	3.41	3.18	3.02	2.92	2.77	2.60	2.42	2.21
14	4.60	3.74	3.34	3.11	2.96	2.85	2.70	2.53	2.35	2.13
15	4.54	3.68	3.29	3.06	2.90	2.79	2.64	2.48	2.29	2.07
16	4.49	3.63	3.24	3.01	2.85	2.74	2.59	2.42	2.24	2.01
17	4.45	3.59	3.20	2.96	2.81	2.70	2.55	2.38	2.19	1.96
18	4.41	3.55	3.16	2.93	2.77	2.66	2.51	2.34	2.15	1.92
19	4.38	3.52	3.13	2.90	2.74	2.63	2.48	2.31	2.11	1.88
20	4.35	3.49	3.10	2.87	2.71	2.60	2.45	2.28	2.08	1.84
21	4.32	3.47	3.07	2.84	2.68	2.57	2.42	2.25	2.05	1.81
22	4.30	3.44	3.05	2.82	2.66	2.55	2.40	2.23	2.03	1.78
23	4.28	3.42	3.03	2.80	2.64	2.53	2.38	2.20	2.00	1.76
24	4.26	3.40	3.10	2.78	2.62	2.51	2.36	2.18	1.98	1.73
25	4.24	3.38	2.99	2.76	2.60	2.49	2.34	2.16	1.96	1.71
26	4.22	3.37	2.98	2.74	2.59	2.47	2.32	2.15	1.95	1.69
27	4.21	3.35	2.96	2.73	2.57	2.46	2.30	2.13	1.93	1.67
28	4.20	3.34	2.95	2.71	2.56	2.44	2.29	2.12	1.91	1.65
29	4.18	3.33	2.93	2.70	2.54	2.43	2.28	2.10	1.90	1.64
30	4.17	3.32	2.92	2.69	2.53	2.42	2.27	2.09	1.89	1.62
40	4.08	3.23	2.84	2.61	2.45	2.34	2.18	2.00	1.79	1.51
60	4.00	3.15	2.76	2.52	2.37	2.25	2.10	1.92	1.70	1.39
120	3.92	3.07	2.68	2.45	2.29	2.17	2.02	1.83	1.61	1.25
∞	3.84	2.99	2.60	2.37	2.21	2.09	1.94	1.75	1.52	1.00

Answers

Exercise 8(A)

29. $z = 2.83$; significant
30. $z = 4.5$; the coin is not fair
31. $z = 1.79$; claim cannot be supported.
32. No, since z (= 1.40) < z_α (= 1.645).
34. No; (16.7, 27.0)
35. difference due to sampling fluctuations
36. $z = 0.725$; not significant
37. $z = 6.49$; Yes; Yes.
38. Yes, since z (= 1.80) > $z_{.05}$ (= 1.645)
39. $z = 2.56$; the difference cannot be hidden

40. $z\ (= 0.104) < z_{.05}\ (= 1.645)$; the machine has improved
41. $z\ (= 3.17) > z_{.05}\ (= 1.96)$; difference significant.
42. $|z| = 1.02$; the claim is valid
43. $z = 3.12$; No
44. No, since $z = 5 : 96.8$
45. Claim cannot be true as $z = 1.89$ and $z_{0.05} = 1.645$
46. Yes, since $z\ (= 1.6) < z_{0.05}\ (= 1.645)$
47. 41
48. (57.2, 58.8) and (57.0, 59.0)
49. No, since $z = 2.58$
50. No, since $z = 8.82$
51. No, since $z = 1.32$
52. Yes, since $z\ (= 4.78) > z_{1\%}\ (= 2.33)$
53. Yes, at 5% level, since $|z|\ (= 1.937) > z_{0.05}\ (= 1.645)$; and No, at 1% level, since $|z|\ (= 1.937) < |z_\alpha|\ (= 2.33)$
54. (1.98, 6.02)
55. H_0: $\mu_1 - \mu_2 = 35$ accepted, as $z = 1.90$
57. Yes, as $z = 1.71$
58. Yes, as $|z|\ (= 2.5) > z_{1\%}\ (= 2.33)$
59. Yes, as $z = 1.70$
60. No, as $z = 3.61$

Exercise 8(B)

16. $|t| = 0.62$; Yes; $83.66 < \mu < 110.74$
17. $|t| = 2.67$; (99.76, 107.74) and (98.22, 109.29)
18. $t = 2.5$; the campaign was successful
19. $|t| = 1.44$; $t_{0.1} = 1.90$; $\bar{x}$ is not less than μ
20. $|t| = 0.26$; machine is reliable
21. (11.82, 12.38); (11.69, 12.51)
22. $t = -1.067$; No
23. $|t| = 0.424$; No
24. $t = 4.33$; Yes
25. $|t| = 1.85$; Not significant
26. $t = 4.0$; coaching was effective
27. $F = 1.057$; Not significant
28. The populations have the same variance
29. No, though the difference between variances is not significant, the difference between the mean is significant.
30. Yes, as the differences between the means and between the variance are not significant.

Exercise 8(C)

11. $\chi^2 = 5.2$; $\nu = 9$; digits randomly drawn

12. $\chi^2 = 4.17$; $\nu = 6$; accidents occur uniformly.
13. $\chi^2 = 12.9$; $\nu = 5$; equal probability hypothesis is refuted.
14. $\chi^2 = 1.68$; $\nu = 5$; the demand does not depend on the day of the week
15. $\chi^2 = 4.5$; $\nu = 2$; genetic theory may be correct
16. $\chi^2 = 3.54$; $\nu = 3$; coins are true
17. $\chi^2 = 2.76$; $\nu = 4$; dice are unbiased
18. E_i: 7, 26, 37, 24, 6; $\chi^2 = 0.06$; $\nu = 1$; binomial fit is good. (E_i: 1, 5, 11, 18, 21, 19, 13, 8, 3, 1)
19. $\chi^2 = 11.30$; $\nu = 4$; binomial fit is not satisfactory;
20. E_i: 301, 362, 217, 87, 26, 6, 1; $\chi^2 = 5.40$; $\nu = 4$; Poisson fit is good
21. E_i: 121, 61, 15, 3, 0 ; $\chi^2 = 0.99$; $\nu = 1$; Poisson fit is good
22. $x^2 = 3.84$; $\nu = 8$; Normal fit is good
23. E_i: 2, 5, 13, 25, 37, 42, 36, 23, 12, 4, 1; $\chi^2 = 1.68$; $\nu = 4$; Normal fit is good
24. $\chi^2 = 0.85$; $\nu = 1$; no association between treatment and after-effect
25. $\chi^2 = 176.12$; $\nu = 1$; coaching was effective
26. $\chi^2 = 10.09$; $\nu = 1$; some relation between area and voting preference
27. $\chi^2 = 2.48$; $\nu = 1$; no, as there is no relation between area and sex of ownership
28. $\chi^2 = 3.52$; $\nu = 2$; do not differ significantly
29. $\chi^2 = 373.40$; $\nu = 4$; preference for type of music influenced by age
30. $\chi^2 = 35.42$; $\nu = 6$; the two criteria are not independent

9 Queueing Theory

There are many situations in daily life when a queue is formed. For example, machines waiting to be repaired, patients waiting in a Doctor's room, cars waiting at a traffic signal and passengers waiting to buy tickets in counters form queues. Queue is formed if the service required by the customer (machine, patient, car, etc.) is not immediately available, that is, if the current demand for a particular service exceeds the capacity to provide the service.

Queues may be decreased in size or prevented from forming by providing additional service facilities which results in a drop in the profit. On the other hand, excessively long queues may result in lost sales and lost customers. Hence the problem of interest is how to achieve a balance between the cost associated with long waiting (queues) and the cost associated with the prevention of waiting in order to maximise the profits. As queueing theory provides an answer to this problem, it has become a topic of interest. Before we consider the solutions of queueing problems, we shall consider the general framework of a queueing system.

Although there are many types of queueing systems, all of them can be classified and described according to the following characteristics:

1. The input (or arrival) pattern

The input describes the manner in which the customers arrive and join the queueing system. It is not possible to observe and control the actual moment of arrival of a customer for service. Hence the number of arrivals in one time period or the interval between successive arrivals is not treated as a constant, but a random variable. So the mode of arrival of customers is expressed by means of the probability distribution of the number of arrivals per unit of time or of the inter-arrival time.

We shall deal with only those queueing systems in which the number of arrivals per unit of time has a poisson distribution with mean λ. In this case, the time between consecutive arrivals has an exponential distribution with mean $\frac{1}{\lambda}$ [Refer to Property 4 of the poisson process discussed in the Chapter 6].

Further the input process should specify the number of queues that are permitted to form, the maximum queue length and the maximum number of customers requiring service, viz., the nature of the source (finite or infinite) from which the customers emanate.

2. The service mechanism (or pattern)

The mode of service is represented by means of the probability distribution of the number of customers serviced per unit of time or of the inter-service time. We shall deal with only those queueing systems in which the number of customers serviced per unit of time has a Poisson distribution with mean μ or equivalently the inter-service time (viz. the time to complete the service for a customer) has an exponential distribution with mean $\frac{1}{\mu}$.

Further the service process should specify the number of servers and the arrangement of servers (in parallel, in series, etc.), as the behaviour of the queueing system depends on them also. The following figures represent the framework of queueing systems in which only one queue is permitted to form:

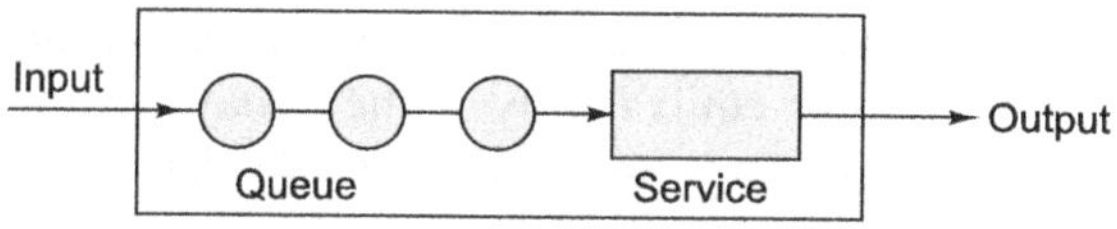

Fig. 9.1 Single server queueing system

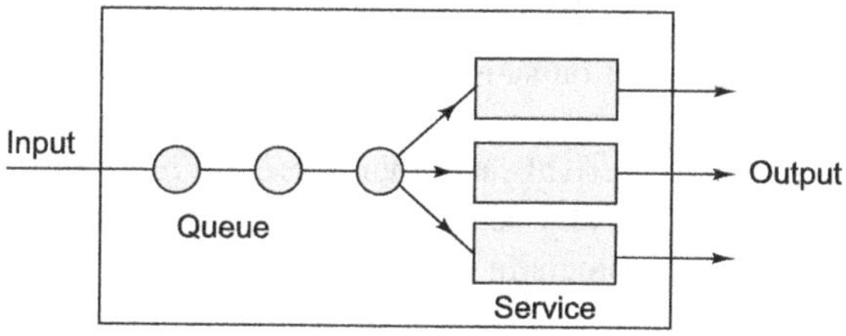

Fig. 9.2 Multiple servers (in parallel) queueing system

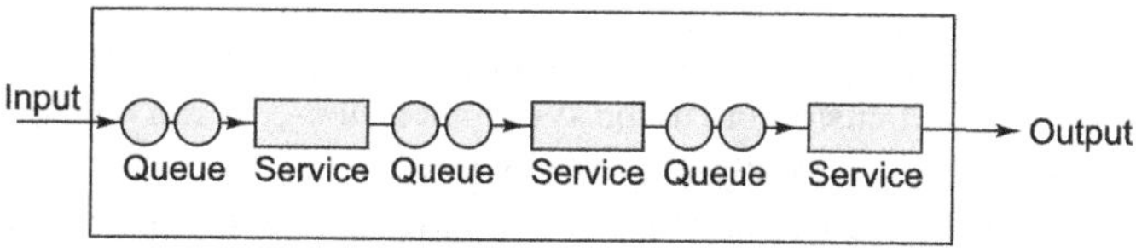

Fig. 9.3 Multiple servers (in series) queueing system

3. The queue discipline

The queue discipline specifies the manner in which the customers form the queue or equivalently the manner in which they are selected for service, when a queue has been formed. The most common discipline is the *FCFS* (First Come First Served) or *FIFO* (First in First out) as per which the customers are served in the strict order of their arrival. If the last arrival in the system is served first, we have the *LCFS* or *LIFO* (last in First Out) discipline. If the service is given in random order, we have the *SIRO* discipline. In the queueing systems which we deal with, we shall assume that service is provided on the *FCFS* (First come First served) basis.

Symbolic Representation of a Queueing Model

Usually a queueing model is specified and represented symbolically in the form $(a/b/c):(d/e)$, where a denotes the type of distribution of the number of arrivals per unit time, b the type of distribution of the service time, c the number of servers, d the capacity of the system, viz., the maximum queue size and e the queue discipline.

Accordingly, the first four models which we will deal with will be denoted by the symbols $(M/M/1):(\infty/FIFO)$, $(M/M/s):(\infty/FIFO)$, $(M/M/1):(k/FIFO)$ and $(M/M/s):(k/FIFO)$.

In the above symbols the letter 'M' stands for Markov' indicating that the number of arrivals in time t and the number of completed services in time t follow Poisson process which is a continuous time Markov chain.

Difference Equations Related to Poisson Queue Systems

If the characteristics of a queueing system (such as the input and output parameters) are independent of time or equivalently if the behaviour of the system is independent of time, the system is said to be in *steady-state*. Otherwise it is said to be in *transient-state*.

Let $P_n(t)$ be the probability that there are n customers in the system at time t $(n > 0)$. Let us first derive the differential equation satisfied by $P_n(t)$ and then deduce the difference equation satisfied by P_n(probability of n customers at any time) in the steady-state.

Let λ_n be the average arrival rate when there are n customers in the system (both waiting in the queue and being served) and let μ_n be the average service rate when there are n customers in the system.

Note *The system being in steady-state does not mean that the arrival rate and service rate are independent of the number of customers in the system.*

Now $P_n(t + \Delta t)$ is the probability of n customers at time $t + \Delta t$.

The presence of n customers in the system at time $t + \Delta t$ can happen in any one of the following four mutually exclusive ways:

(i) Presence of n customers at t and no arrival or departure during Δt time.

(ii) Presence of $(n - 1)$ customers at t and one arrival and no departure during Δt time.

(iii) Presence of $(n + 1)$ customers at t and no arrival and one departure during Δt time.

(iv) Presence of n customers at t and one arrival and one departure during Δt time (since more than one arrival/departure during Δt is ruled out)

$$\therefore \quad P_n(t + \Delta t) = P_n(t)(1 - \lambda_n \Delta t)(1 - \mu_n \Delta t) + P_{n-1}(t)\,\lambda_{n-1}\Delta t\,(1 - \mu_{n-1}\Delta t) + P_{n+1}(t)(1 - \lambda_{n+1}\Delta t)\,\mu_{n+1}\Delta t + P_n(t)\cdot \lambda_n \Delta t \cdot \mu_n \Delta t$$

[since P(an arrival occurs during Δt time) $= \lambda\,\Delta t$ etc.]

i.e., $P_n(t + \Delta t) = P_n(t) - (\lambda_n + \mu_n) P_n(t)\, \Delta t + \lambda_{n-1} P_{n-1}(t)\, \Delta t + \mu_{n+1} P_{n+1}(t)$, on omitting terms containing $(\Delta t)^2$ which is negligibly small.

$$\therefore \quad \frac{P_n(t + \Delta t) - P_n(t)}{\Delta t} = \lambda_{n-1} P_{n-1}(t) - (\lambda_n + \mu_n) P_n(t) + \mu_{n+1} P_{n+1}(t) \quad (1)$$

Taking limits on both sides of (1) as $\Delta t \to 0$, we have

$$P_n'(t) = \lambda_{n-1} P_{n-1}(t) - (\lambda_n + \mu_n) P_n(t) + \mu_{n+1} P_{n+1}(t) \quad (2)$$

Equation (2) does not hold good for $n = 0$, as $P_{n-1}(t)$ does not exist. Hence we derive the differential equation satisfied by $P_0(t)$ independently. Proceeding as before,

$$P_0(t + \Delta t) = P_0(t)(1 - \lambda_0 \Delta t) + P_1(t)(1 - \lambda_1 \Delta t)\, \mu_1 \Delta t,$$

[by the possibilities (i) and (iii) given above and as no departure is possible when $n = 0$]

$$\therefore \quad \frac{P_0(t + \Delta t) - P_0(t)}{\Delta t} = -\lambda_0 P_0(t) + \mu_1 P_1(t) \quad (3)$$

Taking limits on both sides of (3) as $\Delta t \to 0$, we have

$$P_0'(t) = -\lambda_0 P_0(t) + \mu_1 P_1(t) \quad (4)$$

Now in the steady-state, $P_n(t)$ and $P_0(t)$ are independent of time and hence $P_n'(t)$ and $P_0'(t)$ become zero. Hence the differential equations (2) and (4) reduce to the difference equations

$$\lambda_{n-1} P_{n-1} - (\lambda_n + \mu_n) P_n + \mu_{n+1} P_{n+1} = 0 \quad (5)$$

and

$$-\lambda_0 P_0 + \mu_1 P_1 = 0 \quad (6)$$

Values of P_0 and P_n for Poisson Queue Systems

From Equation (6) derived above, we have

$$P_1 = \frac{\lambda_0}{\mu_1} P_0 \quad (7)$$

Putting $n = 1$ in (5) and using (7), we have

$$\mu_2 P_2 = (\lambda_1 + \mu_1) P_1 - \lambda_0 P_0$$

$$= (\lambda_1 + \mu_1) \frac{\lambda_0}{\mu_1} P_0 - \lambda_0 P_0 = \frac{\lambda_0 \lambda_1}{\mu_1} P_0$$

$$\therefore \quad P_2 = \frac{\lambda_0 \lambda_1}{\mu_1 \mu_2} P_0 \quad (8)$$

Successively putting $n = 2, 3, \ldots$ in (5) and proceeding similarly, we can get

$P_3 = \dfrac{\lambda_0 \lambda_1 \lambda_2}{\mu_1 \mu_2 \mu_3} P_0$ etc.

Finally

$$P_n = \frac{\lambda_0 \lambda_1 \lambda_2 \cdots \lambda_{n-1}}{\mu_1 \mu_2 \mu_3 \cdots \mu_n} \cdot P_0, \text{ for } n = 1, 2, \ldots \quad (9)$$

Since the number of customers in the system can be 0 or 1 or 2 or 3 etc., which events are mutually exclusive and exhaustive, we have $\sum\limits_{n=0}^{\infty} P_n = 1$.

i.e., $$P_0 + \sum_{n=1}^{\infty}\left(\frac{\lambda_0\lambda_1\cdots\lambda_{n-1}}{\mu_1\mu_2\cdots\mu_n}\right)P_0 = 1$$

$$\therefore \quad P_0 = \frac{1}{1+\sum\limits_{n=1}^{\infty}\left(\frac{\lambda_0\lambda_1\cdots\lambda_{n-1}}{\mu_1\mu_2\cdots\mu_n}\right)} \qquad (10)$$

Equations (9) and (10) will be used to derive the important characteristics of the four queueing models.

Characteristics of Infinite Capacity, Single Server Poisson Queue Model I [M/M/1]: (∞/FIFO) model], when $\lambda_n = \lambda$ and $\mu_n = \mu$ $(\lambda < \mu)$

1. *Average number L_s of customers in the system:* Let N denote the number of customers in the queueing system (i.e., those in the queue and the one who is being served).

   N is a discrete random variable, which can take the values 0, 1, 2, ..., ∞ such that $P(N = n) = P_n = \left(\frac{\lambda}{\mu}\right)^n P_0$, from Equation (9) of the previous discussion.

   From Equation (10), we have

$$P_0 = \frac{1}{1+\sum\limits_{n=1}^{\infty}\left(\frac{\lambda}{\mu}\right)^n} = \frac{1}{\sum\limits_{n=0}^{\infty}\left(\frac{\lambda}{\mu}\right)^n} = 1-\frac{\lambda}{\mu}$$

$$\therefore \quad P_n = \left(\frac{\lambda}{\mu}\right)^n\left(1-\frac{\lambda}{\mu}\right)$$

Now $$L_s = E(N) = \sum_{n=0}^{\infty} n \times P_n$$

$$= \left(\frac{\lambda}{\mu}\right)\left(1-\frac{\lambda}{\mu}\right)\sum_{n=1}^{\infty} n\left(\frac{\lambda}{\mu}\right)^{n-1}$$

$$= \frac{\lambda}{\mu}\left(1-\frac{\lambda}{\mu}\right)\left(1-\frac{\lambda}{\mu}\right)^{-2}, \text{ by binomial summation}$$

$$= \frac{\frac{\lambda}{\mu}}{1-\frac{\lambda}{\mu}} = \frac{\lambda}{\mu-\lambda} \qquad (1)$$

2. *Average number L_q of customers in the queue or Average length of the queue:*

If N is the number of customers in the system, then the number of customers in the queue is $(N - 1)$

$$\therefore \quad L_q = E(N-1) = \sum_{n=1}^{\infty} (n-1)P_n$$

$$= \left(1-\frac{\lambda}{\mu}\right)\sum_{n=1}^{\infty} (n-1)\left(\frac{\lambda}{\mu}\right)^n$$

$$= \left(\frac{\lambda}{\mu}\right)^2\left(1-\frac{\lambda}{\mu}\right)\sum_{n=2}^{\infty} (n-1)\left(\frac{\lambda}{\mu}\right)^{n-2}$$

$$= \left(\frac{\lambda}{\mu}\right)^2\left(1-\frac{\lambda}{\mu}\right)\left(1-\frac{\lambda}{\mu}\right)^{-2}$$

$$= \frac{\left(\frac{\lambda}{\mu}\right)^2}{1-\frac{\lambda}{\mu}} = \frac{\lambda^2}{\mu(\mu-\lambda)} \qquad (2)$$

3. *Average number L_w of customers in nonempty queues*

$L_w = E\{(N-1)/(N-1) > 0\}$, since the queue is non-empty

$$= \frac{E(N-1)}{P(N-1>0)} = \frac{\lambda^2}{\mu-\lambda} \times \frac{1}{\sum_{n=2}^{\infty} P_n}$$

$$= \frac{\lambda^2}{\mu(\mu-\lambda)} \times \frac{1}{\sum_{n=2}^{\infty}\left(\frac{\lambda}{\mu}\right)^n\left(1-\frac{\lambda}{\mu}\right)}$$

$$= \frac{\lambda^2}{\mu(\mu-\lambda)} \times \frac{1}{\left(\frac{\lambda}{\mu}\right)^2\left(1-\frac{\lambda}{\mu}\right)\sum_{n=0}^{\infty}\left(\frac{\lambda}{\mu}\right)^n}$$

$$= \frac{\mu}{\mu-\lambda} \times \frac{1}{\left(1-\frac{\lambda}{\mu}\right)\left(1-\frac{\lambda}{\mu}\right)^{-1}} = \frac{\mu}{\mu-\lambda} \qquad (3)$$

4. *Probability that the number of customers in the system exceeds k*

$$P(N>k) = \sum_{n=k+1}^{\infty} P_n = \sum_{n=k+1}^{\infty}\left(\frac{\lambda}{\mu}\right)^n\left(1-\frac{\lambda}{\mu}\right)$$

$$= \left(\frac{\lambda}{\mu}\right)^{k+1}\left(1-\frac{\lambda}{\mu}\right)\sum_{n=k+1}^{\infty}\left(\frac{\lambda}{\mu}\right)^{n-(k+1)}$$

$$= \left(\frac{\lambda}{\mu}\right)^{k+1}\cdot\left(1-\frac{\lambda}{\mu}\right)\sum_{n=0}^{\infty}\left(\frac{\lambda}{\mu}\right)^{n}$$

$$= \left(\frac{\lambda}{\mu}\right)^{k+1}\cdot\left(1-\frac{\lambda}{\mu}\right)\left(1-\frac{\lambda}{\mu}\right)^{-1} = \left(\frac{\lambda}{\mu}\right)^{k+1} \tag{4}$$

5. *Probability density function of the waiting time in the system.*
Let W_s be the continuous random variable that represents the waiting time of a customer in the system, viz, the time between arrival and completion of service.

Let its pdf be $f(w)$ and let $f(w/n)$ be the density function of W_s subject to the condition that there are n customers in the queueing system when the customer arrives,

$$\text{Then} \quad f(w) = \sum_{n=0}^{\infty} f(w/n)\, P_n \tag{5}$$

Now $f(w/n)$ = pdf of sum of $(n + 1)$ service times (one part-service time of the customer being served + n complete service times)

= pdf of sum of $(n + 1)$ independent random variables, each of which is exponentially distributed with parameter μ

$= \dfrac{\mu^{n+1}}{n!} e^{-\mu w} w^n; w > 0$ which is the pdf of Erlang distribution.

[$\because$ The mgf of the exponential distribution (μ) is $\left(1-\dfrac{t}{\mu}\right)^{-1}$ and hence the mgf of the sum of $(n + 1)$ independent exponential (μ) variables is $\left(1-\dfrac{t}{\mu}\right)^{n+1}$, which is the mgf of Erlang distribution with parameters μ and $(n + 1)$] (refer to Erlang distribution in chapter 5)

$$\therefore \quad f(w) = \sum_{n=0}^{\infty} \frac{\mu^{n+1}}{n!} e^{-\mu w} w^n \left(\frac{\lambda}{\mu}\right)^{n}\left(1-\frac{\lambda}{\mu}\right), \text{ by} \tag{5}$$

$$= \mu e^{-\mu w}\left(1-\frac{\lambda}{\mu}\right)\sum_{n=0}^{\infty}\frac{1}{n!}(\lambda w)^n$$

$$= \mu\left(1-\frac{\lambda}{\mu}\right) e^{-\mu w} e^{\lambda w}, \text{ by exponential summation}$$

$$= (\mu - \lambda)\, e^{-(\mu-\lambda)w} \tag{6}$$

which is the pdf of an exponential distribution with parameter $(\mu - \lambda)$.

6. *Average waiting time of a customer in the system:*

   W_s follows an exponential distribution with parameter $(\mu - \lambda)$.

$$\therefore \quad E(W_s) = \frac{1}{\mu - \lambda} \tag{7}$$

   ($\because$ the mean of an exponential distribution is the reciprocal of its parameter).

7. *Probability that the waiting time of a customer in the system exceeds t*

$$P(W_s > t) = \int_t^{\infty} f(w)\, dw$$

$$= \int_t^{\infty} (\mu - \lambda)\, e^{-(\mu - \lambda)w}\, dw$$

$$= [-e^{-(\mu - \lambda)w}]_t^{\infty} = e^{-(\mu - \lambda)t} \tag{8}$$

8. *Probability density function of the waiting time W_q in the queue:*

   W_q represents the time between arrival and reach of service point.

Let the pdf of W_q be $g(w)$ and let $g(w/n)$ be the density function of W_q subject to the condition that there are n customers in the system or there are $(n - 1)$ customers in the queue apart from one customer receiving service. Now $g(w/n)$ = pdf of sum of n service times [one residual service time + $(n - 1)$ full service times]

$$= \frac{\mu^n}{(n-1)!}\, e^{-\mu w}\, w^{n-1};\ w > 0$$

$$\therefore \qquad g(w) = \sum_{n=1}^{\infty} \frac{\mu^n}{(n-1)!}\, e^{-\mu w}\, w^{n-1} \left(\frac{\lambda}{\mu}\right)^n \left(1 - \frac{\lambda}{\mu}\right)$$

$$= \lambda\left(1 - \frac{\lambda}{\mu}\right) e^{-\mu w} \sum_{n=1}^{\infty} \frac{1}{(n-1)!}\, (\lambda w)^{n-1}$$

$$= \frac{\lambda}{\mu}\, (\mu - \lambda)\, e^{-\mu w}\, e^{\lambda w}$$

$$= \frac{\lambda}{\mu}\, (\mu - \lambda)\, e^{-(\mu - \lambda)w};\ w > 0 \tag{9}$$

and
$$g(w) = 1 - \frac{\lambda}{\mu}, \text{ when } w = 0$$

Note *1. W_q is a continuous random variable in $w > 0$ and it takes the value 0 with a non-zero probability.*

2. Wq does not follow an exponential distribution.

9. *Average waiting time of a customer in the queue.*

$$E(W_q) = \frac{\lambda}{\mu}(\mu-\lambda)\int_0^\infty w\, e^{-(\mu-\lambda)w}\, dw$$

$$= \frac{\lambda}{\mu}\int_0^\infty x\, e^{-x}\, \frac{dx}{\mu-\lambda}$$

$$= \frac{\lambda}{\mu(\mu-\lambda)}\left[x(-e^{-x}) - e^{-x}\right]_0^\infty$$

$$= \frac{\lambda}{\mu(\mu-\lambda)} \qquad (10)$$

10. *Average waiting time of a customer in the queue, if he has to wait*

$$E(W_q/W_q > 0) = \frac{E(W_q)}{P(W_q > 0)}$$

$$= \frac{E(W_q)}{1 - P(W_q = 0)}$$

$$= \frac{E(W_q)}{1 - P\,(\text{no customer in the queue})}$$

$$= \frac{E(W_q)}{1 - P_0}$$

$$= \frac{\lambda}{\mu(\mu-\lambda)} \times \frac{\mu}{\lambda} \qquad \left(\because P_0 = 1 - \frac{\mu}{\lambda}\right)$$

$$= \frac{1}{\mu-\lambda} \qquad (11)$$

Relations Among $E(N_s)$, $E(N_q)$, $E(W_s)$ and $E(W_q)$

(i) $E(N_s) = \dfrac{\lambda}{\mu-\lambda} = \lambda E(W_s)$ $\qquad [\because E(N_s) = L_s]$

(ii) $E(N_q) = \dfrac{\lambda^2}{\mu(\mu-\lambda)} = \lambda E(W_q)$ $\qquad [\because E(N_q) = L_q]$

(iii) $E(W_s) = E(W_q) + \dfrac{1}{\mu}$

(iv) $E(N_s) = E(N_q) + \dfrac{\lambda}{\mu}$

Note *1. If any one of the quantities $E(N_s)$, $E(N_q)$, $E(W_s)$ and $E(W_q)$ is known, the other three can be found out using the relations given above.*

2. The above relations, called Little's formulas hold good for the models with infinite capacity, but with a slight modification for the models with finite capacity.

Characteristics of Infinite Capacity, Multiple Server Poisson Queue Model II [M/M/s]: (∞/FIFO) model], When $\lambda_n = \lambda$ for all $n(\lambda < s\mu)$

1. *Values of P_0 and P_n:*

For the Poisson queue system, P_n is given by

$$P_n = \frac{\lambda_0 \lambda_1 \lambda_2 \cdots \lambda_{n-1}}{\mu_1 \mu_2 \mu_3 \cdots \mu_n} \times P_0,\ n \geq 1, \tag{1}$$

where
$$P_0 = \left[1 + \sum_{n=1}^{\infty} \left(\frac{\lambda_0 \lambda_1 \cdots \lambda_{n-1}}{\mu_1 \mu_2 \cdots \mu_n}\right)\right]^{-1} \tag{2}$$

If there is a single server, $\mu_n = \mu$ for all n. But there are s servers working independently of each other.

If there be less than s customers, i.e., if $n < s$, only n of the s servers will be busy and the others idle and hence the mean service rate will be $n\mu$

If $n \geq s$, all the s servers will be busy and hence the mean service rate $= s\mu$.

i.e.,
$$\mu_n = \begin{cases} n\mu, & \text{if } 0 \leq n < s \\ s\mu, & \text{if } n \geq s \end{cases} \tag{3}$$

Using (3) in (1) and (2), we have

$$P_n = \frac{\lambda^n}{1\mu \cdot 2\mu \cdot 3\mu \cdots n\mu},\ P_0,\ \text{if } 0 \leq n < s$$

$$= \frac{1}{n!}\left(\frac{\lambda}{\mu}\right)^n P_0,\ \text{if } 0 \leq n < s \tag{4}$$

and
$$P_n = \frac{\lambda^n}{\{1\mu \cdot 2\mu \cdots (s-1)\mu\}\{s\mu \cdot s\mu \cdots (n-s+1)\text{ times}\}} P_0$$

$$= \frac{\lambda^n}{(s-1)!\mu^{s-1}(s\mu)^{n-s+1}} \cdot P_0$$

$$= \frac{1}{s!\, s^{n-s}}\left(\frac{\lambda}{\mu}\right)^n P_0,\ \text{if } n \geq s \tag{5}$$

Now P_0 is given by $\sum_{n=0}^{\infty} P_n = 1$

i.e.
$$\left[\sum_{n=0}^{s-1}\frac{1}{n!}\left(\frac{\lambda}{\mu}\right)^n+\sum_{n=s}^{\infty}\frac{1}{s!\,s^{n-s}}\left(\frac{\lambda}{\mu}\right)^n\right]P_0=1$$

i.e.,
$$\left[\sum_{n=0}^{s-1}\frac{1}{n!}\left(\frac{\lambda}{\mu}\right)^n+\frac{s^s}{s!}\sum_{n=s}^{\infty}\left(\frac{\lambda}{\mu s}\right)^n\right]P_0=1$$

i.e.,
$$\left[\sum_{n=0}^{s-1}\frac{1}{n!}\left(\frac{\lambda}{\mu}\right)^n+\frac{s^s}{s!}\left(\frac{\lambda}{\mu s}\right)^s\frac{1}{1-\dfrac{\lambda}{\mu s}}\right]P_0=1$$

i.e.,
$$\left[\left\{\sum_{n=0}^{s-1}\frac{1}{n!}\left(\frac{\lambda}{\mu}\right)^n\right\}+\left\{\frac{1}{s!\left(1-\dfrac{\lambda}{\mu s}\right)}\left(\frac{\lambda}{\mu}\right)^s\right\}\right]P_0=1$$

or
$$P_0=\frac{1}{\left\{\displaystyle\sum_{n=0}^{s-1}\frac{1}{n!}\left(\frac{\lambda}{\mu}\right)^n\right\}+\left\{\dfrac{1}{s!\left(1-\dfrac{\lambda}{\mu s}\right)}\left(\dfrac{\lambda}{\mu}\right)^s\right\}} \tag{6}$$

2. *Average number of customers in the queue or average queue length*

$$L_q=E(N_q)=E(N-s)=\sum_{n=s}^{\infty}(n-s)P_n$$

$$=\sum_{x=0}^{\infty}x\,P_{x+s}$$

$$=\sum_{x=0}^{\infty}x\times\frac{1}{s!\,s^x}\left(\frac{\lambda}{\mu}\right)^{s+x}\cdot P_0$$

$$=\frac{1}{s!}\left(\frac{\lambda}{\mu}\right)^s P_0\sum_{x=0}^{\infty}x\left(\frac{\lambda}{\mu s}\right)^x$$

$$=\frac{1}{s!}\left(\frac{\lambda}{\mu}\right)^s\frac{\lambda}{\mu s}\cdot P_0\frac{1}{\left(1-\dfrac{\lambda}{\mu s}\right)^2}$$

$$= \frac{1}{s \cdot s!} \cdot \frac{\left(\frac{\lambda}{\mu}\right)^{s+1}}{\left(1 - \frac{\lambda}{\mu s}\right)^2} P_0 \qquad (7)$$

3. *Average number of customers in the system*
   By Little's formula (iv),

$$E(N_s) = E(N_q) + \frac{\lambda}{\mu}$$

$$= \frac{1}{s \cdot s!} \cdot \frac{\left(\frac{\lambda}{\mu}\right)^{s+1}}{\left(1 - \frac{\lambda}{\mu s}\right)^2} P_0 + \frac{\lambda}{\mu} \qquad (8)$$

Result (8) can also be directly derived by using the definition $E(N_s) = \sum_{n=0}^{\infty} n\ P_n$.

4. *Average time a customer has to spend in the system*
   By Little's formula (i)

$$E(W_s) = \frac{1}{\lambda}\ E(N_s)$$

$$= \frac{1}{\mu} + \frac{1}{\mu} \cdot \frac{1}{s \cdot s!} \cdot \frac{\left(\frac{\lambda}{\mu}\right)^{s}}{\left(1 - \frac{\lambda}{\mu s}\right)^2} \cdot P_0 \qquad (9)$$

5. *Average time a customer has to spend in the queue*
   By Little's formula (ii),

$$E(W_q) = \frac{1}{\lambda}\ E(N_q)$$

$$= \frac{1}{\mu} \cdot \frac{1}{s \cdot s!} \cdot \frac{\left(\frac{\lambda}{\mu}\right)^{s}}{\left(1 - \frac{\lambda}{\mu s}\right)^2} \cdot P_0 \qquad (10)$$

6. *Probability that an arrival has to wait*
   Required probability = Probability that there are s or more customers in the system
   i.e., $P(W_s > 0) = P(N \geq s)$

$$= \sum_{n=s}^{\infty} P_n = \sum_{n=s}^{\infty} \frac{1}{s!s^{n-s}} \left(\frac{\lambda}{\mu}\right)^n P_0$$

$$= \frac{1}{s!}\left(\frac{\lambda}{\mu}\right)^s \cdot P_0 \sum_{n=s}^{\infty} \left(\frac{\lambda}{\mu s}\right)^{n-s}$$

$$= \frac{\left(\frac{\lambda}{\mu}\right)^s \cdot P_0}{s!\left(1-\frac{\lambda}{\mu s}\right)} \tag{11}$$

7. *Probability that an arrival enters the service without waiting*
   Required probability

$$= 1 - P(\text{an arrival has to wait})$$

$$= 1 - \frac{\left(\frac{\lambda}{\mu}\right)^s \cdot P_0}{s!\left(1-\frac{\lambda}{\mu s}\right)} \tag{12}$$

8. *Mean waiting time in the queue for those who actually wait.*

$$E(W_q/W_s > 0) = \frac{E(W_q)}{P(W_s > 0)}$$

$$= \frac{1}{\mu} \cdot \frac{1}{s \cdot s!} \frac{\left(\frac{\lambda}{\mu}\right)^s}{\left(1-\frac{\lambda}{\mu s}\right)^2} P_0 \times \frac{s!\left(1-\frac{\lambda}{\mu s}\right)}{\left(\frac{\lambda}{\mu}\right)^s \cdot P_0} \quad \text{[using (10) and (11)]}$$

$$= \frac{1}{\mu s\left(1-\frac{\lambda}{\mu s}\right)} = \frac{1}{\mu s - \lambda} \tag{13}$$

9. *Probability that there will be someone waiting*
   Required probability = $P(N \geq s + 1)$

$$= \sum_{n=s+1}^{\infty} P_n = \sum_{n=s}^{\infty} P_n - P(N = s)$$

$$= \frac{\left(\frac{\lambda}{\mu}\right)^s \cdot P_0}{s!\left(1-\frac{\lambda}{\mu s}\right)} - \frac{\left(\frac{\lambda}{\mu}\right)^s \cdot P_0}{s!} \quad \text{[using (10) and (5)]}$$

$$= \frac{\left(\frac{\lambda}{\mu}\right)^s P_0}{s!} \cdot \frac{\left(\frac{\lambda}{\mu s}\right)}{1 - \frac{\lambda}{\mu s}} \tag{14}$$

10. *Average number of customers (in non-empty queues), who have to actually wait.*

$$L_w = E(N_q/N_q \geq 1) = E(N_q)/P(N \geq s)$$

$$= \frac{1}{s \cdot s!} \frac{\left(\frac{\lambda}{\mu}\right)^{s+1} \cdot P_0}{\left(1 - \frac{\lambda}{\mu s}\right)^2} \cdot \frac{s!\left(1 - \frac{\lambda}{\mu s}\right)}{\left(\frac{\lambda}{\mu}\right)^s P_0}$$

$$= \frac{\left(\frac{\lambda}{\mu s}\right)}{1 - \frac{\lambda}{\mu s}} \tag{15}$$

Characteristics of Finite Capacity, Single Server Poisson Queue Model III [(M/M/1): (k/FIFO) Model]

1. *Values of P_0 and P_n*

For the Poisson queue system, $P_n = P(N = n)$ in the steady-state is given by the difference equations

$$\lambda_{n-1} P_{n-1} - (\lambda_n + \mu_n) P_n + \mu_{n+1} P_{n+1} = 0; n > 0$$

and $\quad - \lambda_0 P_0 + \mu_1 P_1 = 0; n = 0$

This model represents the situation in which the system can accommodate only a finite number k of arrivals. If a customer arrives and the queue is full, the customer leaves without joining the queue.

Therefore, for this model,

$$\mu_n = \mu, n = 1, 2, 3, \ldots$$

and
$$\lambda_n = \begin{cases} \lambda, \text{ for } & n = 0, 1, 2, \cdots (k-1) \\ 0, \text{ for } & n = k, k+1, \cdots \end{cases}$$

Using these values in the difference equations given above, we have

$$\mu P_1 = \lambda P_0 \tag{1}$$

$$\mu P_{n+1} = (\lambda + \mu) P_n - \lambda P_{n-1}, \text{ for } 1 \leq n \leq k-1 \tag{2}$$

and $\quad \mu P_k = \lambda P_{k-1}$, for $n = k$ (3) $\quad (\because P_{k+1}$ has no meaning)

From (1), $\quad P_1 = \dfrac{\lambda}{\mu} P_0$

From (2), $\quad \mu P_2 = (\lambda + \mu) \dfrac{\lambda}{\mu} \cdot P_0 - \lambda P_0$

$$\therefore \qquad P_2 = \left(\frac{\lambda}{\mu}\right)^2 P_0 \text{ and so on}$$

In general, $P_n = \left(\frac{\lambda}{\mu}\right)^n P_0$, true for $0 \le n \le k-1$

From (3) $P_k = \frac{\lambda}{\mu} \cdot \left(\frac{\lambda}{\mu}\right)^{k-1} P_0 = \left(\frac{\lambda}{\mu}\right)^k P_0$

Now $\qquad \sum_{n=0}^{k} P_n = 1$

i.e., $\qquad P_0 \sum_{n=0}^{k} \left(\frac{\lambda}{\mu}\right)^n = 1$

i.e., $\qquad P_0 \frac{\left\{1 - \left(\frac{\lambda}{\mu}\right)^{k+1}\right\}}{1 - \frac{\lambda}{\mu}} = 1,$

which is valid even for $\lambda > \mu$

$$\therefore \quad P_0 = \begin{cases} \dfrac{1 - \frac{\lambda}{\mu}}{1 - \left(\frac{\lambda}{\mu}\right)^{k+1}}, \text{ if } \quad \lambda \ne \mu & (4) \\[2ex] \dfrac{1}{k+1}, \text{ if } \lambda = \mu, \text{ since } \lim\limits_{\frac{\lambda}{\mu} \to 1} \left\{\dfrac{1 - \frac{\lambda}{\mu}}{1 - \left(\frac{\lambda}{\mu}\right)^{k+1}}\right\} = \dfrac{1}{k+1} & (5) \end{cases}$$

$$\therefore \quad P_n = \begin{cases} \left(\frac{\lambda}{\mu}\right)^n \cdot \left[\dfrac{1 - \frac{\lambda}{\mu}}{1 - \left(\frac{\lambda}{\mu}\right)^{k+1}}\right], \text{ if } \lambda \ne \mu & (6) \\[2ex] \dfrac{1}{k+1}, \text{ if } \lambda = \mu & (7) \end{cases}$$

2. *Average number of customers in the system*

$$E(N) = \sum_{n=0}^{k} nP_n = \frac{\left(1-\frac{\lambda}{\mu}\right)}{1-\left(\frac{\lambda}{\mu}\right)^{k+1}} \cdot \sum_{n=0}^{k} n\left(\frac{\lambda}{\mu}\right)^n$$

$$= \frac{\left(1-\frac{\lambda}{\mu}\right)\cdot\frac{\lambda}{\mu}}{1-\left(\frac{\lambda}{\mu}\right)^{k+1}} \cdot \sum_{n=0}^{k} \frac{\mathrm{d}}{\mathrm{d}x}(x^n), \text{ where } x = \frac{\lambda}{\mu}$$

$$= \frac{\left(1-\frac{\lambda}{\mu}\right)\cdot\frac{\lambda}{\mu}}{1-\left(\frac{\lambda}{\mu}\right)^{k+1}} \frac{\mathrm{d}}{\mathrm{d}x}\left(\frac{1-x^{k+1}}{1-x}\right)$$

$$= \frac{(1-x)\,x}{1-x^{k+1}} \cdot \left[\frac{(1-x)\left\{-(k+1)\,x^k\right\}+\left(1-x^{k+1}\right)}{(1-x)^2}\right]$$

$$= \frac{x\left[1-(k+1)\,x^k + k\,x^{k+1}\right]}{(1-x)\left(1-x^{k+1}\right)}$$

$$= \frac{x\left(1-x^{k+1}\right)-(k+1)(1-x)\,x^{k+1}}{(1-x)\left(1-x^{k+1}\right)}$$

$$= \frac{x}{1-x} - \frac{(k+1)\,x^{k+1}}{1-x^{k+1}}$$

$$= \frac{\lambda}{\mu-\lambda} - \frac{(k+1)\left(\frac{\lambda}{\mu}\right)^{k+1}}{1-\left(\frac{\lambda}{\mu}\right)^{k+1}}, \text{ if } \lambda \neq \mu \tag{8}$$

and $$E(N) = \sum_{n=0}^{k} \frac{n}{k+1} = \frac{k}{2}, \text{ if } \lambda = \mu \tag{9}$$

3. *Average number of customers in the queue.*

$$E(N_q) = E(N-1) = \sum_{n=1}^{k} (n-1)P_n$$

$$= \sum_{n=0}^{k} n P_n - \sum_{n=1}^{k} P_n$$

$$= E(N) - (1 - P_0) \tag{10}$$

As per Little's formula (iv),

$$E(N_q) = E(N) - \frac{\lambda}{\mu},$$

which is true when the average arrival rate is λ throughout. Now we see that, in step (8), $1 - P_0 \neq \frac{\lambda}{\mu}$, because the average arrival rate is λ as long as there is a vacancy in the queue and it is zero when the system is full.

Hence we define *the overall effective arrival rate*, denoted by λ' or λ_{eff}, by using step (8) and Little's formula as

$$\frac{\lambda'}{\mu} = 1 - P_0 \text{ or } \lambda' = \mu(1 - P_0) \tag{11}$$

Thus, step (8) can be rewritten as

$$E(N_q) = E(N) - \frac{\lambda'}{\mu}, \tag{12}$$

which is the modified Little's formula for this model.

4. *Average waiting times in the system and in the queue:*
   By the modified Little's formulas,

$$E(W_s) = \frac{1}{\lambda'} E(N) \tag{13}$$

and

$$E(W_q) = \frac{1}{\lambda'} E(N_q) \tag{14}$$

where λ' is the effective arrival rate, given by step (9).

Characteristics of Finite Queue, Multiple Server Poisson Queue Model IV [(M/M/s): (k/FIFO) Model]

1. *Values of P_0 and P_n*
   For the Poisson queue system, P_n is given by

$$P_n = \frac{\lambda_0 \lambda_1 \cdots \lambda_{n-1}}{\mu_1 \mu_2 \cdots \mu_n} \cdot P_0, \; n \geq 1, \tag{1}$$

where

$$P_0 = \left\{1 + \sum_{n=1}^{k} \left(\frac{\lambda_0 \lambda_1 \cdots \lambda_{n-1}}{\mu_1 \mu_2 \cdots \mu_n}\right)\right\}^{-1} \tag{2}$$

For this $(M/M/s)$: $(k/FIFO)$ model,

$$\lambda_n = \begin{cases} \lambda, & \text{for } n = 0, 1, 2, \cdots, k-1 \\ 0, & \text{for } n = k, k+1, \cdots \end{cases}$$

$$\mu_n = \begin{cases} n\mu, & \text{for} \quad n = 0, 1, 2, \cdots, s-1 \\ s\mu, & \text{for} \quad n = s, s+1, \cdots \end{cases}$$

Using these values of λ_n and μ_n in (2) and noting that $1 < s < k$, we get

$$P_0^{-1} = \left\{1 + \frac{\lambda}{1!\,\mu} + \frac{\lambda^2}{2!\,\mu^2} + \cdots + \frac{\lambda^{s-1}}{(s-1)!\,\mu^{s-1}}\right\} + \left\{\frac{\lambda^s}{(s-1)!\,\mu^{s-1}\cdot \mu s}\right.$$

$$\left. + \frac{\lambda^{s+1}}{(s-1)!\,\mu^{s-1}\cdot(\mu s)^2} + \cdots + \frac{\lambda^k}{(s-1)!\,\mu^{s-1}\cdot(\mu s)^{k-s+1}}\right\}$$

$$= \sum_{n=0}^{s-1} \frac{1}{n!}\left(\frac{\lambda}{\mu}\right)^n + \frac{\lambda^s}{s!\,\mu^s}\left[1 + \frac{\lambda}{\mu s} + \left(\frac{\lambda}{\mu s}\right)^2 + \cdots + \left(\frac{\lambda}{\mu s}\right)^{k-s}\right]$$

$$= \sum_{n=0}^{s-1} \frac{1}{n!}\left(\frac{\lambda}{\mu}\right)^n + \frac{1}{s!}\left(\frac{\lambda}{\mu}\right)^s \sum_{n=s}^{k} \left(\frac{\lambda}{\mu s}\right)^{n-s} \tag{3}$$

$$P_n = \begin{cases} \dfrac{1}{n!}\left(\dfrac{\lambda}{\mu}\right)^n P_0, & \text{for} \quad n \le s \\ \dfrac{1}{s!\,s^{n-s}} \cdot \left(\dfrac{\lambda}{\mu}\right)^n \cdot P_0, & \text{for} \quad s < n \le k \\ 0, & \text{for} \quad n > k \end{cases} \tag{4}$$

2. *Average queue length or average number of customers in the queue*

$$E(N_q) = E(N - s) = \sum_{n=s}^{k} (n-s)P_n$$

$$= \frac{P_0}{s!}\sum_{n=s}^{k} (n-s)\left(\frac{\lambda}{\mu}\right)^n / s^{n-s} \text{ [using (4)]}$$

$$= \frac{\left(\frac{\lambda}{\mu}\right)^s \cdot P_0}{s!} \sum_{x=0}^{k-s} x \cdot \left(\frac{\lambda}{\mu s}\right)^x$$

$$= \frac{\left(\frac{\lambda}{\mu}\right)^s \cdot P_0\,\rho}{s!} \sum_{x=0}^{k-s} x\,\rho^{x-1} \text{ where } \rho = \frac{\lambda}{\mu s}$$

$$= \left(\frac{\lambda}{\mu}\right)^s \cdot \frac{P_0\,\rho}{s!} \sum_{x=0}^{k-s} \frac{\mathrm{d}}{\mathrm{d}\rho}(\rho^x)$$

$$= \left(\frac{\lambda}{\mu}\right)^s \cdot \frac{P_0\,\rho}{s!} \frac{\mathrm{d}}{\mathrm{d}\rho}\left\{\frac{1-\rho^{k-s+1}}{1-\rho}\right\}$$

$$= \left(\frac{\lambda}{\mu}\right)^s \cdot \frac{P_0\,\rho}{s!}\left[\frac{-(1-\rho)(k-s+1)\,\rho^{k-s} + \left(1-\rho^{k-s+1}\right)}{(1-\rho)^2}\right]$$

$$= \left(\frac{\lambda}{\mu}\right)^s \cdot \frac{P_0\,\rho}{s!}\left[\frac{-(k-s)(1-\rho)\rho^{k-s} - (1-\rho)\rho^{k-s} + 1 - \rho^{k-s+1}}{(1-\rho)^2}\right]$$

$$= P_0\left(\frac{\lambda}{\mu}\right)^s \frac{\rho}{s!}\left[\frac{-(k-s)(1-\rho)\,\rho^{k-s} + 1 - \rho^{k-s}(1-\rho+\rho)}{(1-\rho)^2}\right]$$

$$= P_0 \cdot \left(\frac{\lambda}{\mu}\right)^s \frac{\rho}{s!(1-\rho)^2}\,[1 - \rho^{k-s} - (k-s)(1-\rho)\,\rho^{k-s}],$$

where $\rho = \dfrac{\lambda}{\mu s}$ (5)

3. *Average number of customers in the system*

$$E(N) = \sum_{n=0}^{k} nP_n = \sum_{n=0}^{s-1} nP_n + \sum_{n=s}^{k} nP_n$$

$$= \sum_{n=0}^{s-1} nP_n + \sum_{n=s}^{k} (n-s)\,P_n + \sum_{n=s}^{k} sP_n$$

$$= \sum_{n=0}^{s-1} nP_n + E\,(N_q) + s\left\{\sum_{n=0}^{k} P_n - \sum_{n=0}^{s-1} P_n\right\}$$

$$= E(N_q) + s - \sum_{n=0}^{s-1} (s-n)P_n \left(\because \sum_{n=0}^{k} P_n = 1\right) \quad (6)$$

Obviously $\left\{s - \sum_{n=0}^{s-1} (s-n)P_n\right\} \neq \dfrac{\lambda}{\mu}$, so that step (6) represents Little's formula.

In order to make (6) to assume the form of Little's formula, we define the *over-all effective arrival rate* λ' or λ_{eff} as follows:

$$\frac{\lambda'}{\mu} = s - \sum_{n=0}^{s-1} (s-n)P_n$$

i.e., $$\lambda' = \mu\left[s - \sum_{n=0}^{s-1}(s-n)\,P_n\right] \quad (7)$$

With this definition of λ', step (6) becomes

$$E(N) = E(N_q) + \frac{\lambda'}{\mu} \quad (8)$$

which is the modified Little's formula for this model.

4. *Average waiting time in the system and in the queue:*
   By the modified Little's formulas,

$$E(W_s) = \frac{1}{\lambda'} E(N) \tag{9}$$

and

$$E(W_q) = \frac{1}{\lambda'} E(N_q) \tag{10}$$

where λ' is the effective arrival rate, given by step (7).

Worked Examples 9

Example 1 *Arrivals at a telephone booth are considered to be Poisson with an average time of 12 min. between one arrival and the next. The length of a phone call is assumed to be distributed exponentially with mean 4 min.*

(a) Find the average number of persons waiting in the system.

(b) What is the probability that a person arriving at the booth will have to wait in the queue?

(c) What is the probability that it will take him more than 10 min. altogether to wait for the phone and complete his call?

(d) Estimate the fraction of the day when the phone will be in use.

(e) The telephone department will install a second booth, when convinced that an arrival has to wait on the average for at least 3 min. for phone. By how much the flow of arrivals should increase in order to justify a second booth?

(f) What is the average length of the queue that forms from time to time?

Solution Mean inter-arrival time $= \frac{1}{\lambda} = 12$ min.

Therefore mean arrival rate $= \lambda = \frac{1}{12}$ per minute.

Mean service time $= \frac{1}{\mu} = 4$ min.

Therefore, mean service rate $= \mu = \frac{1}{4}$ per minute.

(a) $E(N) = \frac{\lambda}{\mu - \lambda}$, (by formula (1) of model I)

$$= \frac{\frac{1}{12}}{\frac{1}{4} - \frac{1}{12}} = 0.5 \text{ customer}$$

(b) $P(W > 0) = 1 - P(W = 0)$

$= 1 - P$ (no customer in the system)

$$= 1 - P_0$$

$$= 1 - \left(1 - \frac{\lambda}{\mu}\right) \text{ (by the formula for } P_0 \text{ of model I)}$$

$$= \frac{\lambda}{\mu} = \frac{1/12}{1/4} = \frac{1}{3}$$

(c) $P(W > 10) = e^{-(\mu - \lambda)\times 10}$ [by formula (8) of model I]

$$= e^{-\left(\frac{1}{4} - \frac{1}{12}\right)\times 10}$$

$$= e^{-\frac{5}{3}} = 0.1889$$

(d) $P(\text{the phone will be idle}) = P(N = 0) = P_0$

$$= 1 - \frac{\lambda}{\mu} = \frac{2}{3}$$

$\therefore$ P (the phone will be in use) $= 1 - \frac{2}{3} = \frac{1}{3}$

or the fraction of the day when the phone will be in use $= \frac{1}{3}$.

(e) The second phone will be installed, if $E(W_q) > 3$.

i.e., if $\frac{\lambda}{\mu(\mu - \lambda)} > 3$ [by formula (10) of model I]

i.e., if $\frac{\lambda_R}{\frac{1}{4}\left(\frac{1}{4} - \lambda_R\right)} > 3$,

where λ_R is the required arrival rate.

i.e., if $\lambda_R > \frac{3}{4}\left(\frac{1}{4} - \lambda_R\right)$

i.e., if $\lambda_R > \frac{3}{28}$

Hence the arrival rate should increase by $\frac{3}{28} - \frac{1}{12} = \frac{1}{42}$ per minute, to justify a second phone.

(f) $E(N_q/\text{the queue is always available})$

$$= E(N_q/N_q > 0)$$

$$= E(N_q/N > 1)$$

$$= \frac{E(N_q)}{P(N > 1)} = \frac{E(N_q)}{1 - P_0 - P_1} = \frac{\lambda^2}{\mu(\mu - \lambda)} \times \frac{1}{1 - \left(1 + \frac{\lambda}{\mu}\right)P_0},$$

[by formula (2) of model 1]

$$= \frac{\lambda^2}{\mu(\mu-\lambda)} \cdot \frac{1}{1-\left(1+\dfrac{\lambda}{\mu}\right)\left(1-\dfrac{\lambda}{\mu}\right)}$$

$$= \frac{\lambda^2}{\mu(\mu-\lambda)} \cdot \frac{\mu^2}{\lambda^2} = \frac{\mu}{\mu-\lambda} = \frac{1/4}{1/4-1/12} = 1.5 \text{ persons.}$$

Example 2 *Customers arrive at a one-man barber shop according to a Poisson process with a mean interarrival time of 12 min. Customers spend an average of 10 min in the barber's chair.*

(a) What is the expected number of customers in the barber shop and in the queue?

(b) Calculate the percentage of time an arrival can walk straight into the barber's chair without having to wait.

(c) How much time can a customer expect to spend in the barber's shop?

(d) Management will provide another chair and hire another barber, when a customer's waiting time in the shop exceeds 1.25 h. How much must the average rate of arrivals increase to warrant a second barber?

(e) What is the average time customers spend in the queue?

(f) What is the probability that the waiting time in the system is greater than 30 min?

(g) Calculate the percentage of customers who have to wait prior to getting into the barber's chair.

(h) What is the probability that more than 3 customers are in the system?

Solution $\frac{1}{\lambda} = 12 \qquad \therefore \lambda = \frac{1}{12}$ *per minute*

$\frac{1}{\mu} = 10 \qquad \therefore \mu = \frac{1}{10}$ *per minute*

(a) $E(N_s) = \frac{\lambda}{\mu-\lambda} = \frac{1/12}{1/10-1/12} = 5$ customers [by formula (1) of model I]

$$E(N_q) = \frac{\lambda^2}{\mu(\mu-\lambda)} \qquad \text{[by formula (2) of model I]}$$

$$= \frac{\dfrac{1}{144}}{\dfrac{1}{10}\left(\dfrac{1}{10}-\dfrac{1}{12}\right)} = 4.17 \text{ customers}$$

(b) P(a customer straight goes to the barber's chair)

$= P$ (No customer in the system)

$$= P_0 = 1 - \frac{\lambda}{\mu} = 1 - \frac{\frac{1}{12}}{\frac{1}{10}} = \frac{1}{6}$$

Therefore, percentage of time an arrival need not wait = 16.7.

(c) $E(W) = \dfrac{1}{\mu - \lambda}$ [by formula (7) of model I]

$$= \frac{1}{\frac{1}{10} - \frac{1}{12}} = 60 \text{ min or } 1 \text{ h}$$

(d) $E(W) > 75$, if $\dfrac{1}{\mu - \lambda_r} > 75$

i.e., if $\lambda_r > \mu - \dfrac{1}{75}$

i.e., if $\lambda_r > \dfrac{1}{10} - \dfrac{1}{75}$

i.e., if $\lambda_r > \dfrac{13}{150}$

Hence, to warrant a second barber, the average arrival rate must increase by

$\dfrac{13}{150} - \dfrac{1}{12} = \dfrac{1}{300}$ per minute.

(e) $E(W_q) = \dfrac{\lambda}{\mu(\mu - \lambda)}$

[by formula (10) of model I]

$$= \frac{\frac{1}{12}}{\frac{1}{10}\left(\frac{1}{10} - \frac{1}{12}\right)} = 50 \text{ min}$$

(f) $P(W > t) = e^{-(\mu - \lambda)t}$, [by formula (8) of model I]

$$\therefore \quad P(W > 30) = e^{-\left(\frac{1}{10} - \frac{1}{12}\right) \times 30}$$

$$= e^{-0.5} = 0.6065$$

(g) $P(\text{a customer has to wait}) = P(W > 0)$

$$= 1 - P(W = 0) = 1 - P(N = 0) = 1 - P_0$$

$$= \frac{\lambda}{\mu} = \frac{1/12}{1/10} = \frac{5}{6}$$

$\therefore$ Percentage of customers who have to wait

$$= \frac{5}{6} \times 100 = 83.33$$

(h) $P(N > 3) = P_4 + P_5 + P_6 + \ldots$

$$= 1 - \{P_0 + P_1 + P_2 + P_3\}$$

$$= 1 - \left(1 - \frac{\lambda}{\mu}\right)\left\{1 + \frac{\lambda}{\mu} + \left(\frac{\lambda}{\mu}\right)^2 + \left(\frac{\lambda}{\mu}\right)^3\right\}$$

[since $P_n = \left(\frac{\lambda}{\mu}\right)^n \left(1 - \frac{\lambda}{\mu}\right)$, for $n \geq 0$, for model I]

$$= \left(\frac{\lambda}{\mu}\right)^4 = \left(\frac{5}{6}\right)^4 = 0.4823$$

Example 3 *At what average rate must a clerk in a supermarket work in order to ensure a probability of 0.90 that the customer will not wait longer than 12 min? It is assumed that there is only one counter at which customers arrive in a Poisson fashion at an average rate of 15 per hour and that the length of the service by the clerk has an exponential distribution.*

Solution $\lambda = 15$/hour; $\mu = \mu_R$/hour

$$P\left(W_q \leq \frac{1}{5}\right) = 0.90$$

i.e., $$P\left(W_q > \frac{1}{5}\right) = 0.10$$

i.e., $$\int_{0.2}^{\infty} g(w)\, dw = 0.10$$

i.e., $$\int_{0.2}^{\infty} \frac{\lambda}{\mu}(\mu - \lambda)\, e^{-(\mu - \lambda)w}\, dw = 0.10$$ [by formula (9) of model I]

i.e., $$\left[-\frac{\lambda}{\mu} e^{-(\mu - \lambda)w}\right]_{0.2}^{\infty} = 0.1$$

i.e., $$\frac{15}{\mu_R} e^{-(\mu_R - 15) \times 0.2} = 0.1$$

i.e., $(15 - \mu_R) \times 0.2 = \log (0.1) - \log 15 + \log \mu_R$

i.e., $0.2\,\mu_R + \log \mu_R = 3 + \log 150 \simeq 8$ (1)

Solving (1), we get $\mu_R = 24$ approximately.

That is, the clerk must serve at the rate of 24 customer per hour.

Example 4 *If people arrive to purchase cinema tickets at the average rate of 6 per minute, it takes an average of 7.5 seconds to purchase a ticket. If a person arrives 2 min before the picture starts and if its takes exactly 1.5 min to reach the correct seat after purchasing the ticket,*

(a) Can he expect to be seated for the start of the picture?

(b) What is the probability that he will be seated for the start of the picture?

(c) How early must he arrive in order to be 99% sure of being seated for the start of the picture?

Solution $\lambda = 6\text{/minute}; \mu = 8\text{/minute}$

(a) $$E(W) = \frac{1}{\mu - \lambda} \quad \text{[by formula (7) of model I]}$$

$$= \frac{1}{8-6} = \frac{1}{2} \text{ min}$$

$\therefore$ E(total time required to purchase the ticket and to reach the seat)

$$= \frac{1}{2} + 1\frac{1}{2} = 2 \text{ min}$$

Hence he can just be seated for the start of the picture.

(b) P(total time < 2 min)

$$= P\left(W < \frac{1}{2}\right) = 1 - P\left(W > \frac{1}{2}\right)$$

$$= 1 - e^{-\mu\left(1 - \frac{\lambda}{\mu}\right) \times \frac{1}{2}} \quad \text{[by formula (8) of model I]}$$

$$= 1 - e^{-1} = 0.63$$

(c) $P(W < t) = 0.99$

i.e., $P(W > t) = 0.01$

i.e., $e^{-(\mu - \lambda)t} = 0.1$

i.e., $-2t = \log(0.1) = -2.3$

$\therefore$ $t = 2.3$ min

i.e., P(ticket purchasing time < 2.3) = 0.99

$\therefore$ P[total time to get the ticket and to go to the seat < (2.3 + 1.5)] = 0.99

Therefore the person must arrive at least 3.8 min early so as to be 99% sure of seeing the start of the picture.

Example 5 *A duplicating machine maintained for office use is operated by an office assistant who earns Rs 5 per hour. The time to complete each job varies according to an exponential distribution with mean 6 min. Assume a Poisson input with an average arrival rate of 5 jobs per hour. If an 8-h day is used as a base, determine*

(a) the percentage idle time of the machine,
(b) the average time a job is in the system and
(c) the average earning per day of the assistant.

Solution $\lambda = 5\text{/hour}; \mu = \frac{60}{6} = 10\text{/hour}$

(a) P(the machine is idle) = $P(N = 0) = P_0$

$$= 1 - \frac{\lambda}{\mu} \text{ (by the formula for } P_0 \text{ in model I)}$$

$$= 1 - \frac{5}{10} = \frac{1}{2}$$

$\therefore$ Percentage of idle time of the machine = 50

(b) $E(W) = \dfrac{1}{\mu - \lambda}$ [by formula (7) of model I]

$$= \frac{1}{10 - 5} = \frac{1}{5} \text{ h or 12 min}$$

(c) E(earning per day)

$= E$(number of jobs done/day) × earning per job

$= E$(number of jobs done/day) × E (time in hour/job) × earning/hour

$$= (8 \times 5) \times \frac{1}{5} \times 5 = \text{Rs } 40.$$

Example 6 *The mean rate of arrival of planes at an airport during the peak period is 20 per hour, but the actual number of arrivals in any hour follows a Poisson distribution. The airport can land 60 planes per hour on an average in good weather or 30 planes per hour in bad weather, but the actual number landed in any hour follows a Poisson distribution with respective averages. When there is congestion, the planes are forced to fly over the field in the stack awaiting the landing of other planes that arrived earlier.*

(a) How many planes would be flying over the field in the stack on an average in good weather and in bad weathers?

(b) How long a plane would be in the stack and in the process of landing in good and bad weathers?

(c) How much stack and landing time to allow so that priority to land out of order will have to be requested only 1 in 20 times.

Solution $\lambda = 20$ *per hour*

$$\mu = \begin{cases} 60 \text{ per hour in good weather} \\ 30 \text{ per hour in bad weather} \end{cases}$$

Note *Landing time is service time; the planes flying over the field in the stack are assumed to form the queue.*

(a) $E(N_q)$ = Average number of planes flying over the field = $\dfrac{\lambda^2}{\mu(\mu - \lambda)}$ [by formula (2) of model I]

$$= \begin{cases} \dfrac{20^2}{60(60-20)}, & \text{in good weather} \\ \dfrac{20^2}{30(30-20)}, & \text{in bad weather} \end{cases}$$

$$= \begin{cases} \dfrac{1}{6}, & \text{in good weather} \\ \dfrac{4}{3}, & \text{in bad weather} \end{cases}$$

(b) $E(W)$ = Average time for flying in the stack and for landing

$$= \frac{1}{\mu - \lambda} \quad \text{[by formula (7) of model I]}$$

$$= \begin{cases} \dfrac{1}{40}\text{ h} & \text{or 1.5 min in good weather} \\ \dfrac{1}{10}\text{ h} & \text{or 6 min in bad weather} \end{cases}$$

(c) Let t_R be the maximum stack and landing time to be allowed, beyond which priority out of order is to be requested.

Then $P(W > t_R) = \dfrac{1}{20}$

i.e., $$e^{-\mu\left(1-\frac{\lambda}{\mu}\right)t_R} = 0.05 \quad \text{[by formula (8) of model I]}$$

i.e., $$\begin{cases} e^{-40t_R} = 0.05, & \text{for good weather} \\ e^{-10t_R} = 0.05, & \text{for bad weather} \end{cases}$$

i.e., $$\begin{cases} t_R = 0.075\text{ h or 4.5 min for good weather} \\ \quad\;\; = 0.299\text{ h or 18 min for bad weather} \end{cases}$$

Example 7 *There are three typists in an office. Each typist can type an average of 6 letters per hour. If letters arrive for being typed at the rate of 15 letters per hour,*

(a) What fraction of the time all the typists will be busy?
(b) What is the average number of letters waiting to be typed?
(c) What is the average time a letter has to spend for waiting and for being typed?
(d) What is the probability that a letter will take longer than 20 min waiting to be typed and being typed?

Solution $\lambda = 15/\text{hour}$; $\mu = 6/\text{hour}$; $s = 3$.

Hence this is a problem in multiple server [$(M/M/s)$: $(\infty/FIFO)$] model, i.e., model II.

(a) P(all the typists are busy) $= P(N \geq 3)$

$$= \frac{\left(\frac{\lambda}{\mu}\right)^3 \cdot P_0}{3!\left(1-\frac{\lambda}{3\mu}\right)} \qquad \text{[by formula (11) of model II]}$$

$$= \frac{(2.5)^3 P_0}{6\times\left(1-\frac{2.5}{3}\right)} \qquad (1)$$

Now

$$P_0 = \frac{1}{\left\{\sum_{n=0}^{s-1} \frac{1}{n!}\left(\frac{\lambda}{\mu}\right)^n\right\} + \left(\frac{1}{s!\left(1-\frac{\lambda}{\mu s}\right)} \cdot \left(\frac{\lambda}{\mu}\right)^s\right)}$$

[by formula (6) of model II]

$$= \frac{1}{\left\{1 + 2\cdot5 + \frac{1}{2}\times(2\cdot5)^2\right\} + \left\{\frac{1}{6\times\left(1-\frac{5}{6}\right)} \times (2\cdot5)^3\right\}}$$

$$= \frac{1}{22.25} = 0.0449 \qquad (2)$$

Using (2) in (1), we have $P(N \geq 3) = 0.7016$.
Hence the fraction of the time all the typists will be busy = 0.7016.

(b) $$E(N_q) = \frac{1}{s.s!} \cdot \frac{\left(\frac{\lambda}{\mu}\right)^{s+1} \cdot P_0}{\left(1-\frac{\lambda}{\mu s}\right)^2} \qquad \text{[by formula (7) of model II]}$$

$$= \frac{1}{3\times 6} \times \frac{(2.5)^4}{\left(1-\frac{2.5}{3}\right)^2} \times 0.0449 = 3.5078$$

(c) $E(W) = \frac{1}{\lambda} E(N)$ [by Little's formula (i)]

$$= \frac{1}{\lambda}\left\{E\left(N_q\right) + \frac{\lambda}{\mu}\right\}, \text{ [by Little's formula (iv)]}$$

$$= \frac{1}{15}\{3.5078 + 2.5\} = 0.4005 \text{ h}$$

or 24 min, nearly

(d) $$P(W > t) = e^{-\mu t}\left\{1 + \frac{\left(\frac{\lambda}{\mu}\right)^s \left[1 - e^{-\mu t\left(s-1-\frac{\lambda}{\mu}\right)}\right] P_0}{s!\left(1 - \frac{\lambda}{\mu s}\right)\left(s - 1 - \frac{\lambda}{\mu}\right)}\right\}$$

(This formula has not been derived; it may be assumed.)

$$\therefore \quad P\left(W > \frac{1}{3}\right) = e^{-6 \times \frac{1}{3}} = \left[1 + \frac{(2.5)^3\{1 - e^{(-2\times -0.5)}\} \times 0.0449}{6\left(1 - \frac{2.5}{3}\right)(-0.5)}\right]$$

$$= e^{-2}\left[1 + \frac{0.7016\,(1 - e)}{(-0.5)}\right]$$

$$= 0.4616$$

Example 8 *Given an average arrival rate of 20 per hour, is it better for a customer to get service at a single channel with mean service rate of 22 customers per hour or at one of two channels in parallel with mean service rate of 11 customers per hour for each of the two channels. Assume both queues to be of Poisson type.*

Solution For the single channel service,

$\lambda = 20$/hour and $\mu = 22$/hour.

$$E(W) = \frac{1}{\mu - \lambda} \qquad \text{[by formula (7) of model I]}$$

$$= \frac{1}{2}\text{ h}$$

For the two channel service,

$\lambda = 20$/hour and $\mu = 11$/hour.

$$E(W) = \frac{1}{\mu} + \frac{1}{\mu}\cdot\frac{1}{s\cdot s!}\cdot\frac{\left(\frac{\lambda}{\mu}\right)^s}{\left(1 - \frac{\lambda}{\mu s}\right)^2} \times P_0$$

[by formula (9) of model II]

$$= \frac{1}{11} + \frac{1}{11 \times 2 \times 2} \times \frac{\left(\frac{20}{11}\right)^2}{\left(1 - \frac{20}{22}\right)^2} \times P_0$$

$$= 0.0909 + 9.0909 \times P_0 \qquad (1)$$

Now $$P_0^{-1} = \left\{\sum_{n=0}^{s-1} \frac{1}{n!}\cdot\left(\frac{\lambda}{\mu}\right)^n\right\} + \left\{\frac{1}{s!\left(1-\frac{\lambda}{\mu s}\right)}\cdot\left(\frac{\lambda}{\mu}\right)^s\right\}$$

[by formula (6) of model II]

$$= 1 + \frac{20}{11} + \frac{1}{2\times\frac{1}{11}} \times \left(\frac{20}{11}\right)^2$$

$$= 21$$

$\therefore$ $$P_0 = 0.0476 \tag{2}$$

Using (2) in (1), we have

$$E(W) = 0.5236 \text{ h}$$

As the average waiting time in single channel service is less than that in two channel service, the customer has to prefer the former.

Example 9 *A telephone company is planning to install telephone booths in a new airport. It has established the policy that a person should not have to wait more than 10% of the times he tries to use a phone. The demand for use is estimated to be Poisson with an average of 30 per hour. The average phone call has an exponential distribution with a mean time of 5 min. How many phone booths should be installed?*

Solution $\lambda = 30$/hour and $\mu = 12$/hour

In order that infinite queue may not build up, *the traffic intensity* $\frac{\lambda}{\mu s} < 1$, for multiserver model.

i.e., $$s > \frac{\lambda}{\mu}$$

i.e., $$s > \frac{30}{12} \ (= 2.5)$$

Therefore, the telephone company must install at least 3 booths.

Now we have to find the number s of telephone booths such that

$$P(W > 0) \leq 0.10 \text{ or equivalently}$$
$$P(N \geq s) \leq 0.10$$

i.e., we have to find s such that

$$\frac{\left(\frac{\lambda}{\mu}\right)^s \cdot P_0}{s!\left(1-\frac{\lambda}{\mu s}\right)} \leq 0.10$$ [by formula (11) model II]

This inequation is not easily solvable. Hence we proceed by trials and find out the least value of s that satisfies this inequation.

Let $s = 3$:

Then $$P(W > 0) = \frac{(2.5)^3 \cdot P_0}{6\left(1 - \frac{2.5}{3}\right)} = 15.625\ P_0,$$

where $$P_0 = \left[\left\{\sum_{n=0}^{s-1} \frac{1}{n!}\left(\frac{\lambda}{\mu}\right)^n\right\} + \frac{\left(\frac{\lambda}{\mu}\right)^s}{s!\left(1 - \frac{\lambda}{\mu s}\right)}\right]^{-1}$$

[by formula (6) of model II]

i.e., $$P_0 = \left[\left\{1 + 2.5 + \frac{1}{2} \times (2.5)^2\right\} + \frac{(2.5)^3}{6 \times \left(1 - \frac{2.5}{3}\right)}\right]^{-1}$$

$$= (22.25)^{-1} = 0.0449$$

$\therefore$ $P(W > 0) = 15.625 \times 0.0449 = 0.7022 \not< 0.10$

Let $s = 4$:

Then $$P(W > 0) = \frac{(2.5)^4 \cdot P_0}{24\left(1 - \frac{2.5}{4}\right)} = 4.3403\ P_0,$$

where $$P_0 = \left[\left\{1 + 2.5 + \frac{1}{2} \times (2.5)^2 + \frac{1}{6} \times (2.5)^3\right\} + \frac{(2.5)^4}{24\left(1 - \frac{2.5}{4}\right)}\right]^{-1}$$

$$= 0.0737$$

$\therefore$ $P(W > 0) = 4.3403 \times 0.0737 = 0.3199 \not< 0.10$

Similarly, when $s = 5$, $P(W > 0) = 0.1304 \not< 0.10$.
When $s = 6$, $P(W > 0) = 0.047 < 0.10$.
Hence the number of booths to be installed = 6.

Example 10 *A bank has two tellers working on savings accounts. The first teller handles withdrawals only. The second teller handles deposits only. It has been found that the service time distributions for both deposits and withdrawals are exponential with mean service time of 3 min per customer. Depositors are found to arrive in a Poisson fashion throughout the day with mean arrival rate of 16 per hour. Withdrawers also arrive in a Poisson fashion with mean arrival rate of 14 per hour. What would be the effect on the average waiting time for the customers if each teller could handle both withdrawals and deposits. What would be the effect, if this could only be accomplished by increasing the service time to 3.5 min?*

Solution When there is a separate channel for the depositors, $\lambda_1 = 16$/hour, $\mu = 20$/hour

$$\therefore \quad E(W_q \text{ for depositors}) = \frac{\lambda_1}{\mu(\mu - \lambda_1)} \quad \text{[by formula (10) of model I]}$$

$$= \frac{16}{20\,(20 - 16)} = \frac{1}{5} \text{ h or 12 min}$$

When there is a separate channel for the withdrawers, $\lambda_2 = 14$/hour, $\mu = 20$/hour.

$$\therefore \quad E(W_q \text{ for withdrawers}) = \frac{\lambda_2}{\mu(\mu - \lambda_2)}$$

$$= \frac{14}{20(20 - 14)} = \frac{7}{60} \text{ h or 7 min}$$

If both tellers do both service,

$$s = 2,\ \mu = 20\text{/hour},\ \lambda = \lambda_1 + \lambda_2 = 30\text{/hour}$$

$$\therefore \quad E(W_q \text{ for any customer}) = \frac{1}{\mu} \cdot \frac{1}{s.s!} \cdot \frac{\left(\frac{\lambda}{\mu}\right)^s \times P_0}{\left(1 - \frac{\lambda}{\mu s}\right)^2},$$

[by formula (10) of model II]

$$= \frac{1}{20} \times \frac{1}{2 \times 2} \times \frac{(1.5)^2}{(1 - .75)^2} \times P_0 \qquad (1)$$

$$= 0.45 \times P_0$$

Now

$$P_0 = \left[\left\{\sum_{n=0}^{s-1} \frac{1}{n!} \cdot \left(\frac{\lambda}{\mu}\right)^n\right\} + \frac{1}{s!\left(1 - \frac{\lambda}{\mu s}\right)}\left(\frac{\lambda}{\mu}\right)^s\right]^{-1}$$

[by formula (6) of, model II]

$$= \left[1 + 1.5 + \frac{(1.5)^2}{2 \times 0.25}\right]^{-1}$$

$$= \frac{1}{7} \qquad (2)$$

Using (2) in (1)

$$E(W_q \text{ for any customer}) = 0.45 \times \frac{1}{7} \text{ h or 3.86 min}$$

Hence if both tellers do both types of service, the customers get benefited as their waiting time is considerably reduced.

Now if both tellers do both types of service but with increased service time,

$$s = 2,\ \lambda = 30,\ \mu = \frac{60}{3.5} = \frac{120}{7} \text{ per hour.}$$

$E(W_q$ of any customer)

$$= \frac{7}{120} \times \frac{1}{2 \times 2} \times \frac{(1.75)^2}{\left(1 - \frac{7}{8}\right)^2} \times P_0 = 2.86\, P_0, \text{ where}$$

$$P_0 = \left[1 + 1.75 + \frac{(1.75)^2}{2 \times \frac{1}{8}}\right]^{-1} = \frac{1}{15}$$

$$\therefore \quad E(W_q \text{ of any customer}) = \frac{2.86}{15} \text{ h or } 11.44 \text{ min}$$

If this arrangement is adopted, withdrawers stand to lose as their waiting time is increased considerably and depositors get slightly benefited.

Example 11 *A supermarket has two girls attending to sales at the counters. If the service time for each customer is exponential with mean 4 min and if people arrive in Poisson fashion at the rate of 10 per hour,*

(a) what is the probability that a customer has to wait for service?

(b) what is the expected percentage of idle time for each girl?

(c) if the customer has to wait in the queue, what is the expected length of his waiting time?

Solution $s = 2,\ \lambda = \frac{1}{6}$ per minute, $\mu = \frac{1}{4}$ per minute

(a) P(a customer has to wait for service)

$$= P(N \geq 2) = 1 - P_0 - P_1 \tag{1}$$

$$P_0 = \left[\left\{\sum_{n=0}^{s-1} \frac{1}{n!}\left(\frac{\lambda}{\mu}\right)^n\right\} + \left\{\frac{\left(\frac{\lambda}{\mu}\right)^s}{s!\left(1 - \frac{\lambda}{\mu s}\right)}\right\}\right]^{-1}$$

[by formula (6) of model II]

$$= \left[1 + \frac{2}{3} + \frac{\left(\frac{2}{3}\right)^2}{2 \times \left(1 - \frac{1}{3}\right)}\right]^{-1} = \frac{1}{2} \tag{2}$$

$$P_1 = \frac{\lambda}{\mu} \cdot P_0, \qquad \text{[by formula (4) of model II]}$$

$$= \frac{2}{3} \times \frac{1}{2} = \frac{1}{3} \qquad (3)$$

Using (2) and (3) in (1), we have

$$P(N \geq 2) = 1 - \frac{1}{2} - \frac{1}{3} = \frac{1}{6}$$

(b) Fraction of time when the girls are busy $= \frac{\lambda}{\mu s} = \frac{1}{3}$

$\therefore$ Fraction of time when the girls are idle $= \frac{2}{3}$

$\therefore$ Expected percentage of idle time for each girl $= \frac{2}{3} \times 100 = 67$

(c) $E(W_q / W_s > 0) = \frac{1}{\mu s - \lambda}$ [by formula (13) of model II]

$$= \frac{1}{\frac{1}{4} \times 2 - \frac{1}{6}} = 3 \text{ min}$$

Example 12 *A petrol pump station has 4 pumps. The service times follow the exponential distribution with a mean of 6 min and cars arrive for service in a Poisson process at the rate of 30 cars per hour.*

(a) What is the probability that an arrival would have to wait in line?

(b) Find the average waiting time, average time spent in the system and the average number of cars in the system.

(c) For what percentage of time would a pump be idle on an average?

Solution $s = 4$, $\lambda = 30$/hour, $\mu = 10$/hour

(a) P(an arrival has to wait) $= P(W > 0)$

$$= \frac{\left(\frac{\lambda}{\mu}\right)^s \cdot P_0}{s!\left(1 - \frac{\lambda}{\mu s}\right)} \qquad \text{[by formula (11) of model II]}$$

$$= \frac{3^4 \times P_0}{24 \times \left(1 - \frac{3}{4}\right)} = 13.5 \times P_0 \qquad (1)$$

$$P_0 = \left[\left\{\sum_{n=0}^{s-1} \frac{1}{n!} \cdot \left(\frac{\lambda}{\mu}\right)^n\right\} + \frac{\left(\frac{\lambda}{\mu}\right)^s}{s!\left(1 - \frac{\lambda}{\mu s}\right)}\right]^{-1}$$

[by formula (6) of model II]

$$= \left[\left(1 + 3 + \frac{1}{2} \times 9 + \frac{1}{6} \times 27 \right) + \frac{3^4}{24 \times \left(1 - \frac{3}{4} \right)} \right]^{-1}$$

$$= 0.0377 \qquad (2)$$

Using (2) in (1), $P(W > 0) = 0.5090$

(b) $$E(W_q) = \frac{1}{\mu} \cdot \frac{1}{s \times s!} \times \frac{\left(\frac{\lambda}{\mu} \right)^s}{\left(1 - \frac{\lambda}{\mu s} \right)^2} \times P_0 \text{ [by formula (10) of model II]}$$

$$= \frac{1}{10 \times 4 \times 24} \times \frac{3^4}{\left(1 - \frac{3}{4} \right)^2} \times 0.0377 = 0.0509 \text{ h}$$

or 3.05 min

$$E(W_s) = \frac{1}{\mu} + E(W_q) \text{ [by formulas (9) and (10) of model II]}$$

$$= 6 + 3.05 = 9.05 \text{ min}$$

$$E(N) = \frac{1}{s \times s!} \times \frac{\left(\frac{\lambda}{\mu} \right)^{s+1}}{\left(1 - \frac{\lambda}{\mu s} \right)^2} \times P_0 + \frac{\lambda}{\mu}$$

[by formula (8) of model II]

$$= \frac{1}{4 \times 24} \times \frac{3^5}{\left(1 - \frac{3}{4} \right)^2} \times 0.0377 + 3$$

$$= 4.53 \text{ cars}$$

(c) The fraction of time when the pumps are busy = traffic intensity

$$= \frac{\lambda}{\mu s} = \frac{3}{4}$$

$\therefore$ The fraction of time when the pumps are idle $= \frac{1}{4}$

Therefore, required percentage = 25%

Example 13 *In a single server queueing system with Poisson input and exponential service times, if the mean arrival rate is 3 calling units per hour, the expected service time is 0.25 h and the maximum possible number of calling* *units in the system is 2, find $P_n(n \geq 0)$, average number of calling units in the system and in the queue and average waiting time in the system and in the queue.*

Solution The situation in this problem is one of finite capacity, single server Poisson queue models.

$$\lambda = 3,\ \mu = 4 \text{ and } k = 2$$

As $\lambda \neq \mu$, $P_0 = \dfrac{1 - \dfrac{\lambda}{\mu}}{1 - \left(\dfrac{\lambda}{\mu}\right)^{k+1}}$ [by formula (4) of model III]

$$= \frac{1 - \frac{3}{4}}{1 - \left(\frac{3}{4}\right)^3} = \frac{16}{37} = 0.4324$$

Since $\lambda \neq \mu$, $P_n = \left(\dfrac{\lambda}{\mu}\right)^n \left[\dfrac{1 - \dfrac{\lambda}{\mu}}{1 - \left(\dfrac{\lambda}{\mu}\right)^{k+1}}\right]$ [by formula (6) of model III]

$$= (0.4324)\,(0.75)^n$$

$E(N) = \dfrac{\lambda}{\mu - \lambda} - \dfrac{(k+1)\left(\dfrac{\lambda}{\mu}\right)^{k+1}}{1 - \left(\dfrac{\lambda}{\mu}\right)^{k+1}}$ [by formula (8) of model IV]

$$= \frac{3}{4-3} - \frac{3 \times \left(\frac{3}{4}\right)^3}{1 - \left(\frac{3}{4}\right)^3} = 3 - \frac{81}{37} = \frac{30}{37} \simeq 0.8 \text{ calling unit}$$

$E(N_q) = E(N) - (1 - P_0)$ [by formula (10) of model III]

$$= \frac{30}{37} - \left(1 - \frac{16}{37}\right) = \frac{9}{37} = 0.24 \text{ calling unit}$$

$E(W_s) = \dfrac{1}{\lambda'}\, E(N)$ [by formula (13) of model III]

where $\lambda' = \mu\,(1 - P_0)$, [by formual (11) of model III]

$$= 4\left(1 - \frac{16}{37}\right) = \frac{84}{37}$$

$$\therefore \qquad E(W_s) = \frac{37}{84} \times \frac{30}{37} = \frac{5}{14} \text{ h or } 21.4 \text{ min}$$

$$E(W_q) = \frac{1}{\lambda'} E(N_q) \text{ [by formula (14) of model III]}$$

$$= \frac{37}{84} \times \frac{9}{37} = \frac{3}{28} \text{ h or } 6.4 \text{ min}$$

Example 14 *The local one-person barber shop can accommodate a maximum of 5 people at a time (4 waiting and 1 getting hair-cut). Customers arrive according to a Poisson distribution with mean 5 per hour. The barber cuts hair at an average rate of 4 per hour (Exponential service time).*

(a) What percentage of time is the barber idle?

(b) What fraction of the potential customers are turned away?

(c) What is the expected number of customers waiting for a hair-cut?

(d) How much time can a customer expect to spend in the barber shop?

Solution $\lambda = 5$, $\mu = 4$, $k = 5$

(a) P(the barber is idle) $= P(N = 0)$

$$= P_0 = \frac{1 - \dfrac{\lambda}{\mu}}{1 - \left(\dfrac{\lambda}{\mu}\right)^{k+1}} \qquad \text{[by formula (4) of model III]}$$

$$= \frac{1 - \dfrac{5}{4}}{1 - \left(\dfrac{5}{4}\right)^{6}} = 0.0888$$

$\therefore$ Percentage of time when the barber is idle $\simeq 9$.

(b) P(a customer is turned away) $= P(N > 5)$

$$= \left(\frac{\lambda}{\mu}\right)^5 \cdot \left[\frac{1 - \dfrac{\lambda}{\mu}}{1 - \left(\dfrac{\lambda}{\mu}\right)^{k+1}}\right] \qquad \text{[by formula (6) of model III]}$$

$$= \left(\frac{5}{4}\right)^5 \cdot \left[\frac{1 - \dfrac{5}{4}}{1 - \left(\dfrac{5}{4}\right)^{6}}\right]$$

$$= \frac{3125}{11529} = 0.2711$$

Therefore, 0.2711 × potential customers are turned away.

(c) $E(N_q) = E(N) - (1 - P_0)$

$$= \frac{\lambda}{\mu - \lambda} - \frac{(k+1)\left(\frac{\lambda}{\mu}\right)^{k+1}}{1 - \left(\frac{\lambda}{\mu}\right)^{k+1}} - (1 - P_0),$$

[by formulas (6) and (10) of model III]

$$= \left\{-5 - \frac{6 \times \left(\frac{5}{4}\right)^6}{1 - \left(\frac{5}{4}\right)^6}\right\} - (1 - 0.0888)$$

$$= \frac{6 \times \frac{15625}{4096}}{\frac{11529}{4096}} - 5.9112 \simeq 2.2 \text{ customers}$$

(d) $E(W) = \frac{1}{\lambda'} E(N)$ [by formula (13) of model III]

$$= \frac{1}{\mu(1 - P_0)} \times E(N) = \frac{3.1317}{3.6448} \simeq 0.8592 \text{ h}$$

or 51.5 min

Example 15 *At a railway station, only one train is handled at a time. The railway yard is sufficient only for 2 trains to wait, while the other is given signal to leave the station. Trains arrive at the station at an average rate of 6 per hour and the railway station can handle them on an average of 6 per hour. Assuming Poisson arrivals and exponential service distribution, find the probabilities for the numbers of trains in the system. Also find the average waiting time of a new train coming into the yard. If the handling rate is doubled, how will the above results get modified?*

Solution

(i) $\lambda = 6$ per hour, $\mu = 6$ per hour, $k = 2 + 1 = 3$

Since $\lambda = \mu$, $P_0 = \frac{1}{k+1}$

$$= \frac{1}{4}$$ [by formula (5) of model III]

$P_n = \frac{1}{k+1} = \frac{1}{4}$ for $n = 1, 2, 3$ [by formula (7) of model III]

$E(N) = \frac{k}{2}$ [by formula (9) of model III]

$= 1.5$ trains

$$E(W) = \frac{1}{\lambda'} E(N) \qquad \text{[by formula (13) of model III]}$$

$$= \frac{1.5}{\mu(1 - P_0)} = \frac{1.5}{6 \times \frac{3}{4}} = \frac{1}{3} \text{ h or 20 min}$$

(ii) $\lambda = 6;\ \mu = 12,\ k = 3$

Since $\lambda \neq \mu$, $$P_0 = \frac{1 - \frac{\lambda}{\mu}}{1 - \left(\frac{\lambda}{\mu}\right)^{k+1}} \qquad \text{[by formula (4) of model III]}$$

$$= \frac{1 - \frac{1}{2}}{1 - \left(\frac{1}{2}\right)^4} = \frac{8}{15}$$

$$P_n = \left(\frac{\lambda}{\mu}\right)^n \left\{ \frac{1 - \frac{\lambda}{\mu}}{1 - \left(\frac{\lambda}{\mu}\right)^{k+1}} \right\} \qquad \text{[by formula (6) of model III]}$$

$$= \frac{8}{15} \cdot \left(\frac{1}{2}\right)^n, \text{ for } n = 1, 2, 3.$$

$$E(N) = \frac{\lambda}{\mu - \lambda} - \frac{(k+1)\left(\frac{\lambda}{\mu}\right)^{k+1}}{1 - \left(\frac{\lambda}{\mu}\right)^{k+1}} \qquad \text{[by formula (8) of model III]}$$

$$= 1 - \frac{4 \times \left(\frac{1}{2}\right)^4}{1 - \left(\frac{1}{2}\right)^4} = 1 - \frac{4}{15} = \frac{11}{15} \simeq 0.73 \text{ train}$$

$$E(W) = \frac{1}{\lambda'} E(N) \qquad \text{[by formula (13) of model III]}$$

$$= \frac{1}{\mu(1 - P_0)} \times E(N)$$

$$= \frac{\frac{11}{15}}{12\left(1 - \frac{8}{15}\right)} = \frac{11}{84} \text{ h or 7.9 min}$$

Example 16 *Patients arrive at a clinic according to Poisson distribution at a rate of 30 patients per hour. The waiting room does not accommodate more than 14 patients. Examination time per patient is exponential with mean rate of 20 per hour.*

(a) Find the effective arrival rate at the clinic.
(b) What is the probability that an arriving patient will not wait?
(c) What is the expected waiting time until a patient is discharged from the clinic?

Solution

(a) $\lambda = 30$ per hour, $\mu = 20$ per hour, $k = 14 + 1 = 15$

Since $\lambda \neq \mu$, $$P_0 = \frac{1 - \dfrac{\lambda}{\mu}}{1 - \left(\dfrac{\lambda}{\mu}\right)^{k+1}}$$ [by formula (4) of model III]

$$= \frac{1 - \dfrac{3}{2}}{1 - \left(\dfrac{3}{2}\right)^{16}} = 0.00076$$

Effective arrival rate $\lambda' = \mu(1 - P_0)$ [by formula (11) of model III]

$$= 20 \times (1 - 0.00076)$$
$$= 19.98 \text{ per hour}$$

(b) P(a patient will not wait)

$$= P_0 = 0.00076$$

(c) $$E(N) = \frac{\lambda}{\mu - \lambda} - \frac{(k+1)\left(\dfrac{\lambda}{\mu}\right)^{k+1}}{1 - \left(\dfrac{\lambda}{\mu}\right)^{k+1}}$$

$$= -3 - \frac{16 \times \left(\dfrac{3}{2}\right)^{16}}{1 - \left(\dfrac{3}{2}\right)^{16}} = 13 \text{ patients nearly}$$

$$E(W) = \frac{E(N)}{\lambda'} = \frac{13}{19.98} = 0.65 \text{ h or } 39 \text{ min}$$

Example 17 *A 2-person barber shop has 5 chairs to accommodate waiting customers. Potential customers, who arrive when all 5 chairs are full, leave without entering barber shop. Customers arrive at the average rate of 4 per hour and spend an average of 12 min in the barber's chair. Compute P_0 P_1, P_7, $E(N_q)$ and $E(W)$.*

Solution The situation in this problem is one of finite capacity, multiserver Poisson queue models.

$$\lambda = 4 \text{ per hour}, \mu = 5 \text{ per hour}, s = 2, k = 2 + 5 = 7$$

(a) $$P_0 = \left[\sum_{n=0}^{s-1} \frac{1}{n!} \left(\frac{\lambda}{\mu} \right)^n + \frac{1}{s!} \left(\frac{\lambda}{\mu} \right)^s \sum_{n=s}^{k} \left(\frac{\lambda}{\mu s} \right)^{n-s} \right]^{-1}$$

[by formula (3) of model IV]

$$= \left[\sum_{n=0}^{1} \frac{1}{n!} \left(\frac{4}{5} \right)^n + \frac{1}{2} \cdot \left(\frac{4}{5} \right)^2 \sum_{n=2}^{7} \left(\frac{2}{5} \right)^{n-2} \right]^{-1}$$

$$= \left[1 + \frac{4}{5} + \frac{8}{25} \left\{ 1 + \frac{2}{5} + \left(\frac{2}{5} \right)^2 + \left(\frac{2}{5} \right)^3 + \left(\frac{2}{5} \right)^4 + \left(\frac{2}{5} \right)^5 \right\} \right]^{-1}$$

$$= \left[\frac{9}{5} + \frac{8}{25} \left\{ \frac{1 - (0.4)^7}{1 - 0.4} \right\} \right]^{-1} = 0.4287$$

(b) $$P_n = \frac{1}{n!} \left(\frac{\lambda}{\mu} \right)^n P_0, \text{ for } n \le s \quad \text{[by formula (4) of model IV]}$$

$$\therefore \qquad P_1 = \left(\frac{4}{5} \right) \times 0.4287 = 0.3430$$

(c) $$P_n = \frac{1}{s!.s^{n-s}} \left(\frac{\lambda}{\mu} \right)^n \cdot P_0, \text{ for } s < n \le k \qquad \text{[by formula (4) of model IV]}$$

$$\therefore \qquad P_7 = \frac{1}{2 \times 2^{7-2}} \times \left(\frac{4}{5} \right)^7 \times 0.4287$$

$$= 0.0014$$

(d) $$E(N_q) = P_0 \left(\frac{\lambda}{\mu} \right)^s \cdot \frac{\rho}{s!(1-\rho)^2} [1 - p^{k-s} - (k-s)(1-\rho)\, \rho^{k-s}],$$

where $$\rho = \frac{\lambda}{\mu s} \text{ [by formula (5) of model IV]}$$

$$= (0.4287).\, (0.8)^{2} \cdot \frac{(0.4)}{2 \times (0.6)^2} [1 - (0.4)^5 - 5 \times 0.6 \times (0.4)^5]$$

$$= 0.15 \text{ customer}$$

(e) $$E(N) = E(N_q) + s - \sum_{n=0}^{s-1} (s-n) P_n \qquad \text{[by formula (6) of model IV]}$$

$$= 0.1462 + 2 - \sum_{n=0}^{1} (2-n) P_n$$

$= 2.1462 - (2 \times P_0 + 1 \times P_1)$
$= 2.1462 - (2 \times 0.4287 + 1 \times 0.3430)$
$= 0.95$ customer

$$E(W) = \frac{1}{\lambda'} . E(N) \text{ [by formula (9) of model IV]}$$

where $$\lambda' = \mu \left[s - \sum_{n=0}^{s-1} (s-n) P_n \right] \text{ [by formula (7) of model IV]}$$

$= 4[2 - (2 \times 0.4287 + 1 \times 0.3430)]$
$= 3.1984$

$$\therefore \quad E(W) = \frac{0 \cdot 9458}{3 \cdot 1984} = 0.2957 \text{ h or } 17.7 \text{ min}$$

Example 18 *At a port there ar 6 unloading berths and 4 unloading crews. When all the berths are full, arriving ships are diverted to an overflow facility 20 kms down the river. Tankers arrive according to a Poisson process with a mean of 1 every 2 h. It takes for an unloading crew, on the average, 10 h to unload a tanker, the unloading time following an exponential distribution. Find*

(a) how many tankers are at the port on the average?
(b) how long does a tanker spend at the port on the average?
(c) what is the average arrival rate at the overflow facility?

Solution $\lambda = \frac{1}{2}$ per hour, $\mu = \frac{1}{10}$ per hour, $s = 4$, $k = 6$

(a) $$P_0 = \left[\sum_{n=0}^{s-1} \frac{1}{n!} \left(\frac{\lambda}{\mu} \right)^n + \frac{1}{s!} \left(\frac{\lambda}{\mu} \right)^s \sum_{n=s}^{k} \left(\frac{\lambda}{\mu s} \right)^{n-s} \right]^{-1}$$

[by formula (3) of model IV]

$$= \left[\left(1 + 5 + \frac{1}{2} \times 5^2 + \frac{1}{6} \times 5^3 \right) + \frac{1}{24} \times 5^4 \times \left\{ \left(\frac{5}{4} \right)^0 + \left(\frac{5}{4} \right)^1 + \left(\frac{5}{4} \right)^2 \right\} \right]^{-1}$$

$= 0.0072$

$$E(N_q) = P_0 . \left(\frac{\lambda}{\mu} \right)^s \cdot \frac{\rho}{s!(1-\rho)^2} [1 - \rho^{k-s} - (k-s)(1-\rho)\rho^{k-s}],$$

where $$\rho = \frac{\lambda}{\mu s} \qquad \text{[by formula (5) of model IV]}$$

$$= 0.0072 \times 5^4 \times \frac{1.25}{24 \times (.25)^2}$$

$$[1 - (1.25)^2 - 2 \times (-.25)(1.25)^2]$$

$= 0.8203$ tanker

$$E(N) = E(N_q) + s - \sum_{n=0}^{s-1} (s-n) P_n$$

[by formula (6) of model IV]

$$= 4.8203 - (4P_0 + 3P_1 + 2P_2 + P_3)$$
$$= 4.8203 - \{4 \times .0072 + 3 \times 0.0360 + 2 \times 0.09 + 0.15\}$$
$$= 4.3535 \text{ tankers}$$

(b) $E(W) = \dfrac{1}{\lambda'} E(N)$ [by formula (9) of model IV]

where $\lambda' = \mu \left[s - \sum_{n=0}^{s-1} (s-n) P_n \right]$ [by formula (7) of model IV]

$$= \frac{1}{10} [4 - \{4P_0 + 3P_1 + 2P_2 + P_3\}]$$
$$= \frac{1}{10} [4 - 0.4668] = 0.3533$$

$\therefore$ $E(W) = \dfrac{4.3535}{0.3533} = 12.32 \text{ h}$

(c) When $N = 6$, i.e., when the number of tankers in the port is 6, overflow occurs.

$$P(N = 6) = \frac{1}{s!s^{n-s}} \left(\frac{\lambda}{\mu} \right)^n P_0, \text{ for } n = k$$

[by formula (4) of model IV]

$$= \frac{1}{24 \times 4^2} \times 5^6 \times 0.0072$$
$$= 0.2930$$

Average arrival rate at the overflow facility = (Average arrival rate at the port) $\times$ (Probability that overflow occurs)

$$= \frac{1}{2} \times 0.2930 = 0.586 \text{ per hour}$$

Example 19 *A car servicing station has 2 bays where service can be offered simultaneously. Because of space limitation, only 4 cars are accepted for servicing. The arrival pattern is Poisson with 12 cars per day. The service time in both the bays is exponentially distributed with $\mu = 8$ cars per day per bay. Find the average number of cars in the service station, the average number of cars waiting for service and the average time a car spends in the system.*

Solution *$\lambda = 12$ per day, $\mu = 8$ per day, $s = 2$, $k = 4$*

(a) $P_0 = \left[\sum_{n=0}^{s-1} \dfrac{1}{n!} \left(\dfrac{\lambda}{\mu} \right)^n + \dfrac{1}{s!} \left(\dfrac{\lambda}{\mu} \right)^s \sum_{n=s}^{k} \left(\dfrac{\lambda}{\mu s} \right)^{n-s} \right]^{-1}$

[by formula (3) of model IV]

$$= \left[1 + \frac{1.5}{1} + \frac{1}{2} \times (1.5)^2 \left\{1 + (.75) + (.75)^2\right\}\right]^{-1}$$
$$= 0.1960$$

$$E(N_q) = P_0 \left(\frac{\lambda}{\mu}\right)^s \cdot \frac{\rho}{s!(1-\rho)^2} [1 - \rho^{k-s} - (k-s)(1-\rho)\rho^{k-s}],$$

where $\rho = \dfrac{\lambda}{\mu s}$ [by formula (5) of model IV]

i.e., $$E(N_q) = 0.1960 \times (1.5)^2 \times \frac{0 \cdot 75}{2 \times (0 \cdot 25)^2} \times [1 - (0.75)^2 - 2 \times 0.25 \times (0.75)^2]$$
$$= 0.4134 \text{ car}$$

(b) $E(N)$ = Average number of cars in the service station

$$= E(N_q) + s - \sum_{n=0}^{s-1} (s-n) P_n$$

[by formula (6) of model IV]

$$= 0.4134 + 2 - \sum_{n=0}^{1} (2-n) P_n$$
$$= 2.4134 - (2P_0 + P_1)$$
$$= 2.4134 - (2 \times 0.1960 + 1.5 \times 0.1960)$$
$$= 1.73 \text{ cars}$$

(c) $E(W) = \dfrac{1}{\lambda'} E(N)$ [by formula (9) of model IV]

where $$\lambda' = \mu \left[s - \sum_{n=0}^{s-1} (s-n) P_n\right]$$ [by formula (7) of model IV]

$$= 8[2 - (2P_0 + P_1)]$$
$$= 10.512$$

$$\therefore \quad E(W) = \frac{1.73}{10.512} = 0.1646 \text{ day}$$

Example 20 *A group of engineers has 2 terminals available to aid in their calculations. The average computing job requires 20 min of terminal time and each engineer requires some computation about once every half an hour. Assume that these are distributed according to an exponential distribution. If there are 6 engineers in the group, find*

(a) the expected number of engineers waiting to use one of the terminals and in the computing centre and

(b) the total time lost per day.

$\lambda = 2$ *per hour,* $\mu = 3$ *per hour,* $s = 2$, $k = 6$

Solution

(a) $$P_0 = \left[\sum_{n=0}^{s-1} \frac{1}{n!} \left(\frac{\lambda}{\mu} \right)^n + \frac{1}{s!} \left(\frac{\lambda}{\mu} \right)^s \sum_{n=s}^{k} \left(\frac{\lambda}{\mu s} \right)^{n-s} \right]^{-1}$$

[by formula (3) of model IV]

$$= \left[1 + \frac{2}{3} + \frac{1}{2} \times \left(\frac{2}{3} \right)^2 \left\{ 1 + \frac{1}{3} + \left(\frac{1}{3} \right)^2 + \left(\frac{1}{3} \right)^3 + \left(\frac{1}{3} \right)^4 \right\} \right]^{-1}$$

$= 0.5003$

$$E(N_q) = P_0 \times \left(\frac{\lambda}{\mu} \right)^s \times \frac{\rho}{s!(1-\rho)^2} \left[1 - \rho^{k-s} - (k-s)(1-\rho)\rho^{k-s}\right],$$

where $$\rho = \frac{\lambda}{\mu s}$$ [by formula (5) of model IV]

i.e., $$E(N_q) = 0.5003 \times \left(\frac{2}{3} \right)^2 \times \frac{\left(\frac{1}{3} \right)}{2 \times \left(\frac{2}{3} \right)^2} \left[1 - \left(\frac{1}{3} \right)^4 - 4 \times \frac{2}{3} \times \left(\frac{1}{3} \right)^4 \right]$$

$= 0.0796$

$$E(N) = E(N_q) + s - \sum_{n=0}^{s-1} (s-n)P_n$$ [by formula (6) of model IV]

$$= 0.0796 + 2 - \sum_{n=0}^{1} (2-n)P_n$$

$= 2.0796 - (2P_0 + P_1)$

$$= 2.0796 - \left(2 \times 0.5003 + \frac{2}{3} \times 0.5003 \right)$$

$= 0.75$

(b) $$E(W_q) = \frac{1}{\lambda'} E(N_q)$$ [by formula (10) of model IV]

where $$\lambda' = \mu \left[s - \sum_{n=0}^{s-1} (s-n) P_n \right]$$ [by formula (7) of model IV]

$$= 3 \left[2 - \sum_{n=0}^{1} (2-n) P_n \right]$$

$= 3\,[2 - (2P_0 + P_1)]$

$$= 3\left[2 - \left(2 \times 0.5003 + \frac{2}{3} \times 0.5003\right)\right]$$

$$= 1.9976$$

$$\therefore \qquad E(W_q) = \frac{0.0796}{1.9976} = 0.0398 \text{ h}$$

Every time an engineer approaches the computer centre, he has to lose 0.0398 h by way of waiting.

If the day consists of 8 working hours, he has to approach the centre 16 times.

$\therefore$ Time lost in waiting in a day per engineer

$$= 16 \times 0.0398 = 0.6368 \text{ h}$$

$\therefore$ Total time lost in waiting in a day by all the 6 engineers

$$= 6 \times 0.6368 = 3.82 \text{ h.}$$

Exercise 9

Part A (Short answer questions)

1. What are the characteristics of a queueing system?
2. What do the letters in the symbolic representation $(a/b/c)$: (d/e) of a queueing model represent?
3. What do you mean by transient state and steady-state queueing systems?
4. Write down the difference equations that give the probability that there are n customers $(n \geq 0)$ in a Poisson queueing system in steady-state.
5. Write down the formulas for P_0 and P_n in a Poisson queue system in the steady-state.
6. Give the formulas for the average number of customers (i) in the system, (ii) in the queue and (iii) in the non-empty queues for the $(M/M/1)$: $(\infty/FIFO)$ model.
7. Obtain the variance of queue length for the $(M/M/1)$: $(\infty/FIFO)$ model.
8. In the usual notation of a $(M/M/1)$: $(\infty/FIFO)$ queue system, find $P(N > 2)$, if $\lambda = 12$ per hour and $\mu = 30$ per hour.
9. In the usual notation of a $(M/M/1)$: $(\infty/FIFO)$ queue system if $\lambda = 12$ per hour and $\mu = 24$ per hour, find the average number of customers in the system and in the queue.
10. Give the formulas for the waiting time of a customer in the queue and in the system for the $(M/M/1)$: $(\infty/FIFO)$ model.
11. If a customer has to wait in a $(M/M/1)$: $(\infty/FIFO)$ queue system, what is his average waiting time in the queue, if $\lambda = 8$ per hour and $\mu = 12$ per hour?
12. What is the probability that a customer has to wait more than 15 min to get his service completed in a $(M/M/1)$: $(\infty/FIFO)$ queue system, if $\lambda = 6$ per hour and $\mu = 10$ per hour?
13. Write down the probability density function of the waiting time of a customer in the $(M/M/1)$: $(\infty/FIFO)$ queue system.

14. Write down the Little's formulas that hold good for all the Poisson queue models.
15. Write down the Little's formulas that hold good for the infinite capacity Poisson queue models.
16. What is the probability that there are no customers in the $(M/M/s)$: $(\infty/FIFO)$ queueing system?
17. Write down the formula for P_n in terms of P_0 for the $(M/M/s)$: $(\infty/FIFO)$ queueing system.
18. Give the formulas for the average number of customers in the system and in the queue for the $(M/M/s)$: $(\infty/FIFO)$ queueing model.
19. If there are 2 servers in an infinite capacity Poisson queue system with $\lambda = 10$ per hour and $\mu = 15$ per hour, what is the percentage of idle time for each server?
20. In a 3 server infinite capacity Poisson queue model if $\lambda/s\mu = \frac{2}{3}$, find P_0.
21. In a 3 server infinite capacity Poisson queue model if $\lambda/s\mu = 2/3$ and P_0 1/9 find the average number of customers in the queue and in the system.
22. If $\lambda/s\mu = \frac{2}{3}$ in a $(M/M/s)$: $(\infty/FIFO)$ queue system find the average number of customers in the nonempty queue.
23. What is the probability that an arrival to an infinite capacity 3 server Poisson queue system with $\lambda/s\mu = 2/3$ and $P_0 = 1/9$ will have to wait?
24. What is the probability that an arrival to an infinite capacity 3 server Poisson queue system with $\lambda/s\mu = 2/3$ and $P_0 = 1/9$ enters the service without waiting?
25. Give the formulas for the average waiting time of a customer in the system and in the queue for the $(M/M/s)$:$(\infty/FIFO)$ queueing model.
26. What is the average waiting time of a customer in the 3 server infinite capacity Poisson queue if he happens to wait, given that $\lambda = 6$ per hour and $\mu = 4$ per hour.
27. Give the probability that there is no customer in an $(M/M/1)$:$(k/FIFO)$ queueing system.
28. Write down the probability that there are n customers in an $(M/M/1)$: $(k/FIFO)$ queueing system.
29. If $\lambda = 4$ per hour and $\mu = 12$ per hour in an $(M/M/1)$:$(4/FIFO)$ queueing system, find the probability that there is no customer in the system. If $\lambda = \mu$, what is the value of this probability?
30. Write the formulas for the average number of customers in the $(M/M/1)$: $(k/FIFO)$ queueing system and also in the queue.
31. Define effective arrival rate with respect to an $(M/M/1)$:$(k/FIFO)$ queueing model.
32. How are N_s and N_q related in an $(M/M/1)$:$(k/FIFO)$ queueing model?

33. Write down the Little's formulas for the average waiting time in the system and in the queue for an $(M/M/1)$:$(k/FIFO)$ queueing model.
34. If $\lambda = 3$ per hour, $\mu = 4$ per hour and maximum capacity $k = 7$ in a $(M/M/1)$:$(k/FIFO)$ system, find the average number of customers in the system.
35. Write the formula for the probability that there is no customer in an $(M/M/s)$:$(k/FIFO)$ queue system.
36. Write the formula for the probability that there are n customers in an $(M/M/s)$:$(k/FIFO)$ queueing system.
37. Write down the formula for the average queue length in an $(M/M/s)$:$(k/FIFO)$ queueing model.
38. Define effective arrival rate with respect to an $(M/M/s)$:$(k/FIFO)$ queueing model.
39. How are N_s and N_q related in an $(M/M/s)$:$(k/FIFO)$ queueing model?
40. Write down Little's formulas for the average waiting time in the system and in the queue for an $(M/M/s)$:$(k/FIFO)$ queueing model.

Part B

41. Arrivals at a telephone booth are considered to be Poisson with an average time of 10 min between one arrival and the next. The length of a phone call is assumed to be distributed exponentially with mean 3 min.
    (a) Find the average number of persons waiting in the system.
    (b) What is the probability that a person arriving at the booth will have to wait in the queue?
    (c) What is the probability that it will take him more than 10 min altogether to wait for phone and complete his call?
    (d) Estimate the fraction of the day when the phone will be in use.
    (e) The telephone department will install a second booth when convinced that an arrival has to wait on the average for at least 3 min for phone. By how much the flow of arrivals should increase in order to justify a second booth?
42. Customers arrive at a one man barber shop according to a Poisson process with a mean interarrival time of 20 min. Customers spend an average of 15 min in the barber's chair.
    (a) What is the expected number of customers in the barber shop in the queue?
    (b) What is the probability that a customer will not have to wait for a hair cut?
    (c) How much can a customer expect to spend in the barber shop?
    (d) Management will put another chair and hire another barber when a customer's average waiting time in the shop exceeds 1.25 h. How much must the average rate of arrivals increase to warrant a second barber?

(e) What is average time customers spend in the queue?

(f) What is the probability that the waiting time in the system is greater than 30 min?

(g) What is the probability that there are more than 3 customers in the system?

43. If customers arrive for service according to a Poisson distribution at the average rate of 5 per day, how fast must they be serviced on the average (assume exponential service time) in order to keep the average number of customers in the system less than 4?

44. Patients arrive at an hospital for emergency service at the rate of one every hour. Currently only one emergency can be handled at a time. Patients spend an average of 20 min for receiving emergency service. How much the average service time need to be decreased to keep the average time to wait and receive the service less then 25 min?

45. A departmental secretary receives an average of 8 jobs per hour. Many are short jobs, while others are quite long. Assume, however, that the time to perform a job has an exponential distribution with a mean of 6 min.

(a) What is the average elapsed time from the time the secretary receives a job until it is completed?

(b) Calculate $E(N)$, $E(W_q)$, $P(W > 2h)$, $P(N > 5)$ and the fraction of time the secretary is busy.

46. A service station expects a customer every 4 min on the average. Service takes, on the average, 3 min. Assume Poisson input and exponential service.

(a) What is the average number of customers waiting for service?

(b) How long can a customer expect to wait for service?

(c) What is the probability that a customer will spend less than 15 min waiting for and getting service?

(d) What is the probability that a customer will spend longer than 10 min waiting for and getting service?

47. A dress shop has 3 sales persons. Assume that arrivals follow Poisson pattern with an average of 10 min between arrivals. Also assume that any salesperson can provide the desired service for any customer. If the time to provide service for a customer is exponentially distributed with a mean of 20 min per customer, calculate $E(N)$, $E(N_q)$ $E(W)$, $E(W_q)$ and P_n for $n = 0$, 1 and 2.

48. If the mean arrival rate is 24 per hour, find from the customer's point of view of the time spent in the system, whether 3 channels in parallel with a mean service rate of 10 per hour is better or worse than a single channel with mean service rate of 30 per hour.

49. Four counters are being run on the frontier of a country to check the passports and necessary papers of the tourists. The tourists choose a counter at random. If the arrivals at the frontier is Poisson at the rate of λ and the

service time is exponential with parameter $\frac{\lambda}{2}$, what is the steady-state average queue at each counter?

50. An insurance company has 3 claim registers in its branch office. People with claims against the company are found to arrive in a Poisson fashion at an average rate of 20/8-h day. The amount of time that an adjuster spends with a claimant is found to have exponential distribution with mean service time 40 min. Claimants are processed in the order of their appearance.
    (a) How many hours a week can an adjuster expect to spend with claimants?
    (b) How much time, on the average, does a claimant spend in the branch office?
51. A telephone exchange has 2 long distance operators. The telephone company finds that during the peak load, long distance calls arrive in a Poisson fashion at an average rate of 15 per hour. The length of service on these calls is approximately exponentially distributed with mean length 5 min.
    (a) What is the probability that a subscriber will have to wait for his long distance call during the peak hours of the day?
    (b) If the subscribers will wait and are serviced in turn, what is the expected waiting time?
52. A petrol pump station has 4 pumps. The service times follow the exponential distribution with a mean of 6 min and cars arrive for service in a Poisson process at the rate of 30 cars per hour.
    (a) What is the probability that an arrival would have to wait in line?
    (b) Find the average waiting time in the queue, average time spent in the system and the average number of cars in the system.
    (c) For what percentage of time would a pump be idle on an average?
53. A one-person barber shop has 6 chairs to accommodate people waiting for a hair cut. Assume that castomers who arrive when all the 6 chairs are full leave without entering the barber shop. Customers arrive at the average rate of 3 per hour and spend an average of 15 min in the barber's chair.
    (a) What is the probability that a customer can get directly into the barber's chair upon arrival?
    (b) What is the expected number of customers waiting for a hair cut?
    (c) How much time can a customer expect to spend in the barber shop?
    (d) What fraction of potential customers are turned away?
54. Assume that the goods trains are coming in a yard at the rate of 30 trains per day and suppose that the inter-arrival times follow an exponential distribution. The service time for each train is assumed to be exponential with an average of 36 min. If the yard can admit 9 trains at a time, calculate the probability that the yard is empty and the average queue length.

55. A car park contains 5 cars. The arrival of cars is Poisson at a mean rate of 10 per hour. The length of time each car spends in the car park has negative exponential distribution with mean of 2 min. How many cars are in the car park on an average and what is the probability of a newly arriving customer finding the car park full and leaving to park his car elsewhere.
56. A stenographer is attached to 5 officers for whom she performs stenographic work. She gets calls from the officers at the rate of 4 per hour and takes on the average 10 min to attend to each call. If arrival rate is Poisson and service time is exponential find (a) the average waiting time for an arriving call (b) the average number of waiting calls and (c) the average time an arriving call spends in the system.
57. A 2-person barber shop has 5 chairs to accommodate waiting customers. Potential customers, who arrive when all the 5 chairs are full, leave without entering the barber shop. Customers arrive at the average rate of 3.7634 per hour and spend an average of 15 min in the barber's chair. Compute P_0, P_1, P_7, $E(N_q)$ and $E(W)$.
58. A barber shop has 2 barbers and 3 chairs for waiting customers. Assume that customers arrive in Poisson fashion at a rate of 5 per hour and that each barber services customers according to exponential distribution with mean of 15 min. Further, if a customer arrives and there are no empty chairs in the shop, he will leave. Find the steady-state probabilities. What is the probability that the shop is empty? What is the expected number of customers in the shop?
59. An automobile inspection station has 3 inspection stalls. Assume that cars wait in such a way that when a stall becomes vacant, the car at the head of the line pulls up to it. The station can accommodate at most 4 cars waiting (7 in the station) at one time. The arrival pattern is Poisson with a mean of 1 car every minute during the peak hours. The service time is exponential with mean 6 min. Find the average number of customers in the system during peak hours, the average waiting time and the average number of cars per hour that cannot enter the house because of full capacity.
60. A mechanic repairs 4 machines. The mean time between service requirements is 5 h for each machine and forms an exponential distribution. The mean repair time is 1 h and also follows the same distribution pattern. Machine downtime costs Rs. 25 per hour and the machanic costs Rs 55 per day.
    (a) Find the expected number of machines under service and in waiting.
    (b) Determine the expected downtime cost per 8-h day?
    (c) Would it be economical to engage 2 mechanics, each repairing only 2 machines?

Answers

Exercise 9

11. 5 min 12. 0.3679 19. 50% 20. $\frac{1}{9}$ 21. $\frac{8}{9}, \frac{26}{9}$

22. 2 23. $\frac{4}{9}$ 24. $\frac{5}{9}$ 26. 10 min 29. $\frac{81}{121}, \frac{1}{5}$

34. 2.11
41. (a) 0.43 (b) 0.3 (c) 0.097 (d) 0.3
    (e) 0.16 person per minute
42. (a) 3; 2.25 (b) 0.25 (c) 1 h (d) 3.2 per hour
    (e) $\frac{3}{4}$ h (f) 0.61 (g) 0.32
43. 3.84 h/service
44. 17.65 min per patient
45. (a) 30 min (b) 4; 24 min; 0.0183; 0.2621; 0.8
46. (a) 2.25 (b) 9 min (c) 0.7135 (d) 0.8465
47. 2.1739; 0.1739; 21.739 min; 1.739 min; $P_0 = 0.1304$, $P_1 = 0.2608$; $P_2 = 0.2602$
48. Single channel is better
49. $\frac{4}{23}$
50. 22.2 h; 49 min
51. (a) 0.48 (b) 3.2 min
52. (a) 0.3826 (b) 3.05 min; 9.054 min; 4.53 cars; (c) 24.98%
53. (a) 0.2778 (b) 1.3878 (c) 43.8 min (d) 3.7%
54. 0.28; 1.55
55. 0.49, 0.0027
56. (a) 12.45 min (b) 0.79 customer (c) 22.42 min
57. 0.36133; 0.33996; 0.00368; 0.2457; 19 min
58. $P_n = (0.56)(0.625)^n$, for $2 \le n \le 5$; $P_1 = 0.35$; $P_0 = 0.28$; 2.956
59. 6.06 cars; 12.3 min; 30.4 cars
60. (a) 1 (b) Total cost = Rs 255
    (c) Total cost with 2 machines = Rs. 270; Use of 2 machines is not economical.

10 Design of Experiments

By 'experiment', we mean collection of data (which usually consist of a series of measurements of some feature of an object) for a scientific investigation, according to certain specified sampling procedures. Statistics provides not only the principles and the basis for the proper planning of the experiments but also the methods for proper interpretation of the results of the experiment.

In the beginning, the study of the design of experiments was associated only with agricultural experimentation. The need to save time and money has led to a study of ways to obtain maximum information with the minimum cost and labour. Such motivations resulted in the subsequent acceptance and wide use of the design of experiments and the related analysis of variance techniques in all fields of scientific experimentation. In this chapter we consider some aspects of experimental design briefly and analysis of data from such experiments using analysis of variance techniques.

Aim of the Design of Experiments

A statistical experiment in any field is performed to verify a particular hypothesis. For example, an agricultural experiment may be performed to verify the claim that a particular manure has got the effect of increasing the yield of paddy. Here the quantity of the manure used and the amount of yield are the two variables involved directly. They are called *experimental variables*. Apart from these two, there are other variables such as the fertility of the soil, the quality of the seed used and the amount of rainfall, which also affect the yield of paddy. Such variables are called *extraneous variables*. The main aim of the design of experiments is to control the extraneous variables and hence to minimise the experimental error so that the results of the experiments could be attributed only to the experimental variables.

Basic Principles of Experimental Design

In order to achieve the objective mentioned above, the following three principles are adopted while designing the experiments—(1) randomisation, (2) replication and (3) local control.

1. *Randomisation* As it is not possible to eliminate completely the contribution of extraneous variables to the value of the response variable (the amount of yield of paddy), we try to control it by randomisation. The group of experimental units (plots of the same size) in which the manure is used is called the *experimental group* and the other group of plots in which the manure is not used and which will provide a basis for comparison is called the *control group*. If any information regarding the extraneous variables and the nature and magnitude of their effect on the response variable in question is not available, we resort to randomisation. That is, we select the plots for the experimental and control groups in a random manner, which provides the most effective way of eliminating any unknown bias in the experiment.

2. *Replication* In a comparative experiment, in which the effects of different manures on the yield are studied, each manure is used in more than one plot. In other words, we resort to replication which means repetition. It is essential to carry out more than one test on each manure in order to estimate the amount of the experimental error and hence to get some idea of the precision of the estimates of the manure effects.

3. *Local Control* To provide adequate control of extraneous variables, another essential principle used in the experimental design is the local control. This includes techniques such as grouping, blocking and balancing of the experimental units used in the experimental design. By *grouping*, we mean combining sets of homogeneous plots into groups, so that different manures may be used in different groups. The number of plots in different groups need not necessarily be the same. By *blocking*, we mean assigning the same number of plots in different blocks. The plots in the same block may be assumed to be relatively homogeneous. We use as many manures as the number of plots in a block in a random manner. By *balancing*, we mean adjusting the procedures of grouping, blocking and assigning the manures in such a manner that a balanced configuration is obtained.

Some Basic Designs of Experiment

1. Completely Randomised Design (C.R.D.)

Let us suppose that we wish to compare 'h' treatments (use of 'h' different manures) and there are n plots available for the experiment.

Let the ith treatment be replicated (repeated) n_i times, so that $n_1 + n_2 + \ldots + n_h = n$.

The plots to which the different treatments are to be given are found by the following randomisation principle. The plots are numbered from 1 to n serially. n identical cards are taken, numbered from 1 to n and shuffled thoroughly. The numbers on the first n_1 cards drawn randomly give the numbers of the plots to which the first treatment is to be given. The numbers on the next n_2 cards drawn at random give the numbers of the plots to which the second treatment is to be given and so on. This design is called a completely randomised design, which is used when the plots are homogeneous or the pattern of heterogeneity of the plots is unknown.

Analysis of Variance (ANOVA) The analysis of variance is a widely used technique developed by R.A. Fisher. It enables us to divide the total variation (represented by variance) in a group into parts which are ascribable to different factors and a residual random variation which could not be accounted for by any of these factors. The variation due to any specific factor is compared with the residual variation for significance by applying the F-test, with which the reader is familiar. The details of the procedure will be explained in the sequel.

Analysis of Variance for One Factor of Classification Let a sample of N values of a given random variable X (representing the yield of paddy) be subdivided into 'h' classes according to some factor of classification (different manures).

We wish to test the null hypothesis that the factor of classification has no effect on the variable, viz., there is no difference between various classes, viz., the classes are homogeneous. Let x_{ij} be the value of the jth member of the ith class, which contains n_i members. Let the general mean of all the N values be $\bar{x}$ and the mean of n_i values in the ith class be $\bar{x}_i$.

$$\text{Now} \quad \sum_i \sum_j (x_{ij} - \bar{x})^2 = \sum_i \sum_j \left\{(x_{ij} - \bar{x}_i) + (x_i - \bar{x})\right\}^2$$

$$= \sum_i \sum_j (x_{ij} - \bar{x}_i)^2 + \sum_i \sum_j (\bar{x}_i - \bar{x})^2$$

$$+ 2 \sum_i \sum_j (x_{ij} - \bar{x}_i)\,(\bar{x}_i - \bar{x})$$

$$= \sum_i \sum_j (x_{ij} - \bar{x}_i)^2 + \sum_i \sum_{j=1}^{n_i} (\bar{x}_i - \bar{x})^2$$

$$+ 2 \sum_i (\bar{x}_i - \bar{x}) \sum_{j=1}^{n_i} (x_{ij} - \bar{x}_i)$$

$$= \sum_i \sum_j (x_{ij} - \bar{x}_i)^2 + \sum_i n_i (\bar{x}_i - \bar{x})^2$$

$$\left[\because \sum_{j=1}^{n_i} (x_{ij} - \bar{x}_i) = \text{sum of the deviations of } n_i \text{ values of the } x_{ij} \text{ in the } i^{\text{th}} \text{ class from their mean } \bar{x}_i = 0\right]$$

i.e., $Q = Q_2 + Q_1$, say, where

$Q_1 = \sum_i n_i (\bar{x}_i - \bar{x})^2$ = sum of the squared deviations of class means from the general mean (variation between classes)

$$Q_2 = \sum_i \sum_j (x_{ij} - \bar{x}_i)^2 = \text{sum of the squared deviations of variates from the}$$

corresponding class means (variation within classes) and Q = total variation.

Since $Q_2 = Q - Q_1$, viz., the variation Q_2 within classes is got after removing the variation Q_1 between classes from the total variation Q, Q_2 is the residual variation.

If s^2 is the variance of a sample of size n drawn from a population with variance σ^2, then it is known from the theory of estimation that $\left(\frac{ns^2}{n-1}\right)$ is an unbiased estimate of σ^2.

i.e., $$E\left(\frac{ns^2}{n-1}\right) = \sigma^2.$$

Since the items in the ith class with variance $\frac{1}{n_i}\sum_{j=1}^{n_i}(x_{ij} - \bar{x}_i)^2$ may be considered as a sample of size n_i drawn from a population with variance σ^2,

$$E\left\{\frac{n_i}{n_i - 1} \cdot \frac{1}{n_i}\sum_{j=1}^{n_i}(x_{ij} - \bar{x}_i)^2\right\} = \sigma^2$$

i.e., $$E\left[\sum_j (x_{ij} - \bar{x}_i)^2\right] = (n_i - 1)\,\sigma^2$$

$\therefore$ $$E\left[\sum_i \sum_j (x_{ij} - \bar{x}_i)^2\right] = \sum_{i=1}^{h}(n_i - 1)\,\sigma^2$$

i.e., $$E(Q_2) = (N - h)\,\sigma^2 \text{ or } E\left\{\frac{(Q)_2}{N - h}\right\} = \sigma^2$$

i.e., $\frac{Q_2}{N-h}$ is an unbiased estimate of σ^2 with $(N - h)$ degrees of freedom.

Now if we consider the entire group of N items with variance $\frac{1}{N}\sum_i \sum_j (x_{ij} - \bar{x})^2$ as a sample of size N drawn from the same population,

$$E\left\{\frac{N}{N - 1} \cdot \frac{1}{N}\sum_i \sum_j (x_{ij} - \bar{x})^2\right\} = \sigma^2$$

i.e., $$E\left(\frac{Q}{N-1}\right) = \sigma^2$$

i.e., $\frac{Q}{N-1}$ is an unbiased estimate of σ^2 with $(N - 1)$ degrees of freedom.

Now $\quad Q_1 = Q - Q_2$

$\therefore \quad E(Q_1) = E(Q) - E(Q_2)$

$$= (N-1)\sigma^2 - (N-h)\sigma^2$$

$$= (h-1)\,\sigma^2 \text{ or } E\left(\frac{Q_1}{h-1}\right) = \sigma^2$$

i.e., $\frac{Q_1}{h-1}$ is also an unbiased estimate of σ^2 with $(h-1)$ degrees of freedom.

If we assume that the sampled population is normal, then the estimates $\frac{Q_1}{h-1}$ and $\frac{Q_2}{N-h}$ are independent and hence the ratio $\frac{Q_1/(h-1)}{Q_2/(N-h)}$ follows a F-distribution with $(h-1, N-h)$ degrees of freedom or the ratio $\frac{Q_2/(N-h)}{Q_1/(h-1)}$ follows a F-distribution with $(N-h, h-1)$ degrees of freedom. Choosing the ratio which is greater than one, we employ the F-test.

If the calculated value of $F < F_{5\%}$, the null hypothesis is accepted, viz., different treatments do not contribute significantly different yields.

These results are displayed in the form of a table, called the ANOVA table, as given below:

Table 10.1 *ANOVA table for one factor of classification*

Source of variation (S.V.)	*Sum of squares (S.S.)*	*Degrees of freedom (d.f.)*	*Mean square (M.S.)*	*Variance ratio (F)*
Between classes	Q_1	$h-1$	$Q_1/(h-1)$	$\frac{Q_1/(h-1)}{Q_2/(N-h)}$
				(OR)
Within classes	Q_2	$N-h$	$Q_2/(N-h)$	$\frac{Q_2/(N-h)}{Q_1/(h-1)}$
Total	Q	$N-1$	—	—

Note *For calculating Q, Q_1, Q_2, the following computational formulas may be used:*

$$Q = N\left\{\frac{1}{N}\sum\sum x_{ij}^2 - \bar{x}^2\right\}$$

$$= N\left\{\frac{1}{N}\sum\sum x_{ij}^2 - \left(\frac{1}{N}\sum\sum x_{ij}\right)^2\right\}$$

$$= \sum\sum x_{ij}^2 - \frac{T^2}{N}, \text{ where } T = \sum\sum x_{ij}$$

Similarly, for the ith class,

$$\sum_j (x_{ij} - \bar{x}_i)^2 = \sum_j x_{ij}^2 - \frac{T_i^2}{n_i}, \text{ where } T_i = \sum_j x_{ij}.$$

$$\therefore \qquad Q_2 = \sum_i \sum_j (x_{ij} - \bar{x}_i)^2 = \sum_i \sum_j x_{ij}^2 - \sum_i \frac{T_i^2}{n_i}$$

Hence

$$Q_1 = Q - Q_2$$

$$= \sum_i \frac{T_i^2}{n_i} - \frac{T^2}{N}$$

2. *Randomised Block Design (R.B.D.)*

Let us consider an agricultural experiment using which we wish to test the effect of 'k' fertilizing treatments on the yield of a crop. We assume that we know some information about the soil fertility of the plots. Then we divide the plots into 'h' blocks, according to the soil fertility, each block containing 'k' plots. Thus the plots in each block will be of homogeneous fertility as far as possible.

Within each block, the 'k' treatments are given to the 'k' plots in a perfectly random manner, such that each treatment occurs only once in any block. But the same 'k' treatments are repeated from block to block. This design is called Randomised Block Design.

Analysis of Variance for two Factors of Classification Let the N variate values $\{x_{ij}\}$ (representing the yield of paddy) be classified according to two factors. Let there be 'h' rows (blocks) representing one factor of classification (soil fertility) and 'k' columns representing the other factor (treatment), so that $N = hk$.

We wish to test the null hypothesis that the rows and columns are homogeneous viz., there is no difference in the yields of paddy between the various rows and between the various columns.

Let x_{ij} be the variate value in the ith row and j^{th} column.

Let $\bar{x}$ be the general mean of all the N values, $\bar{x}_{i*}$ be the mean of the k values in the ith row and $\bar{x}_{*j}$ be the mean of the h values in the jth column.

Now $$x_{ij} - \bar{x} = (x_{ij} - \bar{x}_{i*} - \bar{x}_{*j} + \bar{x}) + (\bar{x}_{i*} - \bar{x}) + (\bar{x}_{*j} - \bar{x})$$

$$\therefore \sum\sum (x_{ij} - \bar{x})^2 = \sum\sum (x_{ij} - \bar{x}_{i*} - \bar{x}_{*j} + \bar{x})^2 + \sum\sum (\bar{x}_{i*} - \bar{x})^2$$
$$+ \sum\sum (\bar{x}_{*j} - \bar{x})^2 + 2\sum\sum (x_{ij} - \bar{x}_{i*} - \bar{x}_{*j} + \bar{x})(\bar{x}_{i*} - \bar{x})$$
$$+ 2\sum\sum (x_{ij} - \bar{x}_{i*} - \bar{x}_{*j} + \bar{x})(\bar{x}_{*j} - \bar{x})$$
$$+ 2\sum\sum (\bar{x}_{i*} - \bar{x})(\bar{x}_{*j} - \bar{x}) \qquad (1)$$

Now, the fourth member in the R.H.S. of (1)

$$= 2\sum_i (\bar{x}_{i*} - \bar{x}) \sum_{j=1}^{k} (x_{ij} - \bar{x}_{i*} - \bar{x}_{*j} + \bar{x})$$

$$= 2\sum_i (\bar{x}_{i*} - \bar{x})(k\,\bar{x}_{i*} - k\,\bar{x}_{i*} - k\,\bar{x} + k\,\bar{x})$$
$$= 0$$

Similarly, the last two members in the R.H.S. of (1) also become zero each.

Also $\sum_i \sum_j (\bar{x}_{i*} - \bar{x})^2 = k \sum_i (\bar{x}_{i*} - \bar{x})^2 = Q_1$, say

$$\sum_i \sum_j (\bar{x}_{*j} - \bar{x})^2 = h \sum_j (\bar{x}_{*j} - \bar{x})^2 = Q_2, \text{ say}$$

Let $Q = \Sigma\Sigma(x_{ij} - \bar{x})^2$ and

$$Q_3 = \Sigma\Sigma(x_{ij} - \bar{x}_{i*} - \bar{x}_{*j} + \bar{x})^2$$

Using all these in (1), we get

$Q = Q_1 + Q_2 + Q_3$, where

Q = total variation.

Q_1 = sum of the squares due to the variations in the rows,

Q_2 = that in the columns and

Q_3 = that due to the residual variations.

Proceeding as in one factor of classification, we can prove that $\frac{Q_1}{h-1}, \frac{Q_2}{k-1}, \frac{Q_3}{(h-1)(k-1)}$ and $\frac{Q}{hk-1}$ are unbiased estimates of the population variance σ^2 with degrees of freedom $h - 1$, $k - 1$, $(h - 1)(k - 1)$ and $(hk - 1)$ respectively. If the sampled population is assumed normal, all these estimates are independent.

$\therefore \frac{Q_1/(h-1)}{Q_3'/(h-1)(k-1)}$ follows a F-distribution with $\{h - 1, (h - 1)(k - 1)\}$ degrees of freedom and $\frac{Q_2/(k-1)}{Q_3/(h-1)(k-1)}$ follows a F-distribution with $\{k - 1, (h - 1)(k - 1)\}$ degrees of freedom. Then the F-tests are applied as usual and the significance of difference between rows and between columns is analysed.

Table 10.2 *The ANOVA table for the two factors of classifications*

S.V.	*S.S.*	*d.f.*	*M.S.*	*F*
Between rows	Q_1	$h - 1$	$Q_1/(h-1)$	$\left[\frac{Q_1/(h-1)}{Q_3/(h-1)(k-1)}\right]^{\pm 1}$
Between columns	Q_2	$k - 1$	$Q_2/(k-1)$	$\left[\frac{Q_2/(k-1)}{Q_3/(h-1)(k-1)}\right]^{\pm 1}$
Residual	Q_3	$(h-1)(k-1)$	$Q_3/(h-1)(k-1)$	—
Total	Q	$hk - 1$	—	—

Note *The following working formulas that can be easily derived may be used to compute Q, Q_1, Q_2 and Q_3:*

1. $Q = \sum\sum x_{ij}^2 - \frac{T^2}{N}$, where $T = \sum\sum x_{ij}$
2. $Q_1 = \frac{1}{k}\sum T_i^2 - \frac{T^2}{N}$, where $T_i = \sum_{j=1}^{k} x_{ij}$
3. $Q_2 = \frac{1}{h}\sum T_j^2 - \frac{T^2}{N}$, where $T_j = \sum_{i=1}^{h} x_{ij}$
4. $Q_3 = Q - Q_1 - Q_2$

It may be verified that $\sum_i T_i = \sum_j T_j = T$.

3. Latin Square Design (L.S.D.)

We consider an agricultural experiment, in which n^2 plots are taken and arranged in the form of an $n \times n$ square, such that the plots in each row will be homogeneous as far as possible with respect to one factor of classification, say, soil fertility and plots in each column will be homogeneous as far as possible with respect to another factor of classification, say, seed quality.

Then n treatments are given to these plots such that each treatment occurs only once in each row and only once in each column. The various possible arrangements obtained in this manner are known as Latin squares of order n. This design of experiment is called the Latin Square Design.

Analysis of Variance for Three Factors of Classifications Let the $N (= n^2)$ variate values $\{x_{ij}\}$, representing the yield of paddy, be classified according to three factors. Let the rows, columns and letters stand for the three factors, say soil fertility, seed quality and treatment respectively.

We wish to test the null hypothesis that the rows, columns and letters are homogeneous, viz., there is no difference in the yield of paddy between the rows (due to soil fertility), between the columns (due to seed quality) and between the letters (due to the treatments).

Let x_{ij} be the variate value corresponding to the ith row, j^{th} column and k^{th} letter.

Let $\bar{x} = \frac{1}{n^2}\sum\sum x_{ij}$, $\bar{x}_{i*} = \frac{1}{n}\sum_j x_{ij}$, $\bar{x}_{*j} = \frac{1}{n}\sum_i x_{ij}$ and $\bar{x}_k$ be the mean of the values of x_{ij} corresponding to the kth treatment.

Now $x_{ij} - \bar{x} = (\bar{x}_{i*} - \bar{x}) + (\bar{x}_{*j} - \bar{x}) + (\bar{x}_k - \bar{x}) + (x_{ij} - \bar{x}_{i*} - \bar{x}_{*j} - \bar{x}_k + 2\bar{x})$

$$\therefore \quad \sum\sum (x_{ij} - \bar{x})^2 = n\sum_i (\bar{x}_{i*} - \bar{x})^2 + n\sum_j (\bar{x}_{*j} - \bar{x})^2 + n\sum_k (\bar{x}_k - \bar{x})^2 + \sum_i\sum_j (x_{ij} - \bar{x}_{i*} - \bar{x}_{*j} - \bar{x}_k + 2\bar{x})^2$$

($\because$ all the product terms vanish as in two factor classification)

i.e., $$Q = Q_1 + Q_2 + Q_3 + Q_4$$

As before we can prove that $\frac{Q_1}{n-1}, \frac{Q_2}{n-1}, \frac{Q_3}{n-1}, \frac{Q_4}{(n-1)(n-2)}$ and $\frac{Q}{n^2-1}$ are unbiased estimates of the population variance σ^2 with degrees of freedom $n-1, n-1, n-1, (n-1)(n-2)$ and (n^2-1) respectively.

If the sampled population is assumed normal, all these estimates are independent.

$\therefore$ Each of $\frac{Q_1/(n-1)}{Q_4/(n-1)(n-2)}, \frac{Q_2/(n-1)}{Q_4/(n-1)(n-2)}$ and $\frac{Q_3/(n-1)}{Q_4/(n-1)(n-2)}$ follows a F-distribution with $\{(n-1), (n-1)(n-2)\}$ degrees of freedom.

Then the F-tests are applied as usual and the significance of differences between rows, columns and treatments is analysed.

Table 10.3 *The ANOVA table for three factors of classification*

S.V.	*S.S.*	*d.f.*	*M.S.*	*F*
Between rows	Q_1	$n-1$	$Q_1/(n-1) = M_1$	$\left(\frac{M_1}{M_4}\right)^{\pm 1}$
Between columns	Q_2	$n-1$	$Q_2/(n-1) = M_2$	$\left(\frac{M_2}{M_4}\right)^{\pm 1}$
Between letters	Q_3	$n-1$	$Q_3/(n-1) = M_3$	$\left(\frac{M_3}{M_4}\right)^{\pm 1}$
Residual	Q_4	$(n-1)(n-2)$	$Q_4/(n-1)(n-2) = M_4$	—
Total	Q	n^2-1	—	—

Note *The following working formulas may be used to compute the Q's:*

1. $Q = \sum\sum x_{ij}^2 - \frac{T^2}{n^2}$, *where* $T = \sum\sum x_{ij}$
2. $Q_1 = \frac{1}{n}\sum T_i^2 - \frac{T^2}{n^2}$, *where* $T_i = \sum_{j=1}^{n} x_{ij}$
3. $Q_2 = \frac{1}{n}\sum T_j^2 - \frac{T^2}{n^2}$, *where* $T_j = \sum_{i=1}^{n} x_{ij}$
4. $Q_3 = \frac{1}{n}\sum T_k^2 - \frac{T^2}{n^2}$, *where* T_k *is the sum of all* x_{ij}*'s receiving the* k^{th} *treatment.*
5. $Q_4 = Q - Q_1 - Q_2 - Q_3$.

Also $T = \sum_i T_i = \sum_j T_j = \sum_k T_k$

Comparison of RBD and LSD

1. The number of replications of each treatment is equal to the number of treatments in LSD, whereas there is no such restrictions on treatments and replication in RBD.
2. LSD can be performed on a square field, while RBD can be performed either on a square field or a rectangular field.
3. LSD is known to be suitable for the case when the number of treatments is between 5 and 12, whereas RBD can be used for any number of treatments.
4. The main advantage of LSD is that it controls the effect of two extraneous variables, whereas RBD controls the effect of only one extraneous variable. Hence the experimental error is reduced to a larger extent in LSD than in RBD.

Note on Simplification of Computational Work

The variance of a set of values is independent of the origin and so a shift of origin does not affect the variance calculations. Hence in analysis of variance problems, we can subtract a convenient number from the original values and work out the problems with the new values obtained. Also since we are concerned with the variance ratios, change of scale also may be introduced without affecting the values of F.

Worked Examples 10

Example 1 *A completely randomised design experiment with 10 plots and 3 treatments gave the following results:*

Plot No.	:	*1*	*2*	*3*	*4*	*5*	*6*	*7*	*8*	*9*	*10*
Treatment	:	*A*	*B*	*C*	*A*	*C*	*C*	*A*	*B*	*A*	*B*
Yield	:	*5*	*4*	*3*	*7*	*5*	*1*	*3*	*4*	*1*	*7*

Analyse the results for treatment effects.

Solution Rearranging the data according to the treatments, we have the following table:

Treatment	*Yield from plots* (x_{ij})				T_i	T_i^2	n_i	$\frac{T_i^2}{n_i}$
A	5	7	3	1	16	256	4	64
B	4	4	7	–	15	225	3	75
C	3	5	1	–	9	81	3	27
		Total			$T = 40$	–	$N = 10$	166

$$\sum\sum x_{ij}^2 = (25 + 49 + 9 + 1) + (16 + 16 + 49) + (9 + 25 + 1)$$
$$= 84 + 81 + 35 = 200$$

$$Q = \sum\sum x_{ij}^2 - \frac{T^2}{N} = 200 - \frac{40^2}{10} = 200 - 160 = 40$$

$$Q_1 = \sum \frac{T_i^2}{n_i} - \frac{T^2}{N} = 166 - 160 = 6$$

$$\therefore \quad Q_2 = Q - Q_1 = 40 - 6 = 34$$

ANOVA table

S.V.	*S.S.*	*d.f.*	*M.S.*	F_0
Between classes (treatments)	$Q_1 = 6$	$h - 1 = 2$	3.0	$\frac{4.86}{3.0}$
Within classes	$Q_2 = 34$	$N - h = 7$	4.86	= 1.62
Total	$Q = 40$	$N - 1 = 9$	–	–

From the F-table, $F_{5\%}\ (v_1 = v_2 = 2) = 19.35$

We note that $F_0 < F_{5\%}$

Let H_0 : The treatments do not differ significantly.

$\therefore$ The null hypothesis is accepted.

i.e., the treatments are not significantly different.

Example 2 *The following table shows the lives in hours of four brands of electric lamps:*

Brand

A : 1610, 1610, 1650, 1680, 1700, 1720, 1800

B : 1580, 1640, 1640, 1700, 1750

C : 1460, 1550, 1600, 1620, 1640, 1660, 1740, 1820

D : 1510, 1520, 1530, 1570, 1600, 1680

Perform an analysis of variance and test the homogeneity of the mean lives of the four brands of lamps.

Solution We subtract 1640 (= the average of the extreme values) from the given values and work out with the new values of x_{ij}

Brand				*Lives of lamps* (x_{ij})					T_i	n_i	$\frac{T_i^2}{n_i}$
A	– 30	– 30	10	40	60	80	160	–	290	7	12014
B	– 60	0	0	60	110	–	–	–	110	5	2420
C	– 180	– 90	– 40	– 20	0	20	100	180	– 30	8	113
D	– 130	– 120	– 110	– 70	– 40	40	–	–	– 430	6	30817
								Total	– 60	26	45364

$$\sum\sum x_{ij}^2 = (900 + 900 + 100 + 1600 + 3600 + 6400 + 25600)$$
$$+ (3600 + 0 + 0 + 3600 + 12100)$$
$$+ (32400 + 8100 + 1600 + 400 + 0 + 400 + 10000 + 32400)$$
$$+ (16900 + 14400 + 12100 + 4900 + 1600 + 1600)$$
$$= 39100 + 19300 + 85300 + 51500 = 195200$$

$$Q = \sum\sum x_{ij}^2 - \frac{T^2}{N} = 1{,}95{,}200 - 138 = 1{,}95{,}062$$

$$Q_1 = \sum \frac{T_i^2}{n_i} - \frac{T^2}{N} = 45{,}364 - 138 = 45{,}226$$

$$Q_2 = Q - Q_1 = 1{,}95{,}062 - 45{,}226 = 1{,}49{,}836$$

ANOVA table

S.V.	*S.S.*	*d.f.*	*M.S.*	F_0
Between brands	$Q_1 = 45{,}226$	$h - 1 = 3$	15,075	$\frac{15{,}075}{6{,}811}$
Within brands	$Q_2 = 1{,}49{,}836$	$N - h = 22$	6,811	= 2.21
Total	$Q = 1{,}95{,}062$	$N - 1 = 25$	–	–

From the F-tables, $F_{5\%}\,(v_1 = 3, v_2 = 22) = 3.06$

$$F_0 < F_{5\%}$$

Hence the null hypothesis H_0, namely, the means of the lives of the four brands are homogeneous, is accepted viz., the lives of the four brands of lamps do not differ significantly.

Note *We could have used a change of scale also. viz., we could have made the change*

$$New\ x_{ij} = \frac{old\ x_{ij} - 1640}{10}$$ *and simplified the numerical work still further.*

Example 3 *A car rental agency, which uses 5 different brands of tyres in the process of deciding the brand of tyre to purchase as standard equipment for its fleet, finds that each of 5 tyres of each brand last the following number of kilometres (in thousands):*

Tyre brands

A	*B*	*C*	*D*	*E*
36	*46*	*35*	*45*	*41*
37	*39*	*42*	*36*	*39*
42	*35*	*37*	*39*	*37*
38	*37*	*43*	*35*	*35*
47	*43*	*38*	*32*	*38*

Test the hypothesis that the five tyre brands have almost the same average life.

Solution We shift the origin to 40 and work out with the new values of x_{ij}.

Tyre brand	x_{ij}					T_i	n_i	$\frac{T_i^2}{n_i}$	$\sum_{j=1}^{5} x_{ij}^2$
A	−4	−3	2	−2	7	0	5	0	82
B	6	−1	−5	−3	3	0	5	0	80
C	−5	2	−3	3	−2	−5	5	5	51
D	5	−4	−1	−5	−8	−13	5	33.8	131
E	1	−1	−3	−5	−2	−10	5	20	40
	Total					−28	25	58.8	384

$$T = \sum_i T_i = -28; \quad \sum\sum x_{ij}^2 = \sum_i \left(\sum_j x_{ij}^2 \right) = 384$$

$$Q = \sum\sum x_{ij}^2 - \frac{T^2}{N} = 384 - \frac{(-28)^2}{25} = 352.64$$

$$Q_1 = \sum \frac{T_i^2}{n_i} - \frac{T^2}{N} = 58.8 - 31.36 = 27.44$$

$$Q_2 = Q - Q_1 = 352.64 - 27.44 = 325.20$$

ANOVA table

S.V.	*S.S.*	*d.f.*	*M.S.*	F_0
Between tyre brands	$Q_1 = 27.44$	$h - 1 = 4$	6.86	$\frac{16.26}{6.86}$
Within tyre brands	$Q_2 = 325.20$	$N - h = 20$	16.26	$= 2.37$
Total	$Q = 352.64$	$N - 1 = 24$	–	–

From the F-tables, $F_{5\%}$ $(v_1 = 20, v_2 = 4) = 5.80$

$$F_0 < F_{5\%}$$

Hence 'H_0 : the five tyre brands have almost the same average life) is accepted viz., the five tyre brands do not differ significantly in their lives.

Example 4 *In order to determine whether there is significant difference in the durability of 3 makes of computers, samples of size 5 are selected from each make and the frequency of repair during the first year of purchase is observed. The results are as follows :*

Makes

A	*B*	*C*
5	*8*	*7*
6	*10*	*3*
8	*11*	*5*
9	*12*	*4*
7	*4*	*1*

In view of the above data, what conclusion can you draw?

Solution

Make	x_{ij}					T_i	n_i	$\frac{T_i^2}{n_i}$	$\sum_{j=1} x_{ij}^2$
A	5	6	8	9	7	35	5	245	255
B	8	10	11	12	4	45	5	405	445
C	7	3	5	4	1	20	5	80	100
					Total	100	15	730	800

$$T = \sum T_i = 100 \,;\ \sum\sum x_{ij}^2 = 800;\ N = \sum n_i = 15$$

$$Q = \sum\sum x_{ij}^2 - \frac{T^2}{N} = 800 - \frac{100^2}{15} = 133.33$$

$$Q_1 = \sum \frac{T_i^2}{n_i} - \frac{T^2}{N} = 730 - 666.67 = 63.33$$

$$Q_2 = Q - Q_1 = 70$$

ANOVA table

S.V.	*S.S.*	*d.f.*	*M.S.*	F_0
Between makes	$Q_1 = 63.33$	$h - 1 = 2$	31.67	$\frac{31.67}{5.83}$
Within makes	$Q_2 = 70$	$N - h = 12$	5.83	5.43
Total	$Q = 133.33$	$N - 1 = 14$	–	–

From the F-tables, $F_{5\%}$ $(v_1 = 2, v_2 = 12) = 3.88$

$$F_0 > F_{5\%}$$

Hence the null hypothesis (H_0: the 3 makes of computers do not differ in the durability) is rejected.

viz., there is significant difference in the durability of the 3 makes of computers.

Example 5 *Three varieties of a crop are tested in a randomised block design with four replications, the layout being as given below: The yields are given in kilograms. Analyse for significance*

C48	*A51*	*B52*	*A49*
A47	*B49*	*C52*	*C51*
B49	*C53*	*A49*	*B50*

Solution Rewriting the data such that the rows represent the blocks and the columns represent the varieties of the crop (as assumed in the discussion of analysis of variance for two factors of classification), we have the following table:

Crops

Blocks	*A*	*B*	*C*
1	47	49	48
2	51	49	53
3	49	52	52
4	49	50	51

We shift the origin to 50 and work out with the new values of x_{ij}.

Crops

Blocks	*A*	*B*	*C*	T_i	$T^2{}_i/k$	$\sum_j x_{ij}^2$
1	−3	−1	−2	−6	36/3 = 12	14
2	1	−1	3	3	9/3 = 3	11
3	−1	2	2	3	9/3 = 3	9
4	−1	0	1	0	0/3 = 0	2
T_j	−4	0	4	$T = 0$	$\sum \frac{T_j^2}{k} = 18$	36
$T^2{}_j/h$	$\frac{16}{4} = 4$	$\frac{0}{4} = 0$	$\frac{16}{4} = 4$	$\sum \frac{T_i^2}{h} = 8$		
$\sum_i x_{ij}^2$	12	6	18	36		

$$Q = \sum \sum x_{ij}^2 - \frac{T^2}{N} = 36 - \frac{0^2}{12} = 36$$

$$Q_1 = \frac{1}{k} \sum T_i^2 - \frac{T^2}{N} = 18 - 0 = 18$$

$$Q_2 = \frac{1}{h} \sum T_j^2 - \frac{T^2}{N} = 8 - 0 = 8$$

$$Q_3 = Q - Q_1 - Q_2 = 36 - 18 - 8 = 10$$

ANOVA table

S.V.	*S.S.*	*d.f.*	*M.S.*	F_0
Between rows (blocks)	$Q_1 = 18$	$h - 1 = 3$	6	$\frac{6}{1.67} = 3.6$
Between columns (crops)	$Q_2 = 8$	$k - 1 = 2$	4	$\frac{4}{1.67} = 2.4$
Residual	$Q_3 = 10$	$(h - 1)(k - 1) = 6$	1.67	–
Total	Q = 36	$hk - 1 = 11$	–	–

From F-tables, $F_{5\%}$ ($v_1 = 3, v_2 = 6$) = 4.76 and $F_{5\%}$ ($v_1 = 2, v_2 = 6$) = 5.14.

Considering the difference between rows, we see that F_0 (= 3.6) < $F_{5\%}$ (= 4.76).

Hence the difference between the rows is not significant. (H_0 is accepted) viz., the blocks do not differ significantly with respect to the yield.

Considering the difference between columns, we see that F_0 (= 2.4) < $F_{5\%}$ (= 5.14)

Hence the difference between the columns is not significant. (H_0 is accepted) viz., the varieties of crop do not differ significantly with respect to the yield.

Example 6 *Five breeds of cattle B_1, B_2, B_3, B_4, B_5 were fed on four different rations R_1, R_2, R_3, R_4. Gains in weight in kg over a given period were recorded and given below:*

	B_1	B_2	B_3	B_4	B_5
R_1	1.9	2.2	2.6	1.8	2.1
R_2	2.5	1.9	2.3	2.6	2.2
R_3	1.7	1.9	2.2	2.0	2.1
R_4	2.1	1.8	2.5	2.3	2.4

Is there a significant difference between (i) breeds and (ii) rations?

Solution We effect the change of origin and scale using $y_{ij} = \dfrac{x_{ij} - 2}{1/10} = 10(x_{ij} - 2)$

and work out with y_{ij} values.

	B_1	B_2	B_3	B_4	B_5	T_i	$\dfrac{T_i^2}{k}$	$\sum_j y_{ij}^2$
R_1	−1	2	6	−2	1	6	7.2	46
R_2	5	−1	3	6	2	15	45.0	75
R_3	−3	−1	2	0	1	−1	0.2	15
R_4	1	−2	5	3	4	11	24.2	55
T_j	2	−2	16	7	8	$T = 31$	$\sum \dfrac{T_i^2}{k} = 76.6$	191
T_j^2/h	1	1	64	12.25	16	$\sum T_j^2/h = 94.25$		
$\sum_i y_{ij}^2$	36	10	74	49	22	191		

$$Q = \sum\sum y_{ij}^2 - \frac{T^2}{N} = 191 - \frac{(31)^2}{20} = 142.95$$

$$Q_1 = \frac{1}{k}\sum T_i^2 - \frac{T^2}{N} = 76.6 - 48.05 = 28.55$$

$$Q_2 = \frac{1}{h}\sum T_j^2 - \frac{T^2}{N} = 94.25 - 48.05 = 46.20$$

$$Q_3 = Q - Q_1 - Q_2 = 142.95 - (28.55 + 46.20) = 68.20$$

ANOVA table

S.V.	S.S.	d.f.	M.S.	F_0
Between rows (rations)	$Q_1 = 28.55$	$h - 1 = 3$	9.52	9.52/5.68 = 1.68
Between Cols. (breeds)	$Q_2 = 46.20$	$k - 1 = 4$	11.55	11.55/5.68 = 2.03
Residual	$Q_3 = 68.20$	$(h - 1)(k - 1) = 12$	5.68	–
Total	$Q = 142.95$	$hk - 1 = 19$	–	–

From the F-tables, $F_{5\%}$ $(v_1 = 3, v_2 = 12) = 3.49$ and $F_{5\%}$ $(v_1 = 4, v_2 = 12) = 3.26$

With respect to the rows, $F_0(= 1.68) < F_{5\%}$ $(= 3.49)$

With respect to the columns, F_0 $(= 2.03) < F_{5\%}$ $(= 3.26)$

Hence the difference between the rations and that between the breeds are not significant.

Example 7 *The following data represent the number of units of production per day turned out by 5 different workers using 4 different types of machines:*

Machine Type

		A	B	C	D
	1	44	38	47	36
	2	46	40	52	43
Workers:	3	34	36	44	32
	4	43	38	46	33
	5	38	42	49	39

(a) Test whether the five men differ with respect to mean productivity.

(b) Test whether the mean productivity is the same for the four different machine types.

Solution We subtract 40 from the given values and work out with new values of x_{ij}.

Machine Type

Worker	A	B	C	D	T_i	T_i^2/k	$\sum_j x_{ij}^2$
1	4	– 2	7	– 4	5	6.25	85
2	6	0	12	3	21	110.25	189
3	– 6	– 4	4	– 8	– 14	49.00	132
4	3	– 2	6	– 7	0	0	98
5	– 2	2	9	– 1	8	16.00	90
T_j	5	– 6	38	– 17	$T = 20$	$\sum \frac{T_i^2}{k} = 181.5$	594
T_j^2/h	5	7.2	288.8	57.8	$\sum T_j^2/h = 358.8$		
$\sum_i x_{ij}^2$	101	28	326	139	594		

$$Q = \sum\sum x_{ij}^2 - \frac{T^2}{N} = 594 - \frac{400}{20} = 574$$

$$Q_1 = \sum \frac{T_i^2}{k} - \frac{T^2}{N} = 181.5 - 20 = 161.5$$

$$Q_2 = \sum \frac{T_j^2}{h} - \frac{T^2}{N} = 358.8 - 20 = 338.8$$

$$Q_3 = Q - Q_1 - Q_2 = 574 - (161.5 + 338.8) = 73.7$$

ANOVA table

S.V.	*S.S.*	*d.f.*	*M.S.*	F_0
Between rows (workers)	$Q_1 = 161.5$	$h - 1 = 4$	40.375	$\frac{40.375}{6.142} = 6.57$
Between Cols. (machines)	$Q_2 = 338.8$	$k - 1 = 3$	112.933	$\frac{112.933}{6.142} = 18.39$
Residual	$Q_3 = 73.7$	$(h - 1)(k - 1) = 12$	6.142	–
Total	$Q = 574$	$hk - 1 = 19$	–	–

From the F-tables, $F_{5\%}(v_1 = 4, v_2 = 12) = 3.26$

and $F_{5\%}(v_1 = 3, v_2 = 12) = 3.49$

With respect to the rows, $F_0(= 6.57) > F_{5\%}\ (= 3.26)$

With respect to the columns, $F_0(= 18.39) > F_{5\%}\ (= 3.49)$

Hence the 5 workers differ significantly and the 4 machine types also differ significantly with respect to mean productivity.

Example 8 *Four doctors each test four treatments for a certain disease and observe the number of days each patient takes to recover. The results are as follows (recovery time in days)*

Treatment

Doctor	*1*	*2*	*3*	*4*
A	*10*	*14*	*19*	*20*
B	*11*	*15*	*17*	*21*
C	*9*	*12*	*16*	*19*
D	*8*	*13*	*17*	*20*

Discuss the difference between (a) doctors and (b) treatments.

We subtract 15 from the given values and work out with the new values of x_{ij}.

Doctor	Treatment 1	2	3	4	T_i	$\frac{T_i^2}{k}$	$\sum_j x_{ij}^2$
A	−5	−1	4	5	3	2.25	67
B	−4	0	2	6	4	4.00	56
C	−6	−3	1	4	−4	4.00	62
D	−7	−2	2	5	−2	1.00	82
T_j	−22	−6	9	20	$T = 1$	$\sum \frac{T_i^2}{k} = 11.25$	267
T_j^2/h	121	9	20.25	100	$\sum T_j^2/h = 250.25$		
$\sum_i x_{ij}^2$	126	14	25	102	267		

$$Q = \sum\sum x_{ij}^2 - \frac{T^2}{N} = 267 - \frac{1}{16} = 266.94$$

$$Q_1 = \sum \frac{T_i^2}{k} - \frac{T^2}{N} = 11.25 - 0.0625 = 11.19$$

$$Q_2 = \sum \frac{T_j^2}{h} - \frac{T^2}{N} = 250.25 - 0.0625 - 250.19$$

$$Q_3 = Q - Q_1 - Q_2 = 266.94 - 261.38 = 5.56$$

ANOVA table

S.V.	*S.S.*	*d.f.*	*M.S.*	F_0
Between rows (doctors)	$Q_1 = 11.19$	$h - 1 = 3$	3.73	$\frac{3.73}{0.62} = 6.02$
Between cols. (treatments)	$Q_2 = 250.19$	$k - 1 = 3$	83.40	$\frac{83.40}{0.62} = 134.52$
Residual	$Q_3 = 5.56$	$(h - 1)(k - 1) = 9$	0.62	–
Total	$Q = 266.94$	$hk - 1 = 15$	–	–

From the F-tables, $F_{5\%}$ $(v_1 = 3, v_2 = 9) = 3.86$

Since $F_0 > F_{5\%}$ with respect to rows and columns, the difference between the doctors is significant and that between the treatments is highly significant.

Example 9 *The following data resulted from an experiment to compare three burners B_1, B_2 and B_3. A Latin square design was used as the tests were made on 3 engines and were spread over 3 days.*

	Engine 1	*Engine 2*	*Engine 3*
Day 1	$B_1 - 16$	$B_2 - 17$	$B_3 - 20$
Day 2	$B_2 - 16$	$B_3 - 21$	$B_1 - 15$
Day 3	$B_3 - 15$	$B_1 - 12$	$B_2 - 13$

Test the hypothesis that there is no difference between the burners.

Solution We subtract 16 from the given values and work out with new values of x_{ij}.

	E_1	E_2	E_3	T_i	$\frac{T_i^2}{n}$	$\sum_j x_{ij}^2$
D_1	$0(B_1)$	$1(B_2)$	$4(B_3)$	5	8.33	17
D_2	$0(B_2)$	$5(B_3)$	$-1(B_1)$	4	5.33	26
D_3	$-1(B_3)$	$-4(B_1)$	$-3(B_2)$	–8	21.33	26
T_j	–1	2	0	$T = 1$	$\sum T_i^2/n = 35$	69
T_j^2/n	0.33	1.33	0	$\sum T_i^2/n = 1.66$		
$\sum_i x_{ij}^2$	1	42	26	69		

Rearranging the data values according to the burners, we have

Burner		x_k		T_k	T_k^2/n
B_1	0	– 1	– 4	– 5	8.33
B_2	1	0	– 3	– 2	1.33
B_3	4	5	– 1	8	21.33
	Total		$T = 1$	$\sum \frac{T_k^2}{n} = 31$	

$$Q = \sum\sum x_{ij}^2 - \frac{T^2}{N} = 69 - \frac{1}{9} = 68.89$$

$$Q_1 = \frac{1}{n}\sum T_i^2 - \frac{T^2}{N} = 35 - \frac{1}{9} = 34.89$$

$$Q_2 = \frac{1}{n}\sum T_j^2 - \frac{T^2}{N} = 1.67 - \frac{1}{9} = 1.56$$

$$Q_3 = \frac{1}{n}\sum T_k^2 - \frac{T^2}{N} = 31 - \frac{1}{9} = 30.89$$

$$Q_4 = Q - Q_1 - Q_2 - Q_3 = 1.55$$

ANOVA table

S.V.	S.S.	d.f.	M.S.	F_0
Between rows (days)	$Q_1 = 34.89$	$n - 1 = 2$	17.445	$\frac{17.445}{0.775} = 22.51$
Between Cols. (engines)	$Q_2 = 1.56$	$n - 1 = 2$	0.780	$\frac{0.780}{0.775} = 1.01$
Between letters (burners)	$Q_3 = 30.89$	$n - 1 = 2$	15.445	$\frac{15.445}{0.775} = 19.93$
Residual	$Q_4 = 1.55$	$(n - 1)(n - 2) = 2$	0.775	–
Total	$Q = 68.89$	$n^2 - 1 = 8$	–	–

From the F-tables, $F_{5\%}$ $(v_1 = 2, v_2 = 2) = 19.00$

Since $F_0(= 19.93) > F_{5\%}$ $(= 19.00)$ for the burners, there is significant difference between the burners.

Incidentally, since $F_0 > F_{5\%}$ for the rows, the difference between the days is significant and since $F_0 < F_{5\%}$ for the columns, the difference between the engine is not significant.

Example 10 *Analyse the variance in the following Latin square of yields (in kgs) of paddy where A, B, C, D denote the different methods of cultivation*

D122	*A121*	*C123*	*B122*
B124	*C123*	*A122*	*D125*
A120	*B119*	*D120*	*C121*
C122	*D123*	*B121*	*A122*

Examine whether the different methods of cultivation have given significantly different yields.

Solution We subtract 120 from the given values and work out with the new values of x_{ij}.

i \ j	1	2	3	4	T_i	T_i^2 / n	$\sum_j x_{ij}^2$
1	D2	A1	C3	B2	8	16	18
2	B4	C3	A2	D5	14	49	54
3	A0	B-1	D0	C1	0	0	2
4	C2	D3	B1	A2	8	16	18
T_j	8	6	6	10	$T = 30$	$\sum T_i^2 / n = 81$	92
T_j^2 / n	16	9	9	25	$\sum T_j^2 / n = 59$		
$\sum_i x_{ij}^2$	24	20	14	34	92		

Rearranging the data according to the letters, we have

Letter			x_k		T_k	T_k^2 / n
A	1	2	0	2	5	6.25
B	2	4	- 1	1	6	9.00
C	3	3	1	2	9	20.25
D	2	5	0	3	10	25.00
		Total			30	60.50

$$Q = \sum\sum x_{ij}^2 - \frac{T^2}{N} = 92 - \frac{30^2}{16} = 35.75$$

$$Q_1 = \frac{1}{n}\sum T_i^2 - \frac{T^2}{N} = 81 - 56.25 = 24.75$$

$$Q_2 = \frac{1}{n}\sum T_j^2 - \frac{T^2}{N} = 59 - 56.25 = 2.75$$

$$Q_3 = \frac{1}{n}\sum T_k^2 - \frac{T^2}{N} = 60.50 - 56.25 = 4.25$$

$$Q_4 = Q - Q_1 - Q_2 - Q_3 = 35.75 - (24.75 + 2.75 + 4.25)$$
$$= 4.0$$

ANOVA table

S.V.	S.S.	d.f.	M.S.	F_0
Between rows	$Q_1 = 24.75$	$n - 1 = 3$	8.25	
Between columns	$Q_2 = 2.75$	$n - 1 = 3$	0.92	
Between letters	$Q_3 = 4.25$	$n - 1 = 3$	1.42	$\frac{1.42}{0.67} = 2.12$
Residual	$Q_4 = 4.0$	$(n - 1)(n - 2) = 6$	0.67	–
Total	$Q = 35.75$	$n^2 - 1 = 15$	–	–

From the F-tables, $F_{5\%}$ $(v_1 = 3, v_2 = 6) = 4.76$.

Since $F_0 (= 2.12) < F_{5\%}$ $(= 4.76)$ with respect to the letters, the difference between the methods of cultivation is not significant.

Exercise 10

Part A (Short answer questions)

1. What do you mean by the term 'experiment' in Design of experiments?
2. What motivated the adoption of design of experiments technique in scientific problems?

3. What is the aim of the design of experiments?
4. Distinguish between experimental and extraneous variables.
5. Name the basic principles of experimental design.
6. What do you mean by experimental group and control group?
7. What are the techniques frequently used in the local control of extraneous variables?
8. Name three basic designs of experiment.
9. What do you mean by analysis of variance.
10. Explain completely randomised design briefly.
11. Write down the format of the ANOVA table for one factor of classification.
12. Explain randomised block design briefly.
13. Write down the format of the ANOVA table for two factors of classification.
14. Explain Latin square design briefly.
15. Is a 2×2 Latin square design possible? Why?
    [**Hint** : No, as the degree of freedom for the residual variation is zero]
16. Write down the format of ANOVA table for three factors of classification.
17. Compare RBD and LSD.
18. What is the main advantage of LSD over RBD?
19. What is the total number of all possible Latin squares of order 3?
20. What is the total number of all possible Latin squares of order 4?

Part B

21. The following tables gives the yields of wheat from 16 plots, all of approximately equal fertility, when 4 varieties of wheat were cultivated in a completely randomised fashion. Test the hypothesis that the varieties are not significantly different.

Plot No. :	1	2	3	4	5	6	7	8	9	10
Variety :	A	B	D	C	B	C	A	D	B	D
Yield :	32	34	29	31	33	34	34	26	36	30

Plot No. :	11	12	13	14	15	16
Variety :	A	C	B	A	B	C
Yield :	33	35	37	35	35	32

22. A random sample is selected from each of 3 makes of ropes and their breaking strength (in certain units) are measured with the following results:

    I : 70, 72, 75, 80, 83
    II : 60, 65, 57, 84, 87, 73
    III : 100, 110, 108, 112, 113, 120, 107

    Test whether the breaking strengths of the ropes differ significantly.
23. The weights in gm of a number of copper wires, each of length 1 metre, were obtained. These are shown classified according to the dye from which they come:

D_1 : 1.30, 1.32, 1.36, 1.35, 1.32, 1.37
D_2 : 1.28, 1.35, 1.33, 1.34
D_3 : 1.32, 1.29, 1.31, 1.28, 1.33, 1.30
D_4 : 1.31, 1.29, 1.33, 1.31, 1.32
D_5 : 1.30, 1.32, 1.30, 1.33

Test the hypothesis that there is no difference between the mean weights of wires coming from different dyes.

24. It is suspected that four machines used in a canning operation fills cans to different levels on the average. Random samples of cans produced by each machine were taken and the fill (in ounces) was measured. The results are tabulated below:

Machine

A	*B*	*C*	*D*
10.20	10.22	10.17	10.15
10.18	10.27	10.22	10.27
10.36	10.26	10.34	10.28
10.21	10.25	10.27	10.40
10.25	–	–	10.30

Do the machines appear to be filling the cans at different average levels?

25. Different numbers of leaves were taken from each of 6 trees and their lengths measured. The following are the lengths in millimetres:

Tree	Lengths							
1	82	87	86	90	81	84		
2	85	84	91	92	88			
3	92	90	84	86	88	93	89	90
4	80	86	87	81	82	82		
5	87	86	88	90	85	86	87	
6	90	86	84	85	85	86	87	84	87

Can all these leaves be regarded as having come from the same species of trees?

26. There are 3 typists working in an office. The times (in minutes) they spend for tea-break in addition to the allowed lunch tea break are observed and noted below:

A : 25 18 30 32 35 37 19
B : 24 22 26 28 30 32 28 26
C : 28 20 27 19 29 35 30 23 27 32

Can the difference in average times that the 3 typists spend for tea break be attributed to chance variation?

27. Four machines *A*, *B*, *C*, *D* are used to produce a certain kind of cotton fabric. 4 Samples with each unit of size 100 square metres are selected from the outputs of the machines at random and the number of flaws in each 100 square metres are counted, with the following results:

A	B	C	D
8	6	14	20
9	8	12	22
11	10	18	25
12	4	9	23

Do you think that there is a significant difference in the performance of the four machines?

28. The following table shows the yield (in certain units) of lima beans on 20 plots of land subject to 4 different treatments, 5 plots per treatment. Set up an analysis of variance table to test the significance of the differences between the yields due to different treatments.

T_1 :	26.3	30.0	54.2	25.7	52.4
T_2 :	18.5	21.1	29.3	17.2	12.4
T_3 :	36.9	21.8	24.0	18.5	10.2
T_4 :	39.8	28.7	21.2	39.4	29.0

29. To test the significance of the variation of the retail prices of a certain commodity in the 4 principal cities Mumbai, Kolkata, Delhi and Chennai, 7 shops were chosen at random in each city and the prices (in Rs.) observed were as follows:

Mumbai :	100,	97,	91,	87,	87,	81,	79
Kolkata :	102,	100,	98,	97,	94,	86,	80
Delhi :	106,	102,	98,	86,	86,	84,	84
Chennai :	97,	95,	94,	92,	90,	86,	82

Do the data indicate that the prices in the 4 cities are significantly different?

30. Steel wire was made by 4 manufacturers A, B, C and D. In order to compare their products, 10 samples were randomly drawn from a batch of wires made by each manufacturer and the strength of each piece of wire was measured. The (coded) values are given below:

A : 55, 50, 80, 60, 70, 75, 40, 45, 80, 70
B : 70, 80, 85, 105, 65, 100, 90, 95, 100, 70
C : 70, 60, 65, 75, 90, 40, 95, 70, 65, 75
D : 90, 115, 80, 70, 95, 100, 105, 90, 100, 60

Carry out an analysis of variance and give your conclusions.

31. A randomised block experiment was laid out (with 4 blocks, each block containing 4 plots) to test 4 varieties of manure A, B, C, D and the yields per acre are given as below. Test for the significance of the difference among the 4 varieties of manure.

Block I	A155	B152	C157	D156
Block II	B152	C150	D156	A154
Block III	C156	D153	A161	B162
Block IV	D153	A154	B156	C155

32. The following table gives the gains in weights of 4 different types of pigs fed on 3 different rations over a period. Test whether
    (i) the difference in the rations significant
    (ii) the 4 types of pigs differ significantly in gaining weight.

	Types of pig			
Ration	*I*	*II*	*III*	*IV*
A	13.8	15.7	16.0	20.2
B	8.7	11.8	9.0	12.9
C	12.0	16.5	13.3	12.5

33. Four experiments determine the moisture content of samples of a powder, each observer taking a sample from each of six consignments. The assessments are given below:

	Consignment					
Observer	1	2	3	4	5	6
1	9	10	9	10	11	11
2	12	11	9	11	10	10
3	11	10	10	12	11	10
4	12	13	11	14	12	10

Perform an analysis of variance on these data and discuss whether there is any significant difference between consignments or between observers.

34. In order to compare three burners B_1, B_2 and B_3, one observation is made on each burner on each of four successive days. The data are tabulated below:

	B_1	B_2	B_3
Day 1	21	23	24
Day 2	18	17	23
Day 3	18	21	20
Day 4	17	20	22

Perform an analysis of variance on these data and find whether the difference between (i) the days and (ii) the burners significant at 5% LOS.

35. A company appoints 4 salesmen *A*, *B*, *C* and *D* and observes their sales in 3 seasons summer, winter and monsoon. The figures (in lakhs of Rs) are given in the following table :

	Salesmen			
Season	*A*	*B*	*C*	*D*
Summer	36	36	21	35
Winter	28	29	31	32
Monsoon	26	28	29	29

Carry out an analysis of variance.

36. The following data represent the numbers of units of production per day turned out by 4 different workers using 5 different types of machines:

	Machine type				
Worker	*A*	*B*	*C*	*D*	*E*
1	4	5	3	7	6
2	6	8	6	5	4
3	7	6	7	8	8
4	3	5	4	8	2

On the basis of this information, can it be concluded that (i) the mean productivity is the same for different machines (ii) the workers do not differ with regard to productivity?

37. The number of automobiles arriving at 4 toll gates were recorded for a 2 hours time period (10 A.M. to 12 noon) for each of six working days. The data are as follows :

Day	Gate 1	Gate 2	Gate 3	Gate 4
Mon	200	228	212	301
Tues	208	230	215	305
Wed	225	240	228	288
Thur	223	242	224	212
Fri	228	210	235	215
Sat	220	208	245	200

Determine whether the rate of arrival (i) is the same at each toll gate (ii) differs significantly during the six days or not.

38. The following table gives the number of refrigerators sold by 4 salesmen in 3 months:

	Salesman			
Months	*I*	*II*	*III*	*IV*
May	50	40	48	39
June	46	48	50	45
July	39	44	40	39

Determine whether (i) there is any difference in average sales made by the four salesmen (ii) the sales differ with respect to different months.

39. Four different drugs have been developed for a certain disease. These drugs are used in 3 different hospitals and the results, given below, show the number of cases of recovery from the disease per 100 people who have taken the drugs :

	D_1	D_2	D_3	D_4
H_1	19	8	23	8
H_2	10	9	12	6
H_3	11	13	13	10

What conclusions can you draw based on an analysis of variance ?

40. The following table gives the additional hours of sleep due to 3 soporofic drugs *A*, *B*, *C* tried on one patient each from 4 different age groups. Examine whether age has got any significant effect on the gain in sleep. Also examine whether the 3 drugs are similar in their effects or not

	Age group			
Drug	30–40	40–50	50–60	60–70
A	2.0	1.2	1.0	0.3
B	1.1	0.8	0.0	– 0.1
C	1.5	1.3	0.9	0.1

41. The following table gives the results of experiments on 4 varieties of a crop in 5 blocks of plots. Prepare the ANOVA table to test the significance of the difference between the yields of the 4 varieties :

Variety	B_1	B_2	B_3	B_4	B_5
A	32	34	33	35	37
B	34	33	36	37	35
C	31	34	35	32	36
D	29	26	30	28	29

42. In the table given below are the yields of 6 varieties of a crop in a 4 replicate RBD experiment. Analyse the data:

	Varieties					
Replicates	1	2	3	4	5	6
1	18.5	15.7	16.2	14.1	13.0	13.6
2	11.7	14.25	12.9	14.4	14.9	12.5
3	15.4	14.6	15.5	20.3	18.4	21.5
4	16.5	18.6	12.7	15.7	16.5	18.0

43. Analyse the variance in the following Latin square:

A8	C18	B9
C9	B18	A16
B11	A10	C20

44. Analyse the variance in the following Latin square:

20 B	17 C	25 D	34 A
23 A	21 D	15 C	24 B
24 D	26 A	21 B	19 C
26 C	23 B	27 A	22 D

45. A varietal trial was conducted on wheat with 4 varities *A*, *B*, *C*, *D* in a Latin square design. The plan of the experiment and the per plot yield are given below.

C25	B23	A20	D20
A19	D19	C21	B18
B19	A14	D17	C20
D17	C20	B21	A15

Analyse the data and interpret the result.

46. The following is the Latin square layout of a design when 4 varieties of seeds are tested. Set up the analysis of variance table and state your conclusions.

A105	B95	C125	D115
C115	D125	A105	B105
D115	C95	B105	A115
B95	A135	D95	C115

47. The table given below shows the yield of a certain crop in kgs per plot. The letters *A*, *B*, *C*, *D* refer to 4 different manurial treatments. Carry out an analysis of variance.

A260	B300	C335	D371
B280	A300	D300	C410
D320	C345	B340	A254
C372	D395	A290	B328

48. The following results were obtained in a textile experiment to compare the effects of sizing treatments *A*, *B*, *C*, *D* on the number of warps breaking per hour. Is the difference between the treatments significant?

Period	Loom 1	Loom 2	Loom 3	Loom 4
1	A 54	B 31	C 70	D 45
2	B 59	A 23	D 100	C 22
3	C 40	D 41	B 74	A 33
4	D 83	C 29	A 100	B 28

49. An agricultural experiment on the Latin square plan gave the following results for the yield of wheat per acre, letters corresponding to varieties.

A16	B10	C11	D9	E9
E10	C9	A14	B12	D11
B15	D8	E8	C10	A18
D12	E6	B13	A13	C12
C13	A11	D10	E7	B14

Discuss the variation of yield with each of the factors corresponding to the rows and columns.

50. The following is a Latine square design of five treatments:

A13	B9	C21	D7	E6
D9	E8	A15	B7	C16
B11	C17	D8	E10	A17
E8	A15	B7	C10	D7
C11	D9	E8	A11	B15

Analyse the data and interpret the results.

Answers

Exercise 10

19. 12   20. 576

21. $Q_1 = 46.08$, $Q_2 = 73.67$, $F_0 = 2.50$, $F_{5\%} = 3.49$;
Difference between varieties not significant.

22. $Q_1 = 5838.4$, $Q_2 = 1126$, $F_0 = 38.89$, $F_{5\%} = 3.68$; Breaking strengths of ropes differ significantly.

23. $Q_1 = 35.98$, $Q_2 = 99.38$, $F_0 = 1.81$, $F_{5\%} = 2.87$; Mean weights of wires do not differ significantly.

24. $Q_1 = 44.44$, $Q_2 = 696$, $F_0 = 2.98$, $F_{5\%} = 3.35$; No, the machines appear to fill at same level.

25. $Q_1 = 151.95$, $Q_2 = 255$, $F_0 = 4.17$, $F_{5\%} = 2.50$; Leaves have not come from the same species.

26. $Q_1 = 2.52$, $Q_2 = 29.27$, $F_0 = 11.62$, $F_{5\%} = 19.45$; Difference may be attributed to chance variation.

27. $Q_1 = 540.65$, $Q_2 = 85.75$, $F_0 = 25.21$, $F_{5\%} = 3.49$; Performances of the machines differ significantly.

28. $Q_1 = 34845.93$, $Q_2 = 10032.78$, $F_0 = 3.47$, $F_{5\%} = 3.24$; Treatments give significantly different yields.

29. $Q_1 = 94.97$, $Q_2 = 1446.03$, $F_0 = 1.9$, $F_{5\%} = 8.64$; Prices do not differ significantly.

30. $Q_1 = 5151$, $Q_2 = 8348$, $F_0 = 7.41$, $F_{5\%} = 8.60$; Strengths of wire do not differ significantly.

31. $Q_1 = 42.75$, $Q_2 = 6.75$, $Q_3 = 96.25$, $F_0 = 4.75$, $F_{5\%} = 8.82$; Difference between manures is not significant.

32. $Q_1 = 3393.59$, $Q_2 = 878.44$, $Q_3 = 344.36$,
F_0 (rows) $= 9.85$ and $F_{5\%} = 5.14$,
F_0 (columns) $= 2.55$ and $F_{5\%} = 4.76$;
Difference between rations significant. Difference between pigs is not significant.

33. $Q_1 = 13.13$, $Q_2 = 9.71$, $Q_3 = 13.12$,

F_0 (rows) = 5.03 and $F_{5\%}$ = 3.29,
F_0 (columns) = 2.23 and $F_{5\%}$ = 5.05;
Difference between observers is significant; Difference between consignments is not significant.

34. Q_1 = 22.00, Q_2 = 28.17, Q_3 = 14.50,
F_0 (rows) = 3.03 and $F_{5\%}$ = 4.76,
F_0 (columns) = 5.83 and $F_{5\%}$ = 5.14%;
Difference between days is not significant; Difference between burners is significant.

35. Q_1 = 32, Q_2 = 42, Q_3 = 136,
F_0 (rows) = 1.42 and $F_{5\%}$ = 19.33,
F_0 (columns) = 1.62, and $F_{5\%}$ = 8.94;
Differences between seasons and between salesmen are not significant.

36. Q_1 = 22.0, Q_2 = 12.8, Q_3 = 30.0;
F_0 (rows) = 2.93 and $F_{5\%}$ = 3.49;
F_0 (columns) = 1.28 and $F_{5\%}$ = 3.26;
Differences between the workers and between machine types are not significant.

37. Q_1 = 2279.83, Q_2 = 1470.05, Q_3 = 820.12,
F_0 (rows) = 1.80 and $F_{5\%}$ = 4.64,
F_0 (columns) = 1.79 and $F_{5\%}$ = 3.29;
Differences between the days and between the gates are not significant.

38. Q_1 = 109.5, Q_2 = 42.0, Q_3 = 64.5,
F_0 (rows) = 5.09 and $F_{5\%}$ = 5.14,
F_0 (columns) = 1.30 and $F_{5\%}$ = 4.76;
Differences between the months and between salesmen are not significant.

39. Q_1 = 55.17, Q_2 = 113.0, Q_3 = 89.5,
F_0 (rows) = 1.85 and $F_{5\%}$ = 5.14;
F_0 (columns) = 2.52 and $F_{5\%}$ = 4.76;
Differences between the hospitals and between the drugs are not significant.

40. Q_1 = 98.17, Q_2 = 341.58, Q_3 = 25.17,
F_0 (rows) = 11.69 and $F_{5\%}$ = 5.14,
F_0 (columns) = 27.11 and $F_{5\%}$ = 4.76;
Age has significant effect on the gain in sleep; Drugs differ significantly in their effect.

41. Q_1 = 134.0, Q_2 = 21.7, Q_3 = 29.5,
F_0 (row) = 18.16 and $F_{5\%}$ = 3.49,
F_0 (columns) = 2.21 and $F_{5\%}$ = 3.26;
Difference between the yields of 4 varieties is significant.

42. Q_1 = 56.76, Q_2 = 12.58, Q_3 = 80.18,
F_0 (rows) = 3.30 and $F_{5\%}$ = 3.34,
F_0 (columns) = 2.27 and $F_{5\%}$ = 4.65;
Difference between the varieties is significant.

43. Q_1 = 11.56, Q_2 = 68.23, Q_3 = 29.56, Q_4 = 68.21,

F_0 (rows) = 5.90, F_0 (cols.) = 1, F_0 (letters) = 2.31,
$F_{5\%}$ (for all) = 19.0;
The differences between rows, between columns and between letters are not significant.

44. $Q_1 = 34.19, Q_2 = 22.69, Q_3 = 141.19, Q_4 = 96.87$, F_0 (rows) = 1.42 and $F_{5\%}$ = 8.94; F_0 (columns) = 2.14 and $F_{5\%}$ = 8.94; F_0(letters) = 2.91 and $F_{5\%}$ = 4.76; Differences between rows, between columns and between letters are not significant.
45. $Q_1 = 46.5, Q_2 = 7.5, Q_3 = 48.5, Q_4 = 10.5$,
F_0 (rows) = 8.86, F_0 (columns) = 1.43, F_0 (letters) = 9.24, $F_{5\%}$ = 4.76. Difference between varieties is significant.
46. $Q_1 = 2, Q_2 = 4, Q_3 = 22, Q_4 = 60$, F_0 (rows) = 15, F_0 (columns) = 7.5, F_0 (letters) = 1.36, $F_{5\%}$ = 8.94; Difference between rows is significant, but differences between columns and between letters are not significant.
47. $Q_1 = 2540.5, Q_2 = 2853.75, Q_3 = 18690, Q_4 = 7515.75$, F_0 (rows) = 1.48, and $F_{5\%}$ = 8.94; F_0 (columns) = 1.32 and $F_{5\%}$ = 8.94; F_0 (letters) = 4.97 and $F_{5\%}$ = 4.76; Differences between rows and between columns are not significant, but difference between treatments is significant.
48. $Q_1 = 376, Q_2 = 8184, Q_3 = 1547.5, Q_4 = 284.5$
F_0 (rows) = 2.64, F_0 (columns) = 57.53, F_0 (letters) = 10.88; $F_{5\%}$ (for all) = 4.76; Difference between periods is not significant; Differences between looms and between treatments are significant.
49. $Q_1 = 2.16, Q_2 = 66.56, Q_3 = 122.56, Q_4 = 5.28$, F_0 (rows) = 1.2, F_0 (columns) = 37.8, F_0 (letters) = 69.6, $F_{5\%}$ (for all) = 3.26; Difference between rows is not significant, but differences between columns and between varieties are significant.
50. $Q_1 = 26, Q_2 = 34, Q_3 = 224.4, Q_4 = 103.6$, F_0 (rows) = 1.33 and $F_{5\%}$ = 5.91, F_0 (columns) = 1.02 and $F_{5\%}$ = 5.91, F_0 (letters) = 6.50 and $F_{5\%}$ = 3.26; Differences between rows and between columns are not significant, but difference between treatments is significant.

Appendix: Important Formulae

1. If A (an event) $\subset$ (sample space),

$$P(A) = \frac{n(A)}{n(S)} = \frac{\text{No. of outcomes favourable to } A}{\text{Exhaustive no. of outcome in } S}$$

2. $$P(A \cup B) = \begin{cases} P(A) + P(B) - P(A \cap B) & \text{if } A \text{ and } B \text{ are any 2 events} \\ P(A) + P(B) & \text{if } A \text{ and } B \text{ are mutually exclusive} \end{cases}$$

3. $$P(A \cap B) = \begin{cases} P(A) \cdot P(B/A) & \text{if } A \text{ and } B \text{ are any 2 associated events} \\ P(A) \cdot P(B) & \text{if } A \text{ and } B \text{ are independent} \end{cases}$$

4. If $B_1, B_2 \ldots, B_n$ be a set of exhaustive and mutually exclusive events and A is another event,

   (a) $$P(A) = \sum_{i=1}^{n} P(B_i) \cdot P(A/B_i)$$

   (b) $$P(B_i/A) = \frac{P(B_i) \cdot P(A/B_i)}{\sum_{i=1}^{n} P(B_i) \cdot P(A/B_i)}; \; i = 1, 2, \ldots, n$$

5. $P(r$ succeses in n Bernoulli's trials)
   $= P_n(r) = nC_r P^r q^{n-r}; r = 0, 1, 2, \ldots, n$

6. $$\sum_{r=r_1}^{r_2} nC_r p^r q^{n-r} \simeq \int_{Z_1}^{Z_2} \phi(z)\, dz, \; z_1 = \frac{r_1 - np - \frac{1}{2}}{\sqrt{npq}} \text{ and } z_2 = \frac{r_2 - np + \frac{1}{2}}{\sqrt{npq}}$$

7. $$P_n(r_1, r_2, \ldots, r_k) = \frac{n!}{r_1!\, r_2! \cdots r_k!} p_1^{r_1} \cdot p_2^{r_2} \cdots p_k^{rk},$$

$$P(A_i) = p_i, \sum_{i=1}^{k} p_i = 1 \text{ and } \sum_{i-1}^{k} r_i = n$$

8. If $\{p_i\}$ is the probability mass function and $F(x)$ is the distribution function of a discrete RV X,

$$F(x) = \sum_{x_i \leq x} p_i$$

9. If $f(x)$ is the probability density function and $F(x)$ is the distribution function of a continuous RV X,

   (i) $P(a \leq X \leq b) = \int_a^x f(x)\,dx = F(b) - F(a);$

   (ii) $F(x) = \int_{-\infty}^{b} f(x)\,dx;$

   (iii) $\frac{d}{dx} F(x) = f(x)$

10. If $f\{x, y)$ is the joint probability density function of (X, Y),

   (i) $P\{(X, Y) \in D\} = \iint_D f(x, y)\,dx\,dy;$

   (ii) $F\{(x, y) = \int_{-\infty}^{y} \int_{-\infty}^{x} f(x, y)\,dx\,dy$

11. Marginal probability distribution of X and Y are $\{x_i, p_{i*}\}$, $i = 1, 2, 3, \ldots$ where $p_{i*} = \sum_j P_{ij}$ and $\{y_j, p_{*j}\}$, $j = 1, 2, 3, \ldots$ where $p_{*j} = \sum_i p_{ij}$ respectively.

12. Conditional probability distributions of X given $Y = y_j$ and of Y given $X = x_i$ are respectively $\left\{x_i, \frac{p_{ij}}{p_{*j}}\right\}$, $i = 1, 2, 3, \ldots$ and $\left\{y_j, \frac{p_{ij}}{p_{i*}}\right\}$, $j = 1, 2, 3, \ldots$

13. Marginal density functions of X and Y are respectively

$$f_X(x) = \int_{-\infty}^{\infty} f(x, y)\,dy \text{ and } f_Y(y) = \int_{-\infty}^{\infty} f(x, y)\,dx$$

14. Conditional density functions of X given Y and of Y given X are respectively

$$f_{X/Y}(x) = \frac{f(xy)}{f_Y(y)} \text{ and } f_{Y/X}(y) = \frac{f(xy)}{f_X(x)}$$

15. X and Y are independent if (*i*) $p_{ij} = p_{i*}\ p_{*j}$ (discrete case) and
    (ii) $f(x, y) = f_X(x) \cdot f_Y(y)$ (continuous case)
16. If $f_X(x)$ and $f_Y(y)$ are density functions of X and Y where $Y = g(x)$,

$$f_Y(y) = \begin{cases} f_X(x)\left|\dfrac{dx}{dy}\right| & \text{if } x = g^{-1}(y) \text{ is single valued} \\ \sum_i f_X(x_i)\left|\dfrac{dx_i}{dy}\right| & \text{if } x = g^{-1}(y) \text{ is many valued} \end{cases}$$

17. If $Z = X + Y$, where X and Y are independent,

$$f_Z(z) = \begin{cases} \int_{-\infty}^{\infty} f_X(z-y) \cdot f_Y(y)\,dy \\ \int_0^z f_X(z-y) \cdot f_Y(y)\,dy \end{cases}$$

$z > 0$, if $f_X(x) = 0$ for $x < 0$ and $f_Y(y) = 0$ for $y < 0$

18. If $Z = XY$, where X and Y are independent,

$$f_Z(z) = \int_{-\infty}^{\infty} \frac{1}{|y|} f_X\left(\frac{z}{y}\right) \cdot f_Y(y)\,dy$$

19. If $Z = \dfrac{X}{Y}$, where X and Y are independent,

$$f_Z(z) = \int_{-\infty}^{\infty} |y|\, f_X(yz) \cdot f_Y(y)\,dy$$

20. $f_{ZW}(z, w) = |J|\, f_{XY}(x, y)$, where $J = \begin{vmatrix} \dfrac{\partial x}{\partial z} & \dfrac{\partial x}{\partial w} \\ \dfrac{\partial y}{\partial z} & \dfrac{\partial y}{\partial w} \end{vmatrix}$

21. $E(X) = \begin{cases} \sum_i x_i p_i & \text{if } X \text{ is discrete} \\ \int_{R_X} x f(x)\,dx & \text{if } X \text{ is continuous} \end{cases}$

22. Var $(X) = E(X^2) - E^2(X)$, where $E(X^2) = \begin{cases} \sum x_i^2 p_i \\ \int_{R_X} x^2 f(x)\, dx \end{cases}$

23. $\mu_n = \mu'_n - nC_1 \mu'_{n-1} d + nC_2 \mu'_{n-2} d^2 - \ldots + (-1)^n d^n$, where $\mu_n = E\{X - E(X)\}^n$, $\mu'_n = E\{X - a\}^n$ and $d = E(X) - a$.

24. $E\{g(X, Y)\} = \begin{cases} \sum_i \sum_j g(x_i, y_j)\, p_{ij}, \text{ if } (x, y) \text{ is discrete} \\ \iint_{R_{XY}} g(x, y)\, f(x, y)\, dx\, dy \ (x, y) \text{ is continuous} \end{cases}$

25. If X and Y are independent, $E(XY) = E(X) \cdot E(Y)$

26. $C_{XY} = E(XY) - E(X) \cdot E(Y)$

27. $\rho_{XY} = C_{XY} / \sqrt{\text{var}(X) \cdot \text{var}(Y)}$

28. $E(X/Y) = \begin{cases} \sum_i x_i\, p_{ij} / p_{*j} & \text{for discrete case} \\ \int_{R_X} x f_{X/Y}(x)\, dx & \text{for continuous case} \end{cases}$

29. $E(Y/X) = \begin{cases} \sum_j y_j\, p_{ij} / p_{i*} & \text{for discrete case} \\ \int_{R_Y} y f_{Y/X}(y)\, dy & \text{for continuous case} \end{cases}$

30. Moment Generating Function of X is

$$M(t) = \begin{cases} \sum_i e^{tx_i} p_i & \text{for discrete case} \\ \int_{R_X} e^{tx} f(x)\, dx & \text{for continuous case} \end{cases}$$

31. Characteristic functions of X is

$$\phi(w) = \begin{cases} \sum_r e^{iw_r} p_r & \text{for discrete case} \\ \int_{R_X} e^{iwx} f(x)\, dx & \text{for continuous case} \end{cases}$$

32. $E(X^n) = \begin{cases} \text{coefficient of } \dfrac{t^n}{n!} \text{ in the expansion of } M(t) \\ \text{coefficient of } \dfrac{i^n w^n}{n!} \text{ in the expansion of } \phi(w) \end{cases}$

33. $E(X^n) = \left[\dfrac{d^n}{dt^n} M(t)\right]_{t=0}$ or $\dfrac{1}{i^n}\left[\dfrac{d^n}{dw^n}\phi(w)\right]_{w=0}$

34. $M_{aX+b}(t) = e^{bt} \cdot M_X(at)$

35. $\phi_{aX+b}(w) = e^{ibw}\,\phi_X(w)$

36. Tchebycheff's inequality is

(i) $P\{|X-\mu| \geq c\} \leq \dfrac{\sigma^2}{c^2}$ or $P\{|X-\mu| \leq c\} \geq 1 - \dfrac{\sigma^2}{c^2}, c > 0$

Where $\mu = E(X)$ and $\sigma^2 = \text{Var}(X)$

(ii) $P\left\{\left|\dfrac{X-\mu}{k}\right| \geq \sigma\right\} \leq \dfrac{1}{k^2}$ or $P\left\{\left|\dfrac{X-\mu}{k}\right| \leq \sigma\right\} \geq 1 - \dfrac{1}{k^2}, k > 0$

37. If $E(X_i) = \mu_i$ and $\text{Var}(X_i) = \sigma_i^2$ = and if $S_n = \sum_{i=1}^{n} X_i$, then S_n follows

$N\left\{\sum_{i=1}^{n}\mu_i, \sqrt{\sum_{i=1}^{n}\sigma_i^2}\right\}$, as $n \to \infty$, where $\{X_i\}$ is a sequence of independent RVs.

38. If $E(X_i) = \mu$ and $\text{Var}(X_i) = \sigma^2$ and $S_n = \sum_{i=1}^{n} X_i$ then S_n follows $N\{n\mu, \sigma\sqrt{n}\}$

as $n \to \infty$, and $\overline{X} = \dfrac{1}{n} S_n$ follows $N\left\{\mu, \dfrac{\sigma}{\sqrt{n}}\right\}$ as $n \to \infty$, where $\{X_i\}$ is a sequence of independent and identically distributed RVs.

39. For the binomial distribution $B(n, p)$,

(i) $P\{X = r\} = nC_r p^r q^{n-r}; r = 0, 1, 2, \ldots, n; p + q = 1$

(ii) $E\{X\} = np$

(iii) $\text{Var}(X) = npq$

(iv) $\mu_{k+1} = pq\left[\dfrac{d\mu_k}{dp} + nk\mu_{k-1}\right]$

(v) $M(t) = (q + pe^t)^n$ and $\phi(w) = (q + pe^{iw})^n$

40. For the Poisson distribution $P(\lambda)$,

(i) $P\{X = r\} = \dfrac{e^{-\lambda} \cdot \lambda^r}{r!}$; $r = 0, 1, 2, \ldots, \infty$

(ii) $E\{X\} = \lambda = \text{Var}(X) = \mu_3$

(iii) $\mu_{k+1} = \lambda\left[\dfrac{d\mu_k}{d\lambda} + k\mu_{k-1}\right]$

(iv) $M(t) = e^{\lambda(e^t - 1)}$ and $\phi(w) = e^{\lambda(e^{iw} - 1)}$

41. For the geometric distribution $G_1(p)$,

(i) $P\{X = r\} = q^r p$; $r = 0, 1, 2, \ldots, \infty$; $p + q = 1$

(ii) $E(X) = \dfrac{q}{p}$

(iii) $\text{Var}(X) = \dfrac{q}{p^2}$

42. For the geometric distribution $G_2(p)$,

(i) $P\{X = r\} = q^{r-1} p$; $r = 1, 2, \ldots, \infty$; $p + q = 1$

(ii) $E(X) = \dfrac{1}{p}$

(iii) $\text{Var}(X) = \dfrac{q}{p^2}$

43. For the hypergemetric distribution $H(N, k, n)$,

(i) $P\{X = r\} = \dfrac{kc_r \cdot (N - k)\, C_{n-r}}{NC_n}$; $r = 0, 1, 2, \ldots, n$

(ii) $E(X) = \dfrac{nk}{N}$ or np; where $p = \dfrac{k}{N}$

(iii) $\text{Var}(X) = \dfrac{nk(N - k)(N - n)}{N^2(N - 1)}$ or $npq\left(\dfrac{N - n}{N - 1}\right)$,

where $p = \dfrac{k}{N}$ and $q = 1 - p$

44. For the uniform (rectangular) distribution $U(a, b)$,

(i) $f\{x\} = \dfrac{1}{b - a}$; $b > a$; $a < x < b$

(ii) $E(X) = \dfrac{1}{2}(b + a)$

(iii) $\text{Var}(X) = \frac{1}{12}(b-a)^2$

(iv) $$\mu_k = \begin{cases} 0 & \text{, if } k \text{ is odd} \\ \frac{1}{k+1}\left(\frac{b-a}{2}\right)^k & \text{, if } k \text{ is even} \end{cases}$$

45. For the (negative) exponential distribution $E(\lambda)$,

(i) $f\{x) = \lambda e^{-\lambda x}$, $x \geq 0$; $\lambda > 0$

(ii) $E\{X) = \frac{1}{\lambda}$

(iii) $\text{Var}(X) = \frac{1}{\lambda^2}$

(iv) μ'_k (about the origin) $= E(X^k) = \frac{k!}{\lambda^k}$

46. For the Erlang or General Gamma distribution $Er(\lambda, k)$,

(i) $f\{x) = \frac{\lambda^k x^{k-1} e^{-\lambda x}}{\Gamma(k)}$, $x \geq 0$; $\lambda > 0$; $k > 0$

(ii) $E(X) = \frac{k}{\lambda}$

(iii) $\text{Var}(X) = \frac{k}{\lambda^2}$

(iv) $M(t) = \left(1 + \frac{t}{\lambda}\right)^{-k}$ and $\phi(w) = \left(1 + \frac{iw}{\lambda}\right)^{-k}$

(v) $\mu'r = E(X^r) = \frac{1}{\lambda^r} \cdot \frac{\Gamma(k+r)}{\Gamma(k)}$

47. For the Weibull distribution $W(\alpha, \beta)$,

(i) $f\{x) = \alpha\beta x^{\beta-1} \cdot e^{-\alpha x\beta}$; $x > 0$; $\alpha, \beta > 0$

(ii) $E(X) = \alpha^{-1/\beta} \cdot \Gamma\left(\frac{1}{\beta} + 1\right)$

(iii) $\text{Var}(X) = \alpha^{-2/\beta}\left[\Gamma\left(\frac{2}{\beta} + 1\right) - \left\{\Gamma\left(\frac{1}{\beta} + 1\right)\right\}^2\right]$

(iv) $\mu_r' = E(X^r) = \alpha^{-r/\beta}\,\Gamma\left(\frac{r}{\beta}+1\right)$

48. For the normal (Gaussian) distribution $N(\mu, \sigma)$,

(i) $f(x) = \dfrac{1}{\sigma\sqrt{2\pi}}\, e^{-(x-\mu)^2/2\sigma^2}\quad -\infty < x < \infty$

(ii) $E(X) = \mu =$ Median $=$ Mode

(iii) $\text{Var}(X) = \sigma^2$

(iv) $M(t) = e^{\mu t + \frac{1}{2}\sigma^2 t^2}$ and $\phi(w) = e^{i\mu w - \frac{1}{2}\sigma^2 w^2}$

(v) $\mu_{2n+1} = 0$ and $\mu_{2n} = 1.\,3.\,5 \ldots (2n-1)\,\sigma^{2n}$

(vi) Mean deviation $= \sqrt{\dfrac{2}{\pi}}\,\sigma \simeq \dfrac{4}{5}\sigma$

(vii) Quartile deviation $= 0.674\,\sigma \simeq \dfrac{2}{3}\sigma$.

49. For the bivariate normal distribution $N(0, 0; \sigma_x, \sigma_y; r)$

(i) $f(x, y) = \dfrac{1}{2\pi\sigma_x\sigma_y\sqrt{1-r^2}}$

$$\exp\left\{-\frac{1}{2(1-r^2)}\left(\frac{x^2}{\sigma_x^2} - \frac{2rxy}{\sigma_x\sigma_y} + \frac{y^2}{\sigma_y^2}\right)\right\};\ -\infty < x, y < \infty$$

(ii) $f_X(x) = \dfrac{1}{\sigma_x\sqrt{2\pi}} \exp\left(-x^2/2\sigma_x^2\right),\ -\infty < x < \infty$

(iii) $f_Y(y) = \dfrac{1}{\sigma_y\sqrt{2\pi}} \exp\left(-y^2/2\sigma_y^2\right),\ -\infty < y < \infty$

(iv) $f_{X/Y}(x) = \dfrac{1}{\sqrt{2\pi}\left(\sigma_x\sqrt{1-r^2}\right)} \exp\left\{-\dfrac{1}{2\sigma_x^2(1-r^2)}\left(x - \dfrac{r\sigma_x y}{\sigma_y}\right)^2\right\}$,

$-\infty < x < \infty$

(v) $f_{Y/X}(y) = \dfrac{1}{\sqrt{2\pi}\left(\sigma_y\sqrt{1-r^2}\right)} \exp\left\{-\dfrac{1}{2\sigma_y^2(1-r^2)}\left(y - \dfrac{r\sigma_y}{\sigma_x}x\right)^2\right\}$,

$-\infty < y < \infty$.

50. For two-tailed large sample tests,

$$|Z_{1\%}| = 2.58, |z_{2\%}| = 2.33, |z_{5\%}| = 1.96 \text{ and } |Z_{10\%}| = 1.645$$

51. For one-tailed large sample tests,

$$z_{1\%} = \pm 2.33,\ Z_{2\%} = \pm 2.05,\ z_{5\%} = \pm 1.645 \text{ and } z_{10\%} = \pm 1.28$$

52. In large samples, to test the significance of the difference between

(i) sample proportion p and population propotion P, the test statistic

$$z = \frac{p - P}{\sqrt{pq/n}} \text{ (size of the sample} = n)$$

(ii) two sample proportions p_1 and p_2,

$$z = \frac{P_1 - P_2}{\sqrt{\hat{P}\hat{Q}\left(\frac{1}{n_1} + \frac{1}{n_2}\right)}}, \text{ where } \hat{P} = \frac{n_1 p_1 + n_2 p_2}{n_1 + n_2} = \text{ and } \hat{Q} = 1 - \hat{P}$$

(iii) sample mean $\bar{x}$ and population mean, μ,

$$z = \begin{cases} \dfrac{\bar{X} - \mu}{\sigma/\sqrt{n}}, & \text{if } \sigma \text{ is known} \\ \dfrac{\bar{X} - \mu}{s/\sqrt{n}}, & \text{if } \sigma \text{ is not known (S.D. of sample} = s) \end{cases}$$

(iv) two sample means $\bar{x}$ and $\bar{x}_2$,

$$z = \begin{cases} \left(\bar{X}_1 - \bar{X}_2\right) \Big/ \sqrt{\dfrac{\sigma_1^2}{n_1} + \dfrac{\sigma_2^2}{n_2}} & \text{if } \sigma \neq \sigma_2 \text{ and } \sigma_1 \text{ and } \sigma_2 \text{ are known} \\ \left(\bar{X}_1 - \bar{X}_2\right) \Big/ \sigma\sqrt{\dfrac{1}{n_1} + \dfrac{1}{n_2}} & \text{if } \sigma_1 = \sigma_2 \text{ and } \sigma \text{ is known} \\ \left(\bar{X}_1 - \bar{X}_2\right) \Big/ \sigma\sqrt{\dfrac{s_1^2}{n_1} + \dfrac{s_2^2}{n_2}} & \text{if } \sigma_1 \neq \sigma_2 \text{ and not known} \\ \left(\bar{X}_1 - \bar{X}_2\right) \Big/ \hat{\sigma}\sqrt{\dfrac{1}{n_1} + \dfrac{1}{n_2}} & \text{if } \sigma_1 = \sigma_2 \text{ is not known} \\ \text{(or)} \left(\bar{X}_1 - \bar{X}_2\right) \Big/ \sqrt{\dfrac{s_1^2}{n_2} + \dfrac{s_2^2}{n_1}} & \text{and } \hat{\sigma}^2 = \dfrac{n_1 s_1^2 + n_2 \sigma_2^2}{n_1 + n_2} \end{cases}$$

(v) Sample SDs and population SDσ, $z = \dfrac{s - \sigma}{\sigma / \sqrt{2n}}$

(vi) Two sample SDs s_1 and s_2,

$$z = \frac{s_1 - s_2}{\hat{\sigma}\sqrt{\frac{1}{2n_1} + \frac{1}{2n_2}}} \text{ or } \frac{s_1 - s_2}{\sqrt{\frac{s_1^2}{2n_2} + \frac{s_2^2}{1n_1}}}$$

$$\hat{\sigma}^2 = \frac{n_1 s_1^2 + n_2 s_2^2}{n_1 + n_2}$$

53. 95% confidence limits for

(i) the population proportion are $\left(p \mp 1.96\sqrt{\frac{pq}{n}}\right)$

(ii) the population mean are $\left(\bar{x} \mp 1.96\frac{\sigma}{\sqrt{n}}\right)$ or $\left(\bar{x} \mp 1.96\frac{s}{\sqrt{n}}\right)$ according

as σ is known or not known.

54. In small samples, to test the significance of the difference between

(i) sample mean $\bar{x}$ and population mean μ, the test statistic

$$t = \frac{\bar{x} - \mu}{s/\sqrt{n-1}}; \text{ degrees of freedom } \nu = n - 1.$$

(ii) two sample means $\bar{X}_1$ and $\bar{X}_2$, $t = \dfrac{\bar{X}_1 - \bar{X}_2}{\sqrt{\left(\frac{n_1 s_1^2 + n_2 s_2^2}{n_1 + n_2 - 2}\right)\left(\frac{1}{n_1} + \frac{1}{n_2}\right)}}$;

$\nu = n_1 + n_2 - 2$

(iii) two population variances σ_1^2 and σ_2^2, $F = \dfrac{\hat{\sigma}_1^2}{\hat{\sigma}_2^2} = \dfrac{n_1\, s_1^2/(n_1 - 1)}{n_2\, s_2^2/(n_2 - 1)}$;

$\nu_1 = n_1 - 1$ and $\nu_2 = n_2 - 1$, provided $F > 1$.

55. For χ^2-test of goodness of fit,

$$\chi^2 = \sum_{i=1}^{n} (O_i - E_i)^2/E_i; \ \nu = \text{number of independent frequencies.}$$

56. For χ^2-test of independence of attributes or for the $(m \times n)$ contingency table,

$$\chi^2 = \sum_{i=1}^{m}\sum_{j=1}^{n} (O_{ij} - E_{ij})^2/E_{ij},\ \nu = (m-1)(n-1), \text{ where } E_{ij} = \frac{O_{i*} \cdot O_{*j}}{N}.$$

57. For a random process $\{X(t)\}$,

(i) Mean $\mu(t) = E\{X(t)\}$

(ii) Autocorrelation $R_{XX}(t_1, t_2) = E\{X(t_1) \cdot X(t_2)\}$

(iii) Autocovariance $C_{XX}(t_1, t_2) = R_{XX}(t_1, t_2) - \mu(t_1) \cdot \mu(t_2)$

(iv) Correlation coefficient $\rho_{XX}(t_1, t_2) = \dfrac{C(t_1 t_2)}{\sqrt{C(t_1, t_1) \cdot C(t_2, t_2)}}$

58. For two random processes $\{X(t)\}$ and $\{Y(t)\}$,

(i) Cross-correlation $R_{XY}(t_1, t_2) = E\{X(t_1) \cdot Y(t_2)\}$

(ii) Cross-covariance $C_{XY}(t_1, t_2) = R_{XY}(t_1, t_2) - \mu_X(t_1) \cdot \mu_Y(t_2)$

(iii) Cross-correlation coefficient

$$\rho_{XY}(t_1, t_2) = \frac{C_{XY}(t_1, t_2)}{\sqrt{C_{XX}(t_1, t_1) \cdot C_{YY}(t_2, t_2)}}$$

59. Density function of Wiener process is

$$f_{X(t)}(x) = \frac{1}{\sqrt{2\pi \alpha t}} e^{-x^2/2\alpha t} \quad -\infty < x < \infty$$

60. Time average of a random process $\{X(t)\}$ over $(-T, T)$ is

$$\bar{X}_T = \frac{1}{2_T} \int_{-T}^{T} X(t)\, dt$$

61. If $R_{XX}(\tau)$ is the autocorrelation function of $\{X(t)\}$,

(i) $\mu_X^2 = \lim_{\tau \to \infty} \{R_{XX}(\tau)\}$

(ii) $E\{X^2(t)\} = R_{XX}(0)$

62. If $\bar{X}_T$ is the time average of the process $\{X(t)\}$ over $(-T, T)$,

(i) $\operatorname{Var}(\bar{X}_T) = \dfrac{1}{4T^2} \displaystyle\int_{-T}^{T} \int_{-T}^{T} C(t_1, t_2)\, dt\, dt$

(ii) $\operatorname{Var}(\bar{X}_T) = \dfrac{1}{2} \displaystyle\int_{0}^{2T} C(\tau) \left\{1 - \frac{|\tau|}{2T}\right\} d\tau$ if $\{X(T)\}$ is a stationary process.

63. If $R_{XX}(\tau)$ and $S_{XX}(\omega)$ are the autocorrelation function and power spectral density of a stationary process,

(i) $S_{XX}(\omega) = \displaystyle\int_{-\infty}^{\infty} R_{XX}(\tau)\, e^{-i\omega\tau} d\tau$

(ii) $R_{XX}(\tau) = \dfrac{1}{2\pi} \displaystyle\int_{-\infty}^{\infty} R_{XX}(\omega)\, e^{i\tau\omega} d\omega$

64. If $Y(t) = \int_{-\infty}^{\infty} h(u)X(t-u)\mathrm{d}u$

$S_{YY}(\omega) = |H(\omega)|^2 S_{XX}(\omega)$.

65. The nth order density of a Gaussian process $\{X(t)\}$ is

$$\frac{1}{(2\pi)^{1/2}|\Lambda|^{1/2}} \exp\left\{-\frac{1}{2|\Lambda|}\sum_{i=1}^{n}\sum_{j=1}^{n}|\Lambda|_{ij}(x_i-\mu_i)(x_i-\mu_j)\right\},$$

where $\mu_i = E\{X(t_i)\}$ and Λ is the nth order square matrix (λ_{ij}),

where $\lambda_{ij} = C\{X(t_i), X(t_j)\}$ and $|\Lambda|_{ij}$ = cofactor of λ_{ij} in $|\Lambda|$.

66. If $\{N(t)\}$ is white noise,

(i) $S_{NN}(\omega) = \dfrac{N_0}{2}$

(ii) $R_{NN}(\tau) = \dfrac{N_0}{2}\,\delta(\tau)$

67. If $\{N(t)$ is a band-limited white noise,

(i) $S_{NN}(\omega) = \begin{cases} \dfrac{N_0}{2}, & \text{for } |\omega| \le \omega_B \\ 0, & \text{elsewhere} \end{cases}$

(ii) $R_{NN}(\tau) = \dfrac{N_0\omega_B}{2\pi}\left(\dfrac{\sin\omega_B\tau}{\omega_B\tau}\right)$

68. Probability law for Poisson process $\{X(t)\}$ is $P\{X(t) = n\} = \dfrac{e^{-\lambda t}\cdot(\lambda t)^n}{n!}$; $n = 0, 1, 2, \ldots, \infty$

69. If $\{p_{ij}\}$ is the transition probability matrix of a homogeneous Markov chain $\{X_n\}$, $\{p_{ij}^{(n)}\} = (p_{ij})^n$.

For the ($M/M/1$):($\infty/F1F0$) queueing model,

70. $P_n = \left(\dfrac{\lambda}{\mu}\right)^n\left(1-\dfrac{\lambda}{\mu}\right) n \ge 0$

71. $L_s = E$(number of customers in the system) $= \dfrac{\lambda}{\mu-\lambda}$

72. $L_q = E$(number of customers in the queure) $= \dfrac{\lambda^2}{\mu(\mu-\lambda)}$

73. $L_w = E$(number of customers in the non-empty queues) $= \dfrac{\mu}{\mu-\lambda}$

74. $P(N_s > k)$ = Probability that the number of customers in the system exceeds $k = \left(\frac{\lambda}{\mu}\right)^{k+1}$

75. $E(W_s)$ = Average waiting time of a customer in the system = $\frac{1}{\mu - \lambda}$

76. $P(W_s > t) = e^{-(\mu - \lambda)t}$

77. $E(W_q)$ = Average waiting time of a customer in the queue = $\lambda/\mu\,(\mu - \lambda)$.

78. $E(W_q/W_q > 0)$ = Average waiting time of a customer in the queue, if he has to wait = $\frac{1}{\mu - \lambda}$

Little's formulas valid for $(M/M/1)$:$(\infty/F1F0)$ and $(M/M/s)$:$(\infty/F1F0)$ models.

79. $L_S = \lambda \cdot E(W_s)$

80. $L_q = \lambda \cdot E(W_q)$

81. $L_S = \frac{\lambda}{\mu} + L_q$

82. $E(W_s) = \frac{1}{\mu} + E(W_q)$

For the $(M/M/s)$:$(\infty/F1F0)$ queueing model,

83. $$P_n = \begin{cases} \frac{1}{n!}\left(\frac{\lambda}{\mu}\right)^n \cdot P_o & \text{if } 0 \le n < s \\ \frac{1}{s!\,s^{n-s}} \cdot \left(\frac{\lambda}{\mu}\right)^n \cdot P_o & \text{if } n \ge s, \text{ where} \end{cases}$$

$$P_0^{-1} = \sum_{n-0}^{s-1} \frac{1}{n!}\left(\frac{\lambda}{\mu}\right)^n + \left\{\frac{1}{s!\left(1 - \frac{\lambda}{\mu s}\right)}\left(\frac{\lambda}{\mu}\right)^s\right\}$$

84. $$L_q = E(N_q) = \frac{1}{s.s!} \cdot \frac{\left(\frac{\lambda}{\mu}\right)^{s+1}}{\left(1 - \frac{\lambda}{\mu s}\right)^2} \cdot P_o$$

85. $$L_s = E(N_S) = L_q + \frac{\lambda}{\mu} = \frac{1}{s.s!} \cdot \frac{\left(\frac{\lambda}{\mu}\right)^{s+1}}{\left(1 - \frac{\lambda}{\mu s}\right)^2} \cdot P_o + \frac{\lambda}{\mu}$$

86. $E(W_s) = \dfrac{1}{\lambda} E(N_s) = \dfrac{1}{\mu} + \dfrac{1}{\mu} \cdot \dfrac{1}{s.s!} \cdot \dfrac{\left(\dfrac{\lambda}{\mu}\right)^s}{\left(1 - \dfrac{\lambda}{\mu s}\right)^2} \cdot P_o$

87. $E(W_q) = \dfrac{1}{\lambda} E(N_q) = \dfrac{1}{\mu} \cdot \dfrac{1}{s.s!} \cdot \dfrac{\left(\dfrac{\lambda}{\mu}\right)^s}{\left(1 - \dfrac{\lambda}{\mu s}\right)^2} \cdot P_o$

88. (i) $P(W_s > 0)$ = Probability that an arrival has to wait = $\dfrac{\left(\dfrac{\lambda}{\mu}\right)^s \cdot P_o}{s!\left(1 - \dfrac{\lambda}{\mu s}\right)}$.

(ii) P(an arrival enters the service without waiting) = $1 - P(W_s > 0)$.

89. $E(W_q/W_s > 0)$ = Average waiting time in the queue for the customers who actually wait = $1/\mu s - \lambda$.

90. $E(N_q/N_q \geq 1)$ = Average number of customers in the queue who have to actually wait = $\dfrac{\dfrac{\lambda}{\mu s}}{1 - \dfrac{\lambda}{\mu s}}$.

For the $(M/M/1)$:$(k/F1F0)$ queueing model,

91. $P_n = \begin{cases} \left(\dfrac{\lambda}{\mu}\right)^n \left\{\dfrac{1 - \dfrac{\lambda}{\mu}}{1 - \left(\dfrac{\lambda}{\mu}\right)^{k+1}}\right\} & \text{if} \quad \lambda \neq \mu \quad n \geq 0 \\ \dfrac{1}{k+1} & \text{if} \quad \lambda = \mu \quad n \geq 0 \end{cases}$

92. $L_s = E(N_s) = \begin{cases} \dfrac{\lambda}{\mu - \lambda} - \dfrac{(k+1)\left(\dfrac{\lambda}{\mu}\right)^{k+1}}{1 - \left(\dfrac{\lambda}{\mu}\right)^{k+1}} & \text{if } \lambda \neq \mu \\ \dfrac{k}{2} & \text{if } \lambda = \mu \end{cases}$

93. Overall effective arrival rate $\lambda' = \mu(1 - P_0)$.

94. $L_q = E(N_q) = L_s - \dfrac{\lambda'}{\mu}$ or $L_s - (1 - P_0)$

95. (i) $E(W_s) = \dfrac{1}{\lambda'} L_s$

(ii) $E(W_q) = \dfrac{1}{\lambda'} L_q$

For the (*M*/*M*/*s*):(*k*/*F*1*F*0) queueing model,

96. $$P_n = \begin{cases} \dfrac{1}{n!}\left(\dfrac{\lambda}{\mu}\right)^n \cdot P_o & \text{for } n \le s \\ \dfrac{1}{s!\,s^{n-s}} \cdot \left(\dfrac{\lambda}{\mu}\right)^n \cdot P_o & \text{for } s \le n \le k \\ 0 & \text{for } n > k, \text{ where} \end{cases}$$

$$P^{-1}{}_0 = \sum_{n=0}^{s-1} \frac{1}{n!}\left(\frac{\lambda}{\mu}\right)^n + \frac{1}{s!} \cdot \left(\frac{\lambda}{\mu}\right)^s \sum_{n=s}^{k} \left(\frac{\lambda}{\mu s}\right)^{n-s}$$

97. $L_q = E(N_q) = P_o \cdot \left(\dfrac{\lambda}{\mu}\right)^s \cdot \dfrac{p}{s!(1 - p)^2} \{1 - \rho^{k-s} - (k - s)(1 - \rho)\,\rho^{k-s}\}$

where $\rho = \dfrac{\lambda}{\mu s}$

98. Overall effective arrival rate

$$\lambda' = \mu\left[s - \sum_{n=0}^{s-1} (s - n)P_n\right]$$

99. $L_s = E(N_s) = L_q + \dfrac{\lambda'}{\mu}$ or $L_q + s - \displaystyle\sum_{n=0}^{s-1} (s - n)P_n$.

100. (i) $E(W_s) = \dfrac{1}{\lambda'} L_s$

(ii) $E(W_q) = \dfrac{1}{\lambda'} L_q$.

Index

M

N

O

P

Q

R

S

T

U

V

W

Y

Made in United States
Orlando, FL
04 November 2026

90208561R00341